S0-CCM-543

The
Science
of
Flight

The Science of Flight

Pilot-oriented Aerodynamics

W. N. Hubin

 IOWA STATE UNIVERSITY PRESS / AMES

W. N. Hubin is Professor of Physics at Kent State University, Kent, Ohio. He has been flying since 1962, has logged over 4000 flight hours, and holds a Commercial certificate with Instrument, Flight Instructor Airplane, Flight Instructor Instrument, and Multiengine ratings. He is a member of The American Physical Society, The American Association of Physics Teachers, the American Institute of Aeronautics and Astronautics, the Experimental Aircraft Association, and the International Aerobatic Club.

© 1992, W. N. Hubin. Previous editions and printings © 1979, 1980, 1981, 1983, 1984, 1987, 1988
All rights reserved

Authorization to photocopy items for internal or personal use, or the internal or personal use of specific clients, is granted by Iowa State University Press, provided that the base fee of $.10 per copy is paid directly to the Copyright Clearance Center, 27 Congress Street, Salem, MA 01970. For those organization that have been granted a photocopy license by CCC, a separate system of payments has been arranged. The fee code for users of the Transactional Reporting Service is 0-8138-0398-5/92 $.10.

⊗ Printed on acid-free paper in the United States of America

First Iowa State University Press edition, 1992

Library of Congress Cataloging-in-Publication Data

Hubin, Wilbert N.
    The science of flight : pilot-oriented aerodynamics / W. N. Hubin.—1st ed.
        p.   cm.
    Includes index.
    ISBN 0-8138-0398-5 (acid-free paper)
    1. Aerodynamics.   I. Title.
TL570.H79 1992
    629.132′3—dc20   92–13229

Hear them call and cry,
Wheeling and gliding,
Off on the wing,
As swift as an arrow in flight!

They sail up to the clouds
Where the sun is riding,
Forever climbing
To the heav'nly height.

May life never confine them,
Free may they wander,
Searching in the clouds
For the heaven they know is there.

How like my heart
They pursue a vision of rapture
Forever climbing
In the golden air!

Storms may beset them,
The light'ning flash may blind them,
Wind may delay them,
The thunder may roar!

They never waver,
They never falter,
As on they journey
Over sea and shore!

Onward they fly
To lands beyond our knowing
Striving to follow
Their own secret star.

Ah, you gypsies of heav'n,
My heart is with you,
A dream bids us to wander,
Afar, afar, a star, a star![1]

[1]"Bird Song," from *Pagliacci* by Ruggiero Leoncavallo. From *A Treasury of Grand Opera*, Henry W. Simon, editor. Translation by George Mead. Reprinted by permission of Simon & Schuster, Inc. ©1946, 1973 by Simon & Schuster, Inc.

Contents

Pre-Takeoff

"Fly it by the numbers!" Every aspiring expert or professional pilot hears this imperious command sometime during his or her training. Although a simple aircraft will sometimes suffer a ham-handed pilot, complex aircraft simply must be flown more precisely. It is not enough to choose a climb or approach speed on the basis that it is "not too fast" or "not too slow." Rather, there is a specific number that should appear opposite that little airspeed pointer needle for each weight, wind condition, and performance requirement. And while a nosewheel-type aircraft usually suffers a pilot who fails to point the aircraft in the direction it is moving on touchdown, a tailwheel-type aircraft immediately informs the pilot of the imprecision by heading for the weeds.

Likewise, the knowledge of a pilot or other aviation enthusiast regarding **how** an airplane flies should far exceed the "thrust forward, lift up, drag rearward, and weight down" level. And understanding how a wing lifts should rise far above the "lower pressure on the top, higher pressure on the bottom" level—if one is to know where frost or ice is most critical, where the wing must be strongest, or why one airfoil is better than another. The knowledge that humidity lowers aircraft performance is at about the same level as knowing that more power increases performance—useful but not useful enough, especially when making critical go/no-go decisions. **Aerodynamic knowledge needs numbers too!**

But does numerical (quantitative) knowledge always require proficiency in computer programming and partial differential equations? No! The numerical examples and problems presented in the following pages assume only good high school proficiency in algebra and some trigonometry, and a review is presented in an appendix. Some previous study of physics, at least at the high school level, is highly desirable but physics principles are introduced and summarized before they are used. Finally, a scientific pocket calculator will handle the calculational needs of the problems quite nicely.

This is not to imply that you **have** to work the numerical problems to derive interesting or useful or significant information from the material presented herein. The numerical examples, by themselves, can be quite instructive, and you will discover an introduction to a wide range of concepts and terms, many of which are not readily available elsewhere. The review questions (with answers) are recommended to you at this level of reading. Yet you are strongly encouraged to invest the considerable additional time and effort required to solve a representative number of numerical problems; the greater study this encourages and the deeper insights that result will provide a plentiful reward.

The Science of Flight begins in Chapter 1 with a little additional justification for studying aerodynamics as well as a first look at control surface terminology (especially for non-pilots), aircraft configurations, and performance terminology. The second chapter introduces vectors and demonstrates their application to the determination of ground speed under **windy** conditions, for either cruise or climb or descent. Readers comfortable with vectors are still encouraged to look at the ground speed equations and the determination of climb or glide **angles** when the wind is either a headwind or a tailwind—subjects of great operational importance. Chapter 3 introduces the concept of the **moment** of a force, especially as applied to the balance of aircraft. Readers familiar with moment determination are still encouraged to inspect the discussion of how a little trim tab can control the location and effect of a much larger surface.

Chapters 4 through 6 discuss the static and dynamic properties of air; Chapter 7 treats the important subject of airspeeds. Chapter 8 introduces airfoils by showing how surface pressures lead naturally to airfoil coefficients, while Chapters 9 through 12 develop airfoil ideas through historical comparisons and application to wings. Chapters 13 and 14 develop the idea of replacing the drag of a complete aircraft by the drag of a flat plate, pointed the wrong way, and make applications to various aspects of aircraft performance. Many types of flight maneuvers, from zero-*g* flight to an emergency glide after takeoff, to aerobatic maneuvers, are included in Chapters 15 and 16. Chapter 17 introduces concepts of stability and control, including a simplified numerical analysis. Wind tunnels and computational fluid dynamics as examples of simulations to the actual air flow around an aircraft are presented in Chapter 18. Then, to bring it all to a happy ending, Chapter 19 shows how the preceding chapters relate to the design of aircraft.

The 1970s and 1980s were exciting times for airplane watchers. The whole history of flying was reenacted before our eyes: first hang gliders, then powered hang gliders, then wheeled, powered hang gliders, and finally "ultralight" aircraft. Within this same time frame, homebuilt aircraft embraced new composite materials, explored the use of new airfoils and new aerodynamic devices such as winglets and vortilons and new configurations such as canard-type and joined-wing aircraft, and cruise speeds were extended to over 300 miles per hour.

The Federal Aviation Administration (FAA) has determined that legitimate ultralights (basically single-seat aircraft with an empty weight of less than 254 pounds, a maximum level speed of 62.5 miles per hour, and a minimum speed of 28 miles per hour or less, as defined by Federal Air Regulation Part 103) do not have to be certificated and do not require certificated pilots. The ultralight industry, providing both kits for building and completed aircraft, has narrowed and matured greatly from its heyday in the 1970s. The limited utility of ultralights is balanced by their short field requirements, low speed maneuverability ("fun quotient"), and relatively low cost; expansion into more sophisticated forms places them in price and performance competition with 2-place used aircraft and homebuilts. (Other countries such as England use different terminology—e.g., "microlight"—and have differing definitions but similarly permit the flight of simple aircraft by pilots with lesser qualifications than previously.)

As long as at least **half** of the work is accomplished by the builder for educational purposes, aircraft that do not have the speed and weight and passenger-carrying limitations of ultralights can be built and **certificated** in the United States under the "amateur-built" subheading within the Experimental category. The builder becomes the manufacturer and is responsible for determining that the aircraft is controllable and can be safely flown. After (usually) 40 hours of satisfactory test flying in a designated flight test area, almost all restrictions are removed except the prohibition against using the aircraft for any commercial purposes whatsoever. Either the builder or a certificated aircraft and powerplant (A&P) mechanic may thereafter perform the annual inspection required of all certificated aircraft.

The Experimental Aircraft Association (EAA) was formed in 1952 by Paul Poberezny with the goal of providing encouragement and assistance to members who were building aircraft for certification within this Experimental/Amateur Built category. The organization has grown to include hundreds of thousands of members worldwide and its members are currently sending new aircraft into the skies at a faster rate than the established commercial manufacturers. Because the FAA requires only that builders follow accepted aircraft construction and assembly practices, it has been possible for Amateur-Built aircraft to try new designs, new aerodynamic ideas, and new construction materials—from the stunning around-the-world-unrefueled Voyager to superlative aerobatic mounts to unlimited racers. At first EAA members built mostly single-seat biplanes or small 2-seat speedsters, but currently 2- and 4-place, single- and multiengine, and 300+ mile/hr aircraft are being built. (Good designs—or least good-looking designs—often result in the builder going into the business of providing plans and hard-to-build parts to those who would duplicate his success.) Because this writer has had the opportunity to assist in the building of two homebuilt aircraft (a Quickie and a Long-EZ) as well as flying a number of different types, you can expect to find some references to homebuilt aircraft in the pages that follow. One important caveat: Because the builder is the manufacturer, there is no requirement that any homebuilt aircraft has been built just like any other version of the same basic design, and any comments regarding homebuilt aircraft in this text, strictly speaking, apply only to those examples the author has studied or flown.

Of those aircraft designers who have offered plans and builder support to others, Burt Rutan stands out for the influence of his diverse canard-type aircraft on aircraft design, for his championing of composite construction (especially epoxy-saturated fiberglass cloth over rigid foam), and for the quality of his builder support. Included in his designs for the homebuilder have been the 2-seat VariViggen, the VariEze and the larger, longer-range Long-EZ, the single-seat Quickie, and the twin-engine, 4-seat Defiant. Designs of his which have not been offered to homebuilders include the Quickie-like "Biplane" Racer, the short-takeoff-and-landing (STOL) Grizzly, and the world-famous Voyager, piloted by his brother, Dick, and Jean Yeager. Burt no longer sells plans for his designs, instead concentrating on the design and development of prototype and proof-of-concept aircraft. Included among them are the NASA AD-1 Skew Wing proof-of-concept aircraft, the Next Generation Trainer for Fairchild Republic, the Beech Starship prototype, and the Pond, unlimited class, racer. New homebuilders still have many other designs from which to choose, though, and they offer performance at each extreme of the possible range of propeller-driven speeds. And yes, homebuilt jets may be next.

The author wishes to gratefully acknowledge the large debt this text owes to other texts and research papers. The numerous references at the end of many of the chapters often indicate sources for some of the material in that chapter as well as suggested further reading. Too, the students over the past twenty years who have taken the KSU aerodynamics course on which this text is based have contributed immeasurably to the development of text material, questions, and problems. Professor Ed Gelerinter, who has shared the teaching responsibility for this course, has provided many suggestions for improvements over the years and has provided valuable assistance in error detection. Marion Bury, Terry Brumbaugh, and Joe Sawyer assisted in the gathering of flight data so that a few shots could be fired in the eternal struggle between theory and experiment. Bob Schimel has contributed smoothed and professionally inked figures. In any event, however, I bear the responsibility for all errors in the text and will greatly appreciate any and all assistance in ferreting them out.

I would also like to thank the staff at Iowa State University Press for its work in getting this book into print, especially Bob Cook for technical assistance in the production department; Bob Campbell for his design; Linda Ross for overall proofing; and Bill Silag for moving the project through the editorial stages.

Kent State University, 1992   **W. N. Hubin**

The
Science
of
Flight

Chapter 1

Some Reasons and Some Terminology

Most gulls don't bother to learn more than the simplest facts of flight—how to get from shore to food and back again. . . . For this gull, though, it was not eating that mattered, but flight. . . . "I don't mind being bone and feathers, mom. I just want to know what I can do in the air and what I can't, that's all. I just want to know." [1]

A. The Why of Aerodynamics

Of the many mechanical devices that have fascinated adventurous men and women in this the twentieth century, perhaps none have had a stronger hold that flying and airplanes. Those so afflicted naturally search out and absorb all available information about flight and flight vehicles—and soon this leads to the *practice* of flight. The aim of this text is to lead the curious student of flight, whether presently a pilot or not, to a higher level of knowledge and understanding about the air and about flying machines, in emulation of that remarkable sea gull, JLS.

We begin in this chapter with the labels that have grown up along with the airplane to describe its various parts; we end the chapter with the terms that are used to summarize performance levels. Most of these labels and terms are as old as manned flight but some are creations of our own generation. In succeeding chapters our study will encompass the properties of the air and just how (and how well) our mechanical birds are able to speed and maneuver in the freedom of three-dimensional space.

For us, "higher" knowledge will include an appreciation for the shape and function of every part of an airplane, an appreciation for the effect on aircraft performance of various changes in the air or the airplane, an appreciation of the challenge that still lies ahead in the never-ending search for mastery of the air. Some of this higher knowledge can be obtained from careful reading and studying of texts such as this. Yet real progress in science came about only when it became **quantitative**. It is not enough to know that the air flow over a wing can separate from the surface and thereby produce turbulent eddies and a dramatic reduction in lift; we must be able to predict the speeds at which this occurs so that an optimum design can be produced for each type of aircraft. Again, it is not enough to realize that air pressures can damage an airplane; we must know the critical flight configurations that can produce such air pressures. And so you are strongly urged to go on to that level of understanding that is granted only to those who slog through numerical computations. You can do a great deal with basic physics and the mathematical tools of algebra and trigonometry; much of this book is devoted to proving that. The physical concepts and principles that are needed, things like the laws of statics and dynamics, will be introduced as they are needed. With inexpensive and powerful scientific calculators, attention can be focused on the *significance* of a numerical answer rather than on the calculation process. Very few of these calculations are difficult or long, particularly for one who has already studied physics, but the answers are now wonderfully meaningful in the real world of airplanes. Numerous examples will be supplied to get you started and on the right track.

If you are now a pilot, you are promised the opportunity to become a better and safer one, for good judgment must build on solid facts and principles and not on the fables and phantasmagoria that we call "hangar talk."

B. Fuselage, Empennage, and Such

The main body of an aircraft, exclusive of its wings, tail section, and powerplant, is known as its **fuselage** (Fig. 1.1). As with many of our common aeronautical terms, the name reflects one of the many French contributions to the birth of aviation. The original French word referred probably to the shape of a typical airplane body when it is without its wings and tail. (Another little-known French connection: "hangar" is from a French word for "shed.")

[1]Reprinted with permission of Macmillan Publishing Company from *Jonathan Livingston Seagull* by Richard Bach. Copyright © 1970 by Richard D. Bach and Leslie Parrish-Bach.

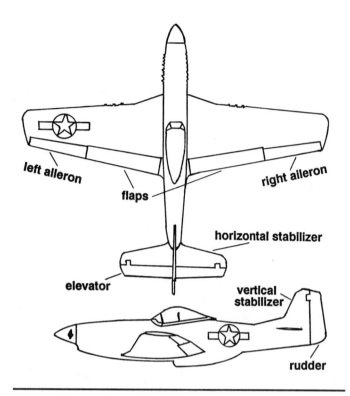

Fig. 1.1 The North American P-51 Mustang, a WW II fighter plane

The **wing** or **wings** provide most of the lift for most flying machines. They do this by giving the air a downward momentum—i.e., lift is obtained by pushing *down* on the air and getting the equal and opposite push promised by Newton's third law, in return. A good wing is one that does this pushing efficiently. The remaining plane surfaces are devoted to the stability and control of the wing-fuselage combination. As with the wing, the control surfaces function by deflecting the air and thereby receiving an opposite push.

To keep the aircraft flying straight into the air, an additional fixed vertical surface (**vertical fin** or **vertical stabilizer**) and an additional horizontal surface are attached to the fuselage at the tail because a fuselage and a wing, by themselves, are normally unstable. That is, if the nose is deflected upward momentarily, perhaps by a gust of wind, it will tend to continue until it is 90° to the direction of flight. Similarly, if the fuselage should "yaw" to the right or left, it will try to go even farther until it is again 90° to the direction of flight. What a ride that would be! The purpose of the tail on an aircraft is similar, then, to the purpose of feathers on the tail of an arrow. If the tail surfaces are made too large or are located too far back, on the other hand, the aircraft will be reluctant to do even normal maneuvering. Optimum sizes and shapes will be discussed later.

The movable vertical surface attached to the vertical stabilizer is the **rudder**. The label is an unfortunate choice of words because it implies a steering function that does not exist. Instead, the rudder is best thought of as a control surface that permits the pilot to place the nose where he wants it relative to his own right or left side. In a normal ("coordinated") turn the rudder is used to make the nose turn an amount appropriate to

the speed and the angle of bank while *entering* or *exiting* a banked turn, as will be discussed in quantitative terms later. In the aerobatic maneuver called a *slow roll*, though, the rudder is used much more generally to prevent the direction of the aircraft's longitudinal axis from changing during the roll; specifically, while passing through the 90° bank point, the rudder is used to keep the nose the proper distance above the horizon, first with rudder deflection one way and then the other way. It is to be noted that the rudder's function *relative to the pilot* has not changed at all, but its function relative to the ground and relative to the horizon has changed completely.

Normally the rudder is controlled via foot pedals and cables or rods, and the pedals are mechanically interconnected so that pushing away on one pedal causes the hinged rudder surface to extend toward the right side of the aircraft, thereby causing a new side force on the right side of the tail that will cause the nose to start to move toward the pilot's right side. (In fact, in the first few decades of aircraft-making, the rudder cables often were attached to the end of a horizontal **rudder bar** that was pivoted at the center, and this was also operated by the feet. This is especially reasonable if individual wheel brakes aren't needed; since they usually are, individual pedals and toe brakes are the norm now.)

The traditional horizontal tail consists of a fixed **horizontal stabilizer** and a movable **elevator**. The most common aircraft wing, by itself, even without the fuselage, is also *unstable*; if a gust of wind turns the nose up, for example, the tendency is again for it to just keep on going up. The horizontal stabilizer thus imparts an overall stability to the aircraft—giving it a tendency to return to its equilibrium position following the occurrence of a vertical disturbance. Payment for this service is exacted in the form of increased drag, particularly because the horizontal tail normally has a *down* load on it, forcing the wings to lift *more* than the aircraft's weight.

The elevator is like the rudder except that it works in an up-down sense (relative to the pilot) rather than in a side-to-side sense. The elevator (*very* poorly named) does *not* do any elevating. Instead, its function is to control the angle of the wings relative to the oncoming air. Within certain aerodynamic limits, the pilot can think of the elevator as a control which moves the aircraft nose either up or down relative to his view directly ahead as the control wheel is pulled toward him or pushed away. Thus in level flight the elevator can be used to place the aircraft nose higher above the horizon (and at **cruise** speeds, this does indeed cause at least momentary elevation—i.e., a gain in altitude) but, contrariwise, in a 90° bank the elevator becomes the control that allows the nose to be swept along the horizon and, even more strangely, in inverted flight the elevator again places the nose in the proper position relative to the horizon but now the *directions* of control movement are *reversed*, with forward movement moving the nose higher *above* the horizon.

Aerobatics, the art of making a properly constructed aircraft into a three-dimensional ballet dancer or heavy metal gymnast, is not for every pilot or passenger. It is a necessary part of the training of every military fighter pilot and it should be included to some extent in the curriculum for every professional pilot, for it is the ultimate training for the recovery from the unusual attitudes that can be caused by severe weather or by turbulence from other aircraft. If it is really to contribute to safety, though, it must be learned under the tutelage of experienced instructors, it must be practiced only in aircraft that are designed for it, and it must be done at high altitudes. To learn one aerobatic maneuver, such as a loop, is to have just enough knowledge to be very dangerous; the important thing is to gain an orientation in space for *every* possible attitude of the aircraft as well as an instinctive ability to make the aircraft do whatever is desired throughout these attitudes. The aerobatically untrained pilot, no matter how many thousands of flight hours are in the logbook, will almost inevitably attempt to recover from inverted flight by completing a half-loop—and lose many hundreds or thousands of feet of altitude, probably saying "bye-bye" to the never-exceed speed of the aircraft, and probably placing excess stress on the wings and tail; in contrast, an aerobatically trained pilot knows how to recover from any attitude with minimum loss of altitude (slow roll!), with minimum airspeed gain, and with minimum stress on the aircraft. For our purposes here, though, it should be noted that aerobatics stimulates some of the most interesting aerodynamic questions and leads to some of the most valuable quantitative questions.

In supersonic flight the control surfaces can no longer affect the air before it gets to them and the result is a great loss in control effectiveness, particularly because a *shock wave* forms in front of the control surface. One early fix for the elevator was to combine it with the stabilizer surface ahead of it to form a single, all-moving control surface known as a **slab tail** (on high speed aircraft) or (more commonly, nowadays) as a **stabilator**. There is not much new under the sun, of course. The 1909 Demoiselle, designed and flown by the French-Brazilian inventor Santos Dumont, used a *single* movable unit for the *whole* tail, mounting it in a universal joint. Some current light aircraft also use stabilators, with mixed reviews from aerodynamicists and pilots.

The stabilizing and control surfaces at the tail of an aircraft are collectively referred to as the **empennage**. (Think *em peh nahzh'* when pronouncing it.) The word derives from the French word for the feathers on an arrow. Americans, not very good at French, tend to use the term **tail feathers** as a colloquial substitute.

The third of the three aerodynamic controls gives the pilot power over roll and bank, a freedom possessed by no earthbound vehicle. Most aircraft use two hinged control surfaces attached to the rear of the outer portions of the wing; these are the **ailerons**. They are mechanically connected so that one always moves in an opposite direction from the other, although not necessarily through the same amount of travel. The result is a change in the average curvature (*camber*) of the two wings which in turn causes a difference in the air pressures around the two wings and a resulting difference in the lift produced by

the two wings, and this induces a rolling acceleration. (The same result was achieved in the first few years of flight by **wing warping**—making the wing flexible enough that it could be differentially cambered by pulling on the rear half with cables.)

Alternately, a few aircraft have used spoilers to produce this differential air pressure and differential lift (Fig. 1.2). When the pilot wishes to begin a roll to his left, for example, he will cause the left wing spoiler to come up; this "spoils" the air flow over the top of the wing so that there is less upward force on that wing, and this causes that wing to dip relative to the other. Spoilers do not tend to be as affective as ailerons and often haven't produced a linear response, causing pilots to complain about a lack of good control "feel," and so they are not used very much even though they leave the whole wing free for other nice things such as high lift flaps for landing.

Spoilers are used in concert on both wings to quickly "dump" the lift after landing, especially for airline transports or soaring aircraft (sailplanes).

The rolling control is fundamentally different from the rudder and elevator controls in that it is a *rate* control rather than a proportional control. When the rudder or elevator surface is deflected, the aircraft tends to stabilize at a new heading or attitude relative to the horizon. However, the rolling control determines the *rate* at which the bank angle is to change. When the control is neutral (ailerons with the same position relative to the wing, or spoilers flush with the wings), the aircraft tends to **stay** at its present angle of bank, whether that bank is 0°, 30°, or 90°. When the control is offset from its neutral position, the degree of deflection determines the rate at which the bank angle is changing. Thus in the aerobatic *slow roll*, the ailerons are held deflected to keep the roll going but must be neutralized to momentarily stop the roll, as in a "4 point slow roll," an "8 point slow roll," etc.

The **flaps** are used to increase the average camber (curvature) of the wing airfoil (cross-sectional shape). They differ from ailerons in that both the flaps deflect together and they usually deflect only *down*, although some sailplanes and experimental designs have a "**cruise flap**" position in which flaps are slightly up. Employment of flaps by the pilot produces an increase in lift and drag which can safely reduce the approach and landing speed for the aircraft, so the primary use of flaps is during the landing approach. Aircraft optimized for short runways sometimes have their ailerons connected to the flap control in such a way that both ailerons deflect partially downward when the flaps are used (while retaining most of their differential action). Ailerons that assist the flaps in this manner

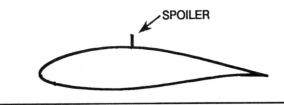

Fig. 1.2 A spoiler in the fully extended position above a wing

are known as **flaperons**. As mentioned, a real advantage of spoilers for roll control is that they free the whole trailing edge for use as a flap surface. On a flying wing or delta wing aircraft, the elevators must provide the aileron function as well; the resulting control surface is called an **elevon**.

Heavy transport aircraft usually have leading edge flaps as well as trailing edge flaps. Both sets of flaps extend forward as well as deflecting downward, so both the camber and the total area of the wing are increased.

In 1903 the Wright brothers made the world's first sustained and controlled flights with a heavier-than-air machine. Their design had the elevator (which they called the rudder!) in *front* of the wing (Fig. 1.3). This configuration has had a small but faithful following throughout the history of flight. Such an aircraft is referred to as a **canard** aircraft and the surface itself is known as the *canard*. Canard is another French word, literally meaning *duck*, which arose because of the supposed duck-like appearance of such an aircraft while in flight. Now the word is used to refer to any horizontal surface *ahead* of the largest horizontal surface (the "wing"), whether or not it contains the elevators.

The elevator on a canard is *lowered* to *increase* its lift during the takeoff roll; with a conventional configuration, though, the

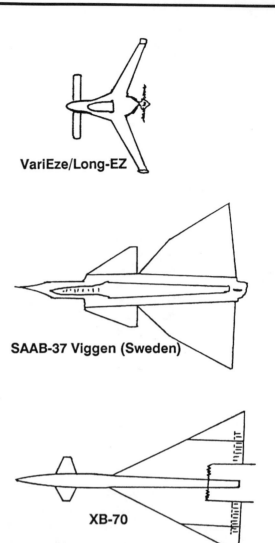

VariEze/Long-EZ

SAAB-37 Viggen (Sweden)

XB-70

Fig. 1.4 Modern canard designs (flying to the left!)

elevator is raised to *decrease* its lift (or increase its negative lift) so that the lift of the wing will increase.

Usually the wing area is considerably greater than the canard surface area and does most of the lifting (but the Quickie is an exception with essentially tandem wings). Aircraft designer Burt Rutan became interested in the canard layout for its potentially safer low speed characteristics. His designs, especially the 2-seat homebuilt VariEze/Long-EZ, the Beechcraft Starship, and the famous round-the-world-unrefueled Voyager, have made canard aircraft an increasingly familiar sight for many people. In addition, in high speed aircraft a canard is often used together with a conventional tail to provide additional elevator authority. Figure 1.4 depicts three modern canard designs.

The ailerons (or spoilers) and the elevator (or stabilator) are controlled by the pilot through either mechanical or electromechanical linkages to a **control stick** or control **wheel**. Transport and cross-country aircraft normally use a control wheel; military, sport, and aerobatic aircraft utilize a stick

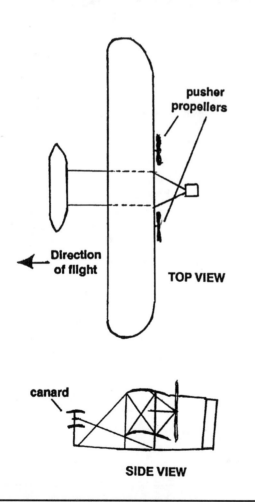

pusher propellers

Direction of flight

TOP VIEW

canard

SIDE VIEW

Fig. 1.3 The Wright Brothers' 1903 Wright Flyer I

SOME REASONS AND SOME TERMINOLOGY
7

mounted ahead of the pilot on the cockpit floor. The stick makes it easier to utilize full control travel but has the disadvantage of impeding entry, especially for an aviatrix in formal wear. Moving the stick (or turning the wheel) to the right tends to roll the plane toward the pilot's right; moving the stick (or wheel) back or forward tends to move (*pitch*) the nose up or down, respectively, relative to the pilot.

The newest military aircraft and the European Airbus A320 (and later model) airliners are using electromechanical linkages so that the pilot generates only electrical *control signals* when moving the control stick, rather than directly moving the surface through cables or control rods. Large aircraft have been using hydraulically boosted controls for some time, but the difference here is that a computer is interposed so that the inputs from the pilot are only requests to the computer. Also, the computer can be programmed to minimize the effect of wind gusts by adjusting the wing lift, without pilot input ("**active control**"), and maneuverability can be enhanced by letting the computer provide the stability. These kinds of aircraft are said to have **fly-by-wire** control systems; they are the norm for current fighter aircraft and may become the norm for transport aircraft. Because muscular effort is adjustable with such a system, it is feasible to save room in the cockpit by transforming the stick into a little miniature version that protrudes from the forward end of an arm rest and is controlled by the wrist and finger tips. Small homebuilt aircraft as well are using these space-saving **side-stick** controllers. In any case, the pilot must be given control *feel* so that the amount of force applied to the stick is directly related to the resulting aerodynamic forces.

In military aircraft at least, **fly-by-light** control systems will probably end up replacing the fly-by-wire systems. Fly-by-light systems use *photons* transmitted over fiber-optic paths, rather than electrons that have been jolted into travelling over copper paths, to convey the pilot's wishes to the computer and to the control surface motors. The potential benefits include immunity to electromagnetic damage, less susceptibility to battle damage, and greater signal density.

A forward extension of the wing along the fuselage produces beneficial air flow effects for high speed, swept-wing aircraft (Fig. 1.5); this extension is known as a **strake**, presumably because of its resemblance to the strake on a boat.

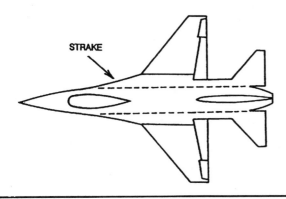

Fig. 1.5 General Dynamics F-16 Air Combat Fighter

It can be considered to be a highly swept forewing. Rutan's Long-EZ and Starship designs use strakes for fuel, baggage, and some additional lifting area. On high speed fighter aircraft such as the F-16, the strakes have a very sharp leading edge that generates "**vortex lift**" during maneuvering.

The most striking of recent wing designs is the fitting of tilted wingtip extensions known as **winglets** (Figs. 1.4a and 1.6). Conceived during a fuel crisis when flight efficiency suddenly became very important, winglets increase the efficiency of the wing by controlling the flow of air around the tip of the wing between lower and upper surfaces. In addition, canard-type aircraft such as the homebuilt Long-EZ and the Beechcraft Starship use winglets to provide the vertical surface area needed for yaw stability as well as a home for the rudders.

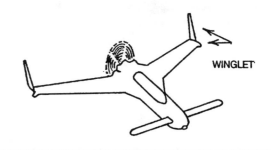

Fig. 1.6 The VariEze in flight

Four other devices are commonly used to control or modify the air flow over the wing. One is the **slot** (Fig. 1.7 and Pic. 1.1), which is a fixed or automatic gap allowing the upper surface air

Fig. 1.7 Cross-section of a wing with a slot

to mix with lower surface air. A second device is the (Pic. 1.2) **vortex generator**, a vane (perhaps 1" by 3", depending on the aircraft size) placed with others as a group on the upper surface of the wing and set at an angle to the direction of flight. The vortex generator has become popular as a band-aid for many aerodynamic ills on both slow and high speed aircraft; it can be thought of as a little wing mounted at 90° to the big wing and at an angle of 20° or so to the oncoming air. The slot is beneficial at slow speeds while vortex generators give extra *rotational* energy to the local air flow and are used to solve problems associated with premature separation and eddy generation in both low and high speed aircraft.

Fences and **vortilons** are plane surfaces that are oriented vertically on the upper middle and lower front, respectively, on the wing of swept-wing aircraft; they are designed to inhibit the tendency of upper-surface air to flow to the wingtip, especially at the slow speeds used for landing. The vortilon (Pic. 1.3) is

Pic. 1.1 A Swift, showing off its slotted wing

Pic. 1.2 A hangared Learjet displays two rows of vortex generators and spoilers.

Pic. 1.3 The Long-EZ's vortilons, as viewed from the cockpit

the newer device and, because it has less drag, is easier to install, looks better to most observers, and is often more effective, will probably prevail in future designs. There will be more about these devices in later chapters.

Some unfamiliar shapes are taking to the skies these days. The Quickie is a very small tandem-winged aircraft that was offered to homebuilders in the late 1970s, in single and 2-seat versions. The 1980s Long-EZ of Fig. 1.9, also an Experimental/Amateur-Built aircraft, is a very successful canard design which has inspired a whole new generation of canard aircraft. A newer design of conventional configuration but also with composite construction is the very speedy Glasair of Fig. 1.10.

Figure 1.11 presents a 3-view of the Beechcraft Starship I, certified in 1988 as a twin-turboprop, 8-seat, 400 miles per hour business aircraft. In Figure 1.12 and scheduled to be certificated in 1989 is the Starship's competition, the Piaggio P180. The Piaggio shows the recent trend toward 3-surface aircraft, attempting to take advantage of the best features of both conventional and canard configurations. The lifting canard on the Piaggio allowed the designers to move the wing back behind the passenger compartment, but the elevator is left on the horizontal tail surface. (Burt Rutan's recent "Catbird" and Advanced Tactical Transport (AT[3]) prototype are also 3-surface designs.)

Figure 1.13 depicts a research aircraft designed to test the "joined-wing" concept of Julian Wolkovitch, an idea which is said to yield a weight saving up to 40% over conventional configurations.

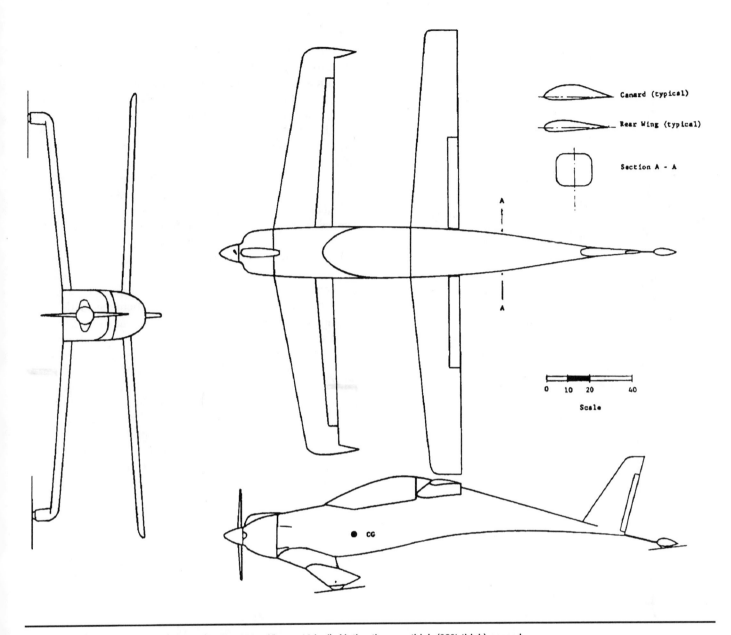

Canard (typical)

Rear Wing (typical)

Section A - A

0 10 20 40
Scale

Fig. 1.8 The tandem-winged Quickie (1 seat, 100 mi/hr on 18 hp!). Notice the very thick (20% thick) canard.

C. Performance Parameters: Wing, Span, and Power Loading

Attempts to quantify the performance characteristics of aircraft have led to the identification of a number of key variables or *parameters*. One of these is **wing loading**, defined as the ratio of an aircraft's gross (or maximum) weight to its wing area. (Here, wing area refers to the area of the shadow that would be cast on the ground by a sun that is directly overhead; thus it is the projected wing area, including an extension of the wing through the fuselage, rather than the wing area involved in recovering one or both surfaces.) The usual unit of wing area in the United States is square feet (ft²).

The wing loading of an aircraft and the shape of its wing together determine the aircraft's minimum level flight speed.

Therefore wing loading is of great importance in determining the climb, approach, and landing speeds. Ultralight aircraft, possessing very low wing loadings, can land at bicycle or jogging speeds while jet transport aircraft possess a very high wing loading and land at much faster speeds.

The effect of a wind gust or air turbulence on the flight path (and on passenger discomfort) is *inversely* proportional to the wing loading. Therefore a jet transport bounces around much *less* in bumpy air than does an aircraft with a low wing loading, independent of the actual weights. Wing loading can be thought of as a measure of how *hard* each little bit of the wing has to work, in pushing down the air, to keep the machine flying.

The usual unit for wing loading in the United States is pounds of weight per square foot of surface area. Typical numerical values: 4 lb/ft² for ultralights, 10 lb/ft² for 2-seat

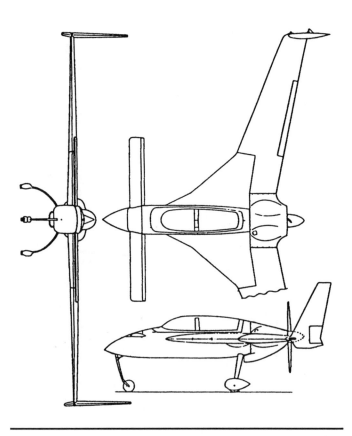

Fig. 1.9 The Rutan Long-EZ (2 seats, 170 mi/hr on 108 hp (Ref. 11.78)

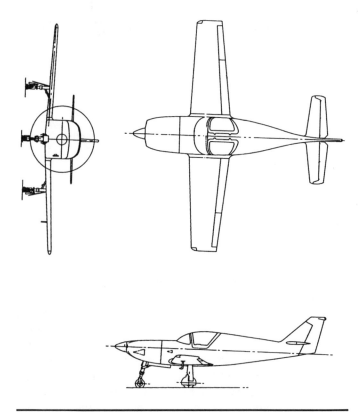

Fig. 1.10 The Glasair III: 2 seats, 280 mi/hr on 300 hp (courtesy of Stoddard-Hamilton Aircraft Inc.)

trainers, 20 lb/ft² for high performance single-engine aircraft, 40 lb/ft² for high performance twin-engine aircraft, 80 lb/ft² for jet fighters, and 120 lb/ft² for jet transports.

Span loading, the ratio of an aircraft's gross weight to its wing *span* (tip to tip), is a parameter than strongly influences the *efficiency* of the wing of an aircraft, and is therefore important in determining the climb rate and the minimum power requirement of the machine. New aircraft designs such as Piper Aircraft's Malibu tend to have longer wing spans than older designs, for this reason. In a piquant sense you can consider the span loading to be a measure of the size and the displaced weight of air created by an aircraft as it bores a hole in the sky. Typical numerical values: 19 lb/ft for ultralights, 50 lb/ft for 2-seat training aircraft, 100 lb/ft for high performance singles, 200 to 300 lb/ft for high performance twins, and 2000 to 3500 lb/ft for jet transports. (Incidentally, the term "span loading" is also commonly used to refer to the spreading out of the payload along the span of a wing to minimize bending loads at the fuselage — especially for a flying wing without a fuselage!)

The third performance parameter that we will introduce here is the **power loading,** the ratio of an aircraft's gross weight to its rated power, usually given in units of pounds of ~~horse~~ weight ~~power~~ per horsepower. (For jet-powered aircraft, though, the appropriate power parameter is the thrust-to-weight ratio.) Typical numerical values: 16 lb/hp for ultralights, 15 lb/hp for 2-seat trainers, 12 or 13 lb/hp for high performance singles, 6 or 7 lb/hp for high performance twins, 0.3 lb of thrust to 1 lb of weight for a jet transport, and 1.1 lb of thrust per lb of weight for a jet fighter. The power loading of an aircraft is the best indicator of the acceleration and rate-of-climb capability of the machine, with the lower values indicating greater performance for power loading. Jet fighters whose thrust capability exceeds their weight can *accelerate* while climbing vertically; they are more like rocket ships than aircraft then.

Pic. 1.4 The classic proportions of the famed North American P-51 Mustang, a high wing-loading, low power-loading fighter aircraft of WW II

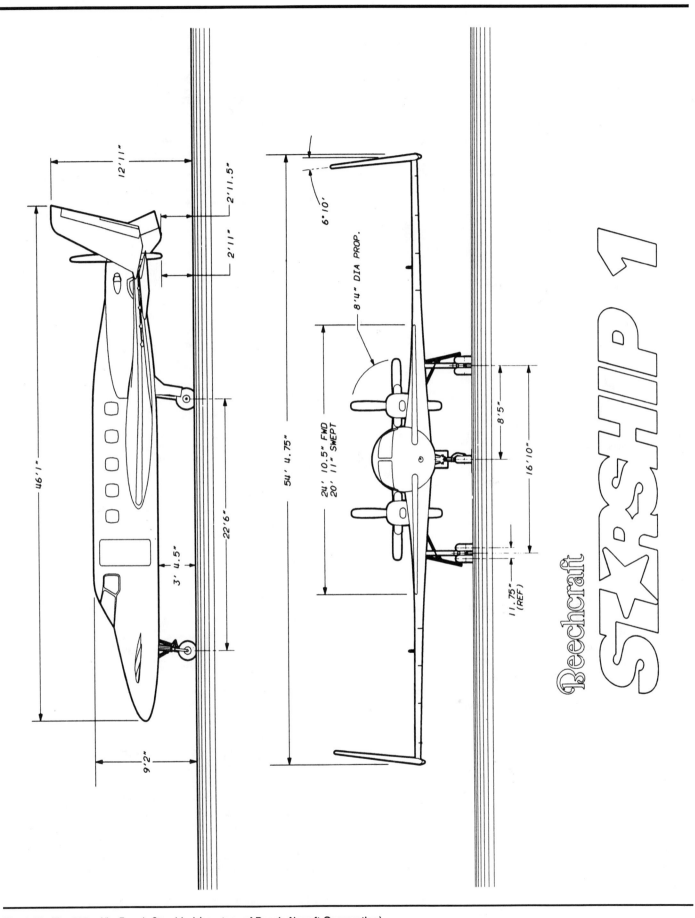

Fig. 1.11 The 380 mi/hr Beech Starship I (courtesy of Beech Aircraft Corporation)

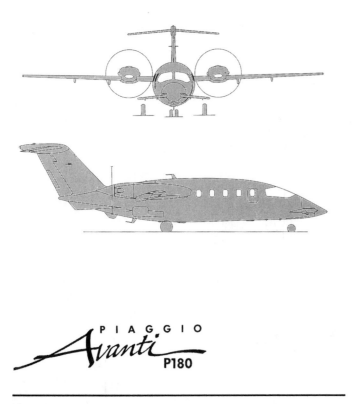

P I A G G I O
Avanti
P180

Fig. 1.12 The Piaggio Avanti P180 3-surface business turboprop
(courtesy of Rinaldo Piaggio)

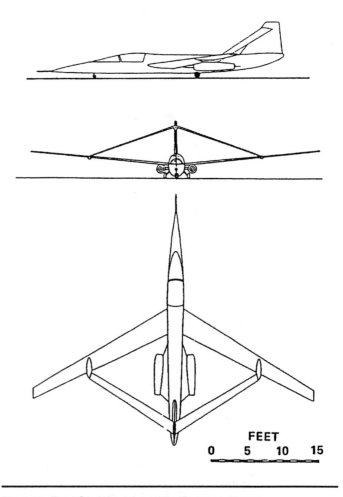

FEET
0   5   10   15

Fig. 1.13 The ACA JW-1 Joined Wing Research Aircraft

Review Questions

1. How do the wings support an airplane on the "thin" air? That is, what is the physical principle involved? (See the discussion of wings.)

2. What is the function of the vertical fin?

3. A pilot begins a slow roll to the right. At 90° of bank she notices that the aircraft nose is too far below the horizon. This means that not enough ___ rudder pressure has been used. (right or left)

4. The pilot in the previous question notices about the same time that the point ahead, which should be in view, is instead hidden behind the aircraft nose. This means that not enough ___ elevator pressure has been used. (forward or back)

5. What parts of an aircraft are included in the empennage?

6. Why do supersonic aircraft need a stabilator?

7. In what direction(s) do the ailerons deflect when a pilot initiates a roll into a right bank?

8. How is the aileron control fundamentally different from the other two primary controls? Compare the position of the controls in a level, 30° bank with that in level flight.

9. Give two effects of deploying flaps.

10. In a slow roll to the right, what is the direction of control wheel/stick deflection and aileron deflection just when the aircraft is passing through the inverted position?

11. What geometrical property of a wing is changed when an aileron on that side is deflected downward? In what direction (increasing or decreasing) does it change?

12. Name one advantage of using spoilers rather than ailerons for roll control.

13. What is a canard surface?

14. What is a strake?

15. What is the purpose of winglets, generally speaking?

16. What is a fly-by-wire control system?

17. Compare the position of spoiler and aileron surfaces while a roll to the left is in progress, from a right bank. Also make this comparison when the aircraft is in a stable 40° bank.

18. The performance parameter that is a measure of how hard the wing has to work to keep an aircraft in the air is the ___ loading.

19. The performance parameter that is closely related to the climb and acceleration capability of an aircraft is the ___ loading.

20. The performance parameter that is closely related to the potential efficiency of an aircraft's wing is the ___ loading.

Chapter 2

Distances, Velocities, and Times

In the first decade of the twentieth century, speed became an American sport. . . . By late 1905, the Wrights were flying their third Flyer at 37 to 38 mph almost routinely. . . . Moseley and the Verville racer won the Pulitzer Trophy Race of 1920 by a margin of several minutes over the runner-up, averaging 157 mph for the 116 miles . . . in the Cleveland National Air Races of 1932 . . . Doolittle and his GeeBee R-1 number 11 . . . tore up the course with an average of 296 mph. . . . In July (1953), activity resumed with an improved F-86D Sabre . . . (Barnes) was clocked at 715.751 mph . . . . in July 27, 1976, an SR-71A reconnaissance plane was flown . . . for an average speed of 2193.64 mph. [1]

A. Distances

Though the idea of going faster isn't what inspired the first successful flying machines, the goddess of speed soon showed her allure and even now (it is said) she holds power over the young. In this chapter we look at some of the basic calculations of flying time, the very quantity that the peripatetic airman seeks to reduce by going ever faster. We begin in this section with a study of distances as vectors.

How far is it to gay Paree? Well, do you wish to have the answer in feet, statute miles, or nautical miles: Since the nautical mile (nm) is winning out in the world of aeronautics, we will use it as our primary unit of distance for the travel plans of this chapter. Conversion to a different system, though, is never farther away than your calculator:

$$
\begin{array}{c}
1 \text{ ft} = 1.646 \times 10^{-4} \text{ nm} \\
1 \text{ nm} = 6080 \text{ ft} = 1.151 \text{ mi} = 1.853 \text{ km} \\
1 \text{ mi} = 0.868 \text{ nm} \\
1 \text{ km} = 0.540 \text{ nm}
\end{array}
$$

A metric unit (1 km = 1 kilometer = 1000 meters) has been given here for reference although the nautical mile appears to be strong enough to resist the metrication tide. Too, the metric system, because of its unfamiliarity, tends to hinder our understanding of the **significance** of a numerical calculation.

[1]Berliner, Don, *Victory over the Wind* (New York: Van Nostrand Reinhold Co. Inc., 1983).

Suppose that we wish to fly from Los Angeles to New York, either direct or by way of Miami (Fig. 2.1). (The directions given on the figure are all referenced to the geographical north pole. It can be seen that the curvature of the earth's surface is being ignored because only on a plane or flat surface do the interior angles of a triangle add up to 180°.)

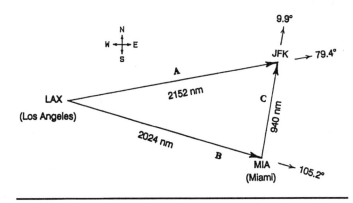

Fig. 2.1 A cross-country odyssey

It is clear that either route takes us to JFK airport. But the distance via Miami (2024 nm + 940 nm = 2964 nm) is much greater than the direct route. The point of this is that distances always have a **direction** as well as a **magnitude** (or size) associated with them and their "addition" does not follow the simple additive rules of pure numbers.

Any quantity that always has a magnitude and a direction associated with it is known as a **vector**. Velocity and force are examples of other vectors that will play an important role in the following pages. Temperature and density are examples of quantities or properties that are not vectors.

Let us call the directed distance from LAX to JFK the vector **A**, the directed distance from LAX to MIA the vector **B**, and the directed distance from MIA to JFK the vector **C**. (Throughout this text we will use **boldface** type whenever the vector nature of a vector quantity is being considered; however, an arrow over the letter is usually used when writing on a blackboard.)

By virtue of its definition, a vector quantity can appear on one side of an equation **only** if the other side is also a vector quantity. (To do otherwise would be to violate the spirit of the equals sign! The double line is sacred here, just as on the highway.) Thus the equation

  A = B + C

is true for the vectors in Fig. 2.1. (Adding vector **C** to vector **B** means that the tail of **C** is touching the head of **B**.) But note well that there is **no** equality for the magnitudes of these vectors:

  A ≠ B + C   (2152 nm ≠ 2024 nm + 940 nm)

It should be apparent from the diagram that there are an infinite number of pairs of vectors **B** and **C** which add vectorially to equal **A**; this is just another way of saying that there are an infinite number of indirect ways to go from LAX to JFK. The equals sign in the equation is telling us that the end result is just the same whether the route is direct or indirect.

The only time we can add (or subtract) vector quantities as if they were ordinary numbers is when they are in the same (or opposite) direction. Clearly, if we walk 5 miles east and then 3 miles west, we end up 2 miles east from our starting point. The vector diagram, and its vector equivalent, would look like this:

N
└─► E

5 nm
3 nm

Equals

N
└─► E

2 nm

The conventional way to handle the problem of adding vectors in the same or opposite direction is to arbitrarily designate one direction as **positive** and the other as **negative**. So we'll choose to call travel toward the east as positive, to the west as negative. Then

  Total Distance Traveled = +5 nm − 3 nm = **+2 nm**

This little example is the key to computing the end result of combining one distance with another distance in a different direction (or even the addition of any number of distances and directions). Suppose that we know only vectors **B** and **C** in Fig. 2.1 and wish to find the resultant vector **A**; that is, suppose we know that we traveled 2024 nm from LAX on a course of 105.2° and then traveled 940 nm on a course of 9.9° and now wish to calculate where we are relative to LAX. Our method will be to

Pic. 2.1 The Cessna 172, a popular 4-place cross-country machine, has flown as far east as Moscow.

(a) replace vectors **B** and **C** with the east-west and north-south vectors to which they are precisely equivalent, **(b)** separately add (or subtract, as appropriate) the east-west and north-south vectors, and **(c)** reconstruct the vector **A** from these two resulting vectors using the theorem of Pythagoras.

First we determine how we could have gone from LAX to MIA by first traveling directly east and then traveling directly south. This is what is usually referred to as "finding the east-west and north-south **components**" of a vector distance.

In Fig. 2.2 the east-west component of **B** is labelled B_X and the north-south component of **B** is labelled B_Y. We arbitrarily (but in accordance with geometric convention) consider north and east to be the positive directions; therefore B_X will be a positive quantity and B_Y will be a negative quantity. By virtue of our choice of two other vectors that are at right angles to each other to equal **B**, we have formed a right triangle, and this permits the use of standard trigonometric relationships.

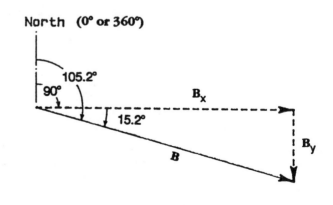

Fig. 2.2 Equivalent perpendicular components of the vector **B**

Referring again to Fig. 2.2,

$$B_X = B \cos(15.2°)$$
$$= (2024 \text{ nm})(0.965)$$
$$= 1953 \text{ nm}$$
$$B_Y = -B \sin(15.2°)$$
$$= -(2024 \text{ nm})(0.2622)$$
$$= -531 \text{ nm}$$

Similarly, for vector **C** (Fig. 2.3), we can write

$$C_X = C \sin(9.9°)$$
$$= (940 \text{ nm})(0.1719)$$
$$= 162 \text{ nm}$$
$$C_Y = C \cos(9.9°)$$
$$= (940 \text{ nm})(0.985)$$
$$= 926 \text{ nm}$$

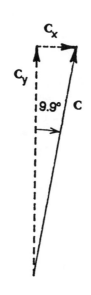

Fig. 2.3 North and east components of vector **C**

Ponder the geometry a few minutes (with the help of Fig. 2.4) and you'll probably soon agree that we can get to JFK by **first** traveling east a distance equal to (B_X + C_X) and **then** traveling north a **net** distance equal to (B_Y + C_Y).

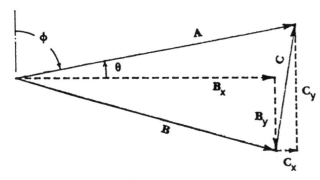

Fig. 2.4 Adding distance vectors **B** and **C** to obtain the equivalent distance vector **A**

Therefore (B_X + C_X) and (B_Y + C_Y) must be the east-west and north-south components, respectively, of vector **A**:

$$A_X = B_X + C_X = 1953 \text{ nm} + 162 \text{ nm} = 2115 \text{ nm}$$
$$A_Y = B_Y + C_Y = -531 \text{ nm} + 926 \text{ nm} = 395 \text{ nm}$$

The Pythagorean theorem tells us that, for a right triangle, the (length)² of the long side (the hypotenuse) is just equal to the sum of the squares of the other two sides. This theorem, with a little algebra, gives us

$$A = \sqrt{A_X^2 + A_Y^2}$$
$$= \sqrt{\left(2115 \text{ nm}\right)^2 + \left(395 \text{ nm}\right)^2}$$
$$= \mathbf{2152 \text{ nm}}$$

and, from the definition of the tangent of an angle,

$$\theta = \arctan(A_Y / A_X)$$
$$= \arctan(395/2115) = \arctan(0.1868)$$
$$= \mathbf{10.6°}$$

The angle θ (the Greek letter **theta**) gives the direction of **A** relative to east; therefore the direction of **A** relative to north is

$$\phi = 90° - \theta = 90° - 10.6° = 79.4°$$

where ϕ is the Greek letter phi.

Now compare this **A** and its direction with Fig. 2.1 to see that we have truly succeeded in our task of recreating the vector **A** from vectors **B** and **C**.

Are there any other ways that this could have been accomplished? Yes! For one thing, we could have chosen to determine components along perpendicular axes other than east-west and north-south directions. (But our choice was a good one because we knew the direction of **B** and **C** relative to north to begin with.)

Second, we could have chosen a drawing scale (1" = 100 nm perhaps) and with a protractor and ruler drawn a scaled **B** (a line 20.24" long in the 105.2° direction, using the suggested scale) and added to its end the scaled vector **C** (then a line 9.40" long in the 9.9° direction). An accurate ruler would then have shown a scaled distance (about 21.52") representing the 2152 nm for the vector sum and the protractor would have given us the angle of 79.4° relative to north for **A**. But this graphical method is much too slow and requires large sheets of paper for accuracy, so the trigonometric method along with calculators or programmed calculator/computers is certainly preferred in practice.

The component-trigonometric method can be used to add **any** number of vectors together. As an example of the addition of three vectors, suppose that our trip was LAX → MIA → JFK → CLE, where CLEveland is 383 nm from JFK on a 289.0° course (Fig. 2.5).

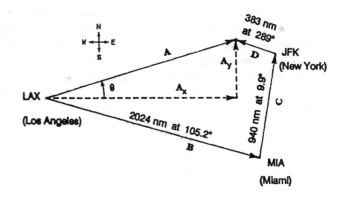

Fig. 2.5 Adding distance vectors **B** and **C** and **D** to obtain the equivalent distance vector **A**

How far and in what direction from our starting point have we traveled? In other words, what is **A**? (These calculations are appropriate for flight planning and for en route calculations when the ground is not visible.) Again we first add all the east-west components separately and then add the north-south components.

$$A_X = B_X + C_X + D_X$$
$$= 1953 \text{ nm} + 162 \text{ nm} + (-383 \text{ nm})(\cos 19.0°)$$

$$= 1753 \text{ nm}$$
$$A_Y = B_Y + C_Y + D_Y$$
$$= -531 \text{ nm} + 926 \text{ nm} + (383 \text{ nm})(\sin 19.0°)$$
$$= 520 \text{ nm}$$
$$A = \sqrt{A_X^2 + A_Y^2} = \textbf{1828 nm}$$
$$\theta = \arctan(A_Y / A_X) = \arctan(0.297) = \textbf{16.5°}$$
Course relative to north $= 90° - \theta = \textbf{73.5°}$

Note that Fig. 2.5 makes it clear that **D** has a north component and a west component so they must be considered as positive and negative components respectively.

B. Horizontal Velocities and Times

Now for the vector that lures airplane buffs—velocity! In this section we will derive formulas that will allow us to predict the ground path and ground speed of our aircraft when it is subjected to a known wind of arbitrary strength and direction.

The *velocity* of an object is defined as the rate at which its position is **changing**. If an object's location is changing, you'll have no problem agreeing that it has a non-zero velocity. But the magnitude or strength of that velocity vector depends on how rapidly its position vector, measured from any arbitrary point, is changing. In other words, we must divide a position change (a distance moved) by the time it took to make that change and this will give us the *average* velocity during that time interval. Therefore we define

$$\text{average velocity} \equiv \frac{\text{vector distance moved}}{\text{time interval required}} \quad (2.1)$$

(Mathematical notation: In this equation three parallel lines, $\equiv$, are used as a symbol meaning equal by *definition*.)

Note that (**a**) a particular time interval is stated or implied whenever the term *average velocity* is used, (**b**) average velocity is a **vector** whose direction is the *same* direction as the distance vector connecting starting and ending points for the stated time interval but whose numerical value (magnitude) is obtained by dividing that distance vector by the time interval, and (**c**) the units of average velocity must be a length divided by a time unit since units must be carried along in any numerical operation.

Whenever we wish to refer only to the **magnitude** of velocity, without reference to its direction, we will be careful to use the word **speed**. This distinction, although not maintained in everyday speech, is a useful and conceptually important one.

Commonly used units for velocity or speed include nautical mi/hr, feet/sec, and feet/min. The first unit is usually replaced by **knot** (here abbreviated as **kt**); this makes it easier to say and write the unit but does tend to obscure the fact that the unit represents a distance divided by a time. We will generally write the second unit as **ft/s** and the third unit as **ft/min**; common (but less explicit) usage is to write them as FPS and FPM respectively.

In that first trip, from LAX to JFK via MIA, in the previous section, suppose that the first leg took 2 hours and 12 minutes [or 2 hr + 12/60 hr = 2.20 hrs] and the second leg took

Pic. 2.2 This Luscombe 8A is an example of an efficient 1940s cross-country design; it was among the first to use a metal semi-monocoque fuselage rather than fabric-covered steel tube.

precisely the **same** amount of time (Fig. 2.6). Then we can calculate the **average** speed for each of these legs:

average speed (1st leg) = (2024 nm)/(2.20 hr)
$$= \textbf{920 kt} \quad (\text{LAX} \rightarrow \text{MIA})$$
average speed (2nd leg) = (940 nm)/(2.20 hr)
$$= \textbf{427 kt} \quad (\text{MIA} \rightarrow \text{JFK})$$

The average speed for the **whole** trip was
$$\text{average speed}_{\text{TOTAL TRIP}} = \frac{(2024 \text{ nm} + 940 \text{ nm})}{(2.20 \text{ hr} + 2.20 \text{ hr})}$$
$$= \textbf{674 kt}$$

It's neater to work in symbols. We use V_{AV} for the average speed, d_{TOT} for the total distance traveled, and t_{TOT} for the total time involved. Then

$$V_{AV} = \frac{d_{TOT}}{t_{TOT}} \quad (2.2)$$

and, by algebra,

$$t_{TOT} = \frac{d_{TOT}}{V_{AV}} \quad (2.3)$$

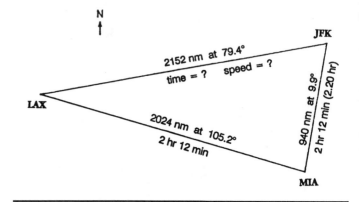

Fig. 2.6 Speeds and times for the odyssey

Vectors Adding are referenced to x & y axies

This, of course, is just the relationship that is used to calculate travel time before any sort of trip. The hard part is determining the correct value of V_{AV}!

The trip from LAX to JFK took 4.40 hours at an average speed of 674 kt. If we had flown directly (refer to Fig. 2.6 again) and maintained the same average speed, it would have taken us only

$$t_{TOT} = (2152 \text{ nm})/(674 \text{ nm/hr}) = 3.19 \text{ hr}$$
$$= 3 \text{ hr} + (0.19 \text{ hr})(60 \text{ min/hr})$$
$$= \textbf{3 hr 12 min}$$

...but it was worth it to go via Florida!

On the other hand, if we had gone directly from LAX to JFK in the same time of 2 hrs 12 min, our average speed would have been

$$V_{AV} = (2152 \text{ nm})/(2.20 \text{ hr}) = \textbf{978 kt}$$

Do you realize the significance of this computation? It is that, as far as our end points are concerned, precisely the same result is obtained from either (**a**) flying at a speed of 920 kt on a heading of 105.2° for 1 hr 12 min and then flying at a speed of 427 kt on a heading of 9.9° for the same time or (**b**) flying at a speed of 978 kt on a heading of 79.4° for just the same 1 hr 12 min. **Velocities are vectors that add just as do distances!** (Fig. 2.7)

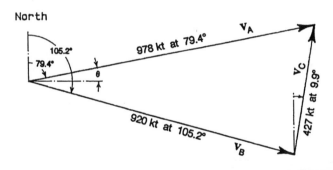

Fig. 2.7 Adding velocity vectors to obtain an equivalent velocity

These velocity vectors were obtained by dividing each of the previous distance vectors by the same value of time, so all we really did was scale the previous vector diagram. (Graphically this could have been from 1" = 100 nm to 1" = [(100 nm) / (2.20 hr)] = 45.45 kt.) As another indication that vectors V_B and V_C really do add up to vector V_A, we can use the method of components:

$$V_{Ax} = V_{Bx} + V_{Cx}$$
$$= (920 \text{ kt})[\cos(15.2°)] + (427 \text{ kt})[\sin(9.9°)]$$
$$= 888 \text{ kt} + 73 \text{ kt}$$
$$= 961 \text{ kt}$$

$$V_{Ay} = V_{By} + V_{Cy}$$
$$= (920 \text{ kt})[\sin(15.2°)] + (427 \text{ kt})[\cos(9.9°)]$$
$$= -241 \text{ kt} + 421 \text{ kt}$$
$$= 180 \text{ kt}$$

$$V_A = \sqrt{V_{Ax}^2 + V_{Ay}^2} = \sqrt{(961)^2 + (180)^2} = \textbf{978 kt}$$

$$\theta = \arctan\left(\frac{V_{Ay}}{V_{Ax}}\right) = \arctan(0.187) = \textbf{10.6°}$$

(so V_A has an angle of [90°–10.6°] = 79.4° relative to north)

Instead of going all the way to MIA, we could have gone a few miles in that direction and then a few miles in the MIA → JFK direction, so that our path would have been a zigzag one (Fig. 2.8).

Fig. 2.8 A different route between Los Angeles and New York

In the limit of smaller and smaller increments of zig and zag, the direct route can be considered to be the result of simultaneous motion in the LAX → MIA and the MIA → JFK directions. Similarly, the 978 kt velocity vector direct to JFK can be considered to be the result of simultaneous velocities in the LAX → MIA and the MIA → JFK directions. This should make it clear that the **end result** is the same if we do any one of the following:

(**a**) fly direct from LAX → JFK at an average speed of 978 kt for 1 hr 12 min, or

(**b**) fly direct to MIA at an average speed of 920 kt for 1 hr 12 min and then direct to JFK at an average speed of 427 kt for 1 hr 12 min, or

(**c**) fly direct to MIA at an average speed of 920 kt for 1 hr 12 min and then remain stationary in a body of air (wind) that is moving toward JFK at 428 kt for the same 1 hr 12 min, or

(**d**) point (or head) the airplane directly toward MIA and travel through the air that is around us at 920 kt while that body of air is itself moving (really blowing) at 428 kt in the 9.9° direction, for a total time of 1 hr 12 minutes.

This last interpretation of the addition of vector velocities is the basis for dead-reckoning calculations. There are **two** situations of particular interest: (**a**) in a preflight mode we wish to calculate the proper heading and the resulting speed with respect to the ground (ground speed) for our aircraft, for a given destination and for the (forecast!) winds aloft, and (**b**) in an inflight mode we have held a heading for a measured length of time and wish to determine the actual wind velocity from our present location.

Both of these problems can be accomplished either graphically or with mechanical or electronic calculators. The procedures required are now accessible to us, using the vector concepts presented earlier along with trigonometry. Even if you don't wish to follow all the details of the algebra in the next page or so, you should follow the general steps of the derivation and make some use of the resulting equations; this will make future use of a flight computer much more meaningful. It will be shown shortly that the preflight problem can be solved through equations (2.7) through (2.9) while the in-flight problem requires equations (2.10) through (2.13).

Suppose that we wish to fly direct from LAX to JFK and need to correct for a south-southwest (SSW) wind V_W (Fig. 2.9). We know that our still-air cruising speed is V_{AA} (where the subscript AA means aircraft relative to air), the wind speed is V_W, the wind direction is θ_W (relative to true north), and our desired course direction is θ_C (relative to true north). The problem is to find the direction that we should head the aircraft θ_H (also relative to true north) and the magnitude of our resultant aircraft speed V_{AG} over the ground.

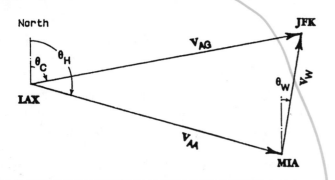

Fig. 2.9 Flying east while fighting a south-southwest wind

First, note that two of the interior angles of the vector triangle can be given in terms of θ_H and θ_W (Fig. 2.10).

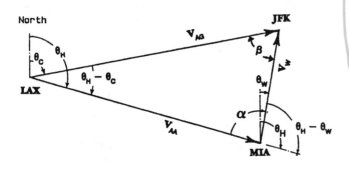

Fig. 2.10 Geometry of the windy trip from Los Angeles to New York

The angle on the left is just $(\theta_H - \theta_C)$. The angle in the lower right corner, which we designate by the first letter of the Greek alphabet α (alpha), requires a little more work (algebra and geometry). Note from the figure that we can write

$$\alpha + (\theta_H - \theta_W) = 180°$$

or $\qquad \alpha = 180° - \theta_H + \theta_W \qquad (2.4)$

But the interior angles of the triangle must also add to 180°, so

$$(\theta_H - \theta_C) + (180° - \theta_H + \theta_W) + \beta = 180°$$

or $\qquad \beta = \theta_C - \theta_W \qquad (2.5)$

Next we use the law of sines:

$$\frac{\sin(\beta)}{V_{AA}} = \frac{\sin(\theta_H - \theta_C)}{V_W} \qquad (2.6)$$

Using algebra to combine Equations (2.5) and (2.6) yields

$$\frac{\sin(\theta_C - \theta_W)}{V_{AA}} = \frac{\sin(\theta_H - \theta_C)}{V_W}$$

which can be solved for θ_H :

$$\sin(\theta_H - \theta_C) = (V_W / V_{AA}) [\sin(\theta_C - \theta_W)]$$

or $\qquad \theta_H - \theta_C = \arcsin [(V_W / V_{AA}) \{\sin(\theta_C - \theta_W)\}]$

and therefore:

$$\boxed{\theta_H = \theta_C + \arcsin \left[\left(\frac{V_W}{V_{AA}}\right) \sin \left(\theta_C - \theta_W\right) \right]} \qquad (2.7)$$

An expression for the aircraft's speed relative to the ground follows rather directly from the law of cosines:

$$V_{AG}^2 = V_{AA}^2 + V_W^2 - 2 V_{AA} V_W \cos(\alpha)$$
$$= V_{AA}^2 + V_W^2 - 2 V_{AA} V_W$$
$$- 2 V_{AA} V_W [\cos(180) \cos(\theta_W - \theta_H)$$
$$+ \sin(180°) \sin(\theta_W - \theta_H)]$$
$$= V_{AA}^2 + V_W^2 - 2 V_{AA} V_W [-\cos(\theta_W - \theta_H)]$$
$$= V_{AA}^2 + V_W^2 + 2 V_{AA} V_W [\cos(\theta_W - \theta_H)]$$

where we have substituted for α from Eq. 2.4 and used a trigonometric identity for the cosine of the sum of two angles, considering 180° as one angle and $(-\theta_H + \theta_W) = (\theta_W - \theta_H)$ as the other angle. The magnitude of our aircraft's ground speed should then be the square root of the right side of this equation, but mathematically this can be **either** a positive or a negative number:

$$\boxed{V_{AG} = \pm\sqrt{V_{AA}^2 + V_W^2 + 2 V_{AA} V_W [\cos(\theta_W - \theta_H)]}} \qquad (2.8)$$

(Note that it will be necessary to solve Eq. 2.7 for the proper aircraft heading angle, θ_H, before this equation can be solved for the ground speed.)

The square root of a positive number can be negative. Is it possible in this case that the negative choice represents a physically meaningful answer? **Yes!** If the wind is too strong for the speed of the aircraft, Eq. 2.7 will tell us the heading that will make the aircraft back directly away from the destination!

Let's try Eqs. 2.6 and 2.8 for the situation of Figs. 2.7 and 2.9.

Input Conditions:

$$V_{AA} = 920 \text{ kt}, \theta_C = 79.4°, \quad V_W = 427 \text{ kt}, \theta_W = 9.9°$$

Substituting in Eq. 2.7 gives us

$$\theta_H = \theta_C + \arcsin[(V_W / V_{AA})\{\sin(\theta_C - \theta_W)\}]$$
$$= 79.4° + \arcsin [(427/920) \{ \sin(79.4° - 9.9°) \}]$$
$$= 79.4° + \arcsin(0.435) = 79.4° + 25.8°$$
$$= \mathbf{105.2°}$$

which indeed agrees with the aircraft heading shown in Fig. 2.7. The ground speed is obtained from Eq. 2.8:

$$V_{AG} = \sqrt{V_{AA}^2 + V_W^2 + 2\,V_{AA}\,V_W\,[\cos(\theta_W - \theta_H)]}$$
$$= \sqrt{920^2 + 427^2 + (2)(920)(427)\,[\cos(9.9° - 105.2°)]}$$
$$= \mathbf{978\ kt}$$

which is again the correct value.

How about a simple check to verify what happens when the wind is too strong for us? Suppose (a) our plane flies at 70 kt, (b) we wish to fly a course of 79.4°, and (c) the wind is right on our nose at 80 kt ($\theta_W = 259.4°$). We already know that our ground speed must be 10 kt backwards if we head directly into this wind, but let's see what Eqs. 2.7 and 2.8 tell us.

$$\theta_H = 79.4° + \arcsin[(80/70)\{\sin(79.4° - 259.4°)\}]$$
$$= 79.4° + \arcsin[(80/70)\{\sin(-180°)\}]$$
$$= 79.4° + \arcsin(0)$$
$$= \mathbf{79.4°}$$

$$V_{AG} = (-)\sqrt{70^2 + 80^2 + 2(70)(80)\cos(259° - 79.4°)}$$
$$= (-)\sqrt{70^2 + 80^2 + 2(70)(80)\cos(180°)}$$
$$= (-)\sqrt{70^2 + 80^2 + 2(70)(80)(-1)}$$
$$= (-)\sqrt{(4900 + 6400 - 11200)}$$
$$= (-1)\sqrt{100}$$
$$= \mathbf{-10\ kt}$$

So our equations give the correct answer if we can find a rule that will tell us when to use the negative sign with Eq. 2.8. Our solution of the wind triangle is not complete until we discover a sure way to predict a negative ground speed like this one! Figure 2.11 should clarify the problem and lead us to a solution. When the wind direction is equal or greater than 90° from our intended course, the wind is certainly going to hurt our ground speed (Figs. 2.11 and 2.12). If our airplane's airspeed is greater than the wind speed (Fig. 2.11), we can think of using the aircraft's air velocity (a) to cancel out the wind's effect ($V_{BUCKING}$) and (b) to produce some ground speed toward the destination with what is left over (V_{EXTRA}). If our aircraft's airspeed is less than the wind's speed, though, we may be able to cancel out only the perpendicular component of the wind's

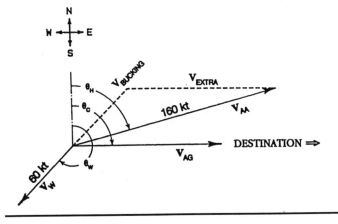

Fig. 2.11 The BAD wind condition

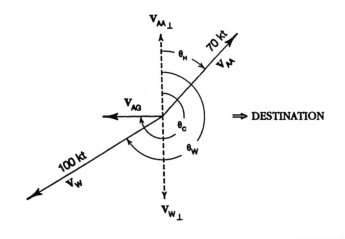

Fig. 2.12 The REALLY BAD wind condition

velocity (Fig. 2.12) and we end up backing away from our intended destination!

So: There may be no heading that will allow us to go either directly toward our destination or directly away from it. This would be the situation if we had a 90° crosswind that was greater than our airspeed. In this case the argument for the arcsine function in Eq. 2.7 will have an absolute value greater than one and the equation will not have a solution (since the sine of any real angle is always between −1 and +1) and we will receive fair warning. (Try $V_W = 100$ kt, $V_{AA} = 80$ kt, $\theta_C = 90°$, $\theta_W = 180°$ and check for yourself.)

Figure 2.13 depicts the situation where we can always make headway but the wind can either hurt or help us. (This assumes that our aircraft speed is greater than the wind's perpendicular velocity component, V_W, as discussed in the previous paragraph.)

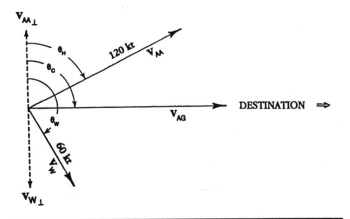

Fig. 2.13 The SOMETIMES GOOD wind condition

A different situation arises en route. Here we typically have experimentally discovered the heading that keeps us on course as well as our ground speed. (The ground speed is obtained from radio navigation aids and/or from using a map and dividing the distance between two identified points by the measured flight time between the points.) With this information we can calculate what the wind aloft really is doing and it may be quite different from the forecast.

Refer to Fig. 2.14. We use the actual direction of flight over the ground, θ_{AG}, rather than the course direction, θ_C, to permit the calculation to apply whether we're on or off course.

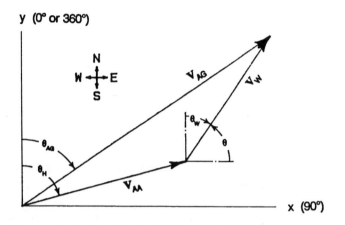

Fig. 2.14 Determining properties of the wind during flight

Since $V_{AG} = V_{AA} + V_W$, by the use of algebra we have
$$V_W = V_{AG} - V_{AA},$$
and we can obtain the wind components along the x and y axes as follows:

$$V_{Wx} = V_{AGx} - V_{AAx} = V_{AG} \sin\theta_{AG} - V_{AA}\sin\theta_H \quad (2.10)$$

$$V_{Wy} = V_{AGy} - V_{AAy} = V_{AG}\cos\theta_G - V_{AA}\cos\theta_H \quad (2.11)$$

Then we obtain the strength of the wind from

$$V_W = \sqrt{V_{Wx}^2 + V_{Wy}^2} \quad (2.12)$$

The angle θ relative to the x axis is obtained from

$$\theta = \arctan\left(\frac{V_{Wy}}{V_{Wx}}\right) \quad (2.13)$$

However, calculators will always return a θ in the range of 0° to 90° if the arctangent is positive and a θ in the range of 0° to –90° if the arctangent is positive. We have to make the check ourselves to see if θ really is in one of the other quadrants. If θ is negative and if V_{Wx} is negative, θ is really in the $\theta = 270°$ to 360° quadrant and we should add 180° to the value calculated from Eq. 2.13. Also, for θ positive and both V_{Wx} and V_{Wy} negative, θ is really in the $\theta = 180°$ to 270° quadrant and we should again add 180°. Finally, we must convert angles measured counterclockwise from the x axis to compass angles measured clockwise with respect to north. The following formulas summarize these steps.

First:

$$\theta \to \theta + 180° \text{ if } \theta \text{ and } V_{Wx} < 0$$

or:

$$\theta \to \theta + 180° \text{ if } \theta > 0 \text{ and } V_{Wx} \text{ and } V_{Wy} < 0 \quad (2.14)$$

and then obtain the wind direction from:

$$\theta_W = -\theta + 90° \quad (2.15)$$
(add 360° if necessary to make angle positive)

We can use the example in Fig. 2.11 (in reverse) to verify these equations:

$\theta_{AG} = 90°$, $\theta_H = 75°$, $V_{AA} = 160$ kt, $V_{AG} = 112$ kt:

$V_{Wx} = (112)(\sin 90°) - (160)(\sin 75°) \quad$ (from Eq. 2.10)
$\quad = -42.5$ kt

$V_{Wy} = (112)(\cos 90°) - (160)(\cos 75°) \quad$ (from Eq. 2.11)
$\quad = -41.4$ kt

$V_W = \sqrt{(-42.5)^2 + (-41.4)^2} \quad$ (from Eq. 2.12)
$\quad = \mathbf{59\ kt}$

$\theta = \arctan[(-41.4)/(-42.5)] = 44° \quad$ (from Eq. 2.13)

$\theta \to \theta + 180° = 224° \quad$ (from Eq. 2.14)

$\theta_W = -224° + 90° = -134° = \mathbf{+226°} \quad$ (from Eq. 2.15)

Note that these are the correct values except for a slight round-off error (because we worked backwards from a rounded-off answer). Note also that we add 360° to the value of θ_W if it turns out to be negative after Eq. 2.15 is used.

C. Vertical Velocities and Times

We have not previously mentioned the important fact that all legitimate flying takes place above the ground, so that a part of each flight is necessarily spent in climbing to and descending from the cruise altitude. In this section we make amends by investigating obstacle clearance and making some total flight time computations.

Pic. 2.3 The PJ-260 is an experimental aircraft that is also a fine sport aerobatic machine.

The first aeronautical problem facing a pilot is that of escaping the ground and clearing any obstacle beyond the end of the runway. The 1977 Cessna T210, a high performance, turbocharged single-engine aircraft, requires a ground roll of 1200 feet and an additional 770 feet to clear a 50 ft obstacle at an optimum climb speed of 75 kt under standard sea level conditions, according to the flight manual. (Government-issue airport trees are all 50 feet in height.) These performance figures are for zero wind. The ground roll and climbout are diagrammed in Fig. 2.15. The angle of climb relative to the ground in no-wind conditions is designated by the Greek letter gamma, γ.

(CLIMB ANGLE EXAGGERATED)

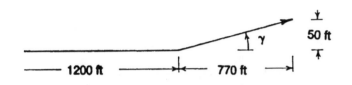

Fig. 2.15 Zero-wind takeoff and climbout for a turbo T-210

The angle of climb is easily calculated:

$$\gamma = \arctan\,[(50\text{ ft})/(770\text{ ft})] = \textbf{3.7°}.$$

The vector diagram for the aircraft velocity under these zero wind conditions is shown in Fig. 2.16. The horizontal component of velocity has been obtained from $\cos\gamma$, and the vertical component from $\sin\gamma$. The conversion of kt to ft/min requires this conversion factor:

$$\boxed{1\text{ kt} = 1\text{ nm/hr}\times6080\text{ ft/nm}\times1\text{ hr/60 min} = 101.3\text{ ft/min}}$$

so that

$$4.84\text{ kt} = 4.84\text{ kt}\times\frac{101.3\text{ ft/min}}{\text{kt}} = \textbf{490 ft/min}.$$

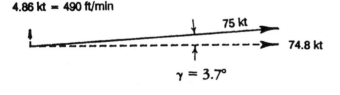

Fig. 2.16 Vector velocities during climbout (no wind)

Now if we have a steady **headwind** component during the climb, V_{HW}, the velocity relative to the ground, V_{AG}, is given by the velocity diagram of Fig. 2.17.

Evidently taking off into a wind gives us a steeper angle of climbout, θ_{AG}, than a climbout under no-wind conditions. In addition, the flight manual specifies a reduction in ground roll

(CLIMB ANGLES GREATLY EXAGGERATED)

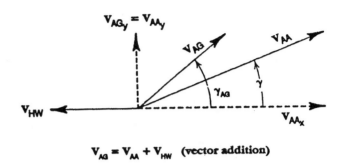

$$V_{AG} = V_{AA} + V_{HW}\quad\text{(vector addition)}$$

Fig. 2.17 Vector velocities for climbout into a 20 kt headwind

of 10% (i.e., multiply by 0.1) for each 10 kt of headwind or an increase of 10% for each 2.5 kt of tailwind.

The key factor to realize is that a pure headwind or tailwind affects **only** the horizontal component of the aircraft's velocity relative to the ground.

EXAMPLE 2.1. Determine the minimum ground roll, the maximum angle of climb, and the minimum total distance required by the 1977 Cessna T210 if it is to clear a 50 ft obstacle under conditions of a 20 kt headwind and standard sea level conditions.

Solution: (a) Ground roll = 1200 ft
$- (0.1\times1200\text{ ft})/(10\text{ kt headwind})\times(20\text{ kt headwind})$
$= 1200\text{ ft} - 240\text{ ft} = \textbf{960 ft}$

(b) We first calculate and add velocity components in the x and y directions (Fig. 2.17).

$$V_{AGx} = V_{AAx} - V_{HW} = 74.8\text{ kt} - 20.0\text{ kt} = 54.8\text{ kt}$$
$$V_{AGy} = V_{AAy} = 4.84\text{ kt}$$

Therefore $\gamma_{AG} = \arctan\,[V_{AGy}/V_{AGx}] = \textbf{5.05°}$

(c) Distance traveled while climbing to 50 feet:

$$x = (50\text{ ft})/[\tan(\gamma_{AG})] = \textbf{566 ft}\quad\text{(see Fig. 2.18)}$$

Therefore, the total distance required to take off and climb 50 feet with a 20 kt headwind is
$$(960\text{ ft} + 566\text{ ft}) = \textbf{1526 ft.}$$

(CLIMB ANGLE EXAGGERATED)

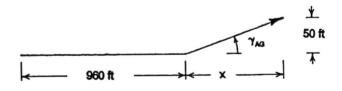

Fig. 2.18 Vector distances from takeoff to 50 ft altitude, with a 20 kt headwind

EXAMPLE 2.2. Repeat the calculations for a 20 kt **tailwind** component (Figs. 2.19 and 2.20).

(CLIMB ANGLES GREATLY EXAGGERATED)

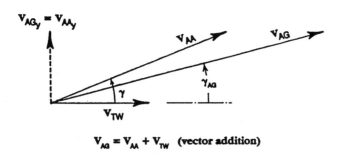

$V_{AG} = V_{AA} + V_{TW}$ (vector addition)

Fig. 2.19 Vector velocities for climbout into a 20 kt tailwind

(CLIMB ANGLE EXAGGERATED)

Fig. 2.20 Vector distances from takeoff to 50 ft altitude, with a 20 kt tailwind

Solution: (a) Ground roll = 1200 ft
+ [(0.1 x 1200 ft)/(2.5 kt tailwind)]×(20 kt tailwind)
= **2160 ft**
(b) $V_{AGx} = V_{AAx} + V_{TW}$ = 74.8 kt + 20.0 kt = 94.8 kt
$V_{AGy} = V_{AAy}$ = 4.84 kt

Therefore γ_{AG} = arctan [V_{AGy} / V_{AGx}] = **2.92°**
(c) Distance traveled while climbing to 50 feet
= x = (50 ft) / [tan (γ_{AG})] = **979 ft**

Therefore the total distance required to take off and climb 50 ft with a 20 kt tailwind is
(2160 ft + 979 ft) = **3139 ft.**

The performance difference between taking off into or with a wind is **spectacular!** Don't check it out in an airplane if at all possible!

Realistically, you **will** encounter some downwind conditions if you fly long enough. In hilly or mountainous terrain, some airports have sloping runways or are even one-way because of higher obstacles in one direction. (In a later chapter we'll be able to estimate where the benefits of taking off downhill outweigh the debits of taking off downwind.) Even in the flatlands, many airports have a "calm wind" runway that is used whenever a "small" downwind component exists even though there may be a much larger downwind component at

higher altitudes during climbout. Finally, in unstable, convective air, aircraft have experienced a measured 10 kt headwind followed by a 15 kt tailwind—during the takeoff roll. Medium-performance aircraft may have no problem with a slight tailwind, but when you appear in your low-performance single on a hot day, or when your twin-engine becomes a single-engine, you may find yourself in exceedingly difficult circumstances!

Another "minor" correction from the real world: the air in motion that we call wind often has a vertical component as well as a horizontal one. Updrafts and downdrafts are caused by variations in the solar heating of the earth's surface, by the thermal interactions between differing air masses, and by obstacles such as trees and buildings. It is these vertical components that present the most challenge to most pilots most of the time. Fortunately, the piston-powered light aircraft which are most affected by downdrafts are normally not flown in strongly convective weather, and also have the advantage of nearly instantaneous throttle response. Nevertheless, there are plenty of circumstances for which the 18 hp Quickie and other under-powered aircraft belong "in the barn" and not in the air.

The wind usually changes direction and increases in speed as the altitude increases. In particular, after a cold front passage there is often a strong west wind on the ground that turns toward the north and becomes much stronger at altitude. If only a north-south runway is available, it is clearly preferable to use the north runway.

If there is a 90° crosswind on takeoff, climbout performance is actually slightly better than for no-wind conditions because the aircraft's down-the-runway speed component is reduced by having to "crab" into the wind to track down the runway; on climbout a low ground speed is desirable! In calculating takeoff distance, though, a conservative approach is to only use the component of the wind that is down the runway.

Now that we've cleared the trees it is time to begin the en route climb to our cruising altitude. The 1977 T210 at maximum weight can maintain a climb rate of around 620 ft/min up to about 16,000 ft (thank you, turbocharging) at an indicated airspeed of 115 kt. (To keep the arithmetic reasonable we will ignore changes in wind velocity with altitude, the variation in climb rate with altitude, and the difference between indicated and true airspeed. The errors resulting from making these approximations can be minimized as much as desired by re-doing the calculations for every 100 ft of climb or so—which is small potatoes for a computer.)

EXAMPLE 2.3. How long would it take the 1977 T210 to climb from 1000 ft to 12,000 ft and how much ground would be covered during the climb? (Fig. 2.21)

Solution: (a) Time to climb
= vertical distance/vertical airspeed
= 11,000 ft/(620 ft/min) = 17.7 min = 0.3 hr
(b) Vertical component of airspeed = V_Y = 620 ft/min
= [620 ft/min]×[(1 kt)/(101.3 ft/min)] = 6.1 kt

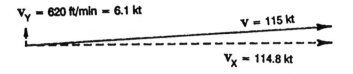

$v_Y = 620$ ft/min $\doteq 6.1$ kt

$V = 115$ kt

$v_X \doteq 114.8$ kt

Fig. 2.21 Velocity vectors for the en route climb

Horizontal component of airspeed = V_X
$$= \sqrt{115^2 - 6.1^2} = 114.8\text{ kt} \cong 115\text{ kt}$$

Horizontal distance covered over ground
$$= (115\text{ kt})(0.3\text{ hr}) = \textbf{34.5 nm}$$

Evidently the angle of climb is so shallow that, in this case, we can consider V_X to be **equal** to V in magnitude, with negligible error.

These calculations were for zero-wind conditions. If there is a wind, Eqs. 2.7 and 2.8 must be solved to find the velocity over the ground and this speed is used to obtain the horizontal distance traveled during the climbout. (See the next example.)

A descent from cruising altitude is usually made using a speed close to the cruising speed and we will make this assumption when calculating the descent time. For efficient flight, it is most important that a pilot begin his descent well before arriving at the airport! A rate of descent that will provide reasonable passenger comfort is 500 ft/min. Thus a descent from a cruising altitude of 12,000 ft down to an airport at an elevation of 2000 ft should begin about [10,000 ft/(500 ft/min)] or 20 min before the estimated time of arrival.

We are now able to make a complete flight time calculation.

EXAMPLE 2.4. Suppose that we are planning to fly our 1977 T210 from MIA to JFK (Fig. 2.1). The elevation of the Miami airport is 9 ft above sea level and that of the New York is 12 ft above sea level; we can consider them to be at sea level with negligible error. We assume an average wind from 270° at 30 kt on the climb to a cruising altitude of 13,000 ft and an average wind from 300° at 50 kt at our cruising altitude. The T210 can cruise at 181 kt at 13,000 ft and that is our planned en route airspeed. We wish to estimate the **(a)** the time to climb to, and **(b)** descend from, the cruising altitude, and **(c)** the total flight time.

Solution: Equations 2.7 and 2.8 tell us that on climbout our heading will be about
$$\theta_H = 9.9° + \arcsin\,[(30/150)\sin(9.9° - 90°)]$$
$$= 9.9° - 16.3° \cong -6° = \textbf{354°}$$
and our ground speed will be about
$$V_{AG} = \sqrt{105^2 + 30^2 + 2(105)(30)\cos(90° - 354°)}$$
$$= \textbf{106 kt}$$
(a) Time to climb to altitude
$$= 13,000\text{ ft} / [620\text{ ft/min}] = 21\text{ min} \doteq \textbf{0.35 hr}$$
(b) Distance traveled during the climb

$$= 106\text{ kt} \times 0.35\text{ hr} = \textbf{37 nm}$$
Heading at cruise altitude of 13,000 ft = θ_H
$$= 9.9° + \arctan\,[(50/181)\sin(9.9° - 120°)]$$
$$= 9.9° - 15.0° \cong -5° = \textbf{355°}$$
Ground speed at cruising altitude = V_{AG}
$$= \sqrt{181^2 + 50^2 + 2(18)(50)\cos(120° - 355°)}$$
$$= \textbf{158 kt}$$
Descent should begin about (13,000 ft)/(500 ft/min)
$$= \textbf{26 min before ETA}\text{ (Estimated Time of Arrival)}$$
Total flight time
$$= \text{time for climb} + \text{time for cruise and descent}$$
$$= 0.35\text{ hr} + (940\text{ nm} - 37\text{ nm})/158\text{ kt} = \textbf{6.1 hr}$$

Although we have made a number of simplifying assumptions, we have covered the main concepts involved in calculating flight time. Computers, hand-held or desk-top, can be programmed from these concepts; the actual airway routes and aircraft performance can be stored in the memory of the computer, and current wind information can be obtained electronically over telephone lines. The result is that the optimum altitude for a flight as well as flight time can be quickly generated. Airlines commonly utilize this capability now, but private pilots can also obtain it through computer-based, online flight-planning services.

Pic. 2.4 The Boeing A75N-1 Stearman was a standard trainer in World War II; now it is a popular sport aerobatic machine.

D. Precision Power-Off Approach

One of the most important skills a single-engine pilot can acquire is the ability to make a power-off approach to a given spot on the ground. Once this skill is acquired, the possibility of a powerplant failure is much less to be feared. A private pilot is expected to demonstrate this ability by performing the maneuver within a normal traffic pattern, safely arriving near the beginning of a suitable off-airport landing field or chosen airport runway. A commercial pilot is expected to arrive more precisely at the beginning of the intended runway. (Pilots of antique or experimental aircraft are extra-likely to be asked to perform this maneuver for real.)

Figure 2.22 provides a concrete example of the challenge. We propose to extend our downwind leg 0.5 nm past the runway threshold, fly a 0.5 nm base leg, and then a 0.5 nm final approach leg. The problem is to determine how much altitude is required to just make the runway if the power is reduced to idle just opposite the runway threshold. Suppose further that our aircraft is a Cessna 152, a popular 2-place trainer possessing an optimum glide angle of 6.1° at 60 kt. (This 60 kt is the true airspeed under standard sea level and gross weight conditions, which we assume for concreteness. At higher altitudes the optimum glide angle is the same but requires a higher true airspeed. At lower weights the optimum glide angle is again the same but the required speeds are lower.)

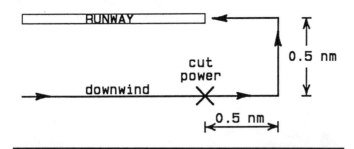

Fig. 2.22 Possible path over the ground during a precision, power-off approach to landing

The joker in this problem, in theory and in practice, is the wind. Each leg in our downward glide will require a wind correction, in general, and will yield a ground speed different than simply the horizontal component of our 60 kt airspeed. Problem 14 asks you to determine the required altitude under conditions of (a) zero wind and (b) a 20 kt wind right down the runway. For the wind down the runway, the wind speed simply adds to or subtracts from the horizontal component of the glide speed on the downwind and final approach legs, so the calculations of Eqs. 2.7 and 2.8 need to be made only for the base leg.

However, the calculation is so important operationally that it is very worthwhile to examine the more general solution — with a wind from every possible direction (Fig. 2.23). First, note that a direct headwind (on final) and a direct tailwind require the same initial altitude; the wind correction angle on the crosswind leg is to the left for a headwind and to the right for a tailwind, but otherwise the downwind and final approach legs are simply exchanged. (Surprised?) Of course the landing with a tailwind is much faster and will require much more runway — but the altitude requirements to get there are the same.

A direct crosswind provides the really critical condition. If the wind is a direct headwind on the base leg, the greatest

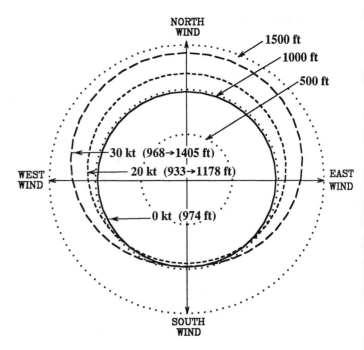

Fig. 2.23 Required initial altitude for the precision approach of Fig. 2.22, for various winds and all wind directions

possible altitude is lost during the approach — about 200 ft more than under no-wind conditions for a 20 kt wind! But if the wind is a direct tailwind on the base leg, a little less altitude than for no-wind conditions is required. The explanation is that a base-leg headwind has the same net effect as a base-leg tailwind as far as the (downwind + final approach) legs are concerned, but the direct headwind really reduces the ground speed and increases the altitude lost on the base leg while the direct tailwind speeds things up and more than compensates for the extra altitude required on the other two legs.

In an actual power-off, forced landing, needless to say, the pilot is encouraged to modify the leg lengths and directions, and to round the corners to whatever extent is needed to make a fully successful approach.

Chapter Summary

In Section A we defined a vector, showed that distance was a vector, and discovered how to determine the end result (a single vector distance from the origin) after moving one vector distance and then moving an additional one or more vector distances (vector addition or subtraction). Our method for obtaining an equivalent resultant vector was to obtain components along two perpendicular axes for all of the vectors being added, add these components to obtain the components of the final resultant vector, use the theorem of Pythagoras to obtain the magnitude of the final resultant vector, and use the arctangent of the components to determine the direction of the final resultant vector.

In Section B we defined the (vector) velocity of an object as the rate of change of its (vector) position with time. This led

Pic. 2.5 A Belgian-designed Stampe SV4c is joined by the PJ-260.

to Eq. 2.3 for the en route time for a trip of given distance and given average speed. We found that vector velocities add just as do vector distances, and that we could interpret the vector sum of two velocities as being the simultaneous application of two distinct aircraft velocities in two reference frames (air and ground). This led to Eqs. 2.7 and 2.8 for an airplane's (vector) velocity relative to the ground when it is subjected to its own velocity relative to the air and a wind velocity of the air relative to the ground. We also derived Eqs. 2.9 and 2.10 which give us the **wind's** velocity based on a knowledge of our velocity with respect to the air and our velocity relative to the ground.

Section C showed that velocities in a vertical plane can be vectorially added also, and this led to the calculation of takeoff angles of climb, en route angles of climb, and total flight times.

Review Questions

1. A vector quantity is a quantity that has both ___ and ___ associated with it.

2. Give three examples of vector quantities.

3. When we say that one vector is equal to the **sum** of two other vectors, we mean that the ___ result or effect is just the same for either choice.

4. The normal addition or subtraction rules are valid for two vectors only if they are ___.

5. If this condition is not satisfied, we must first obtain ___ of the vectors along two arbitrary perpendicular axes.

6. The second step is to do the two sets of subtractions or additions and the third step is to obtain the **magnitude** of the resultant vector by ___.

7. The final step is to obtain the **direction** of the resultant vector from ___.

8. What is the difference between the **speed** and the **velocity** of an aircraft?

9. Give three possible interpretations of the situation where an aircraft's resultant velocity is the vector sum of two other velocities.

10. Looking at Eq. 2.7, note that there will exist no possible heading that will take us to our destination if the argument of the arcsine exceeds 1, since such an angle does not exist. Explain why the probability of this happening depends on (**a**) the ratio of wind speed to airplane airspeed and also (**b**) on how different the wind direction and the course direction are. For what values for (a) and (b), generally speaking, is it most likely that we can't make it to our destination?

11. Specifically, if we wish to fly directly west with a wind directly out of the north, what is the condition on the strength of the wind if success is to greet our efforts? (See Fig. 2.11.)

12. Using the result of question 4, show that if we have a wind direction exactly 90° from our aircraft's **heading**, our ground speed will always be greater than our airspeed. (Draw a little vector diagram of V_{AA}, V_W, and V_{AG} to see why this must be true.)

13. A pilot who takes off downwind is hurting his or her chances of clearing trees at the end of the runway because the ground roll before takeoff is (increased, decreased) and the angle of climbout relative to the ground is (increased, decreased).

14. An en route descent as a pilot approaches the destination airport is usually made at an airspeed close to the ___ speed.

15. Suppose you are living right for a change and you find a direct tailwind of 20 kt at 5000 ft altitude and a direct tailwind of 25 kt at 8000 ft altitude. Under what general conditions would a 5000 ft cruising altitude get you to your destination faster than an 8000 ft cruising altitude? (The optimum climb speed of modern aircraft is less than the cruising speed for this kind of flight.)

Problems

1. Convert a 10,000 ft altitude to its equivalent in nm and km. Answer: 1.645 nm, 3.048 km

2. Convert a 24,500 ft altitude to its equivalent in nm and km.

3. What is the vector sum of a 10 nm displacement to the north followed by a 12 nm displacement to the south?

4. Find the east-west and north-south components of a 1000 nm trek on a heading of 330° relative to 360° as north. (Sketch the action first.) Answer: –500 nm and +866 nm respectively.

5. Find the east-west and north-south components of an 8.4 ft walk on a heading of 120° relative to 360° as north.

6. What is your position and distance relative to your starting point if you travel 100 nm east and then 45 nm south?
Answer: 110 nm at 114°.

7. What is your position and distance relative to your starting point if you travel 374 nm south and then 518 nm west?

8. What is your position and distance relative to your starting point if you travel 218 nm on a course of 037° and then 132 nm on a course of 090°? Answer: 315 nm at 56.5°

9. Determine the proper headings and the resulting ground speed for the course, wind, and aircraft speeds in Table 2.1. Use a negative ground speed to indicate when the wind correction is possible but forward progress is not. (Answers given are rounded off to the nearest whole number.)

INPUT DATA				ANSWERS	
θ_C (deg)	θ_W (deg)	V_{AA} (kt)	V_W (kt)	θ_H (deg)	V_{AG} (kt)
45	225	110	50	45	60
45	90	110	40	30	135
253	170	123	45	274	120
270	360	132	85	230	101
237	330	95	100	(Impossible)	
336	182	98	100	3	–2
335	24	460	80	327	509
23	241	130	65	41	72
207	330	95	100	145	–10
185	300	60	80	(Impossible)	
130	240	115	30	116	101
227	330	95	10	221	92
Table 2.1					

10. Determine the actual wind at your flight altitude if your flight path and aircraft speeds are as given in Table 2.2. (Answers are rounded off to the nearest whole number.)

INPUT DATA				ANSWERS	
θ_{AG} (deg)	θ_H (deg)	V_{AA} (kt)	V_{AG} (kt)	V_W (kt)	θ_W (deg)
124	114	123	99	31	260
253	274	123	120	44	170
19	37	173	35	140	221
203	211	462	500	77	146
203	211	462	449	65	105
342	359	75	62	24	228
Table 2.2					

Pic. 2.6 The classic Piper J-3 Cub takes advantage of its low wing loading to achieve low takeoff and landing speeds; its lack of streamlining also yields very low cruise speeds (70–75 mi/hr on 65 hp).

Pic. 2.7 A modified Great Lakes Trainer, a popular 1930s sport aircraft

11. Your 1977 Cessna 182Q requires a minimum ground roll of 520 ft and then an additional 480 ft to clear a 50 ft obstacle at gross weight, under standard sea level conditions with zero wind, and at a climb speed of about 55 kt.

    a. What is the no-wind climb angle for this kind of maximum performance takeoff? Answer: 5.95°

    b. At what rate is altitude gained after takeoff? Answer: 577 ft/min

    c. What is the climb angle into a 25 kt headwind? Answer: 10.9°

    d. What is the climb angle with a 25 kt tailwind? Answer: 4.1°

    e. What total distance is required to clear a 50 ft tree when taking off into a 25 kt headwind if the ground roll is reduced 10% for each 9 kt of headwind? Answer: 636 ft

12. The 182Q can climb to 6000 ft from sea level with an average rate of climb of 880 ft/min at 76 kt.

    a. How long does it take to make this climb? Answer: 6.8 min

    b. What is the angle of climb under zero-wind conditions? Answer: 6.6°

    c. How much ground is covered during this climb? Answer: 8.6 nm

    d. How much ground is covered if you have an average wind of 25 kt that is exactly perpendicular to your path over the ground during the climb? Answer: 8.1 nm

    e. What is your flight time to an airport 300 nm away if your climb is under the conditions of (d) and if your en route winds are also perpendicular to your path over the ground but are at 38 kt? (The maximum cruise speed at 6000 ft is 140 kt.) Answer: 2 hr 17 min

13. The twin-engine airliner of which you are captain is in a spot of trouble: Both your engines have conked out due to water ingestion as you are penetrating heavy rain. You are at 14,000 ft and the ground elevation is 1300 ft.

You flip rapidly through the pages in the emergency section of your flight manual, finding that at your current weight your best glide angle will be obtained at a speed of 187 kt, yielding a rate of descent of 1340 ft/min. (This is **not** the landing configuration, of course, but with landing gear and flaps up.) (Note: These are the actual performance figures for a DC-9 that did lose both engines while flying through extremely heavy rain on April 4, 1977. A partially successful landing was made on a road some 28.3 nm away.)

    a. How much time do you have before a landing will be made? Answer: 9.5 min

    b. What is the maximum distance that you can glide? Answer: 29.5 nm

    c. What is the angle of the descent? Answer: 4.05°

14. In Fig. 2.22.

    a. How high must you be when you are opposite the beginning of the runway, if you are to just make it to the runway under zero-wind conditions? Neglect turning effects—the extra altitude lost in the banked turns might well compensate for the shortened distance. Answer: 974 ft

    b. How high must you be if you have a direct headwind of 20 kt on your base leg? (Include the effects of the crosswind increasing your glide angle on the downwind and final legs of your pattern.) Answer: 1180 ft

15. The optimum glide capabilities of a glider or an airplane are customarily described by giving the **glide ratio** of the machine. Glide ratio is the maximum ratio of the horizontal distance glided to the vertical height lost, using the same units for the distances, under no-wind conditions. Determine the glide ratio for the DC-9 from the

information of problem 13 and for the Cessna 152 of problem 14. Answer: 14.1 to 1 and 9.4 to 1.

16. An airliner on approach to the Dallas–Fort Worth airport in 1985 found itself in a small-scale downburst of air from a small thunderstorm cell (now called a **microburst**) and was unable to maintain its desired glide angle to the runway. One of the attempts to provide a warning about potentially dangerous wind conditions was in use at the time but did not generate a timely warning. This "wind shear alert" system is designed to activate whenever a **15 kt** vector **difference** between (**a**) a centerfield anemometer (located 500 ft from runway 10 which is, from its designation, aligned at 100° ±5° relative to magnetic north) and (**b**) any or all of five low-level wind shear alert system (LLWSAS) anemometers located on the airport periphery is present, averaged over 30 seconds. Suppose that the centerfield anemometer indicated a wind of 12 kt directly toward an aircraft on runway 10 (therefore a wind **from** a magnetic direction of 100°). What is the minimum northerly wind (**from** 360°) at any of the other anemometers that would trigger this wind shear alert system? Answer: 7.15 kt (Hint: Start by drawing the vector diagram and then use vector subtraction.)

17. In problem 16, suppose that the wind down runway 10 is 8 kt. What is the minimum wind from 190° (a direct crosswind) at any of the other anemometers that would trigger this wind shear alert alarm system? Answer: about 13 kt.

18. Two engines provide a certain safety advantage over a single engine, at least in theory. **Light** twins, however, typically can provide only marginal single-engine climb performance, especially in low density air (high altitudes or high temperatures). The PA-34-200T Seneca II, for example, promises an **optimum** climb rate of only 225 ft/min at 91 kt at maximum weight and sea level standard conditions. Calculate the climb angle and the distance required to clear a 50 ft tree under these very favorable single-engine conditions. Answers: 1.4°, 2048 ft

Pic. 2.8 A Quickie pilot has a good view of the countryside while flying around.

Chapter 3

Force, Mass, and Moments

While machine was outside of building we found center of gravity with man on to be very close to 24" from front edge and about halfway between man and engine.[1]

A. Forces, Center of Mass, Center of Gravity

In the last chapter we spoke approvingly of speed. In this section we define force, the **cause** of any change of speed that occurs. We also discuss the equilibrium of forces on a stationary aircraft and this will turn out to be good preparation for a later investigation of the dynamic forces on an aircraft in flight.

A **force** is anything that has the potential for changing the **velocity** of an airplane (giving it a speed **or** a direction of motion that it did not previously possess).

The force of gravity is a mysterious, challenging, and essential aspect of our life on earth. For an unknown reason every atom attracts, and is attracted by, every other atom. On a small, atom scale, this force is overwhelmed by much stronger electrical and nuclear forces, but on a large scale it is responsible for holding us, our earth, and the stars together. We thoroughly understand this force only to the extent that we can accurately predict and measure its **magnitude** and **direction** and effects. Yes, the gravitational force, as with all forces, is a vector quantity. This means that **vector addition** will have to be used to determine the total gravitational force on any object because the earth's gravity acts individually on the atoms of the object. The first dividends from this realization will be definitions of **center of mass** (*cm*) and **center of gravity** (*cg*).

Consider a hypothetical point in empty space, inhabited only by a uniform stick and a uniform ball (Fig. 3.1a). Since the objects are assumed to be uniform, their centers of mass are at their geometrical centers, as shown.

The forces that hold together the constituent atoms in these objects and determine the normal atomic separation distances are electrical in nature because the electrical force is enormously stronger than the gravitational force (e.g., a factor of 10^{39} in the hydrogen atom). However, every object normally has just as much positive charge as negative charge so that its net charge is zero and any electrical force on it is negligible compared to the always-attractive gravitational force. Thus, in space isolated from other objects, the only *external* forces acting on our hypothetical objects are the gravitational forces of attraction between the atoms of the stick and the atoms of the ball.

This gravitational force is known to diminish rapidly with increasing separation distance. In fact the gravitational force decreases in proportion to the inverse **square** of the distance separating the two atoms being described. If the separation distance is doubled, the gravitational force is only one-fourth $[(\frac{1}{2})^2]$ of its previous value; if the separation distance is tripled, the gravitational force is reduced to $[(\frac{1}{3})^2]$ or one-ninth of its previous value, and so forth.

This means that the atoms of the ball attract atoms at the near end of the stick much more than they do those atoms at the far end. Figure 3.1b attempts to show this by drawing vectors for the forces between an atom at the center of the ball and two atoms in the stick, one atom one-fourth of the way from the end and one atom one-half of the way from the end.

The center of gravity of the stick is defined as the point location where all the mass of the stick could be concentrated to give exactly the **same** gravitational force on the stick. This

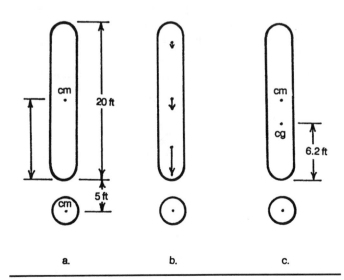

Fig. 3.1 A ball and a stick, stuck out in empty space

[1]Orville Wright, 1903, *The Papers of Wilbur and Orville Wright*, Marvin W. McFarland, Editor (McGraw-Hill Book Company, Inc., 1953).

point is obtained, theoretically, by adding vectorially all the gravitational forces between all the atoms of the ball and all the atoms of the stick. Because of the inequality of these forces, the *cg* of the stick that is calculated in this manner is noticeably closer to the ball than is the *cm* of the stick (see Fig. 3.1c, for a stick with five times the mass of the ball). The closer the two objects come to each other, the greater the distance of separation between the *cm* and the *cg*.

We can guess that our Fig. 3.1 represents only a snapshot in the history of the stick and the ball out there in lonely outer space, for with only the attractive force of gravity present, the objects must be in the process of coming together, much as primeval gases came together to form planets and suns.

As a result of observation and measurement of gravitational and other forces, it has long been realized that no force acts in isolation. In Fig. 3.1, the ball cannot pull on the stick without itself being pulled toward the stick; these forces are **equal** in magnitude and opposite in direction. There are no "unmoved movers" in space! This is just Sir Isaac Newton's **third** "law" of motion; we believe that it is a correct description for all types of forces between any two particles or any two objects. It also is true irrespective of any relative motion. On a more intuitive level, the clue that you are exerting a force on something else comes from the fact that you feel it exerting that equal and opposite force on you.

Consider next the more familiar situation of an airplane and the earth (Fig. 3.2). On the scale of this figure it is not possible to show the airplane as any more than a tiny dot. Thus the resultant force of gravity exerted by all the atoms of the earth on (say) the landing gear atoms is not significantly different from the force on similar atoms in the cabin area or in the tail. Therefore we can consider the *cg* of an aircraft to be **coincident** with its *cm*, a nice simplification. It means that it is possible to accurately predict the *cg* of an aircraft based on the density and shape of its parts, in the design process. (After it is built, the aircraft always must be weighed to determine its empty *cg*, as we'll discuss later.)

The earth is so large and massive relative to aircraft that not only does the force of gravity not vary significantly over distances comparable to the dimensions of a plane, it also does not change very much with normal changes in aircraft altitude. A Concorde SST (supersonic transport) that typically weighs

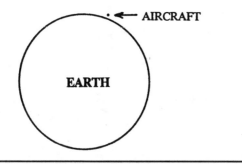

Fig. 3.2 One earth and one plane, out in empty space

300,000 pounds at sea level will weigh about 1700 pounds **less** at its typical cruising altitude of 60,000 ft, a reduction of only 0.5%. Also, the earth is so large and massive compared to airplanes that we don't have to worry about the effects of that equal force of the plane on the earth (the third law again) as the earth pulls or pushes on the plane; a hard landing may shake up the passengers but it won't detectably disturb the earth's orbit.

Next we consider the external forces acting on an airplane that is resting quietly on a level airport parking ramp (Fig. 3.3). Since this machine is neither changing its speed nor its direction of motion, the **vector sum of all the external forces on the airplane must equal zero**. This deduction follows from our earlier definition of a force; the general principle is known as Newton's **first** "law" of motion (non-motion?).

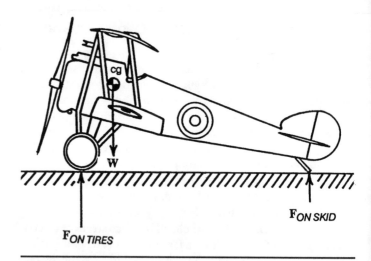

Fig. 3.3 A World War I fighter aircraft, resting

There are some external forces acting on this object and we know one of them: it is the net force of gravity, the vector sum of the attraction of all of earth's atoms for all of the atoms of the airplane. This vector sum, this overall attraction of the airplane to the earth, is by definition the **weight**, *W*, of the airplane. Weight is thus a force vector that acts on all objects associated with the earth; the direction of this vector (as shown by Newton after he had devised the calculus for that purpose) is toward the center of earth. (Actually, the weight vector does not point *exactly* toward the earth's center because the earth is not perfectly spherical and its mass distribution isn't perfectly symmetric about its center. This nicety can be ignored for practical calculations.)

We also know that the point at which the weight vector appears to act is the *cg* of the airplane, represented in the figure by the partially filled-in circle. For many problems involving flight it is quite unnecessary to concern oneself with the actual distribution of the mass in the airplane; instead we simply consider **all the mass to be at the *cg*** and the structure around it to be without mass. Because the *cg* is an *imaginary* point, there is no reason that any of the mass must be at the point; the *cg* of a basketball is at its airy geometrical center, for example. It is when one concerns oneself with the internal strength, the stresses and strains on the airplane structure, that it **is** necessary to consider the actual distribution of mass and the internal forces that develop in response to the external forces. The mass distribution is also important in the determination of the dynamic behavior of the aircraft in maneuvers such as spins.

There must be one or more additional forces acting on this airplane because we know that the vector sum of external forces acting on the machine must be zero, and so far we have discovered just the weight vector. The key point is that we are looking for forces acting **on** the aircraft (external applied forces). Gravity is the only force which we need consider that acts through empty space; all the other external forces arise due to apparent contact of the airplane with adjacent bodies (the earth or molecules of air). If in Fig. 3.3 the wind is not blowing, the only external body that can be exerting a force on the airplane is the ground. Therefore the ground is pushing **up** on the airplane with a total force that is just equal to the weight of the machine. (Because we're focusing on the airplane, we don't have to worry about the fact that the plane is pushing down on the earth, by the third law.)

On level ground the weight vector is perpendicular to the ground, so the vector sum of external forces on the airplane can be zero if the earth pushes exactly up on the wheels and the tail skid. At this point we have no way of predicting how much force is on the main wheel and how much is on the tail skid or (a closely related question) where the cg of the aircraft is located. Suppose, though, that each of the main gear tires has an upward force of 705 pounds on it and the main skid 72 pounds, resisting a total weight of 1482 pounds. These vector forces are shown in two different ways in Figs. 3.4a and 3.4b. In this case the vector sum of the external forces on the plane is evidently zero. (For an aircraft on a level surface with the engine quiet and with no wind, there is no reason to expect any external forces at all in either the nose-to-tail or wingtip-to-wingtip direction, so the vector sums in those directions are trivially zero.)

So far we have been using the pound (lb) as the unit of **force**. We could switch to the metric unit, the newton (1 newton = 0.2247 lb), but the pound wins out for us because pilots and most U.S. aeronautical engineers still work in these units. Most of us haven't developed an intuitive feel for forces in metric units, yet, and helping pilots to obtain a "feel" for why planes do what they do is a major goal of this text.

B. Mass

We have used the terms **mass** and **weight** already, but now it is time to draw a careful distinction between them. In this section we will define mass and indicate how the mass of an object can be determined from a knowledge of its weight.

The weight of an object is the sum of the gravitational forces on its constituent atoms and is usually measured by determining how hard something must push up on the object to keep it stationary. (Note that this is using Newton's third law in a very practical way.) Even on the surface of the earth there is

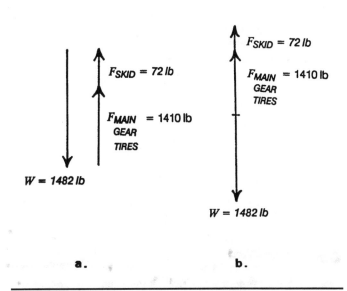

Fig. 3.4 Vector forces on the resting World War I aircraft

a variation in an object's weight of about 0.3% with latitude (because the earth's spin causes a slight equatorial bulge) and about 0.2% because of mountains. At a distance of one radius out from the surface of the earth (3450 nm or 20,000,000 ft), all objects lose 75% of their weight—but only *approach* true weightlessness as they get infinitely far away, despite what you may read in the non-scientific press.

There is a property of objects that doesn't change with position or altitude above the earth! This is the property called inertia. Inertia is defined as the resistance of an object to having its speed (or lack of it) **changed**, or its **direction** of motion **changed** (if it is moving). And mass is the quantitative measure of the amount of inertia possessed by an object.

It is only in scientific or engineering work that one really needs to obtain a numerical value for the mass of an object, so the unit of mass is unfamiliar to most folk. Because mass is **proportional** to weight (with a nearly constant proportionality constant, on the earth), we can use weight as a measure of inertia for purposes of non-technical speech. Everyone knows or can intuitively feel that a 250 lb fullback is going to be tougher to stop than a 170 lb quarterback, if they are going the same speed.

Mass really is a different concept that weight, though; it is **not** a vector because inertial properties are not dependent on direction. Mass also is **not** a force, even apart from its non-directionality.

Experimentally, **all** objects have the same initial acceleration when released from a given point above the ground. That is, the acceleration of a freely-falling object is independent of its weight. (A "freely-falling" object is one for whom the aerodynamic drag force is negligible compared to the weight—typically true only at very low speeds.) The acceleration experienced is called the acceleration of gravity (g) and is a measure of the strength of the gravitational force at that point and therefore has only the same small variation with latitude and elevation as mentioned earlier for the gravitational force.

For those who might need a quick little refresher on what is meant by **acceleration**, let it be noted that acceleration is defined as the **rate** of **change** of the **vector velocity**. Thus an object is accelerating if its **speed** is **changing** or if its **direction of motion** is **changing**. If only its speed is changing, the acceleration is easily calculated by subtracting an initial speed from a later speed and dividing by the time interval. If the acceleration is itself changing, the process yields the *average* acceleration; the instantaneous acceleration is obtained as a limiting value as the time interval is decreased toward zero.

How can it be that a heavy object accelerates and falls exactly like a light object? (Whatever happened to "The heavier they are, the harder they fall" advice?) The answer is that the acceleration of an object depends **(a)** directly on the net external force acting on it (so this tends to make a heavy object fall faster) but **(b)** also inversely on its inertia (and this tends to make a heavy object fall slower); therefore the two effects **cancel** and there is no dependence on weight or mass (so long as other external forces such as air drag are negligible). This is Newton's **second law of motion**. In words, because it is so important,

> The acceleration of an object
> equals
> the vector sum of the external forces acting on it
> divided by its mass.

or, in symbols,

$$a = \frac{\sum F_{\text{EXTERNAL}}}{m} \qquad (3.1)$$

**NEWTON'S SECOND LAW OF MOTION
FOR AN OBJECT OF MASS m**

(Mathematical notation: The capital Greek letter sigma, Σ, means to sum or add up all the items like the variable that follows. Thus, for example

$$\Sigma(\text{pets}) = \text{dog1} + \text{dog2} + \text{cat\#1} + \text{cat\#2} + \text{gerbel}$$

might describe a 5-pet family and

$$\Sigma F_{\text{EXTERNAL, Y DIRECTION}} = F_{\text{UP, BY GROUND}} + F_{\text{DOWN, BY GRAVITY}}$$

would describe a stationary vehicle.)

For a freely-falling object, one for which the aerodynamic drag is negligible, the only external force is by definition the gravitational force, and so we can write

$$a = \frac{W}{m} \qquad \text{(freely-falling object)} \qquad (3.2)$$

Because this acceleration is g and is the same for all objects, the weight must be proportional to mass and the proportionality factor must be g. To prove this statement we write the statement of proportionality in symbols

$$W = mg \qquad (3.3)$$

and use this to substitute for the weight in Eq. 3.2, giving

$$a = g$$

which is the experimental observation.

In other words, we can answer our earlier question by saying that a 20 pound object does indeed have twice as much force pulling it toward the earth as does a 10 pound object, but it also has twice the inertia or resistance to acceleration, and the effects exactly cancel.

The actual value of g varies with latitude in the same proportion as does weight. Its value in English units is about (32.2 ft/s)/s or 32.2 ft/s². This number means that a freely-falling object **changes** the vertical component of its speed by about 32.2 ft/s every second that it is falling. (In earthier units, this is a change in speed of about 22 mi/hr every second.) So if an object with negligible air drag is released from rest (speed = 0), it will have gained a downward speed of 32.2 ft/s (22 mi/hr) after 1 second has elapsed, a speed of 64.4 ft/s (44 mi/hr) after 2 seconds, and so forth. If it is thrown directly upward at 32.2 ft/s, it will have a speed of 0 (highest point in trajectory) after 1 second, a downward speed of 32.2 ft/s after 2 seconds (and be back at the origin), etc. In sum, g is the **rate of change** with **time** of the **vertical** component of the speed of an object that is under the influence of the gravitational force only.

Any object in the earth's atmosphere, and most surely this includes airplanes, are under the influence of at least the external force of air drag in addition to the gravitational force. This means that a freely-falling airplane (vertical dive) will have an acceleration less than g and this acceleration will decrease with each passing second until eventually the air drag force is as large as the gravitational force and there is **no** more acceleration! The speed at which this equality is obtained depends greatly on the aerodynamic cleanliness of the aircraft but in any case is known as the **terminal velocity** (or, better, the terminal speed) of that aircraft. Military aircraft at one time were tested for their terminal speeds—but I recommend to you a calculator and the analytical methods described in a succeeding chapter, instead. However, all objects at the instant of release do have an acceleration equal to g because the air drag force starts from a zero value and builds with increasing speed. It is this initial acceleration that is a measure of the gravitational force and allows us to separate out the contribution of an object's mass to its weight.

We will want to be able to calculate the mass of an object; a formula to do this follows immediately from Eq. 3.3:

$$m = \frac{W}{g} \qquad (3.4)$$

where we are able to drop the vector notation because W and g are in the same direction. The mass calculated from this formula will be in a unit called the **slug** when the weight is expressed in pounds and g is in units of ft/s². Thus a 300,000 pound Concorde has a mass of

$$m = \frac{300,000 \text{ lb}}{32.2 \text{ ft/s}^2} = \textbf{9317 slugs}$$

where we have assumed that g was 32.2 ft/s^2 at the location where that 300,000 lb of weight was determined. (For our work in this text we will always assume that g is equal to 32.2 ft/s^2; the actual variation on the earth's surface is no more than ± 0.2 ft/s^2.)

C. Moments

Now we are ready to return to the problem of determining the distribution of forces on the main gear tires and the nose-wheel or tailwheel. Actually, our primary interest will be in doing what the Wright brothers found necessary on their #1 Flyer: determining the *cg* from a measurement of the forces on the tires.

For equilibrium, it is not enough to require that the external forces on an object have a vector sum of zero. If the forces do not act at the same point on the object, they will tend to produce **rotation**.

The **moment** of a force is a measure of the tendency of that force to produce rotation about the point where the moment is taken. Specifically, we define the moment of a force about a chosen point (x) as the **product** of (a) the distance from that chosen point to the point of application of that force and (b) the component of that force that is perpendicular to the distance. In symbols,

$$\left(\begin{array}{c}\text{moment of a force } F \\ \text{about point } x\end{array}\right) \equiv M_{\text{x}} = d_{\text{x}} \times F_{\perp} \qquad (3.5)$$

DEFINITION OF THE MOMENT OF A FORCE

Only the perpendicular component of the force is to be used because the parallel component cannot produce any tendency toward rotation about the given point.

We define a positive moment as one whose tendency is to cause clockwise rotation **except** for an aircraft in flight, in which case a positive moment is defined as one which tends to pitch the nose up.

A completely equivalent definition of the magnitude of a moment is that it is the product of (a) the perpendicular distance to a line that is extended along the direction of application of the force and (b) the force. In symbols

$$M_{\text{x}} = d_{\perp} \times F \qquad (3.6)$$

EQUIVALENT DEFINITION OF THE MOMENT OF A FORCE

We now demonstrate that these definitions are equivalent.

Suppose that we suspend a 50 pound bag of sand from the propeller of our airplane, as depicted in Fig. 3.5a. Suppose also that the *cg* of this airplane is 6.5 ft behind the propeller and that the plane's longitudinal axis forms a 12° angle with the level ground. What is the moment of this 50 pound force about the *cg* of the aircraft?

Referring to the force diagram in Fig. 3.5b, the definition of Eq. 3.5 yields

$$M_{\text{CG}} = -(6.5 \text{ ft})[50 \text{ lb} \times \cos(12°)] = \textbf{--318 ft-lb}$$

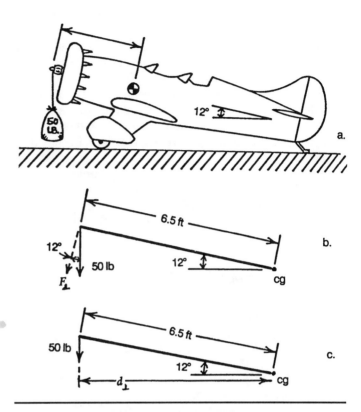

Fig. 3.5 Moments about the *cg* for a stationary aircraft

Note that the moment is negative because the force tends to tip the plane on its nose, which is a counterclockwise rotation. Also note that the **unit** for a moment is just the product of the units of the quantities that are being multiplied.

The second definition of a moment (Fig. 3.5c and Eq. 3.6) yields

$$M_{\text{CG}} = -[6.5 \text{ ft} \times \cos(12°)](50 \text{ lb}) = \textbf{--318 ft-lb}$$

which is the same as the first definition produced.

Feel free to use whichever definition is easier; sometimes it will be one definition and sometimes the other. Practice is necessary to feel comfortable about the moment-taking process. Remember that specifying a moment is **meaningless** unless the point about which the moment is taken is also specified — and remember to connect this with where you intuitively grab a handle being used as a lever..

We can now state a very general and very powerful principle: An object is in **equilibrium** (a) if the vector sum of external forces acting on it is zero **and** (b) if the sum of the moments of the external forces about **any** point is also zero. Just now we are concerned with static equilibrium (objects at rest) but the principle also holds for dynamic equilibrium (constant linear speed in a constant direction and a constant rotational speed with a fixed direction for the axis of rotation). (Note that this means that an aircraft is in dynamic equilibrium for **most** of the time during most flights, since constant-speed climbs and descents are part of dynamic equilibrium! But an aircraft is not in equilibrium during any sort of turn.)

Look now at the Sopwith Camel of Fig. 3.6. Suppose that we know that the *cg* is right above the leading edge of the lower wing (in flight attitude), as shown. What then are the forces exerted by the ground on the main gear tires and the tail skid? (Note that we have drawn the object of interest all by itself and have drawn in only the **external** forces acting on it. This is a good procedure for you to follow, too. Also, the usefulness of the concept of *cg* should be apparent as we work through this example, for we need not concern ourselves with the fact that gravitational forces are acting on every Camel atom but instead can consider the plane to be weightless except for having all its mass at the *cg*. This means that **one** moment calculation describes the rotational tendency for all the uncountable billions of atoms that are holding hands to keep the Camel in one piece!)

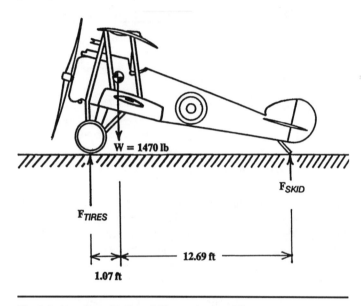

Fig. 3.6 It's a Sopwith Camel (Snoopy's aircraft)!

If we add the moments about the *cg* using the second definition of a moment, we obtain

$$\begin{pmatrix} \text{sum of the moments} \\ \text{about the } cg \end{pmatrix} \equiv \sum M_{\text{CG}}$$
$$= +(1.07 \text{ ft})(F_{\text{TIRES}}) - (12.69 \text{ ft})(F_{\text{SKID}}) + (0 \text{ ft})(1470 \text{ lb})$$
$$= +(1.07 \text{ ft})(F_{\text{TIRES}}) - (12.69 \text{ ft})(F_{\text{SKID}})$$

Note well that the weight **doesn't** contribute a moment (a rotating tendency) about the *cg* because there is a zero distance between the *cg* and the effective point of application of the force of gravity; this is a good reason to customarily take moments about the *cg*. (It is also the reason you customarily pick up heavy objects at their *cg*.) → *No spinning or rotating during lifting*

The principle of equilibrium tells us that this sum must be zero, so by algebra

$$F_{\text{TIRES}} = \left(\frac{12.69}{1.07}\right) F_{\text{SKID}} = 11.9 \, F_{\text{SKID}}$$

The principle of equilibrium also states that the external forces must add to zero, so

$$F_{\text{TIRES}} + F_{\text{SKID}} - W = 0$$

from which, by algebra,

$$F_{\text{TIRES}} + F_{\text{SKID}} = W = 1470 \text{ lb} \qquad (3.7)$$

and solving these equations simultaneously yields

$$F_{\text{TIRES}} = \textbf{1536 lb} \quad \text{and} \quad F_{\text{SKID}} = \textbf{114 lb}$$

Let it be noted that nature forces us to solve simultaneous equations whenever we don't give her enough information. The first equation could not be solved for either of the forces because each depends not only on the relative distances but also on the weight of the aircraft.

The principle of equilibrium also demands that the sum of the moments about **any** point must equal zero. We can test this by taking moments about the point where the main gear tires touch the ground:

$$\sum M_{\text{TIRES}} = (1.07 \text{ ft})(1470 \text{ lb})$$
$$- (1.07 \text{ ft} + 12.69 \text{ ft})(F_{\text{SKID}}) = 0$$

Solving this equation yields $F_{\text{SKID}} = 114$ lb immediately and this value can be combined with Eq. 3.7 to obtain the same value for F_{TIRES} as before.

It should be clear that the *cg* of an airplane must always lie between the main gear and either the nose- or tailwheel. Otherwise the moments about any point will be of the same sign and the taildragger will become a nosedragger and vice versa.

EXAMPLE 3.1. Suppose that the 2400 lb Piper PA-28-180 of Fig. 3.7 has a 250 lb force on its nosewheel and 2150 lb on the main gear tires. Where is the *cg* of this aircraft located?

Solution: Let us use the symbol x to represent the horizontal distance from the nosewheel to the *cg*. Then moments about the nosewheel's contact point are *x Moment ..... distance ... nosewheel = 0 because the nose is i'se at point ... calc mom*

$$\sum M_{\text{NOSEWHEEL}} = (x)(2400 \text{ lb}) - (6.22 \text{ ft})(2150 \text{ lb}) = 0$$

Solving this yields $x = \textbf{5.57 ft}$

Note that we gained no knowledge about the vertical location of the *cg* from this calculation.

EXAMPLE 3.2. The engine in our Cherokee weighs 285 lb and its *cg* is about 5.0 ft ahead of the airplane's *cg*. We propose to substitute a 207 lb turboprop engine so as to have a real going machine. How far forward must the new engine be mounted in order to keep the *cg* of the airplane at the same location as before, while retaining the same 2400 lb gross weight?

Solution: The moment contributed by the old engine was

$$M_{\text{CG, OF OLD ENGINE}} = -(5.0 \text{ ft})(285 \text{ lb}) = 1425 \text{ ft-lb}$$

Therefore the new engine should be mounted a distance

$$\frac{1425 \text{ ft-lb}}{207 \text{ lb}} = \textbf{6.9 ft}$$

ahead of the *cg* to provide the same moments—i.e., 1.9 ft farther forward!

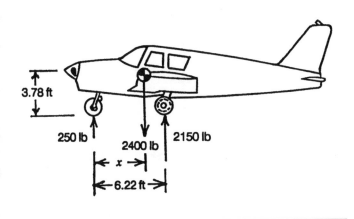

Fig. 3.7 A Piper Cherokee, sitting for its weight and balance check

The calculation of *cg* location is one of the most important things that a homebuilder must do before making a first flight (or even high speed "taxi" tests!). Accurate scales must be placed under each tire after the plane has been "leveled" according to specifications (often using a cockpit longeron member as the level reference). The *cg* location must be determined for two configurations: (a) farthest forward location (usually when there is negligible fuel, no baggage, and no passengers) and (b) farthest rearward location (usually maximum fuel, passengers, and baggage). If this range of *cg* travel does not fall within the range specified by the designer, lead is usually added in the tail or in the nose to correct the problem. Don't laugh! Plenty of aircraft, including production aircraft, fly around very nicely—with lead in their tails!

The pilot-in-command of an airplane, by FAA regulations, must ensure that the weight and *cg* of the aircraft are within the certification specifications before flight is attempted. The empty weight and the empty weight *cg* are listed in the weight and balance papers that must be kept with the plane. Some arbitrary point is indicated as the **datum**, the point about which moments are calculated. The firewall just behind the engine is often chosen as the datum point. Distances from the datum to the fuel, to the seats, and to the baggage compartment are provided so that moments can be calculated for each item. The sum of the weights and the sum of the moments for the actual condition of flight are usually compared with a graph to see if the *cg* and weight fall within acceptable limits (see problem 8, for example).

The author once had the pleasure of discovering a homebuilt, tailwheel-type biplane that had flown for many years (including an aerobatic demonstration on TV!) with an official *cg* located well **forward** of the main gear tires! The datum on this aircraft was well forward of the engine, so the calculated *cg* location had the correct sign, but a calculational error placed the official *cg* in an impossible location—and the error hadn't been detected during either certification or during annual relicensing.

The ultimate system, now in use on some large aircraft, is to have a force sensor on each gear and a readout in the cockpit!

We will be discussing the basis for a range of allowable *cg* locations in the chapter on stability.

It is possible now to understand how a **trim tab** works. In Fig. 3.8 we have a horizontal stabilizer and an elevator that are 15 feet behind the *cg* of the aircraft. The trim tab is adjusted from inside the cockpit to a range of angles relative to the elevator. Normally the trim tab, the elevator, and the stabilizer will all have nearly the same angle with respect to the longitudinal axis of the plane during cruising flight because this is a minimum drag configuration (Fig. 3.8a). However, in the climb phase the aircraft must have a positive angle relative to the oncoming air and for this situation the wing produces a greater negative moment that must be balanced by a greater positive moment produced by the tail. In Fig. 3.8b the pilot must hold constant back pressure on the control wheel because the flowing air tends to straighten out the elevator relative to the stabilizer. In Fig. 3.8c we see that the pilot has adjusted the trim tab. How has this affected the pressure that must be exerted on the control wheel?

It is worthwhile to make some rough numerical estimates. Suppose that a force of 20 lb on the tail is needed to produce the required moment during climb. This force is effectively applied about one-quarter of the way back from the leading edge of the elevator (as we'll see later), as shown in Fig. 3.8b, and this is the force that must be counteracted by the pilot through a constant force on the control wheel.

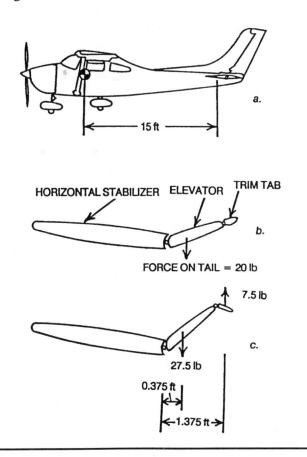

Fig. 3.8 A Cessna 172, sitting for a trim tab inspection

Suppose the trim tab in Fig. 3.8c is contained in the last one-sixth of the elevator. Suppose also that the force due to the air pressure distribution on the main part of the elevator is 27.5 lb and the force on the trim tab due to air pressure is 7.5 lb. Then we still have the net force of 20 lb exerted downward on the tail that we had in Fig. 3.8b, and therefore essentially the same moment about the *cg* of the airplane.

Now we calculate the moment about the elevator hinge line because this tells us how much force must be supplied by the pilot.

$$\sum M_{\text{ELEVATOR HINGE}}$$
$$= (0.375 \text{ ft})(27.5 \text{ lb}) - (1.375 \text{ ft})(7.5 \text{ lb})$$
$$= 0 \text{ ft-lb}$$

Therefore the pilot need exert **no** force on the controls and the plane is said to be "trimmed" for climbing flight. (Note: The angles are exaggerated for illustrative clarity.)

The use of a trim tab at other than cruise speeds conveys a possibly critical disadvantage (question 9).

What we have been discussing in this section is an in-flight adjustable trim tab that maintains its angle relative to the elevator surface as the elevator is rotated up or down about the hinge axis. A ground-adjustable tab, typically just a rectangular surface of sheet metal that can be bent on the ground, is also commonly used to trim an aircraft for cruise, especially for the ailerons and rudders.

Another very important type of tab is one that is connected such that it deflects either opposite or with the direction of motion of the control surface. If the deflection is in the same direction and increases as the deflection of the control surface increases, the tab would be called an anti-servo tab and would serve to increase control surface forces and increase their effectiveness with deflection. If the deflection of the tab is opposite in direction to the deflection of the control surface, the tab would tend to reduce control forces and also decrease their effectiveness, and would be called a servo tab.

By Federal Air Regulations, newly certificated aircraft must be designed so that they can be successfully flown by trim tab **only** if the elevator control linkage should fail.

There are two **experimental** techniques for finding the location of the *cg* of an object that work well with **aircraft models** but not too well with the full-scale version. One is to suspend the object from a string or rope (Fig. 3.9a,b). Then there are only two external forces on the object and the only way that the principle of equilibrium can be satisfied is for the *cg* to be directly **under** the point of attachment. Suspending the object from a second point and drawing a line straight down from the point of attachment a second time will identify the location of the *cg* as being at the intersection of the two lines.

The other direct experimental technique is to find the point for which a single upward force will support the object (Fig. 3.9c). If you try to support the object at a point that is **not** below the *cg*, there will be an unbalanced moment about the point of support and the object will immediately begin to rotate about the point of support. The *cg* again must lie somewhere directly

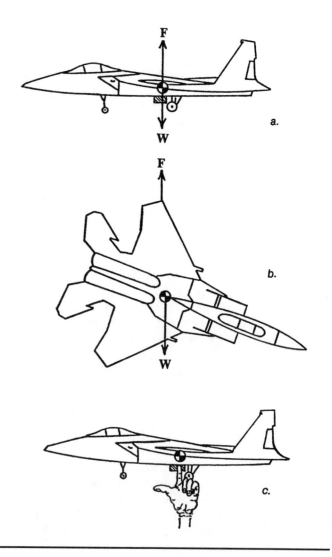

Fig. 3.9 How to determine the *cg* of the F-15A Eagle

above the support point. Modelers typically use this technique, holding a model at about one-third of the way back from the leading edge of the wing to see if the *cg* is about there.

Summary

Section A defined a force as anything that is capable of changing the speed or direction of motion of an object. The gravitational force is a force that acts independently between every two atoms, but we saw that it was possible to replace these innumerable tiny forces with a single force, the weight, acting at a point called the center of gravity. For airplanes it is sufficient to consider the center of gravity to be at the center of mass.

Section B defined mass from a study of the forces on, and the acceleration of, an object subjected only to the gravitational force, using Newton's second law of motion. The mass of an object is calculated by dividing the object's weight by the free-falling acceleration due to gravity, *g*, at that point. The unit of mass in the English or engineering system of units is the slug.

The moment of a force about a given point was defined in Section C as the quantitative measure of the tendency of that force to produce rotation of an object about that point. Two equivalent definitions, for calculation purposes, of the moment of a force were given. The principle of equilibrium was stated: An object is in static or dynamic equilibrium if the vector sums of forces and moments acting on it are separately zero. Following this principle, we were able to find a missing force or a missing distance such as the location of the *cg* for an object that was in equilibrium.

The physical principle behind the trim tab is the balancing of moments about the elevator hinge line (to bring to zero the force the plot must apply) while maintaining the same moment about the *cg*. Finally, we saw that the *cg* of an object must lie directly below the point of attachment when a rope holds the object suspended in air, or directly above the point of contact when a single force supports and balances the object from below.

Review Questions

1. What is meant by the term "center of gravity?"
2. Generally speaking, when are the center of gravity and the center of mass of an object **not** co-located?
3. Aircraft can maintain level flight if they can exert a downward component of force on the air that is equal to their own ___; this follows from the principle of equilibrium for the vertical direction and from Newton's ___ law.
4. The weight of an object can be considered to be a single force that acts at the ___ of the object.
5. How can it be that a giant Boeing 747 airliner and a little Cessna 152 have the same acceleration if they are each dropped in from a foot above the runway?
6. Is terminal velocity an equilibrium situation? Explain.
7. Why must the *cg* of the Camel of Fig. 3.3 be located in back of the point of contact of the main gear tires?
8. Is the moment of the weight about the contact point of the main gear tires in Fig. 3.3 a positive or negative moment?
9. Can you see a possible disadvantage to the use of a trim tab to lessen control forces? (Professional pilots, especially, should!)
10. Does the principle of equilibrium require that the ground **always** exert an upward force equal to the weight of the aircraft when it is on the ground? Explain.
11. For a glider in a vertical dive at terminal velocity, the air drag force on the glider is equal to ___.

Problems

1. Given that the earth has a radius of about 3444 nm, verify the statement in Section A regarding the weight loss of a Concorde SST at 60,000 foot altitude.
Hint: You can use the fact that $[1/(1+x)]^2 = 1 - 2x$ for $x<<1$.
2. What are the masses of the fullback and quarterback described in the text?
3. Determine the mass of the Camel of Figs. 3.3 and 3.4. Answer: 46.0 slugs
4. What is the moment of the force of gravity on the Camel of Fig. 3.6 about the tail skid? Answer: -1.865×10^4 ft-lb

5. In order to turn the loaded Cherokee of Fig. 3.7 by hand, a lineperson exerts a downward force on the tail end of the fuselage, some 7.53 feet behind the main gear tires. How large must this force be? Answer: 207 lb (nosewheel just lifted clear of the ground)
6. Suppose the Cherokee of the preceding problem teeters on its main gear without any force on the rear fuselage when the fuselage has been rotated 19° tail down. What was the **vertical** location of the *cg* before the plane was rotated? Answer: 1.9 ft above the ground
7. The Experimental Aircraft Association (EAA) offers plans for the Acro Sport, a single-seat, aerobatic sport biplane, shown in Fig. 3.10. The datum is the front face of the propeller. The plane in the figure has been leveled to the attitude specified by the designer. There is no fuel, oil, pilot, or baggage in the aircraft when it is weighed. It is found that the left tire is exerting a downward force of 348.2 lb, the right tire 346.8 lb, and the tailwheel 20.5 lb.
   (a) Determine the *cg* location of this empty aircraft.
        Answer: 57.1 inches from the datum
   (b) Next you are to find the most **forward** possible location for the *cg* while in flight. Because the fuel is in the fuselage and is ahead of the wing, full fuel will contribute to a most-forward *cg* location. The same is true for the oil. However, the minimum pilot weight must be considered because the pilot sits behind the empty weight *cg*.
   So determine the *cg* location for a 150 lb pilot, full fuel (20 gallons or 120 lb), and maximum oil in the engine (8 quarts or 15.2 lb). Does the *cg* fall behind the most-forward limit of 60.0 inches specified for this machine? (Hint: Use the concept of Fig. 3.9c. Add the moment due to the empty weight to the moments due to the pilot, gas, and oil to obtain the total clockwise moment about the datum. Forgetting that the upward forces act at three points when on the ground, determine where a single upward force equal to the total weight would produce a balancing counterclockwise moment.) Answer: 60.8 inches
   (c) The most rearward *cg* location in flight would occur with minimum fuel and oil, maximum pilot weight, and maximum baggage. Suppose that a properly frightened 230 lb pilot is flying this machine with 4 quarts of oil and essentially no gas. He does have 25 lb of emergency rations in the baggage compartment. Is the actual *cg* forward of the 66.75 inch limit for most rearward location of the *cg*? Answer: not quite
8. The flight manual for a certain Cessna Super Skymaster push-pull twin contains the allowable *cg* range in the form of the graph of Fig. 3.11. The weight and moment must at all times fall within the enclosed area (envelope) on the graph. The datum is at the nose of the aircraft. Estimate the allowable *cg* range for this machine.
9. Consider a little variation on the trim tab theme of Fig. 3.8. First, suppose that the surface area of the trim tab is ¼ of the total elevator area (instead of the ⅛ assumed in the text). Then the effective force on the part of the elevator that **doesn't** include the trim tab should be about **three** times as great as the force on the trim tab (¾ of total area versus ¼ of total area). You can also assume that the effective force on the trim tab acts at a point about 3.3 times farther from the hinge than does the effective force on the rest of the elevator. (You don't need to know the actual distances because it is the **ratio** of the distances that is important when comparing moments.)
   What now are the forces on (a) the trim tab and (b) the rest of the elevator, if the **net** force on the whole elevator is to be **23 pounds** and if the aircraft is trimmed for this condition (i.e., no pitching moment about the hinge point)?

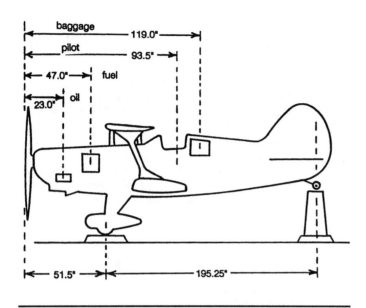

Fig. 3.10 The EAA Acro Sport, at its weigh-in before first flight

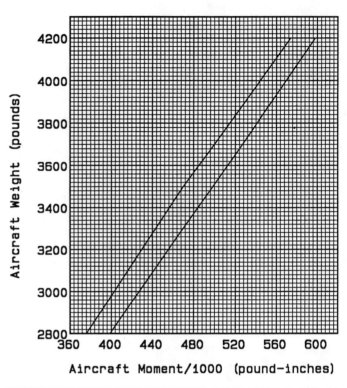

Fig. 3.11 The allowable center of gravity range for a push-pull twin

Procedure: (**1**). Draw a neat sketch of the elevator and its trim tab, including the two forces (using appropriate symbols since they are unknowns) and their moment arms (also as symbols; for example, L and 3.3 L). (**2**). Write an equation for the net force on the whole elevator. (**3**). Write another equation for the pitching moment about the hinge point. (**4**). Solve these two equations simultaneously to find the two forces. (For example, you can solve one of the equations for one of the forces and substitute it in the other.) [Partial answer: the force on the trim tab is 10 lb.]

10. Suppose that the moment arm L (as given in step 2 of the previous problem) has a value of 1.5 ft and the center of gravity of the aircraft is 15.0 feet ahead of the elevator hinge line. What is the moment about the cg due to the forces on the elevator surface? [Answer: +345 ft-lb] (Note that this moment is just **balancing** a moment in the **negative** direction caused by the **wing**, if the aircraft is in a stable climb or descent or cruise.)

11. Suppose the Cherokee of Fig. 3.7 is taxiing over very soft ground, thereby requiring a thrust from the propeller of 300 lb in order to maintain a steady speed. If the thrust force is assumed to act 46 inches above the ground and be directed parallel to the ground and through the *cg*, what is the vertical component of the force on the nosewheel? (Normally the pilot would be holding the wheel back in this situation to reduce the load on the nosewheel, but assume that this is not being done in this instance. It has been said that the load on the nose wheel can be as great as that on the main gear during certain ground operations and this is why Fred Weick wanted a nosewheel tire that was the same size as the main gear tires.) (Note that this is the same "weight transfer" effect that occurs with cars, and the reason that the front brakes need to be more powerful than the rear.) Answer: 436 lb

12. Light tricycle (nosewheel) gear airplanes can be turned by hand by pushing down on the fuselage at a point near the leading edge of the horizontal stabilizer, so that the nosewheel lifts off the ground. It is rather easy to do this with a 2-seat Cessna but noticeably harder with a 4-seater. Suppose that you, a 160 pounder, push down with all your weight on the 4-seat Cessna 172N of Fig. 3.8 at a point just 12.7 ft behind the point where the main gear touches the pavement, and this just barely lifts the nosewheel free of the ground. Assume the point of contact of the nose gear is just 5.50 ft in front of that for the main gear and that the 172 is at its typical empty weight of 1424 lb.
   a. Where is the *cg* located? (Take moments about the point of contact of the main gear on the pavement.) Answer: 1.43 ft in front of the main gear
   b. How much weight is normally borne by the nose gear when it is at its empty weight? (Take moments about the same point.) Answer: 370 lb

13. If an aircraft lands on soft ground or brakes very hard after landing, the rearward force on the tires generates an additional nose-down moment about the *cg* (since the force acts below the *cg*) and this moment must be counteracted by additional upward force by the ground on the poor little nosewheel tire. Suppose, for example, that the C-172 of problem 12, at the same weight (for simplicity), manages to stop in only four plane lengths (about 106 ft) after touching down at 50 mi/hr (73 ft/s). This means that the average deceleration was

$$a = \frac{V_F^2 - V_0^2}{2\,s} = \frac{0^2 - 73^2}{2 \times 106} = -25.1 \frac{\text{ft}}{\text{s}^2}$$

This means, further, that the rearward force of the ground on the airplane must have been (with the help of Eq. 3.4)

$$F_{\text{GROUND}} = m\,a = \left(\frac{W}{g}\right)a = \frac{1424 \text{ lb}}{32.2 \text{ ft/s}^2} \times (25.1 \text{ ft/s}^2) = 1110 \text{ lb}$$

Assuming the *cg* is 3.8 ft above the ground for this airplane, what was the average force on the nosewheel during this rapid stop? (Neglect any aerodynamically-induced moment that the pilot **should** be generating by pulling back on the wheel to provide a maximum down-load on the tail.) Take moments about the *cg* —necessary in this case because the plane is decelerating. Answer: 1137 lb !

Chapter 4

Static Properties of the Atmosphere

"What is essential is invisible to the eye," the little prince repeated, so that he would be sure to remember.[1]

A. Static Properties of Gases

There is hardly a prettier sight than a pair of wings flying in close formation with yours. At such times one is forced to admit that thin air, and not a giant hand from underneath, keeps aircraft suspended above the ground. In this chapter we explore the properties of our planet's gaseous atmosphere and the ways in which these properties are described by pilots and other aeronautical types.

Matter is composed of atoms that are about 5×10^{-8} inches in diameter; this is too small for individual atoms to be seen or even directly photographed. Atoms themselves are mostly empty space. They have a tiny central core of positive nuclear matter (the *nucleus*) which is surrounded by an entourage of negatively charged electrons; the two are attracted to each other by electrical forces. The nucleus contains most of the mass of the atom. Atomic masses range from about 1×10^{-28} slug for the ultralight hydrogen atom to about 2.7×10^{-26} slug for the massive uranium atom. The outermost electrons in the atom are the ones that determine the chemical properties of the atom — i.e., the degree to which it is electrically attracted to, or repelled by, other atoms.

The concept of **temperature** is meaningful only for the description of a group of interacting atoms. This is because we have no hope of following or describing the forces on, or the motions of, individual atoms — there are too many of them! Instead we use the average values for the mechanical properties of the whole group of atoms.

A **solid** is a group of interacting atoms for which the attractive forces are so strong that the atoms cannot change their relative positions without outside intervention. A **liquid** is a group of interacting atoms for which the attractive forces are strong enough to hold the atoms at more or less fixed distances relative to each other but are not strong enough to prevent the

atoms from readily shifting their relative positions. A **gas** is a group of slightly interacting atoms for which the attractive forces are not strong enough to even hold the atoms together.

In each of these three **phases** of matter, at least at the temperatures commonly found on earth, the atoms are in rapid, random motion. However, in a solid the atomic motions are all about fixed points in the solid so that on a large scale the solid holds its shape and its dimensions constant. In a liquid the random motions allow the atoms more relative freedom and so a liquid readily changes its shape but not its volume. In a gas the random motions describe the complete atomic motions except for brief instants when a pair of atoms happen to get close together.

The attractive electrical forces between atoms are the strongest for solids, and the atoms of solids are the closest together. The average atomic separation distance is greater for liquids, in general, and much greater for gases.

Thus in all phases of matter the individual atoms are in constant motion. In a solid the random motions are strongly restricted by the attractive electrical forces of nearby atoms. In a liquid the balance between random motion and attractive forces is more even; in a gas the random motions predominate.

It is not possible for any one atom to hold on to extra energy of motion for very long in any of these phases of matter because, by hypothesis, it is interacting with other atoms. If all the atoms have the same mass, they tend to end up with the same speed, on the average. If the atoms have differing masses, they tend to end up with the same average energy of motion. This energy of motion is called the kinetic energy and its numerical value can be computed from

$$\begin{pmatrix} \text{kinetic energy} \\ \text{of an atom} \end{pmatrix} \equiv \tfrac{1}{2}\,(\text{mass of atom})(\text{speed of atom})^2$$

or, in symbols,

[1]Antoine de Saint-Exupéry, *The Little Prince* (New York: Harcourt Brace Jovanovich, 1943).

$$\boxed{\mathbf{KE} = \tfrac{1}{2}mv^2 \qquad (4.1)}$$

Kinetic Energy

Now we are ready to define temperature. The temperature of a gas is a measure of the average, random, translational kinetic energy of the group of atoms forming the gas. We use the word random to emphasize that we are not talking about motion that is shared by all the atoms; in an aircraft traveling at 500 mi/hr, for instance, the temperature of the gas atoms in the cockpit refers to their kinetic energy relative to the rest of the aircraft and not relative to the earth. We specify translational kinetic energy because only translational speeds cause collisions and heat transfer. In a solid the atoms are not free to translate and for them the temperature is proportional to the vibrational energy of the atoms about their equilibrium positions.

It is an experimental observation that substances in contact eventually reach the same temperature, independent of their phase. The atoms of a hot gas enclosed in a cool box will lose energy to the atoms of the box whenever they interact, for example, and eventually both groups of atoms will have the same average translational kinetic energy. This interchange of energy between interacting types of matter is true only for the translational (linear) kinetic energy of atoms. It is not true for their rotational energy; rotational energy is not transmitted through the interatomic forces of interaction ("collision") and this is why we had to make our definition of temperature refer only to the random translational kinetic energy of the atoms.

Absolute zero is the temperature at which atoms have no removable energy, including removable random kinetic energy. Absolute zero is −459.7° on the Fahrenheit scale and −273.2° on the Celsius scale. (The "degrees" part of these units indicates that the measurement is from a non-zero value — which is the freezing temperature of water in this case.) When describing atoms and not people, it is much more meaningful to use an absolute temperature scale, one that starts at absolute zero rather than using degrees from some arbitrary higher temperature. We shall use the engineering unit of absolute temperature, the **Rankine (R)**, because it is based on the familiar Fahrenheit scale that is still used by many pilots and most American engineers. More importantly, it is also the temperature unit that is consistent with the sl/lb/ft/s system that has already been chosen.

$$\boxed{\textbf{Temperature in R} = \textbf{Temperature in °F} + \textbf{459.7} \quad (4.2)}$$

**CONVERSION OF FAHRENHEIT TEMPERATURE
TO ABSOLUTE TEMPERATURE UNIT**

If the temperature in °F is known within an accuracy of only ±1°, you may as well use 460 as the conversion number. Note also that, because the Rankine is an **absolute** unit just like the ft and the second and the dollar, there is no good reason to use the degree symbol with it. Similar reasoning has caused the Kelvin unit to lose its degree symbol in the past decade or so.

What other average properties of a gas might be significant? Consider a box containing a gas, with both the box atoms and the gas atoms at the same temperature this time. A gas atom that heads toward a wall of the box will continue until its outer electrons are sufficiently repelled by the outer electrons of atoms of the box to cause it to "rebound" from the wall. It never actually "touches" the other atoms and the walls don't move and so the average speed of rebound is just the same as the average speed of approach. (This is called an "elastic" collision.) By Newton's third law an atom of the box cannot exert a force on an atom of the gas without receiving in return an equal and opposite force on it. This force of a gas on enclosing walls can be a very powerful force; if air at the surface of the earth is enclosed in a cubical box 1 foot on a side, the force of the air atoms on each wall will be about 2100 pounds! The only reason boxes don't burst under this force is that normally air molecules are also pushing with an equal force on the opposite side of the wall!

The force exerted by atoms inside a stationary box of gas on a confining wall is always at **right** angles (perpendicular) to the wall. This follows from the fact that, on the average, there will be as many collisions between atoms with a negative sideways component as there will be collisions between atoms with a positive sideways component. Therefore the **average** vector sum of the force components of the gas atoms acting parallel to the wall will be zero.

It is most convenient to describe this "rebounding" force caused by gas atoms in terms of **pressure**. Consider the force that is exerted on an area of wall that is large compared to the size of an atom but otherwise is arbitrarily small compared to an inch. The pressure of a gas at a given point in the gas is defined as the force exerted by the gas atoms on an arbitrarily small area of wall that might be placed there, divided by the area of that wall. In words,

$$\left(\begin{matrix} \text{Pressure of a gas} \\ \text{at a given point} \\ \text{in the gas} \end{matrix} \right) \equiv \frac{\text{Force exerted on a small area of wall}}{\text{the area of the wall}}$$

$$\boxed{p \equiv \frac{F}{A} \qquad (4.3)}$$

DEFINITION OF PRESSURE

The area that is used in such a measurement must be large enough that millions of atoms are involved (so that statistical fluctuations in the random motions are averaged out) but small enough that any average change of pressure with position can be detected. Because of molecular interactions, a gas in a normal container on the earth will attain the **same** pressure throughout the container. As an example, in that one-foot-on-a-side cubical box mentioned earlier, the pressure must have been uniform in value and equal to

$$p = \frac{F}{A} = \frac{2100 \text{ lb}}{1 \text{ ft}^2} = 2100 \frac{\text{lb}}{\text{ft}^2}$$

Because aircraft move relative to the air, varying pressures are generated on the outer surfaces. These pressures result in a net force of the air on the aircraft; for an aircraft in level cruising flight, this force has an upward component (counteracting the force of gravity) and a rearward component (pressure drag, due to greater average pressures **behind** the aircraft wings and fuselage than in front). Air is invisible, but air pressure **can** be measured, and this measurement is the premier method for determining the interactions between the atmosphere and an aircraft. In a wind tunnel, pressures formerly were measured with liquid manometers (see Section D) but now solid state strain gauge transducers (whose electrical resistance depends on pressure) are generally used.

A third fundamental property of gases is **density**. We consider a hypothetical cubical box that is large compared to atomic dimensions so that it contains millions of gas atoms but small enough that we can detect any large-scale variations. Then the density of the gas is defined as the **mass** of the enclosed gas atoms divided by the volume of the enclosing "box":

$$\left(\begin{array}{c}\text{mass density}\\ \text{of a gas}\end{array}\right) \equiv \frac{\text{mass of the gas atoms}}{\text{volume containing the gas}}$$

or, in symbols,

$$\rho \equiv \frac{m}{V} \qquad (4.4)$$

DEFINITION OF DENSITY

We have used the Greek letter rho (ρ) for the density symbol, following conventional practice. Again, a gas in a normal container on earth will always attain a **uniform** density because of force interactions between the atoms; it is only on a microscopic level that density fluctuations can be detected. A cubical box one foot on a side situated on the surface of the earth will contain about 1.44×10^{24} atoms with an average mass of 1.65×10^{-27} slug, so the density of the gas in the box is

$$\rho = \frac{(1.44 \times 10^{24} \text{ atoms})(1.65 \times 10^{-27} \text{ slug/atom})}{1 \text{ ft} \times 1 \text{ ft} \times 1 \text{ ft}}$$

$$= 2.38 \times 10^{-3} \frac{\text{slug}}{\text{ft}^3}$$

These three properties of a gas (temperature, pressure, and density) must surely be related. In particular, we might reasonably expect that the pressure would increase with increasing density (more atoms available to interact with a wall) and with increasing temperature (greater approach and rebound speeds). We won't derive the relationship here, since it is available in most every college physics textbook, but the concept behind the derivation is that the definitions of temperature and density are combined with Newton's second law (Eq. 3.1) to show that pressure is directly **proportional** to density and to absolute temperature:

$$p \propto \rho T$$

(Mathematical notation: $\propto$ is the authorized symbol meaning "**is proportional to.**")

This tells us that doubling the density of a gas or doubling its absolute temperature will each double the pressure of the gas at that point. In order to make calculations of one of these three fundamental properties based on a knowledge of the other two, we must know what the proportionality constant is. If we call this proportionality constant k, then

$$p = \rho kT \qquad (4.5)$$

IDEAL GAS LAW

k has an experimental value of about 1716.56 lb-ft/sl/R for **dry** air within the earth's lower atmosphere. We'll normally round this off to 1716 lb-ft/sl/R in doing calculations.

An important note regarding units: In all of the formulas in this and successive chapters, you should **always** use the following consistent set of engineering units:

Property	Unit
force	lb
distance	ft
speed	ft/s
pressure	lb/ft^2
density	sl/ft^3
temperature	R

CONSISTENT SET OF ENGINEERING UNITS

Any quantity not expressed in this system of units must be converted to this system before being used in a formula or equation; the only exception is when ratios are involved, for then any conversion would apply to both numerator and denominator and no change in the ratio would occur. The happy result of following this rule is that you will always know the proper units for your answer without trying to follow them through the formula along with the numbers; if you are using a formula for density, for instance, you know your answer must have units of sl/ft^3; if you are using a formula for speed, your answer must have units of ft/s, and so forth.

The relationship expressed in Eq. 4.5 is known as the ideal or perfect gas law and is followed rather closely by real gases, including our atmosphere, as long as (**a**) the gas density is not so great that intermolecular forces are significant all the time and not just during "collisions," and (**b**) the gas density is not so small that collisions don't take place often enough to transfer changes in temperature or pressure. The first condition is not a factor for aircraft operating in the earth's atmosphere but the second condition is reached at the very fringes of the earth's atmosphere, say at 3,300,000 ft above the earth's surface (where spacecraft "fly"). Therefore the validity of the ideal gas law can be assumed in aerodynamic calculations for most aircraft.

The gas law is a very nice thing to have around because it allows us to calculate the one gas property that is difficult to measure, the density. You'll get your chances later.

Yet another fundamental property of a gas is its **specific heat**. Specific heat is a measure of how much heat energy is required to raise the temperature of a unit mass of gas. Adding heat energy to a gas, possibly with a moving wall or a paddle wheel, causes the average atomic energy to increase. However, some of the added energy often goes into rotational or vibrational energy (between the atoms of a molecule) and this part does **not** contribute to a rise in temperature since it is not the translational energy that our thermometers measure. The result is that two different gases may have very different values for their specific heats.

To try to obtain a feel for what specific heat is, we might try a little thought experiment. Suppose that we had similar metal containers on top of similar burners on an oven and suppose that we filled each container with the same mass (or weight) of different liquids. If we then monitor the liquid temperatures, the liquid in which the temperatures rises **most** quickly has the **smallest** specific heat and the liquid in which the temperature rises mostly slowly has the largest specific heat. They are all receive the same heat input, but the liquid with a high specific heat is finding places to put it that don't cause the temperature to go up. For a liquid, water has a rather large specific heat and therefore is commonly used as a coolant. Surprisingly, the specific heat of air is relatively close to that of water (one-quarter) even though it would take a far greater volume of air to get the same mass as for a given volume of water.

There are **two** distinct types of specific heats that can be measured with gases. If we add heat energy to a gas while keeping the gas volume constant, we can measure the specific heat at constant volume, c_V.

The specific heat of a gas at constant volume is therefore defined as the heat added to a gas in a constant-volume box divided by the product of the mass of the gas and its resulting temperature change:

$$c_V = \frac{Q \text{ (at constant volume)}}{m \, \Delta T} \qquad (4.6)$$

SPECIFIC HEAT AT CONSTANT VOLUME

(Mathematical notation: the Greek letter capital delta (Δ) is used to indicate that a **change** in a quantity is meant rather than the quantity itself.)

Thus ΔT represents the change in temperature produced by the addition of a measured amount of heat energy, Q. The measured value of specific heat typically depends slightly on the temperature at which it is determined, but normally this temperature dependence is small enough to neglect.

Alternately, if we add heat energy to a gas while allowing the gas to expand as much as is necessary to keep its **pressure** constant, we can measure the specific heat at constant pressure, c_V.

The specific heat of a gas at constant pressure is therefore defined as the heat added to a gas at **constant pressure** divided by the product of the mass of the gas and its resulting temperature change:

$$c_P \equiv \frac{Q \text{ (at constant pressure)}}{m \, \Delta T} \qquad (4.7)$$

SPECIFIC HEAT AT CONSTANT PRESSURE

You yourself have a specific heat which you measure every time you swallow a cold or hot drink. Your specific heat is the heat energy in the drink due to its temperature being different from yours, divided by your mass and your resulting temperature change. Since your volume is effectively constant during the maneuver, you have only one type of specific heat to worry about.

Fortunately for us, it is the **ratio** of c_P to c_V that usually appears in the mathematical descriptions of the properties of **air,** and the ratio is important enough to get its own mathematical symbol, the Greek small letter gamma (γ).

$$\gamma = \frac{c_P}{c_V} \qquad (4.8)$$

SPECIFIC HEAT RATIO

γ has a value of **1.4** for air. (γ doesn't have a unit associated with it because it is a **ratio** of two quantities that are measured with the same unit.)

The final static property of a gas that we will consider is the **speed** at which a pressure disturbance propagates or travels through the gas. Because a pressure disturbance in a certain frequency range is what our ears perceive as **sound**, we are equivalently enquiring as to the **speed of sound in a gas.**

When one part of a gas is given a pressure that is different from the average pressure in the rest of the gas, it will transmit this different pressure by the mechanism of "collisions" with adjacent atoms. If one part of the gas has a higher pressure than the average, for instance, those atoms will transmit a higher average speed to atoms adjacent to them, those atoms will transmit it to their neighbors, etc. We might guess then that the speed of sound would be closely related to the average speed at which the atoms themselves are moving. And because the absolute temperature is proportional to the average random translational kinetic energy of the atoms, which is itself proportional to the **square** of the average atomic speed, we might reasonably expect the speed of sound to be proportional to the **square root** of the absolute temperature. Such is the case.

$$\boxed{V_A = \sqrt{\gamma k T} \qquad (4.9)}$$

SPEED OF SOUND

You may wish to remember the significance of the subscript A by recalling that the speed of sound can also be called the **acoustic** speed.

EXAMPLE 4.1. What is the speed of sound in dry air when the air temperature is 32.0° F?

Solution: from Eq. 4.2, $T = 32.0°F + 459.7 = 491.7$ R

from Eq. 4.9, $V_A = \sqrt{(1.4)(1716)(491.7)}$

$= 1087$ ft/s (741 mi/hr)

For comparison, the average speed of a nitrogen molecule at this temperature is 1613 ft/s; for the heavier oxygen molecule it is 1509 ft/s.

Another comment on units is apropos here. In calculating the speed of sound in the example above, we have followed our earlier rule of substituting quantities with their proper engineering system of units into a formula and simply assuming that the answer has the appropriate engineering system unit. But it is always possible to follow the units through the calculation, if desired. For example, substituting the units for the quantities in the equation for the speed of sound (4.9) yields

$$\sqrt{(lb\ ft\ /\ (sl\ R) \times R} = \sqrt{lb \times ft\ /\ sl}$$
$$= \sqrt{(sl\ ft\ /\ s^2)(ft\ /\ sl)} = ft/s$$

However, in order to obtain the final result in this case it was necessary to realize that the unit **lb** is equivalent to the unit **sl ft/s²**. This isn't obvious but follows directly from Newton's second law (Eq. 3.1) which can be considered to be the definition of the lb as a unit of force. I'll bet you agree now that it is easier to follow the rule of sticking to one consistent system of units and assuming the appropriate unit for the answer rather than trying to derive the unit for the answer each time. Just be very sure you are using a consistent set of units!

B. The Earth's Atmosphere

The "boxes" of the previous section are not needed to contain our planet's atmosphere. Rather, the familiar gravitational force is responsible for maintaining this gaseous sheath around us even though the atoms retain a constant, high speed motion. The **composition** of the atmosphere is very **uniform** up to about 300,000 ft. By mass, the composition is typically 74.7% nitrogen, 22.9% oxygen, 1.3% argon, 1.0% water vapor, and 0.1% carbon dioxide, with traces of other elements. In its lowest level the atmosphere often contains suspended **liquid water** (fog or clouds).

Nitrogen and oxygen are diatomic (two-atom) **molecules**; after a nitrogen or oxygen atom finds another similar atom, its electrons as well as its nucleus are quite happy and have almost zero interest in other atoms. Argon, though, is an inert gas and all its atoms are quite happy in quantity 1. The percentage of water vapor in the lower atmosphere (the relative humidity) varies considerably, of course. Water vapor consists of molecules of 1 atom oxygen, 2 atoms hydrogen. The masses of these atoms and molecules are shown in Table 4.1.

Taking into account the relative abundances of these gases, an average air molecule has a mass of about 29 atomic mass units or 3.3×10^{-27} slug.

Different layers of the earth's atmosphere are given **labels** based on the variation of temperature within the layer. The first layer is the **troposphere** (ground to about 36,000 ft), next is the **stratosphere** (to about 170,000 ft), the **mesosphere** (to about 304,000 ft), the **thermosphere** (to about 1,800,000 ft), and the

exosphere (to perhaps 8,600,000 ft, at which altitude the atmosphere becomes indistinguishable from interstellar space).

ATOM or MOLECULE with chemical symbol	MASS in Atomic Mass Units	Percent of atmosphere by weight
Nitrogen Molecule N_2	28	74.7
Oxygen molecule O_2	32	22.9
Argon atom A	40	1.3
Water vapor molecule H_2O	18	0.1
Table 4.1 Major Constituents of the Earth's Atmosphere		

The boundary between the troposphere and the stratosphere is not determined by altitude but is defined as the height at which the air temperature stops its overall decrease with altitude. It is in this troposphere layer that turbulence, clouds, and other weather phenomena occur. The actual height of the troposphere varies with time of year and with latitude; it is about 26,000 ft at the poles and increases to about 55,000 ft at the equator.

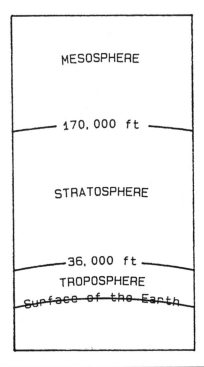

Fig. 4.1 Most aircraft fly in the troposphere, which is only a small part of the earth's atmosphere.

The stratosphere, now visited on a daily basis by jet aircraft, is itself composed of several layers in which the temperature at first is constant with increasing altitude and then **increases** rather rapidly with increasing altitude.

In the exosphere the density of atoms is so low that the atoms are essentially free particles; collisions are very rare and the ideal gas law is not usable.

As far as the composition of the atmosphere is concerned, it remains remarkably uniform up to about the top of the mesosphere. Above that altitude the density of atoms and molecules becomes so low that there are not enough collisions to keep the heavy and light atoms mixed together. Then the proportion of lighter atoms increases with increasing altitude.

It is not difficult to understand why (under static conditions) the **pressure** of the air always **decreases** with increasing altitude. The conclusion follows directly from the principle of static equilibrium (Section 3C.) Consider one more imaginary box somewhere in the lower atmosphere (Fig. 4.2). For static equilibrium (no energy input from the earth or from the sun), the force upward (due to the air pressure of the atoms below the box) must be exactly equal to the total downward force on the box. This downward force is due to the pressure of the atoms above the box on the top of the box **plus** the weight of the air in the box, W. Thus the pressure of the air must always decrease with increasing altitude under equilibrium conditions. If at any time this equality of forces does **not** exist, air will be accelerated upward or downward (resulting in updrafts and downdrafts).

FORCE DOWN DUE TO ATOMS ABOVE

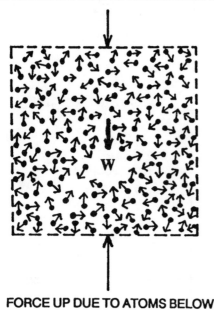

FORCE UP DUE TO ATOMS BELOW

Fig. 4.2 A vertical slice through the troposphere

Normally the density of the air and the temperature of the air also decrease with increasing altitude. However, all too often there exists a **temperature inversion** in the lower troposphere in which the temperature **increases** with altitude. This condition can occur with stable air that is trapped by geographical features (as in the Los Angeles Pacific Coast basin) or with frontal conditions (as when a warm air mass overruns a cooler air mass). The resulting air is stable against vertical motion and tends to collect moisture, dust, and industrial pollutants such as carbon monoxide and sulfur dioxide.

For whatever way in which the temperature may change with altitude in the lower atmosphere, the pressure and density of the air must correspondingly change—and change in such a fashion that $p = \rho kT$ is always satisfied. This tells us that the density of the air must decrease at a greater rate than normal within a temperature inversion, for example.

The Bell X-1A, a rocket-powered research aircraft that in 1947 was the first to exceed the speed of sound in level flight, reached an altitude of 90,000 ft in the 1950s. Current jet fighter aircraft have service ceilings that range from about 50,000 ft to about 75,000 ft. The Lockheed SR-71 surveillance aircraft can operate at ceilings of about 86,000 ft. For contrast, a satellite in a geosynchronous orbit has an altitude of about 118 **million** feet above the earth's surface. (A geosynchronous orbit is one that has a period of 24 hours so that it will appear to be stationary relative to a ground observer. This band of space is rapidly becoming crowded with satellites but pilots are eagerly awaiting the day that a full complement of special transmitters up there will allow them to affordably determine their position within a fraction of a mile, over all of the world and at all altitudes. In the USA, this system is known as the NAVSTAR Global Positioning System (GPS) while the Soviet-built system is Glonass.

C. Standard Atmosphere

Aeronautical designers soon realized that they needed to know the variation with altitude of the properties of the air in order to predict and compare the performance of their airspeed indicators, altimeters, and aircraft themselves. Because the real troposphere is so variable with location on the earth and with time of year, hypothetical or **model** or **standard atmospheres** were devised in the 1920s. In 1952 the International Civil Aviation Organization (ICAO) published a standard or model atmosphere that reconciled U.S. and European models; this standard atmosphere model assigned standard properties to altitudes between −16,400 ft and +65,600 ft. Zero altitude within this model is called **mean sea level** (MSL). A model for even higher altitudes was soon needed and so in 1962 the United States published the U.S. Standard Atmosphere (1962) which essentially agrees with the ICAO Standard Atmosphere within their common range of altitudes but extends the coverage upward to about 2,296,700 ft or 378 nm. Supplements to this 1962 model were published in 1966 and provide model atmosphere for various latitudes and for winter and summer. The most recent revision was completed in 1976; the standard for altitudes above about 164,000 ft was revised, based on newly available rocket and satellite data and improved theory.

The Standard Atmosphere (which will normally be referred to as **StAt** in this text) in the troposphere can be considered as representing average atmospheric properties for the middle latitudes in the spring or fall of the air, except that the air is considered to be perfectly dry (no water vapor or clouds) and completely motionless (no wind and no updrafts or downdrafts!).

The StAt, in the troposphere, is derived by making the following assumptions:

(a) Sea level pressure **is 2116.2 lb/ft²** (29.92" Hg)

(b) Sea level temperature is 518.7 R (59°F)

(c) The gravitational force is such that a freely-falling object has an acceleration of 32.17 ft/s² at SL, and the force decreases with increasing altitude as the inverse square of the distance from the earth's center (assumed to be 3950 statute miles).

(d) The molecular composition (atomic constituents) does not vary from its assumed sea level value.

(e) The air is dry and motionless.

(f) The air obeys the ideal gas law.

(g) The temperature decreases **linearly** with altitude at a rate of –3.566°F per thousand feet.

Although these assumptions are certainly only average values and simplifications, they result in a model atmosphere for which properties can be easily calculated without requiring such sophisticated computational techniques as numerical integration. The result is a completely def ...ed variation of air pressure, density, and temperature with altitude; these are plotted in Fig. 4.5 (next page). Because the equations relating pressure and density to altitude are still rather cumbersome to use, a tabulation of StAt properties at the lower altitudes is presented in Appendix B, which you should use in all numerical problem solving. Included in the tabulation are the speed of sound at each altitude (calculated from Eq. 4.9) and the **viscosity** of the air (needed in the calculation of the dynamic properties of air – as in the next chapter).

D. Pressure Altitude

Every pilot needs to be aware of the fact that his altimeter is nothing more than a device for measuring air pressure (a barometer); it reads feet of altitude rather than a unit of pressure only because it **assumes** that the atmosphere is following the altitude vs. pressure variation given by the StAt. Because air pressure decreases with increasing altitude (under equilibrium conditions, at least; see Section A), the altimeter will tell the pilot when altitude is gained or lost. However, whether the indicated altitude is the same as the true (geometrical) altitude (or even close!) depends on how closely the atmospheric pressure follows the StAt model. As a practical consequence, it is perfectly possible to fly into a 9200 ft mountain while a perfectly accurate altimeter is indicating an "altitude" of 10,000 ft!

The face of a typical altimeter is shown in Fig. 4.3. The sensing element, shown in cross-section in Fig. 4.4, is an evacuated chamber mechanically linked and geared to the dial pointer. The chamber is strong enough to resist collapse from the atmospheric pressure on the outside but is also flexible enough to measurably change its volume when the pressure changes only slightly.

It is no easy task to design a barometer that will accurately measure pressure while operating in the difficult aircraft environment. The design must be immune to the vibrations and accelerations of flight and to changes in ambient temperature.

Fig. 4.3 The face of a typical altimeter

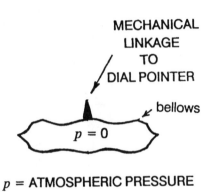

Fig. 4.4 Inside an altimeter lies an aneroid (without liquid) barometer.

The temperature compensation involves the use of metals with low thermal expansion coefficients and a special bimetallic temperature compensator. This temperature compensating aspect of altimeter is **highly** important because there are large differences between summer and winter temperatures and between storage and in-flight temperatures. The indicated altitude must not be affected by these changes. However, there is no way the altimeter can compensate for non-standard changes in pressure with altitude due to temperatures that differ from those assumed by the StAt.

Before takeoff a pilot normally uses the "barometric setting knob" to adjust the pointer of the altimeter so that it indicates an altitude equal to the published airport field elevation. Unless the pressure happens to be just equal to the StAt pressure for that altitude, offset in the altitude versus pressure calibration based on the StAt will be introduced. The window in the altimeter face, the **Kollsman window**, tells the pilot how much offset is being used and permits a correction to be made in flight, based on radioed information; this is necessary whenever flight progresses into a higher or lower pressure area.

Making the in-flight offset correction is particularly important for instrument flight (i.e., flight relying on interior instruments rather than the outside view), for then the pilot relies on

46CHAPTER 4

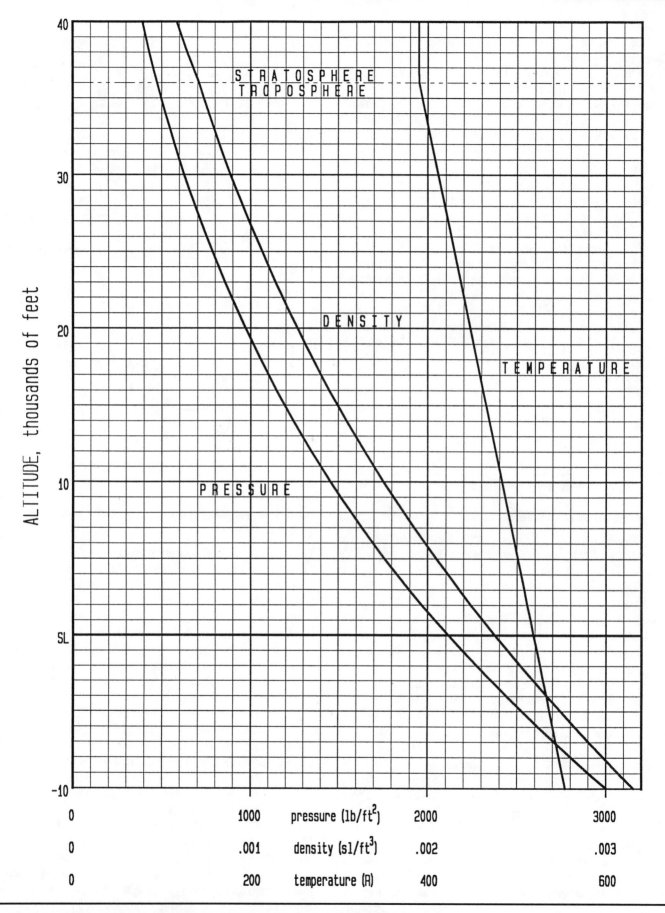

Fig. 4.5 The variation of pressure, density, and temperature with altitude in the Standard Atmosphere

the altimeter for safe clearance from the ground and ground-based obstacles. Adequate separation between en route aircraft under instrument conditions also assumes similar altimeter indications — but not accurate geometrical altitudes — for **all** the aircraft. For this reason, altimeters and their associated pressure lines must be checked every 24 months if the aircraft is used for instrument flight.

The weather over the continental USA is governed by the arrival and departure of air masses possessing differing properties, due mostly to their place of origin. An air mass arriving from the Canadian plains tends to be cool and dry; an air mass arriving from the Gulf of Mexico tends to be warm and wet. The air pressure on the ground near the center of any of these air masses is well above average (a high pressure region) and doesn't change very rapidly. Circulation of air about the high pressure center is clockwise in the northern hemisphere. At the boundary with another air mass (the **front**), there is unstable air, often much precipitation, and air pressures well below average. Circulation of air about a center of low pressure is counterclockwise and the most severe weather is typically close to the center.

The barometric setting knob allows the pilot to compensate for this surface pressure variation so the altimeter always indicates field elevation before takeoff and after landing. This still doesn't mean that the altimeter can be trusted to give accurate heights above the ground, though, because its calibration assumes the StAt variation of pressure with altitude. Therefore the indicated heights are said to be **MSL** (mean sea level) altitudes.

We have to face the prospect of learning to use a different unit of pressure at this point because inches of mercury is the unit for the pressure offset in the Kollsman window. To see how a length can be used to describe pressure (force / area), consider how pressure is often measured (Fig. 4.6). Figure 4.6a shows a U-shaped glass tube that is open to the atmosphere. The tube contains liquid mercury (chemical symbol **Hg**) and the height of the mercury must be the same on both sides because (**a**) the fluid pressure at the left end of the connecting tube must be equal to the fluid pressure at the right end, or else the fluid would be in motion, and because (**b**) equal fluid pressure at the bottom of the tube requires equal heights of mercury since both tubes are open to the atmospheric air pressure.

Suppose now that we connect one tube to a good vacuum pump, reducing the air pressure in the right side of the tube to a negligible value (Fig. 4.6b). Then the atmospheric air pressure on the left side of the U-tube will be unbalanced and the mercury will be forced to move up in the right side and down in the left side. (The mercury is **not** being pulled up in the right tube; instead it is being pushed from the left side. Similarly, if a pressurized aircraft suddenly loses its pressure seal, objects can be pushed out the opening by the escaping air but "suction" is not an adequate description of the effect.)

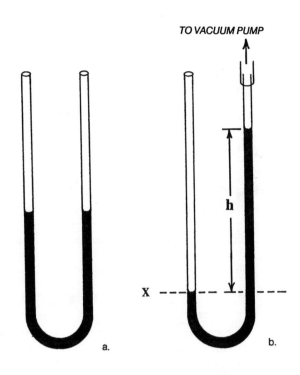

Fig. 4.6 Liquid manometers were the first accurate pressure-measuring devices.

The principle of static equilibrium gives us the relation between the height difference and the atmosphere that we require. First, the pressure at level X must be equal to the atmospheric pressure, p_{AIR}, in both sides of the U-tube, from arguments similar to those presented with Fig. 4.6a. The balance of force on that part of the fluid that is **above** level X in the right side is

$$\sum F_{UP} = \text{force due to pressure at level } X = p_{AIR} A$$
$$= \sum F_{DOWN} = \text{weight of supported fluid}$$
$$= m_{FLUID}\, g = \rho_{FLUID}\, V_{FLUID}\, g = \rho_{FLUID}\, h\, A\, g$$

Therefore $p_{AIR} = \rho_{FLUID}\, h\, g$

and
$$h = \frac{p_{AIR}}{\rho_{FLUID}\, g} \qquad (4.10)$$

BAROMETRIC PRESSURE EXPRESSED AS A HEIGHT OF A LIQUID

In this derivation we have used Eq. 3.3 for weight and Eq. 4.3 for the definition of pressure. V_{FLUID} is the volume of the fluid above level X and A is the cross-sectional area of the tube, so $V_{FLUID} = A \times h$. Note that A does not appear in the final expression for the height (assuming it is constant) because it appears linearly on both sides of the equation.

Evidently the difference in heights is a direct measure of atmospheric air pressure although the correspondence does depend on g (which varies slightly with altitude) and the density of the fluid. Using StAt sea level values and mercury (density of 26.379 sl/ft^3) or water (density of 1.9386 sl/ft^3) in the U-tube, we obtain from Eq. 4.10 the following equivalent descriptions of SL pressures:

$$h = \frac{2116.22 \text{ lb/ft}^2}{(26.379 \text{ sl/ft}^3)(32.174 \text{ ft/s}^2)} = 2.4934 \text{ ft of Hg}$$

$$= \textbf{29.92" of Hg}$$

and

$$h = \frac{2116.22 \text{ lb / ft}^2}{(1.9386 \text{ sl/ft}^3)(32.174 \text{ ft/s}^2)} = 33.9287 \text{ ft of water}$$

$$= \textbf{407.1" of water}$$

Using calculations like these, the conversion factors between various pressure units can be summarized (Table 4.2).

PRESSURE UNIT		PRESSURE UNIT
1 lb/ft^2	=	0.014139 inches of Hg
1 lb/ft^2	=	0.19239 inches of water
1 inch of Hg	=	70.73 lb/ft^2
1 inch of water	=	5.198 lb/ft^2

Table 4.2 Pressure Conversion Constants

When an altimeter is set to 29.92" Hg, it displays the altitude equivalent from the StAt for the actual air pressure at that point. Because weather variations cause offsets from the StAt, it is worthwhile to make the following definition:

> **Pressure altitude is the indicated altitude when an accurate altimeter is set to 29.92" Hg.**

In other words, pressure altitude is the altitude that corresponds to the **actual** pressure according to the StAt correlation between pressure and altitude. In a sense, pressure altitude is just another unit of pressure, one that will not frighten a pilot who doesn't wish to hear about pounds per square ft. Graphically, pressure altitude is the altitude that is read from the StAt graph of pressure versus altitude (Fig. 4.3).

The 29.92" Hg that is in the window does **NOT** mean that the outside air pressure is 29.92" Hg! It only means that the indicated altitude directly correlates with StAt pressure without an offset.

For a SL airport under StAt conditions the pressure will be 2116.2 lb/ft^2; for a 1000 ft airport under StAt conditions the pressure will be 2040.9 lb/ft^2 (Appendix B). This is a pressure difference of 75.3 lb/ft^2 or 1.06" of Hg and illustrates the useful **rule of thumb**:

> **A change of altitude of 1000 ft is approximately equal to a change in pressure of 1" Hg.**

Thus a pilot at a **sea level** airport where the current pressure is 2040.9 lb/ft^2 would have to adjust the aircraft altimeter so that (29.92"–1.06") = 28.86" appeared in the Kollsman window, in order to obtain a display of field elevation.

EXAMPLE 4.2. A pilot is at Kent State University airport (elevation 1150 ft above SL). When set to 29.92", the altimeter indicates 200 ft. What is the current pressure and pressure altitude?

Solution: From the StAt table (Appendix B), the current air pressure is 2101.0 lb/ft^2 for this pressure altitude of 200 ft. (Note that the geometric or true altitude is not involved in the determination of pressure altitude.)

Figure 4.7 presents a worthwhile overview of the dependence of an accurate altimeter's indicated altitude on the actual air pressure and the altimeter setting. Note that the actual air pressure is equal to the altimeter setting only at sea level.

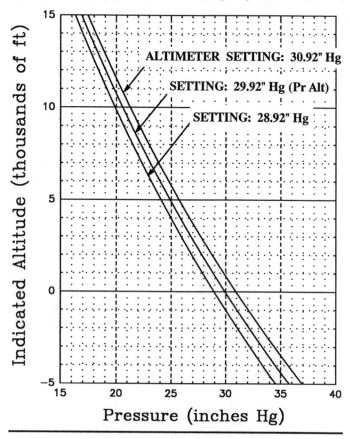

Fig. 4.7 Indicated altitude versus actual air pressure and altimeter setting

E. Density Altitude

The most important atmospheric factor determining the power output of a reciprocating aircraft engine and the overall takeoff and climb performance of an aircraft is the **air density**. The difference in performance due to density variations is truly spectacular; for instance, the distance required to clear a 50 ft obstacle under low density conditions can easily be **two** times the distance under standard SL density conditions.

Unfortunately, a "density meter" does not exist. The only way to obtain the air density is to **calculate** it from the measured air pressure and the measured air temperature, using the ideal gas law. The air pressure is obtained from the aircraft altimeter (set to 29.92") and the air temperature from an outside air temperature (OAT) gauge. Most pilots use a flight calculator (electronic or in the form of a circular slide rule) when they are forced to make this calculation, but this doesn't help to understand what is going on. We'll make the density calculations using a standard calculator and the StAt, which is much more instructive.

Density is usually given in terms of ft of **density altitude** rather than in sl/ft^3 or some other legitimate density unit.

> **Density altitude is the altitude in the StAt that corresponds to the actual air density.**

For example, the air density at 3000 ft in the StAt (see Appendix B) is 0.002175 sl/ft^3; an airport with this value of air density is defined as having a density altitude of 3000 ft **independent** of its pressure altitude **or** its field elevation.

When a manufacturer lists aircraft performance information in terms of altitudes and "standard conditions," the altitudes given are really **density** altitudes.

EXAMPLE 4.3. A pilot is about to take off from an airport that has a listed field elevation of 2830 ft. The Kollsman window shows 28.34" when the altimeter is adjusted to read field elevation. When the altimeter is adjusted to 29.92", it reads 4320 ft. The OAT is 80°F. What is the density altitude at this airport?

Solution:
1. Determine the actual pressure in lb/ft^2 from the pressure altitude, using the StAt tables and interpolating between given values if necessary.
2. Calculate the **absolute** temperature based on the Fahrenheit temperature.
3. Calculate the **density** by using the ideal gas law.
4. Read the altitude corresponding to this density from the StAt tables.

Therefore,

$$p = 1806 \text{ lb/ft}^2, \text{ from the StAt tables}$$

$$T = 80° + 460 = 540 \text{ R, from Eq. 4.2}$$

and finally, from the ideal gas law, Eq. 4.5.

$$\rho = \frac{1806 \text{ lb/ft}^2}{(1716)(540)} = 0.001949 \text{ sl/ft}^3$$

So the density altitude = **6600 ft** (picking the closest density value in the StAt tables in Appendix B).

For those who wish to use programmable calculators or computers, the following steps can be substituted:
(a) Calculate the density of the air, as above.
(b) Calculate the quantity $X = 0.23496 \ln\left(\dfrac{\rho}{0.00237692}\right)$
(c) Calculate the geopotential altitude, h_{POT}, from:

$$h_{POT} = (1.4545 \times 10^5)(1 - e^X)$$

(d) Calculate the density altitude, h_{DEN}, from:

$$h_{DEN} = h_{POT} + (4.795 \times 10^{-8})(h_{POT})^2$$

To check your use of these equations, try the numbers from Ex. 4.3. You should obtain: $X = -0.04664$, $h_{POT} = 6628$ ft, and h_{DEN} = 6628 ft + 2 ft = **6630 ft**. (Note that the e in part (c) is the number base for natural logarithms and, on a typical scientific calculator, the calculation indicated would be made by keying in the value for X and then pressing the keys labelled as INV and (ln x), in that order.)

F. Humidity and Density Altitude

Pilots are warned that humidity raises the density altitude and therefore a decrease in aircraft performance should be expected. In this section the reason for this influence of sticky weather as well as the magnitude of the effect will be discussed.

Water vapor molecules have a mass of 18 atomic mass units (2×1 amu for hydrogen + 1×16 amu for oxygen). Because the average air molecule has a mass of 29 amu (Table 4.1), the water vapor molecule has only **62%** as much mass as the average air molecule.

The pressure due to the motion of gas molecules of **any** type depends on the average translational kinetic energy that they possess, and this in turn depends solely on the **absolute** temperature (Section A). Therefore a given number of water vapor molecules exert the **same** pressure as an equal number of the heavier air molecules **if** they are at the same temperature. It can be deduced then that, when some of the molecules producing a given pressure are water vapor molecules, the **density** of the air is **less** than if all the pressure was being produced by heavier (on the average) dry air molecules. It also means that the gas constant for a gas of purely water vapor molecules, which we will call k_W, must be larger than the gas constant (k) for dry air.

We now derive an equation for the air density when water vapor is present. The basic approach is to realize that the measured air pressure is equal to the **sum** of the individual pressures of the various molecular compounds, which is equivalent to assuming that their interactions with each other don't affect their interactions with solid boundaries. (Chemists refer to this as Dalton's **law** of partial pressures; it follows directly from the assumption that the density is so low that gas molecules are essentially independent particles.) When we combine this law of partial pressures with the ideal gas law and do some algebra, the result will be the useful Eq. 4.16.

Step 1. The observed air pressure, p, is the sum of the separate pressures of the dry air molecules, p_A, and the water vapor molecules, p_W:

$$p = p_A + p_W \tag{4.11}$$

Step 2. The observed density is the sum of the separate densities.

$$\rho = \rho_A + \rho_W \tag{4.12}$$

Step 3. Substitute for the densities, using the ideal gas law and recognizing that the gas constant for water molecules, k_W, must be larger than the gas constant for dry air molecules, k, since the pressure is proportional to T only.

$$\rho = \frac{p_A}{k\,T} + \frac{p_W}{k_W\,T} \tag{4.13}$$

Step 4. Substitute in Eq. 4.13 for p_A, using Eq. 4.11.

$$\rho = \frac{p - p_W}{k\,T} + \frac{p_W}{k_W\,T} \tag{4.14}$$

Step 5. Rearrange algebraically:

$$\rho = \frac{p}{k\,T} - \left(\frac{1}{k} - \frac{1}{k_W}\right)\frac{p_W}{T} \tag{4.15}$$

The first term in Eq. 4.15 is our previous equation for the density of the air with no water vapor present, so the second term must be the reduction in air density due to the fact the lighter water vapor molecules are providing part of the measured air pressure. When we substitute numerical values for k and k_W, the result can be written in a form convenient for computations:

$$\rho = \frac{p}{k\,T} - \left(2.204 \times 10^{-4}\right)\frac{p_W}{T} \tag{4.16}$$

AIR DENSITY WITH THE EFFECT OF HUMIDITY INCLUDED

The **maximum** value for the water vapor pressure at any given temperature is known as the **saturated** water vapor pressure (Table 4.3); it shows a very strong dependence on temperature, as people in temperate zones relearn during every seasonal change. Hot air can hold **ten** times as much water vapor than air close to freezing.

If more water vapor molecules are added to air than the air can hold, some of them will get together and condense into a **liquid**, resulting in the formation of fog or a cloud. The **relative humidity** is defined as the **ratio** of actual water vapor pressure to the maximum possible water vapor pressure for that temperature; it is usually expressed as a percent. Thus 70% relative humidity at 100°F means that the actual water vapor pressure is (using Table 4.3 for the saturated water vapor pressure)

$$p_W = (0.70)(137\ \text{lb/ft}^2) = 96\ \text{lb/ft}^2$$

EXAMPLE 4.4. Determine the density altitude for Example 4.3 except assume 96% relative humidity this time.

Solution: $p_W = (0.96)(73.01\ \text{lb/ft}^2) = 70\ \text{lb/ft}^2$, so the actual density of the air, from Eq. 4.16, is

$$\rho = \frac{1806}{(1716)(540)} - \left(2.204 \times 10^{-4}\right)\left(\frac{70}{540}\right)$$
$$= 0.001949 - 0.000029 = 0.0001920\ \text{sl/ft}^3$$

and then, from the StAt table, density altitude = **7100 ft**

For this example the density altitude is only 500 ft higher for 96% relative humidity than it was for perfectly dry air at the same pressure and temperature. The reduction in performance due to high humidity is at least partly as much psychological as physical!

We can obtain a little bonus from Table 4.3. Flight Service Stations report the humidity to pilots by telling them the **dew**

T (°F)	p_S	T (°F)	p_S	T (°F)	p_S
32	12.75	68	48.83	104	154.1
34	13.83	70	52.29	106	163.4
36	14.98	72	55.96	108	173.3
38	16.21	74	59.86	110	183.6
40	17.53	76	63.99	112	194.5
42	18.94	78	68.37	114	205.9
44	20.45	80	73.01	116	217.9
46	22.07	82	77.93	118	230.5
48	23.79	84	83.12	120	243.7
50	25.64	86	88.62	122	257.6
52	27/61	88	94.43	124	272.1
54	29.72	80	100.6	126	287.4
56	31.96	92	107.1	128	303.3
58	34.36	94	113.9	130	320.0
60	36.90	96	121.1	132	337.5
62	39.62	98	128.7	134	355.8
64	42.50	100	136.7	136	375.0
66	45.57	102	145.2	138	395.0

Table 4.3 Saturated Vapor Pressures for Water, in Units of lb/ft^2

point temperature. This temperature is defined as the temperature at which the present air would become saturated (100% relative humidity) if it were suddenly cooled. It is a meaningful way to give the humidity because it makes it easy to guess if fog is likely to form, as with unstable frontal conditions or with the rapid cooling that often follows the setting of the sun in the fall, especially under a clear sky. (It also makes it unnecessary for pilots to learn a new unit, once again!)

EXAMPLE 4.5. Suppose there is 83% relative humidity at 84°F. What is the dew point temperature?

Solution: Actual vapor pressure $= (0.83)(83.12\ \text{lb/ft}^2)$
$$= 68.99\ \text{lb/ft}^2$$

From Table 4.3, this is the saturated water vapor pressure at a temperature of about **78°F**.

G. True Altitude

It should be obvious by now that the aircraft altimeter is capable of giving only an approximation to the actual or true altitude of the aircraft. A long list of reasons why true and indicated altitude may differ is easy to draw up:

(a) Errors in the altimeter system. Only altimeters used for instrument flight are required to undergo a periodic check, but even for them an error of up to ±80 ft at 10,000 ft altitude is permissible. (Federal Air Regulations Part 43, Appendix E).

(b) A failure by the pilot to reset the altimeter when flying into air that is at a different pressure for the same altitude. This often happens with non-radio aircraft following visual flight rules but must not be allowed to happen on an instrument flight. Particularly hazardous is flying into deteriorating weather because the pressure at a given altitude will be decreasing, and the approach to landing will be under erroneously high altimeter indications if the altimeter hasn't been adjusted.

(c) Lapse rate calibration assumption. The real atmosphere doesn't often follow its model. Temperature inversions, in which warm air is trapped above cooler air, are a fact of flying in basin areas and when warm air pushes cooler air ahead of it. As an example, if the actual rate of temperature change with altitude is positive 2.0°F per thousand ft, the altimeter will indicate about 600 ft lower at 10,000 ft than for a normal lapse rate.

The greater danger lies in flying into an area where the lapse rate is greater than the assumed −3.566°F per thousand ft. For example, if the average temperature change with altitude is −5.0°F per thousand ft from SL, the altimeter will indicate an altitude about 150 ft higher than the 10,000 ft it would indicate with a normal lapse rate.

(d) A non-standard ground temperature. If the pressure for a given altitude is at its standard value but the temperature is higher than standard, the density of the air at that altitude must be less than the standard value (from the ideal gas law, Eq. 4.5). Then it will require a greater altitude change to effect a given pressure change (Eq. 4.10), and the true altitude is greater than the indicated altitude.

Conversely, if the ground temperature is lower than standard, the density of the air will be greater than standard and less altitude change is necessary to obtain a given pressure change, and then the true altitude is **less** than the indicated altitude. Clearly the latter situation is the most hazardous because it presents pilots with potential encounters with ground obstructions.

To see if this effect is significant, though, we need to make some computations. Let us assume that we take off from a SL airport with standard pressure but with an outside air temperature of 100° F, and that we fly to another SL airport, also with standard pressure, with an outside air temperature of 0° F. If we further assume that the temperature variation with altitude is at its standard value of −3.566° F per thousand feet, it can be shown that the true altitude is the indicated altitude multiplied by the ratio of the actual SL air temperature (in R) to the standard SL air temperature (518.7 R) — but note well that this is only true when the takeoff or altimeter-reference altitude is sea level.

$$\begin{pmatrix} \text{true} \\ \text{altitude} \end{pmatrix} = \begin{pmatrix} \text{indicated} \\ \text{altitude} \end{pmatrix} \times \left(\frac{T}{518.7\,R} \right) \qquad (4.17)$$

**CORRECTION FOR NON-STANDARD TEMPERATURE
(FROM SL AIRPORT)**

Thus if our flight is at a constant **indicated** altitude of 10,000 feet, our true altitude near the departure airport will be

True altitude = (10,000 ft) × [(100° + 459.7) / 518.7]
= **10,790 ft**

But at our **destination** airport, our true altitude will be

True altitude = (10,000 ft) × [(0° + 459.7) / 518.7]
= **8863 ft!**

Figure 4.8 provides visual reinforcement of this important result.

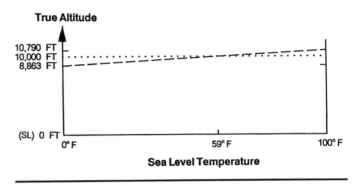

Fig. 4.8 Altimeter error due to a non-standard temperature at a sea level reporting station (dotted line is indicated altitude, dashed line is true altitude)

Because the error is proportional to the altitude above the ground, the difference between indicated and true altitude increases with flight altitude. For example, you would have to climb to an indicated altitude of 32,724 ft in order to clear Mount Everest's 29,002 ft peak under the assumptions of a 0° F temperature at SL. The difference between indicated and true altitude could be even greater if the lapse rate was also non-standard.

Reasons (b) and (d) are the basis for the oft-repeated advice for pilots,

> **from high to low, or from warm to cold, watch out below!**

and now you know the physical basis for the advice. Note, however, that the advice is misleading insofar as it implies that the problem exists only when flight proceeds **into** cold air en route. Actually, the problem exists whenever the temperature is colder than standard, even if that occurs at the departure airport.

On the other hand, the error is **minimized** if the pilot is able to obtain an altimeter error from a high altitude airport, because then the error builds up only **from that altitude** (since the altimeter setting is guaranteed to yield an indicated altitude equal to the true altitude at the point at which it is given).

For the general case, then, Eq. 4.17 is replaced by an equivalent equation that references just the difference between the flight altitude and the takeoff or altimeter reference airport (as might be obtained en route).

$$\begin{pmatrix} \text{true} \\ \text{altitude} \\ \text{difference} \end{pmatrix} = \begin{pmatrix} \text{indicated} \\ \text{altitude} \\ \text{difference} \end{pmatrix} \times \left(\frac{T_{\text{ACTUAL}}}{T_{\text{STANDARD}}} \right) \qquad (4.18)$$

**CORRECTION FOR NON-STANDARD TEMPERATURE
(ABOVE A REPORTING STATION)**

EXAMPLE 4.6. Estimate the true altitude for a flight at an indicated altitude of 10,000 ft if the pilot uses an altimeter setting from a 3000 ft airport with a surface air temperature of –20° F. Assume standard pressure at the surface of the airport and a standard lapse rate.

Solution: From the StAt table, the standard temperature for a 3000 ft airport is 48.3° F or 508.0 R. Therefore the true altitude **above** the airport can be estimated as

$$\begin{pmatrix} \text{true} \\ \text{altitude} \\ \text{difference} \end{pmatrix} = \left(7000 \text{ ft}\right) \times \left(\frac{439.7 \text{ R}}{508.0 \text{ R}}\right) = 6059 \text{ ft}$$

Therefore the estimated true altitude is

true altitude = 3000 ft + 6059 ft = **9059 ft**

compared to the indicated 10,000 ft—which is again a very significant effect.

A pilot flying at constant indicated altitude toward a region where the pressure is lower or the air colder is flying **downhill** and may be pleasantly surprised by the apparent increase in performance of the aircraft!

Flight calculators attempt to calculate true altitude by using the actual outside temperature at the flight altitude and assuming a standard lapse rate down to the reporting station, but without a detailed knowledge of the actual variation in lapse rate, this can only be a approximation, just as we have done.

In sum, for a conservative pilot the altimeter will provide adequate separation from the terrain, particularly when the situations that lead to a too-low reading are kept in mind. The condition for which a great deal of extra altitude must be gained is when approaching a **cold** mountain from the leeward side (at an angle less than 90°, please) because of the altimeter errors and the expectation of strong downdrafts, **especially** when strong winds aloft are present.

Summary

The atomic theory of matter was used to indicate the differences among the solid, liquid, and gas phases of matter in Section A. The meaning of temperature, pressure, and density for a gas were presented, as was the ideal gas law which relates these three properties. Specific heats and the speed of sound in air were presented as related properties.

Section B described properties of the earth's atmosphere and indicated why and how the properties vary with height above the earth's surface. Section C presented a model or standard atmosphere that is widely used as a basis for aircraft instrument design and performance specification.

Section D defined pressure altitude as the altitude in the StAt that corresponds to the actual air pressure. Density altitude is the altitude in the StAt that has the same density as the actual air density; it is the primary indicator of aircraft performance variations with atmospheric properties. Density altitudes were calculated in Section E for dry air and in Section

F for humid air. Finally, the reasons why the aircraft altimeter usually doesn't indicate the true altitude were summarized. The two most important situations are those in which the true altitude is **less** than the indicated altitude, and this occurs whenever flight is made into an area where the pressure is lower (and the altimeter isn't adjusted) or when flight is made into air that is colder and denser than standard; the difference between indicated and true altitude can then be in the thousands of feet.

Symbol Table (in order of introduction)

KE	kinetic energy (of molecules of air)
m	mass (of molecules of air)
v	speed (of molecules of air)
p	pressure (of air)
ρ	density (of air)
V	volume (of air)
k	gas constant for dry air
c_V	specific heat of a gas at constant volume
c_P	specific heat of a gas at constant pressure
γ	ratio of c_P to c_V (for air, value is 1.400)
V_A	speed of sound in air
StAt	U. S. Standard Atmosphere (1976)
g	acceleration due to the gravitational force
h	(a) height above the surface of the earth
	(b) height difference of the liquids in a U-tube manometer
k_W	gas constant for gaseous water molecules

Review Questions

1. Why is temperature a property of **many** atoms or molecules rather than a property of a single particle?

2. What kind of motion is associated with the concept of temperature? (a) linear (b) rotational (c) average group velocity (d) all of these

3. What is the **direction** of the force exerted by a gas in equilibrium on a wall or other bounding surface?

4. Why don't objects or people usually fail structurally under the strong pressures exerted by the air molecules close to the surface of the earth?

5. If you stand on the tips of your shoes, you will increase the ___ on the ground. (a) force (b) pressure (c) both of these

6. The ideal gas law states that the pressure of a gas is directly proportional to its___ and its ___.

7. What are the engineering system units for pressure, density, and temperature? ___, ___, ___

8. Why is density best defined in terms of mass rather than in terms of weight? (Engineers do talk about weight density, but it is a less fundamental property than mass density. Why?)

9. If you could provide a known amount of heat energy to the air trapped in an otherwise empty pressure cooker (fitted with a pressure indicator), you should be able to determine the ___ of the air. (a) weight (b) c_V (c) c_P (d) density

10. The speed of sound is proportional to the ___ of the air molecules.

11. What are the two most abundant molecules in the earth's lower atmosphere? What percent by weight do they alone account for?

12. What property of the air determines the boundary between the troposphere and the stratosphere?

13. Does this boundary change with latitude or with time of year?

14. It follows from the principle of ___ that pressure in the atmosphere decreases with increasing altitude.

15. The StAt is a mid-latitude, spring or fall model of the atmosphere that makes two simplifying assumptions about the composition and motion of the air. What are these assumptions?

16. The easiest altitude for a pilot to determine while in flight is the (**a**) pressure altitude, (**b**) density altitude, (**c**) true altitude.

17. Why must altimeters incorporate temperature compensation?

18. The primary function of the offset feature of altimeters is to allow pilots to correct for (**a**) weather conditions, (**b**) altimeter errors, (**c**) non-standard lapse rates.

19. Why does the barometer of Fig. 4.6b show a greater height difference when water is used than when mercury is used? (This suggests one way to make a more sensitive manometer.)

20. Define pressure and density altitude.

21. The pressure caused by a gas is directly related to the average translational KE of its constituent molecules, so the pressure due to light molecules is ___ that due to heavy molecules. (**a**) the same as, (**b**) greater than, (**c**) less than

22. Define relative humidity and dew point temperature.

23. If the temperature increases with increasing altitude (contrary to the assumption behind the StAt), an altimeter will tend to **indicate** an altitude ___ the true altitude. (**a**) the same as, (**b**) greater than, (**c**) less than

24. If a pilot flies to an airport than is farther from the center of a high pressure region, an unadjusted altimeter will indicate an altitude ___ the true altitude. (**a**) the same as, (**b**) greater than, (**c**) less than

25. Is it possible for the pressure to be the same at two airports with different field elevations? How about for density altitude?

26. The indicated altitude is **greater** than the true altitude (which is bad for obstacle clearance) when the surface temperature is (warmer, cooler) than the standard value for that elevation because then the air is (more, less) dense than standard and less altitude is required to effect a given pressure difference (which is what the altimeter displays as a height difference).

Problems

1. How does the average random translational kinetic energy of air molecules at 100°F compare with its 0°F value? Answer: 22% greater

2. What is the pressure of dry air that has a density of 0.002 sl/ft^3 and a temperature of 50°F? Answer: 1750 lb/ft^2

3. Calculate the speed of sound at 0°F. Answer: 1051 ft/s

4. What is the greatest height to which a (vacuum) pump could draw water? Answer: About 34 ft at SL, less at higher pressure altitudes

5. A certain altimeter, adjusted to 29.92", reads 2600 ft for a field elevation of 3400 ft and an outside air temperature of 63°F.
   (**a**) What is the pressure altitude? Answer: 2600 ft
   (**b**) What is the density altitude? Answer: 3500 ft
   (**c**) What is the reading in the altimeter window after the altimeter is adjusted to read field elevation? (use the rule-of-thumb conversion factor) Answer: 30.72"

6. Some of the highest density altitudes in the continental USA occur on hot days at Denver's Stapleton Airport (elevation 5330 ft). Suppose the sun has heated the air above the concrete runways to 122°F on a scorching July day and the pressure altitude is a rather high 8600 ft. What is the density altitude?

Answer: 14,200 ft (Assume zero relative humidity)

7. What is the density altitude under the conditions of problem 6 but with 95% relative humidity? Answer: Density altitude = 16,000 ft!

8. What is the dew point temperature for problem 7?
Answer: 120°F (!)

9. It is a typical midwinter morning at your home airport in International Falls, Minnesota (elevation 1184 ft). A cold front slipped through during the night and now the temperature is a dry –43°F; the altimeter setting is 30.33" Hg.
   (**a**) What is the pressure altitude?
   (**b**) What is the density altitude? Answer: –6600 ft

10. At sunset on a fine day in the fall, the temperature might be 68°F and the relative humidity 81%. To what temperature will the air have to cool for fog to form? Answer: about 62°F

11. The owner's manual for a 1969 PA-28-140 Cherokee lists a 470 ft/min climb rate at a density altitude of 5000 ft (at a gross weight of 2150 lbs). The manual also shows that each additional 1000 ft in density altitude will cause a reduction of 39.5 ft/min in the climb rate.
   (**a**) What is the climb rate under the conditions of problem 6? Answer: about 110 ft/min
   (**b**) How about under the conditions of problem 7? Answer: about 36 ft/min

12. Your airport is enjoying stable weather for a change as a large cool, dry air mass parks itself directly over your part of the country. You tie down ol' Paint just as the sun goes down and note that ol' Paint's altimeter indicates the field elevation of 1150 ft and the outside air temperature is 83°F. However, a few hours later the air temperature has dropped to 65°F under clear skies. What does ol' Paint's altimeter indicate when you return to the hitching post? (Hint: Inspect Table 4.4.)

13. What is the density altitude at an airport for which the temperature is 78°F, the barometric pressure is 29.92" Hg, and the airport elevation is 2000 ft?

14. What is the density altitude at an airport for which the temperature is 48°F, the barometric pressure is 28.89" Hg, and the airport elevation is 2000 ft?

15. Verify the statement in Sec. G regarding the indicated altitude required to clear Mt. Everest on a day in which SL temperature is 0° F, the lapse rate is standard, and the SL pressure is standard.

16. Estimate the true altitude for flight at an indicated 6000 ft on a day for which the air temperature at a 1200 ft airport with standard pressure is –15° F. Answer: 5350 ft

17. Estimate the true altitude for flight at an indicated 11,000 ft if the altimeter is set based on an airport at an elevation of 5000 ft and with a surface air temperature of 95° F. Assume standard pressure at the reporting airport and a standard lapse rate. Answer: 11,650 ft

18. Suppose the temperature at the Portage County, Ohio, airport (elevation 1200 ft) is 90° F and a flight, at a constant indicated altitude of 8500 ft, is made to Duluth, Minnesota, where the temperature at a 1200 ft elevation is 20° F. If the barometric pressure is the same at both 1200 ft locations, estimate the true cruise altitudes at the beginning and then at the end of the journey. Answers: 9005 ft and 8012 ft

19. Hurricane Gilbert, an extremely powerful 1988 storm with at one time sustained winds over 170 mi/hr, claims the record for the lowest pressure ever measured in the eye of a hurricane: 26.31" Hg.
   (**a**) Determine the SL pressure altitude in Gilbert's eye. (Do not use the "rule of thumb.")
   (**b**) For a SL air temperature of 80° F, what was the density altitude in Gilbert's eye? Answer: 5650 ft!

20. In an aircraft with a **fixed**-pitch propeller, the power is normally set by reference to the tachometer (engine revolutions per minute). In an aircraft with a constant-speed propeller, however, a second instrument, the (intake) manifold pressure gauge, must be consulted as well. Suppose you are flying such an aircraft and note to your surprise that the manifold pressure gauge shows a pressure of 28.0" Hg **before** the engine is started, even though the altimeter shows the correct field elevation when it is set to 29.92" Hg. Why are the two pressure instruments in apparent disagreement? What useful information can you calculate from the given data? Answer: The field elevation and the pressure altitude are about 1800 ft.

21. Blood, which has nearly the same density as water, is pumped from the heart at a maximum pressure of about 4.7" of Hg.

   (a) How high above the heart can blood be raised, based on this typical maximum pressure?

   (b) Under positive accelerations such as caused by abrupt pull-ups and steep turns, the **apparent** weight of the blood in the body increases because the upward force of the seat of the pilot on the pilot increases (to keep the pilot in the airplane!). In level, unaccelerated flight, this upward force is just **equal** to the pilot's weight and affairs are said to be in the "1-*g*" state. Assuming that the pilot's eyes and brain are about 1 ft above the heart (check for yourself), about what sustained positive acceleration will deprive the pilot's brain and eyes of their oxygen? Answer: about 5.3 *g*'s (Perhaps short pilots have an advantage here?)

Atmospheric Data (including cold front passage)				
Day	Time	Pressure inches Hg	Temperature, °F	Relative Humidity
Friday	11 P.M.	30.08	60	
Saturday	6 P.M.	30.31	67	
Saturday	11 P.M.	30.32	53	
Sunday	6 P.M.	30.20	71	43
Sunday	11 P.M.	30.18	56	72
Table 4.4				

References

1. Dommasch, Daniel O., Sherby, Sydney S., and Connolly, Thomas F., *Airplane Aerodynamics*, Pitman Publishing Company, New York, 1967.

2. Andresen, Jack, *Fundamentals of Aircraft Flight and Engine Instruments*, Hayden Book Company, Inc., New York, 1969.

3. *U.S. Standard Atmosphere, 1976*, U.S. National Oceanic and Atmospheric Administration, National Aeronautics and Space Administration, United States Air Force, Government Printing Office, Washington, D.C., October 1976.

Chapter 5

Subsonic Fluid Flow

The discovery of the nature of lift is due to Kutta, a German mathematician, and to Joukowski, a professor of mathematics at the University of Moscow. Joukowski, who continued in Russia the work started in 1877 by Lord Raleigh in England, showed by mathematics that when you subject a body in motion to circulating air, or in other words when you spin it in a wind, you get a subsequent change in motion due to lift.

For almost half a century before Kutta and Joukowski the artillery engineers knew this principle. The called it the Magnus effect, after its early German discoverer. To the artillery it was disturbing because it made spinning shells deviate in path and miss the target. In aircraft design, however, the Magnus effect became the basis for the theory of lift. This is because the wing is so shaped as to cause the same type of circulation as obtained by spinning.[1]

Now these (alternating) vortices (from cylinders in a flow of air) have many physical applications. Shortly after the publication of my paper, Raleigh got the idea that the alternating vortices must give the explanation of the Aeolian harp – the singing wires. Some people will still remember the singing wires of the biplane cellules. The singing comes from the periodic shedding of vortices.

A French naval engineer told me of a case where the periscope of a submarine was completely useless at speeds over 7 knots under water, because the rod of the periscope produced periodic vortices whose frequency at a certain speed was in resonance with the natural vibration of the rod. Radio towers have shown resonant oscillations in natural wind. The galloping motion of power lines also has some connection with the shedding of vortices. The collapse of the bridge over the Tacoma Narrows was also caused by resonance due to periodic vortices. The designer wanted to build an inexpensive structure and used flat plates as side walls instead of trusses. Unfortunately, these gave rise to shedding vortices, and the bridge started torsional oscillations, which developed amplitudes up to 40° before it broke. The phenomenon was a combination of flutter and resonance with vortex shedding. I am always prepared to be held responsible for some other mischief that the Kármán vortices have caused![2]

A. Viscosity and the Boundary Layer

Air possesses a distinct resistance to any sustained relative motion of layers within it. This property of internal friction, perhaps surprising for a gas, is vitally significant in the small-scale interactions of air with the surfaces of aircraft, and these small-scale interactions in turn produce critical effects on large-scale behavior. We begin this study of air in motion by carefully defining the coefficient of viscosity as a measure of the

internal friction and then we describe its variation within the troposphere.

Suppose that we exert a **shear force** on a semi-rigid solid such as a sponge (Fig. 5.1a). The upper surface will move relative to the lower surface as we apply the two opposing forces but, unless the sponge fails structurally, the displacement will stabilize when the external disturbing force is just balanced by a net internal force due to molecular displacements. The system (the solid) changes to a new state of static equilibrium in this process.

Liquids and gases are fundamentally different from solid in their response to shear forces. It is because they share a similar response to shear that gases and liquids are often con-

[1]von Kármán, Theodore, *The Wind and Beyond* (Little, Brown & Co., 1967).
[2]von Kármán, Theodore, *Aerodynamics* (Little, Brown & Co., 1954).

sidered together and given the joint name of **fluid**. Therefore the topics discussed in this chapter fall within the discipline called fluid mechanics.

Suppose that water fills a channel between two walls and that the walls are moved in opposite directions (Fig. 5.1b). It is found that the water molecules next to each wall will be attracted to, and tend to follow, that wall; zero fluid speed is in the middle and intervening layers have intermediate speeds. It requires a **continuous** force to keep the walls moving in opposite directions because the molecules in each layer feel opposing forces on them. Just the same result obtains when a gas such as air is between the channel walls except the force needed to move the walls because of the internal fluid friction is much less. The viscosity of air at SL is only about 1.6% of that for water. Note that, unlike the situation of Fig. 5.1a for a solid, the driving forces on the walls are needed **only** when the walls are **kept** in motion; once the walls stop there is no force trying to return them to their previous position.

For the definition and measurement of fluid viscosity, we might chose to use an apparatus more like that of Fig. 5.1c. It

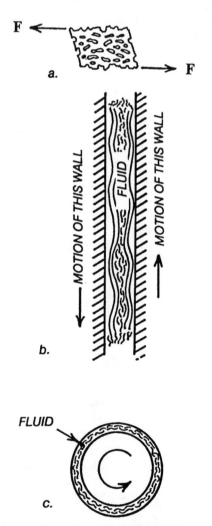

Fig. 5.1 Possible ways to measure the viscosity of a fluid

is necessary to rotate only the inner cylinder because **relative motion** is all that is significant. The distance between the walls should be small because we want the viscosity within a thin layer. Ultimately, our definition is for the limit as the thickness of the layer approaches zero.

Experimentally, the force required to give relative motion to a thin layer of fluid varies directly with the speed variation across the layer, directly as the area of the layer of fluid, and inversely as the thickness of the layer. In symbols,

$$F \propto \frac{(\Delta V) A}{\Delta x}$$

(Mathematical notation: The symbol $\propto$ should be read as "is proportional to." The Greek letter delta, Δ, is used to indicate an increment or change in a variable, so ΔV is the change in fluid speed in a layer of thickness Δx.)

We can rewrite this as

$$\text{shear stress} \equiv \frac{F}{A} \propto \frac{\Delta V}{\Delta s}$$

(Mathematical notation: The three parallel lines should be read as "is defined as.")

In this last expression, (F/A) is the shear stress on the layer and $(\Delta V/\Delta s)$ is the speed gradient within the layer.

The **coefficient of viscosity** for a given fluid is defined as the constant of proportionality for the above equation. That is, it is the ratio of the shear stress to the speed gradient within the layer. We represent this coefficient of viscosity with the Greek letter μ (mu). The units for μ are $(\text{lb/ft}^2)/[(\text{ft/s})/\text{ft}] = [\text{lb-s/ft}^2]$. In words,

$$\binom{\text{coefficient of}}{\text{viscosity}} \equiv \frac{\text{shear stress on fluid layer}}{\text{speed gradient within layer}}$$

In symbols,

$$\frac{F}{A} = \mu \frac{\Delta V}{\Delta s} \quad \text{or} \quad \mu \equiv \frac{\left(\frac{F}{A}\right)}{\left(\frac{\Delta V}{\Delta x}\right)} \qquad (5.1)$$

DEFINITION OF VISCOSITY COEFFICIENT

Readers who have studied the calculus will appreciate the fact that a more precise definition of the coefficient of viscosity is

$$\mu \equiv \frac{\left(\frac{F}{A}\right)}{\left(\frac{dV}{ds}\right)},$$

where (dV/ds) is the rate of change of the fluid speed with distance from the wall, at the wall. (dV/ds) can also be interpreted graphically as the **slope** of the V versus s curve (see Fig. 5.7).

Nothing in world seems to be truly constant. Neither is the "constant" coefficient of viscosity, even though we took the trouble to remove the geometrical dependence in our definition. Although the coefficient of viscosity shows almost no dependence on fluid **pressure**, it **does** depend on the **temperature** of the fluid. Figure 5.3 shows that μ for air decreases by about 17% as the temperature decreases from the SL standard temperature of 59°F to the 30,000 ft standard temperature of –47.8° F. The **SL standard value** of μ is 3.737×10^{-7} lb-s/ft². At other temperatures you can consult the StAt tables (Appendix B) or calculate it from

$$\mu = \frac{(2.27 \times 10^{-8}) \, T^{3/2}}{T + 198.7} \tag{5.2}$$

where T must be in units of R. (Mathematical note: On a calculator, $T^{3/2}$ can be obtained either with $(\sqrt{T})^3$ or with the y^x function key, with $x = 1.5$.)

It shall soon become apparent that in aerodynamics the property of air density often appears at the same time as the coefficient of viscosity, one divided by the other. Therefore it is convenient to define a new quantity, called the **kinematic viscosity**, which is defined as the value of the coefficient of viscosity divided by the density, for a given fluid at a given temperature. The symbol for kinematic viscosity is the Greek letter ν (nu).

$$\nu \equiv \frac{\mu}{\rho} \tag{5.3}$$

DEFINITION OF KINEMATIC VISCOSITY

The variation of ν with altitude in the StAt is also shown in Fig. 5.3. ν varies more rapidly with altitude than does μ because the density of air decreases much faster with altitude than does μ. The **SL standard value** of ν is 1.5723×10^{-4} ft²/s.

EXAMPLE 5.1. Estimate the viscous drag force on a very thin 3 ft×8 ft sheet that is moving slowly and directly through standard SL air (Fig. 5.2). The speed gradient at the sheet for this situation has an average value of 966 ft/s per ft when the sheet is moving at 10 ft/s. (Note well that this implies a layer only $(10 \text{ ft/s})/(966 \text{ ft/s/ft}) = 0.01035$ ft $= 0.124$ inches thick in which the air transitions from the sheet speed to the speed of the surrounding air. This transition region is called the **boundary layer** (often abbreviated as **BL**) and plays the key role in determining overall flow patterns, as we shall see later. When a smooth plate or sheet is moving at this slow a speed relative to the air, the air flow around the plate is rather uniform and very steady.

Solution: From Eq. 5.1,

$$\text{shear stress} = \; = \frac{F}{A} = \mu \left(\frac{\Delta V}{\Delta s}\right) \qquad \text{so}$$

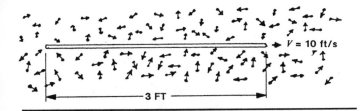

Fig. 5.2 A flat plate moving through a viscous fluid at 10 ft/s

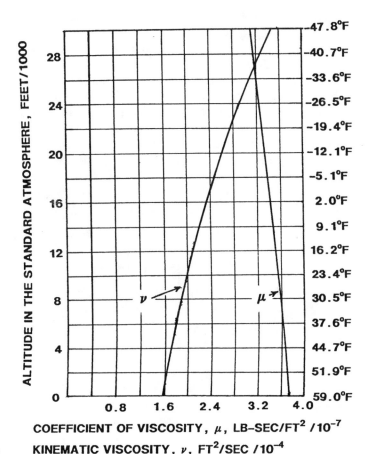

Fig. 5.3 Viscosity variation with altitude in the Standard Atmosphere

$$\text{shear force} = \mu \left(\frac{\Delta V}{\Delta s}\right) A = \left(3.737 \times 10^{-7}\right) \left(966\right) \left(3 \times 8\right)$$
$$= 0.00866 \text{ lb}$$

This is for each side, so the total calculated drag force on the plate due to air viscosity is 2×0.00866 lb = **0.0173 lb**. This is also an experimental value for the drag force. Is it larger or smaller than you expected?

Air does possess very **little** viscosity compared to liquids such as STP® or honey or grape juice. It turns out that it is possible to learn useful information about the air flow outside the BL by making the simplifying assumption that air has **no** viscosity. Aerodynamic theories that make this assumption are characterized as **inviscid flow** theories.

B. Laminar and Turbulent Flow

Undoubtedly you have observed laminar fluid flow. The smoke that leaves the end of a cigarette in a still room has a regular and ordered upward flow (Fig. 5.4); the layers (or **laminas**) of smoke molecules retain their shape and continuity as they move upward. The water that flows out of a faucet is laminar at low speeds. The flow of water past small stones in a gently meandering stream is laminar.

You have also observed turbulent fluid flow. After the insalubrious smoke from that cigarette has travelled upward for a certain distance, it suddenly **transitions** to an irregular flow (represented by the squiggly lines in Fig. 5.4) in which there is much mixing and sharing of energy between adjacent layers. This region of turbulent flow appears to be constantly **changing** as you observe it; i.e., the motion is irregular and unpredictable on a small time scale. In general, the defining characteristics of turbulent flow are disorder, efficient mixing, and vorticity (rotational motion). The overall direction of fluid motion is never in question here, but the small-scale motions are very time dependent.

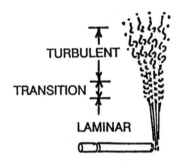

Fig. 5.4 A small smoke source in a still room shows both types of fluid flow.

It is easily deduced from observation that the **transition** from laminar to turbulent flow is associated with (**a**) an increasing interaction distance, or (**b**) an increasing fluid speed. A large rock is more likely to cause a transition to turbulent flow than is a small stone. Rippling water flow is associated with the narrowing of a wide stream into rapids. The water that leaves a faucet in laminar flow at low flow speeds quickly makes the transition to completely turbulent flow as the speed is increased.

The factors that determine **when** a fluid that is in relative motion **parallel** to a bounding surface will make the **transition** to **turbulent** flow were studied by Osborne Reynolds in 1883. He discovered a dimensionless (unit-less) parameter, now universally known as the **Reynolds number** (RN), that is still the **best** predictor we have for this transition.

$$\text{RN} \equiv \frac{\rho V x}{\mu} = \frac{V x}{\nu} \qquad (5.4)$$

DEFINITION OF REYNOLDS NUMBER

In this equation for the Reynolds number, ρ is the density of the fluid, V is the mean speed of the fluid relative to the bounding surface, x is a characteristic length associated with the bounding surface, μ is the coefficient of viscosity of the fluid, and ν is the kinematic viscosity of the fluid. The last form of the Reynolds number, using the kinematic viscosity, justifies our

previous definition of ν because it shows that the total dependence of RN on the properties of the fluid are contained in ν (the remaining dependence being the size of the bounding surface and the fluid speed). Other texts and journals often use **R** or **Re** as the symbol for the Reynolds number.

When the RN for an **airplane** is given, it usually means that the **cruising speed** has been used for V and the average chord (width) of the wing has been used for x when substituting into Eq. 5.4, because of the importance of the wing for lift and drag. However, it is often desirable to deal with fluid properties at a **specific** point on the wing or fuselage (a **local** RN), and then the **distance** from a **leading edge** is used for x.

We can substitute units for the quantities in the second form of the definition to prove the statement that RN is a dimensionless parameter.

$$\text{Units of RN} = \frac{(\text{ft/s})(\text{ft})}{\text{ft}^2/\text{s}} = 1 \qquad (\text{no units})$$

As an aid in getting a feel for the magnitude of the Reynolds number for aircraft, let us determine its value for a length of 1 ft at an air speed of 100 kt under SL standard conditions. Using the sea level value for the kinematic viscosity from the previous section and Eq. 5.4, we have

$$\text{RN per ft} = \frac{(100 \text{ kt} \times 1.689 \text{ ft/s/kt})(1 \text{ ft})}{1.5723 \times 10^{-4} \text{ ft}^2/\text{s}} = 1.07 \times 10^6$$

REYNOLDS NUMBER/FT AT 100 kt, AT SL

Thus an aircraft possessing a wing with a four-foot chord and cruising at 200 kt would have a wing RN of about 8 million; if it lands at 50 kt, it would have a wing RN of about 2 million then. (The conversion factor for kt to ft/s that was used in this calculation will be derived a little bit farther down the road.)

Reynolds studied water flow through pipes, injecting a colored fluid into the water to make it easy to detect the transition to turbulent flow. He found that the **diameter** of the pipe was the most significant dimension, presumably because that determined the speed gradient in the fluid, and therefore he used the pipe diameter for x in the formula. In his experiments, the transition to turbulent flow occurred for RN between about 2000 and 40,000, depending primarily on the smoothness of the flow into the pipe and the smoothness of the pipe; above RN = 40,000, the flow was always turbulent. So we associate low RN (low speeds or small sizes or high viscosity) with **laminar flow** and high RN (high speeds or large sizes or low viscosity) with predominantly **turbulent** flow.

For example, the relative air flow past a settling dust particle in front of your nose is at a very low RN (despite the low viscosity of the air, in this case) and the flow is totally laminar. The RN for aircraft in flight is typically in the millions, though, and we have to struggle to get a significant amount of laminar flow around them.

Notice that the RN does contain the dependence on speed that we anticipated earlier from general observations. The dependence on the type or the properties of the fluid is contained in the **kinematic viscosity** term. A little later, there will be an attempt to give a physical interpretation of the RN.

Einstein's famous principle of relativity reveals (among many other things!) that there is no way to determine absolute motion; only **relative motion** can be measured and therefore only relative motion is significant for science. The immediate application of this to our subject is that we have the choice of considering the fluid to be stationary and the bounding surface in motion **or** we can consider the fluid to be moving relative to a stationary surface. Usually the latter choice is taken because of its easier implementation in a wind tunnel (and perhaps because we like to think of ourselves in the driver seat when talking about airplanes?).

So consider the air to be in motion relative to the flat plate of Fig. 5.5. Following custom, we represent the speed of the plate relative to the undisturbed air by the symbol V_∞. (This is just what we've called the airspeed in earlier work but now we sometimes need the subscript to distinguish this speed from the local speed, which varies greatly in both directions. V_∞ is often called the freestream velocity. The same ∞ subscript is used to refer to other properties of the undisturbed, freestream fluid; thus p_∞ refers to the freestream fluid pressure, etc.

The boundary layer (**BL**) can be defined in general as the region close to a bounding surface in which the **viscosity** of the fluid has a significant effect on the fluid flow. Therefore the flow outside the BL can be considered to be inviscid flow, by definition. The thickness variation of the BL has been measured and theoretically described for flat plates. The same general ideas apply as well to airfoils, so we show this variation (in greatly exaggerated form) in Fig. 5.5 for SL air, a one-foot-long plate, and a V_∞ of 150 ft/s (about 90 kt).

The BL always **begins** with laminar flow. The point (and therefore the local RN) at which the laminar BL transitions to a turbulent one depends a great deal upon (**a**) the smoothness of the plate and (**b**) the degree of (very small scale) inherent turbulence in the airstream. In Fig. 5.5 the transition is shown

at RN $= 5\times10^5$, using the distance from the leading edge as the x in RN, and this is a typical transition RN for a smooth, flat plate.

A turbulent BL is both good news and bad news. The bad news is that there is considerably more energy transfer to the air, which means that the **skin friction** or **viscous drag** is much greater for a turbulent BL than for a laminar one. For both types of BL, there must be a region in which the relative airspeed is zero (at the surface) and where it is (essentially) equal to V_∞. The difference between the two types of BLs is the rate at which the speed changes close to the surface and the total thickness of the BL.

Figure 5.6 depicts the **velocity profile** at two points in the BL for the flat plate of Fig. 5.5: (**a**) before transition, in the laminar BL and (**b**) after transition, in the turbulent BL. Figure 5.7 is an enlargement of these profiles. The length of the arrow in each case is a measure of the air velocity at the point where the arrow begins. Here it is obvious that the turbulent BL has a much greater **speed gradient** (rate of change of speed with distance from the surface) **at** the surface and is much **thicker**. This is the origin of the much greater skin friction drag of a turbulent BL.

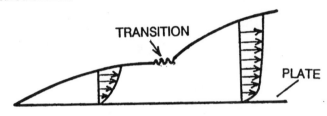

Fig. 5.6 Velocity profiles within a laminar and a turbulent boundary layer

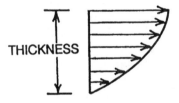

a. LAMINAR BOUNDARY LAYER

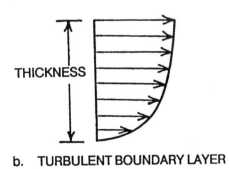

b. TURBULENT BOUNDARY LAYER

Fig. 5.5 A greatly enlarged sketch of the actual boundary layer on a 1 ft flat plate at about 90 kt

Fig. 5.7 Enlarged sketch of the velocity profiles within the boundary layer

What cannot be diagrammed adequately is the **mixing** between layers that occurs in the turbulent BL; the velocities shown are only **average** values because there are significant **instantaneous** perpendicular velocity components and even the parallel component varies significantly over time.

Some conclusions: (**a**) the viscosity of air is solely responsible for the existence of the BL, (**b**) this always results in a viscous or skin friction drag, and (**c**) this drag force is much greater for a turbulent BL than for a laminar one. (We defer to a little later the **good news** about a turbulent BL.)

EXAMPLE 5.2. If the transition RN for a certain airplane wing at a cruise speed of 200 kt at SL is 2.0×10^6, about how far back on the wing would you expect the BL to change from laminar to turbulent?

Solution: First we require the conversion factor so that this airspeed can be expressed in ft/s.

$$1 \text{ kt} = \left(\frac{1 \text{ nm}}{\text{hr}}\right)\left(\frac{6080 \text{ ft}}{\text{nm}}\right)\left(\frac{1 \text{ hr}}{3600 \text{ s}}\right) = 1.689 \text{ ft/s} \quad \text{so}$$

> **1 kt = 1.689 ft/s   and   1 ft/s = 0.592 kt**

CONVERSION BETWEEN UNITS OF ft AND kt

So 200 kt = (200)(1.689) = 338 ft/s and

$$\text{RN} = \frac{V x}{\nu_0} = \frac{(338) x}{1.572 \times 10^{-4}} = 2.00 \times 10^6$$

and solving this equation for x yields: $x = 0.93$ ft

I hope you noticed that I slipped in the use of an "$_0$" subscript to indicate a **SL** standard value for a quantity, in this case the SL kinematic viscosity given earlier.

Figure 5.2 shows that the kinematic viscosity increases with increasing altitude, so x will have to get larger with increasing altitude to maintain the same transition RN. Thus, for a given (true) air speed, the BL tends to be slightly more laminar as the cruising altitude increases. However, most aircraft fly at altitudes where their true airspeed is greater than at lower altitudes, and then the RN may well be greater and the BL more turbulent than at sea level.

It turns out, as will be noted later, that transition to a turbulent BL is **delayed** if the BL air is **accelerating**, as it is initially on the top surface of the wing, but the transition is **encouraged** by a decelerating flow, as on the wing at a point beyond the thickest point.

C. Life Outside the BL: Streamlines and Streamtubes

There are two fundamentally different way to approach the conceptual or mathematical study of fluid flow. **One** approach is to choose a particular particle and follow its path, repeating this for different particles in all different parts of the flow. A **second** approach is to choose an instant of time and study the distribution of velocities at that particular time. The two ap-

proaches give equivalent results **if** the fluid flow is steady; that is, with no time dependence. In this section, then, we consider steady, non-turbulent fluid flow, outside the BL.

Suppose that air is flowing at 150 kt past a "streamlined" body such as the symmetrical wing section or airfoil shown in Fig. 5.8. (For the symmetrical flow depicted, this airfoil is not currently producing any lift.) The lines with arrows represent the direction of the velocity outside the BL at one chosen instant of time; because of the assumption of steady flow, the lines also trace the path that a tiny particle would take if it were to be placed at the beginning of a line. These imaginary lines are called **streamlines**. Notice that the streamlines begin to curve well before the air gets to the airfoil. It is an amazing fact that the air (at 150 kt) is "warned" well ahead of time (by pressure waves from the leading edge) that a wing is coming and is able to begin to alter its path to get around it. Notice also that the streamlines are closer together near the upper and lower surfaces of the airfoil.

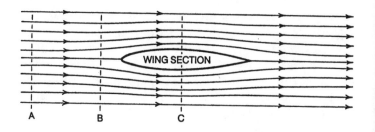

Fig. 5.8 Streamlines around a symmetrical, non-lifting wing section

Our Fig. 5.8 is only a two-dimensional snapshot of the action, of course. We should be thinking of tubes which contain nearly similar fluid speeds; such tubes are (surprise!) called **streamtubes**. These streamtubes must be shrinking in cross-sectional area at the locations where the fluid streamlines are coming closer together. Figure 5.9 illustrates some comparative streamtube sizes by showing a cross-section to the flow of Fig. 5.8 at three different points. (We show circular and elliptical cross-sections because of tradition and the name; it might be better to use square and rectangular cross-sections to make it clear that all the fluid is included, but the problem becomes less

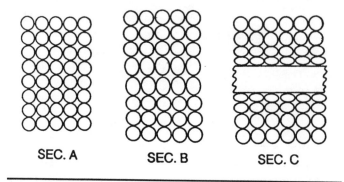

SEC. A   SEC. B   SEC. C

Fig. 5.9 Cross-sections through the streamtubes for the wing of Fig. 5.8

and less noticeable, in any actual analytical work, as the stream-tubes are shrunk to a more realistic size.)

Figure 5.10 is a side view of a streamtube that passes just above the BL on the top surface of the airfoil of Fig. 5.8. We can make these streamtubes as small and as numerous as necessary to make the fluid velocities as constant as desired, at each cross-section.

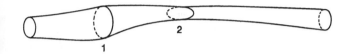

Fig. 5.10 A perspective view of a streamtube, needed for derivations

Our need is for a mathematical description of what changes and what doesn't as we pass from one cross-sectional area in a single streamtube to another cross-sectional area in the same streamtube. Suppose we station our favorite gremlin at cross-section #1 of Fig. 5.10 and ask him or her to please determine how much **mass** of fluid passes that point in the next little time interval, Δt. If the streamtube speed at this point has an average value of V_1, a length of fluid in the streamtube equal to $V_1 \Delta t$ (from the definition of speed, Eq. 2.2) will pass point #1 in the time interval Δt. Thus, for a cross-sectional area A_1 at that point, a **volume** of fluid equal to the product of this length of the fluid and the area of the fluid will pass point #1 in time Δt.

$$\begin{pmatrix} \text{volume of fluid passing point \#1} \\ \text{during time interval } \Delta t \end{pmatrix} = \text{area} \times \text{length}$$
$$= A_1 V_1 \Delta t \qquad (5.5)$$

The **mass** of fluid passing point #1 during this time must be a product of the density of the fluid at that point and the volume of fluid (Eq. 4.4).

$$\begin{pmatrix} \text{mass of fluid passing point \#1} \\ \text{during time interval } \Delta t \end{pmatrix} = \rho_1 A_1 V_1 \Delta t \qquad (5.6)$$

A similar analysis for a second point down the tube gives us:

$$\begin{pmatrix} \text{mass of fluid passing point \#2} \\ \text{during time interval } \Delta t \end{pmatrix} = \rho_2 A_2 V_2 \Delta t \qquad (5.7)$$

We can choose the time interval in this "thought" experiment to be as small as we wish; we choose a time interval that is short enough that (**a**) the average fluid speeds at points 1 and 2 don't change during that time and (**b**) the cross-sectional areas A_1 and A_2 are separately constant for the little volume of fluid being described. Then Eqs. 5.6 and 5.7 must represent the **same** amount of mass because mass cannot be created or destroyed in this fluid flow. (Recall that in Chapter 3 we carefully differentiated between mass and weight, recognizing that mass was a more fundamental property, the inertial property.) Conservation of mass is a principle that appears to be universally true so long as speeds are much less than the speed of light, a limitation not yet in question for anything we fly or shoot. We can write this mass equality in symbols:

$$\rho_1 A_1 V_1 \Delta t = \rho_2 A_2 V_2 \Delta t$$

and finally, dividing through by the time interval,

$$\rho_1 A_1 V_1 = \rho_2 A_2 V_2 \qquad (5.8)$$

CONSERVATION OF MASS, COMPRESSIBLE FLOW

This relationship, really just a statement of mass conservation for a compressible fluid, is known as the continuity equation and tells us that the product of these three fluid properties never changes along a streamtube. If the density decreases along the streamtube, for instance, the product of the area and the speed must increase to exactly compensate.

For many fluids in motion, the change in density of the fluid within a streamtube is negligible (inviscid flow). Then $\rho_1 = \rho_2$ and we obtain from Eq. 5.8 the simpler equation

$$A_1 V_1 = A_2 V_2 \qquad (5.9)$$

CONSERVATION OF MASS, INCOMPRESSIBLE FLOW

You may well wish to ask whether air in motion can ever be considered as an incompressible fluid (that is, of constant density) because gases are known to be relatively easy to compress. The answer is **yes**. At aircraft speeds below about 200 kt or 250 kt, the changes of air density within a streamtube usually **are** negligible. The explanation lies in the fact that, outside the BL, the air is able to sufficiently change its velocity (and therefore the cross-sectional area of each streamtube) that density changes are not induced by the relatively slow passage of an object. We can immediately conclude from Eq. 5.9 and Fig. 5.10 that most of the air above and below and next to the aircraft wing section of Fig. 5.8 is going faster than the freestream fluid speed.

Some of the excitement accompanying compressible air flow will be discussed in the next chapter.

EXAMPLE 5.3. A certain stream of water at one point has a width of 8.7 ft, an average speed of 0.42 ft/s, and a depth of 3.8 ft. About how fast would you expect the water to be flowing at another point where the width is 2.8 ft and the depth is 4.3 ft?

Solution: We can use the incompressible continuity equation (Eq. 5.9) for the description of this stream because we can reasonably assume constant fluid density, no loss of water, and a small BL at this low, meandering speed. We identify the fully-known point with the 1 subscript. Then the symbolic statement of the problem is

$$\rho_1 = \rho_2, \ V_1 = 0.42 \text{ ft/s},$$
$$A_1 = \text{width} \times \text{height} = (8.7 \text{ ft})(3.8 \text{ ft}) = 33.06 \text{ ft}^2,$$
$$A_2 = (2.8 \text{ ft})(4.3 \text{ ft}) = 12.04 \text{ ft}^2, \qquad V_2 = ?$$

Solving Eq. 5.9 for V_2 and substituting in the appropriate values, we obtain

$$V_2 = \frac{A_1 V_1}{A_2} = \frac{(33.06)(0.42)}{12.04} = \textbf{1.15 ft/s}$$

D. The Bernoulli Energy Equation

In this section we introduce the incompressible Bernoulli equation which describes the interdependence of the static **and** dynamic properties of the fluid, somewhat as the ideal gas law (Chapter 4) describes the interdependence of just the intrinsic, static properties of a fluid. The Bernoulli equation is valid whenever no outside energy source or sink is present; it would be unfair, however, to try to apply it to the end of a streamtube in which there was an energy source such as a heated wire.

The Bernoulli equations (one for compressible flow and a much simpler one for incompressible flow) are a mathematical formulation of the principle of **conservation of energy** for fluids in motion. "Energy" is a word that was coined to describe a property of a system that never changes if the system is isolated from outside influences. ("System" here refers to any particle, object, or group of objects.) The discovery that such a property even exists had to wait until it was realized that **heat** is a form of internal kinetic energy. (Recall that absolute temperature is directly proportional to the random translational kinetic energy of the molecules of the air.) The various types of energy include kinetic energy (energy of motion) and potential energy (energy of position, as due to geographical height or compression of a spring). It is the **sum** of all the types of energy that is constant for an isolated system.

Energy is sometimes defined as the ability to do work, and work is done by exerting a force **through** some distance. A gas can do work on another system, for instance a bounding surface like a wing, by pushing that system some distance so that the system acquires a kinetic energy it didn't have previously (or, speaking relatively, a stationary gas can be pushed aside by the wing and thereby exert a reaction force and do work). The principle of conservation of energy states that energy can be **transferred** from one system to another through the mechanism of work, but **no** energy is ever destroyed or created in any interaction.

One simplification that can be made is to realize that the gravitational potential energy of the air molecules is not going to change very much compared to the other energy changes, particularly for air flowing mostly parallel to the ground. The gravitational potential energy of an object is a measure of the work that the gravitational force does on the object when and if it succeeds in reducing the distance of separation (the altitude). An object motionless above the ground has gravitational potential energy; when it is released it exchanges gravitational potential energy for kinetic energy. While it is falling it has large-scale kinetic energy and when it hits the earth that kinetic energy is converted into small-scale kinetic energy (heat energy, as demonstrated by a measurable temperature rise) in both the earth and the object.

The pendulum provides another example of the exchange of gravitational potential energy with kinetic energy except the pendulum doesn't collide with the earth at its lowest point; once a pendulum is started, the bob begins a very regular interchange between gravitational potential energy and large-scale kinetic energy. The total energy (sum of kinetic and potential) remains **constant** except for the small fraction continually being lost to heat energy via (**a**) frictional forces in the pivot and (**b**) viscous drag in the air. Anyway, it is this gravitational potential energy that is not usually important for understanding the air flow about an airplane because (**a**) the flow is usually at nearly constant altitude and (**b**) other energy changes are much larger.

The types of energy that are possessed by a fluid in motion can be determined by asking how such a fluid can do work on a bounding surface. It can do work in two ways: (**a**) through the overall flow speed and large-scale kinetic energy of the fluid (for example, the fluid can push away a bounding surface as its overall speed is reduced) and (**b**) by the force exerted by **individual** molecules as they rebound from a moving surface they are pushing and thereby lose some small-scale kinetic energy. Also, the most useful energy quantity for a fluid under steady flow conditions turns out to be not the total energy of the fluid but rather the **energy density** within a streamtube or streamline.

For the streamtube of Fig. 5.10, the overall kinetic energy of a small volume of fluid passing point #1 is

$$\text{KE}_1 = \tfrac{1}{2} m_1 V_1^2 = \tfrac{1}{2} \rho_1 A_1 V_1 \, \Delta t \, V_1^2 = \tfrac{1}{2} \rho_1 A_1 V_1^3 \, \Delta t$$

in which we have substituted for the mass from Eq. 5.5. The kinetic energy **density** of the fluid at point #1 is the kinetic energy divided by the volume of fluid (Eq. 5.5):

$$\text{KE}_{\text{DENSITY AT PT \#1}} = \frac{\tfrac{1}{2} \rho_1 A_1 V_1^3 \, \Delta t}{A_1 V_1 \, \Delta t} = \tfrac{1}{2} \rho_1 V_1^2 \qquad (5.10)$$

The second energy density factor comes from the random kinetic energy of the fluid molecules. This energy can be considered to be a form of **potential energy**. We obtain this internal fluid energy density by asking how much work would be done on a little volume of fluid at point #1 if the pressure was p_1 on one end and zero on the other end. Work is equal to the product of the force and the distance moved in the direction of the force. The net force is the net pressure times the cross-sectional area of the streamtube (Eq. 4.3) and the length of the fluid element is $V\Delta t$ as before. So

Work done by the pressure of the fluid $= \left(\rho_1 A_1 \right) \left(V_1 \, \Delta t \right)$

Dividing this by the volume from Eq. 5.4 yields the energy density again:

pressure energy density at point #1 $= p_1 \qquad (5.11)$

If no energy is added to the streamtube, the total energy density at point #1 must remain the same as the total energy density at point #2, assuming the validity of the principle of energy conservation. Thus, for the streamtube of Fig. 5.10,

$$\binom{\text{total energy density}}{\text{at point \#1}} = \binom{\text{total energy density}}{\text{at point \#2}}$$

or

$$\begin{pmatrix} \text{pressure} \\ \text{energy} \\ \text{density} \end{pmatrix}_1 + \begin{pmatrix} \text{kinetic} \\ \text{energy} \\ \text{density} \end{pmatrix}_1 = \begin{pmatrix} \text{pressure} \\ \text{energy} \\ \text{density} \end{pmatrix}_2 + \begin{pmatrix} \text{kinetic} \\ \text{energy} \\ \text{density} \end{pmatrix}_2$$

in symbols,

$$p_1 + \tfrac{1}{2}\rho V_1^2 = p_2 + \tfrac{1}{2}\rho V_2^2 \qquad (5.12)$$

**BERNOULLI EQUATION FOR ANY TWO POINTS
ALONG A CONSTANT-ENERGY STREAMTUBE
FOR INCOMPRESSIBLE FLUID FLOW
AND NEGLIGIBLE HEIGHT VARIATION**

It would be highly improper to attempt to apply this equation to first the fluid ahead of a wing and then behind it, because the wing is subtracting energy from the flow (causing the skin friction or viscous drag experienced by the wing). In fact, this is one of the methods used to measure the drag force on a wing section in a wind tunnel.

Because points #1 and #2 are arbitrary points along a streamtube, the Bernoulli equation equivalently states that

pressure energy density + kinetic energy density = constant

This means that, along a streamtube, the energy can change its form, much as it does in a frictionless pendulum, but the **total** energy density remains **constant**.

Some valuable insights into the energy states of fluids can be gained by doing a little thought experiment with air that is trapped in two separate compartments of a box (Fig. 5.11). Suppose that the box is perfectly insulated so no heat energy can enter or leave. Suppose also that the two gases have equal volumes and equal temperatures but that the air in the left compartment has a higher pressure than the one on the right ($p_1 > p_2$). (By both the ideal gas law and a little thought, this evidently means that there are more molecules in the chamber on the left.) If we now release the barrier, it will move to the right until the pressures are equal. From experience and energy considerations, we know that air in the left compartment will be **cooled** by its expansion and the air in the right compartment will be **heated** by its compression. From the energy viewpoint, the air on the left has done **work** on the air in the right compartment by exerting a force on it and moving it into a smaller

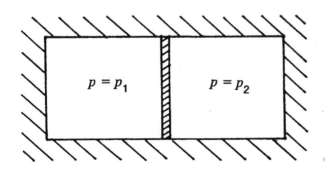

Fig. 5.11 Two compartments of air at the same temperature and volume, different pressures

volume; this work goes into additional heat energy of the air in the right compartment and correspondingly reduces the heat energy of the air in the left compartment. But it is equally valid, and more fun, to look at the event from the molecules' point of view. The molecules of air in the left compartment have been "rebounding" from the barrier for a long time before we let the wall move; each time the molecules of the wall were able to send them back at just the same speed at which they approached (on the average), thanks to intermolecular forces which do their job well before the molecules "hit" each other. But when the barrier is allowed to move, the molecules in the left compartment rebound from the moving wall with a lesser speed, much to their surprise, and therefore the average speed of these molecules, and the temperature of the air, is lowered. The molecules in the right chamber, on the other hand, rebound with a greater speed from the barrier that is suddenly moving toward them, so their average speed and the temperature of the gas **increases**.

The kinetic energy density at any point in a fluid flow is also given the name **local dynamic pressure** and the symbol q.

$$q \equiv \tfrac{1}{2}\rho V^2 \qquad (5.13)$$

(LOCAL) DYNAMIC PRESSURE

However, the dynamic pressure of the **free** (undisturbed) air is usually the most useful and significant property bearing on the aerodynamic properties of the **whole** airplane; when it is necessary to explicitly indicate this dynamic pressure, it is called the **flight dynamic pressure** and given the symbol q_∞.

$$q_\infty \equiv \tfrac{1}{2}\rho V_\infty^2 \qquad (5.14)$$

(FLIGHT) DYNAMIC PRESSURE

The ∞ subscript serves to emphasize the fact that the reference point is the freestream well away from the aircraft, so the speed involved is the true airspeed of the aircraft. (**Note well**: if the local dynamic pressure is **not** of interest, the symbol q by itself refers to the flight dynamic pressure.) To give you some feeling for the significance of flight dynamic pressure, note that it is just the **overpressure**, the amount that the air pressure at a point where the air is brought to a **halt** will exceed the static air pressure; also, multiplied by the cross-sectional area of an obstacle, it will give an idea of the drag force on the obstacle.

Just how potent is this energy of moving air? At SL and 100 kt (168.8 ft/s), the dynamic pressure is

$$\tfrac{1}{2}(0.002377)(100 \times 1.688)^2 = 33.9 \text{ lb/ft}^2$$

DYNAMIC PRESSURE AT SL AND 100 kt

This isn't all that impressive when we recall that standard SL **static** pressure, which is a measure of the work that can be done by an **evacuated** chamber, is 34 times greater, at 2116 lb/ft². But because the dynamic pressure varies as the **square** of the speed, static and dynamic pressure become equal when the speed reaches about 790 kt (1.2 times the speed of sound) under standard SL conditions.

The sum of the pressure and kinetic energy density, which is the total energy density of the fluid, is given the name **total pressure** and the symbol p_T. So we can write

$$p_T \equiv p + \frac{1}{2}\rho V^2 \qquad (5.15)$$

DEFINITION OF TOTAL PRESSURE
(INCOMPRESSIBLE, CONSTANT ENERGY FLUID FLOW)

Thus the Bernoulli equation, applying to any point on a streamtube, can be written as

$$p_T = p + q \qquad (5.16)$$

or, for any point in the **free** airstream,

$$p_T = p_\infty + q_\infty \qquad (5.17)$$

These last two equations just emphasize the fact that the total pressure has a constant value anywhere along a streamline, for constant-energy, incompressible fluid flow. Mostly this is just a matter of definition, for the total pressure is simply the total energy density of the fluid flow.

Evidently the total pressure is quantitatively equal to the static pressure at any point in the airflow where the fluid speed (and therefore the local dynamic pressure) is **zero**, by inspection of Eq. 5.16. Such a point is called a **stagnation point** and such a point always exists at the leading edge of a wing at the point (or in three dimensions, the line) where the flow divides between upper and lower surfaces. The movement of this stagnation point as the angle of the wing relative to the air changes is of great interest, as we'll see shortly.

We can rewrite Eq. 5.17 as

$$q_\infty = p_T - p_\infty \qquad (5.18)$$

This is a more fundamental definition of freestream dynamic pressure than is Eq. 5.14 because this definition is usable for both compressible and incompressible fluid flow. However, dynamic pressure sometimes is taken as $(\frac{1}{2}\rho V_\infty^2)$ even for compressible flow. In any case, the dynamic pressure that is defined by Eq. 5.18 is the quantity that is measured by airspeed indicators (Chapter 7). And in Chapter 8 we will find that the dynamic pressure is the primary air flow property determining the lift and drag of an aircraft.

In the last section it was noted that the incompressible continuity equation, when applied to a streamtube like Fig. 5.10, told us that the airspeed was greatest where the streamtube was smallest in cross-sectional area. Specifically, air **speeds up** when passing above and below an airfoil such as that shown in the figure. Now Eq. 5.15 tells us additionally that, in incompressible fluid flow, the static air pressure must be **less** than p_∞ at such points because that is the only way the total pressure can remain constant.

The venturi tube, often used on older aircraft to drive gyro instruments, presents a good opportunity to demonstrate the use of the energy equations just developed.

Venturi was an Italian physicist who developed the instrument now known by his name; it is shown in cross-section in Fig. 5.12a. The fluid passing through the tube **increases** its speed in the constricted middle of the tube (by the continuity equation) and this results in pressure **less** than p_∞ inside the tube (by the Bernoulli equation). Three U-tube manometers are shown; one end of each manometer tube samples the static air pressure inside the venturi and the other end samples the free airstream pressure, p_∞. The difference in heights of the liquid in each manometer indicates the amount by which p is lower than p_∞ or, equivalently, the amount by which q is greater than q_∞.

Venturi tubes at one time were mounted on the sides of many aircraft; the reduced air pressure near the narrowest part was used to draw air from a static pressure source past a flywheel driving the gyro. (Now almost all gyro flight instruments are electrically driven or obtain a reduced pressure from an engine-driven "vacuum" pump, although a few venturi have returned as back-ups.) Venturi tubes are also used to measure fluid speeds because, as will be shown shortly, the reduction in air pressure in the venturi is a simple function of V_∞ and the geometry of the tube. Finally, the venturi principle is used in carburetors to draw gaseous vapor into the air that is on its way to the cylinders for combustion.

Figure 5.12b presents a qualitative picture of how the static pressure, dynamic pressure, and total pressure change (or don't change) at different points within a streamtube that passes through a venturi tube. At moderate airspeeds, the BL should be thin and the derived properties should be representative of most of the flow inside the tube.

First we apply the continuity equation, Eq. 5.9, and identify the subscript 1 with the nearly-freestream conditions that exist at the opening of the tube and subscript 2 with the conditions that exist at the narrowest part of the tube. Then

$$A_1 V_\infty = A_2 V_2 \qquad (5.19)$$

Second, using Eq. 5.12, we apply the energy conservation condition to these same points:

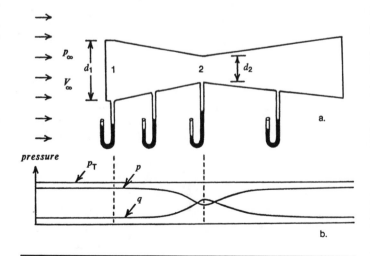

Fig. 5.12 A venturi tube and the pressure variation inside it

$$p_\infty + \frac{1}{2}\rho V_\infty^2 = p_2 + \frac{1}{2}\rho V_2^2 \qquad (5.20)$$

If we are using the venturi tube to measure V_∞, we should eliminate V_2 from Eqs. 5.19 and 5.20 and solve for V_∞. The result of this algebra exercise is

$$V_\infty = \sqrt{\frac{2(p_\infty - p_2)}{\rho\left[\left(\dfrac{A_1}{A_2}\right)^2 - 1\right]}} \qquad (5.21)$$

A manometer at point #2, as shown in Fig. 5.12, will measure the quantity $(p_\infty - p_2)$ in Eq. 5.21. If the density of the fluid and the two cross-sectional areas of the venturi tube are known, Eq. 5.21 is conveniently used to calculate the airspeed before it is influenced by the venturi.

On the other hand, if the venturi tube is being used to produce a lowered fluid pressure, Eqs. 5.19 and 5.20 should be solved for the pressure difference, and performing this algebra yields

$$\begin{pmatrix} \text{pressure} \\ \text{drop} \\ \text{in} \\ \text{venturi} \end{pmatrix} \equiv \Delta p = p_\infty - p_2 = \frac{1}{2}\rho V_\infty^2 \left[\left(\frac{A_1}{A_2}\right)^2 - 1\right] \qquad (5.22)$$

EXAMPLE 5.4. Suppose a venturi tube is connected to a water faucet to make a primitive "suction" pump. Assume that the entrance diameter is 0.50" and the narrowest diameter is 0.35". For a water density of 1.94 sl/ft^3 and an entry water speed of 3.1 ft/s, how much is the pressure at the narrowest point lowered below the entry pressure in the venturi?

Solution: We can use Eq. 5.22 as soon as we figure out the relationship of the cross-sectional areas of the venturi tube to its diameters. The equation we should use is:

$$\text{cross-sectional area} = \frac{1}{4}\pi d^2. \text{ Therefore}$$

$$\frac{A_1}{A_2} = \frac{\frac{1}{4}\pi d_1^2}{\frac{1}{4}\pi d_2^2} = \frac{d_1^2}{d_2^2} = \left(\frac{d_1}{d_2}\right)^2 = \left(\frac{0.50}{0.35}\right)^2 = 2.041$$

Substituting this in Eq. 5.22,

$$\Delta p = \frac{1}{2}(1.94)(3.1)^2[(2.041)^2 - 1] = \textbf{29.5 lb/ft}^2$$

Using a pressure conversion factor from Table 4.2, this pressure drop can also be expressed as 5.7" of water.

E. Circulation and Lift

Experimentalists and theorists had little meaningful communication at the beginning of the twentieth century when the age of manned flight was just dawning. Newton had proposed his laws of motion and a basic understanding of the motions of planets was well in hand. But the huge numbers of molecules involved in fluid motion presented seemingly insurmountable difficulties in the quest for equal precision in fluid mechanics, and so it was that the pioneers of flight developed wings that

worked by experiment and by trial and error rather than from fundamental theories of lift and drag. In this section we spend just a little time looking at the early steps taken by fluid mechanicians toward understanding how a surface can produce lift.

Anyone who has flown a kite or held a hand outside the window of a moving car can testify that a flat plate (or a flat hand) obtains a significant force on it from its interaction with the air. If the plate is pointed directly into the wind, the resultant force of the air is directed just opposite to the direction of the oncoming air. This is a pure **drag** force and we now recognize it as being caused by the viscosity of the air.

On the other hand, if the leading edge of the flat plate is higher than the trailing edge, the resultant force of the air on the plate is tilted **upwards**. We conventionally resolve this resultant aerodynamic force into two mutually perpendicular forces (just as it is often convenient to resolve a distance or velocity vector into components in perpendicular directions), one of which is **parallel** and another which is **perpendicular** to the oncoming air. The one parallel, and opposite in direction, to the oncoming air is reasonably enough called the **drag force** while the perpendicular component is called the **lift force**. (Notice that these definitions make **no** reference to the earth or to the horizon; they apply equally well in upright, inverted, vertical, or banked flight.) Early attempts to predict experimental values of lift and drag theoretically were miserable failures.

Newton's papers suggest that he thought the resultant force of the air on the flat plate was perpendicular to the plate and had a magnitude determined by the equivalent area presented to the oncoming air (Fig. 5.13). Thus the streamlines that were not aimed at the flat plate were assumed to be unaffected by the plate. This is a little like the intuitive understanding of many

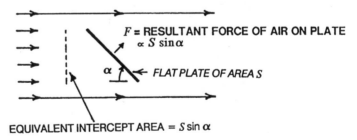

a. ASSUMED FLUID FLOW & FORCE ON PLATE

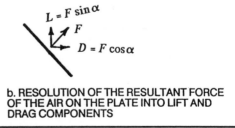

b. RESOLUTION OF THE RESULTANT FORCE OF THE AIR ON THE PLATE INTO LIFT AND DRAG COMPONENTS

Fig. 5.13 The Newtonian theory of the lift and drag created by a flat plate

people yet today that lift is primarily due to the increased pressure on the bottom surface—a sort of "**air surfboard**"; in fact the upper surface usually does **more** than the lower surface by virtue of the reduced pressure there, as we will see in detail in a later chapter.

Next, a famous theory by D'Alembert derived the result that a non-viscous fluid would generate **zero drag** on either a streamlined or unstreamlined object (Fig. 5.14). Evidently the small viscosity of air **is** important, particularly for an unstreamlined object. Even worse, no theory could come close to predicting the experimental lift developed by even a flat or curved plate.

Lord Raleigh tackled the tough theoretical puzzle of lift with his 1878 paper, "On the Irregular Flight of a Tennis Ball." Figure 5.15 illustrates the arguments presented in Raleigh's paper except that the drawing is for a long cylinder of circular cross-section rather than a sphere (so that we can honestly draw the air flow as a predominantly two-dimensional affair).

If the cylinder is moving into the air but is not rotating, the air flow is symmetric above and below it (Fig. 5.15a). If the cylinder is not moving relative to the air but **is** rotating, the viscosity of the air will result in some air circulating around with the cylinder (Fig. 5.15b). If now the cylinder is both rotating and moving into the air, the streamlines should look like a combination of the previous two illustrations (Fig. 5.15c). The air being dragged along by the cylinder will exert viscous forces on the air above the cylinder to speed up the air flow there and will act on the air below it to slow it down, all speeds being relative to the speed of the oncoming air. The converging streamlines above the cylinder, along with the pressure-velocity trade-off expressed by the Bernoulli equations, tell us that the average

pressure above the cylinder must be **less** than the freestream air pressure, p_∞, and the average pressure below the cylinder must be **greater** than p_∞.

This differential pressure must produce a net force **perpendicular** to the direction of motion of the cylinder; in the figure, the cylinder has "back spin" and the upward force (positive lift) causes the trajectory to be higher than it would be with no rotation (as in a baseball pitcher's rising fast ball). Any opposite spin ("top spin") would produce a sinker ball (negative lift force), and a sideways spin produces a curve ball. Actually, the flow behind the cylinder is **not** symmetric with the flow in front of the cylinder, as we have drawn it, and this produces a **pressure drag** that adds to the **viscous drag** (see next section).

Contrary to the Newtonian assumption, the air flow well above and below the cylinder developing lift **is** affected by the

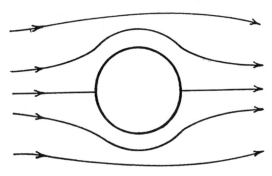

a. PURE TRANSLATION (NO LIFT)

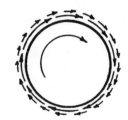

b. PURE CIRCULATION (NO LIFT)

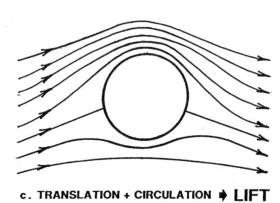

c. TRANSLATION + CIRCULATION ➡ LIFT

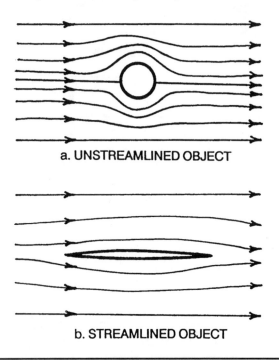

a. UNSTREAMLINED OBJECT

b. STREAMLINED OBJECT

Fig. 5.14 Completely symmetrical flow about a symmetrical object always produces zero pressure drag, in non-viscous flow.

Fig. 5.15 The circulation theory of lift explains how a rotating cylinder produces lift.

cylinder; there is an "upwash" of air before the cylinder and a "downwash" of air behind the cylinder. Note how the stagnation point moves to a slightly lower position when the cylinder develops lift. All of these observations are applicable to wings developing lift as well.

The principle of lift from a rotating cylinder was utilized by Anton Flettner in 1924 to obtain motive power from the wind for a boat without a sail; a large vertical cylinder was mounted on the deck and rapidly rotated. It worked, but mechanical losses in the drive system made it too inefficient to be practical. It will be reinvented momentarily, no doubt.

So now the secret of the airfoil (or flat plate) is revealed: An airfoil produces lift by inducing a "circulation" around it, much as does a rotating cylinder, but with considerably more efficiency. There remains only one more critical assumption to add to this circulation theory before the lift of an airfoil can be quantitatively predicted. This is the *Kutta-Joukowski* condition (proposed independently by the German mathematician M. Wilhelm Kutta and the Russian Nikolai E. Joukowski); it states that the velocity of the air leaving the upper surface will be the **same** as the velocity of the air leaving the lower surface. This condition is equivalent to requiring that the rear **stagnation point** be located at a sharp **trailing edge** of an airfoil. Because only **one** value for the circulation satisfies this condition, the generated lift is uniquely predicted.

The Kutta-Joukowski condition can be justified for a streamlined airfoil by the fact that predicted values are within about 10% of experimental ones, for low and intermediate values of lift at least (for which separation is small and the flow **does** nearly come together at the trailing edge). Figure 5.16 shows the circulation predicted by this theory for an airfoil pointing directly into the oncoming air. This airfoil clearly is developing lift because of the curvature of the upper surface, even though the flat bottom surface points directly toward the oncoming air. Notice the location of the stagnation points.

Fig. 5.16 Streamlines about an airfoil developing lift through circulation

Even though the circulation theory is successful in predicting the lift for low and intermediate values of lift, it cannot predict the loss of lift when the flow over the upper surface of the airfoil breaks away (the "stall") and it cannot predict the **drag** of an airfoil at all.

Nature doesn't allow rotation to be created from nothing, any more than energy can suddenly appear from nowhere. (This is conservation of angular momentum.) When an airplane wing first develops lift on the takeoff roll, or when the angle of the wing changes the circulation and the lift while in flight, there is always formed a **starting vortex** which has an opposite direction from that of the change in circulation (Fig. 5.17), so the net

Fig. 5.17 The starting vortex that is formed when lift is first developed, or increased

circulation is unchanged. This starting vortex trails behind the aircraft and is soon dissipated by the viscous forces in the air.

Theodore von Kármán was a graduate assistant of Ludwig Prandtl, the great Göttingen aerodynamicist who pioneered in the development of **boundary layer theory**. A doctoral student of Prandtl's at that time was engaged in studying the flow of water behind a circular cylinder with the intention of checking boundary layer theory. However, no amount of precision machining or careful alignment could remove the violent oscillations that were observed in the wake.

Finally, von Kármán made a calculation (over a weekend!) and was able to show that only a **periodic** shedding of vortices from alternating top and bottom surfaces of the cylinder was a stable solution for this type of fluid flow (Fig. 5.18). This regular and periodic shedding of vortices occurs only in a particular range of speeds. When the frequency at which the vortices are shed is equal to a natural frequency of an object, there is **resonance** and a maximum energy transfer from fluid to object occurs. (A common example of mechanical resonance is the observation that pushes on a swing or pendulum are effective only if they occur at the proper time with the same frequency as the natural swing frequency.)

Fig. 5.18 Periodic vortex shedding from a cylinder occurs in a critical range of speeds.

Resonance is usually something to be avoided at all costs when it comes to aircraft design because we want nothing to do with maximum energy transfer to the air or, in a bending sense, to a propeller. Avoiding resonance is not always easy because of the large number of natural mechanical and structural frequencies in an aircraft. In the 1950s a couple of wings of Electra airliners separated from their parent fuselages when an unpredicted resonance occurred under certain power settings and airspeeds. Most modern aircraft are limited in their dive speeds by "flutter" considerations. Flutter is an **aeroelastic** phenomenon; it occurs when a frequency of the air flow is in resonance with a natural frequency of the airplane structure (usually a control surface and its supporting surface).

Other applications of this concept of alternating vortices are recited in the quotations that open this chapter. The theory of alternating vortices brought fame and the beginning of a long and productive career for von Kármán; he later settled in the

United States and was probably the best-known American aerodynamicist during that time.

F. Boundary Layer Separation and Pressure Drag

We have seen that the circulation theory is able to predict the low and medium values of lift produced by an airfoil but it is unable to predict the breakdown of that lift at larger values, when the rear stagnation point is no longer at the trailing edge; also it is unable to predict the drag of an airfoil. But now only one more concept remains between us and a qualitative understanding of the lift and drag generated by **both** streamlined and unstreamlined objects. This key concept is the **separation** of the BL from the surface. We shall find that the tendency of a particular BL to separate is intimately related to the laminar or turbulent nature of that BL.

We begin by defining what is meant by the term *fluid particle*. A fluid particle is an imaginary group of fluid molecules; the group is assumed to be large relative to molecular size but small relative to airfoil or wing dimensions. The important thing is that the fluid particle is assumed to be small enough that we can consider it to move and rotate as a single entity. Then we can use mechanical concepts of particle motion to try to understand the overall fluid motion.

Consider the opposing forces acting on a fluid particle that chances to find itself in the BL (Fig. 5.19). On one side of it are molecules closer to the surface, and they have a lower average speed and therefore are tending to slow it down; on the other side of it are molecules trying to speed it up. The net result of these interactions is the viscous retarding force on the fluid particle, and this force is proportional to the viscosity of the fluid, the speed gradient in the BL, and the interacting surface area of the fluid particle (refer to the discussion of viscosity in Section A). The speed gradient should be related to the overall fluid speed, V, divided by some characteristic dimension of the surface, x. So we might suspect that, in the BL,

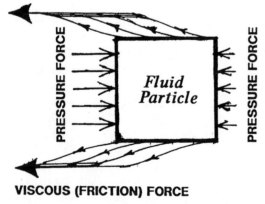

VISCOUS (FRICTION) FORCE

PRESSURE FORCE   *Fluid Particle*   PRESSURE FORCE

VISCOUS (FRICTION) FORCE

Fig. 5.19 These are the forces that are pushing and pulling on a "fluid particle."

$$\left(\begin{array}{c}\text{average viscous retarding force}\\\text{on the fluid particle}\end{array}\right) \propto \mu\left(\frac{V}{x}\right)A$$

where A is the interacting surface area of the fluid particle.

The **driving force** on the fluid particle is the difference in pressure on its front and back surfaces. When the fluid pressure decreases with distance (a favorable, negative pressure gradient), this force will tend to speed up the particle; when the fluid pressure increases with distance (adverse, positive pressure gradient), this force will tend to slow down the fluid particle. We might suspect that this pressure difference is related to the pressure difference between stopped and normally moving fluid; by Bernoulli's theorem this pressure difference is the dynamic pressure and is proportional to ρV^2 for incompressible flow. Therefore the **pressure-induced force** on a fluid particle in the BL might show the following relationship:

$$\left(\begin{array}{c}\text{average pressure force}\\\text{on a fluid particle}\end{array}\right) \propto \rho V^2 A$$

The **ratio** of the driving, pressure-induced force to the viscous retarding force should determine the overall nature of the fluid flow in the BL.

$$\left(\begin{array}{c}\text{nature}\\\text{of the}\\\text{fluid}\\\text{flow}\end{array}\right) \propto \frac{\text{pressure force}}{\text{viscous force}} = \frac{\rho V^2 A}{\mu\left(\frac{V}{x}\right)A} = \frac{\rho V x}{\mu}$$

It helps a **great** deal to know the answer ahead of time in a "derivation" of this sort, but I hope you realize that we now have a rough physical interpretation of the **meaning** of the Reynolds number (Eq. 5.4). If the viscous retarding forces are comparable to the pressure forces, we expect to find minimal acceleration of the fluid particles, a low RN, and laminar flow in the BL; if the pressure forces dominate the viscous forces, we expect to find large accelerations, a large RN, and a transition to turbulent flow in the BL. Stated differently, **low RN** fluid flow is **viscosity-dominated flow** (with lots of laminar flow) and high RN is **pressure-dominated flow** (with mostly turbulent flow).

It is only fair to warn you that this "derivation" of the Reynolds number is not the one presented in most engineering texts. Consider the equation of motion, using Newton's second law (Eq. 3.1), for our fluid particle:

$$\sum F_{\text{EXTERNAL, ON FLUID PARTICLE}} = \left(\begin{array}{c}\text{pressure}\\\text{force}\end{array}\right) + \left(\begin{array}{c}\text{viscous}\\\text{force}\end{array}\right)$$

$$= \left(\begin{array}{c}\text{mass of}\\\text{fluid particle}\end{array}\right) \times \left(\begin{array}{c}\text{acceleration of}\\\text{fluid particle}\end{array}\right)$$

It is common engineering practice to call the (mass $\times$ acceleration) term the **inertial force** on the particle, even though it is **not** a force but the **result** of unbalanced **external** forces. (The "centrifugal" force is a similar fictitious force, being defined as the mass times the **radial** acceleration, for circular motion.) However, this "fictitious force" approach allows one to write Newton's second law in the seductive form

$$\left(\begin{array}{c}\text{pressure}\\\text{force}\end{array}\right) + \left(\begin{array}{c}\text{viscous}\\\text{force}\end{array}\right) - \left(\begin{array}{c}\text{inertial}\\\text{force}\end{array}\right) = 0$$

just as if this were a statics problem where the forces always sum to zero because there is no acceleration! (Note that the pressure force can be positive or negative but the viscous force **always** opposes the motion.) Anyway, in the engineering literature, it is often stated that the Reynolds number is the **ratio** of the viscous force to the inertial force. The derivation of this conclusion does not seem to me to be any more immaculate or convincing than mine, and I find my result more satisfying conceptually, but at least now you know.

If there were no viscous forces, the pressure would increase over the rear part of an object just as much as it decreased over the front part; that is, there would be complete pressure recovery and the net force due to fluid pressures would be zero. This is just the result obtained by D'Alembert's inviscid flow theory and is illustrated by the pressures shown on the center cross-section of a long cylinder in Fig. 5.20a. In reality, viscous forces cause the average pressure on the rear part of an object to be less than those on the front half, and therefore there **always** exists some pressure drag (in addition to viscous drag) on every object (although it can be negligibly small in some cases). The actual magnitude of the average pressure loss, and the importance of the resulting pressure drag, depends on just what happens in the BL! Note that the theoretical, inviscid flow of Fig. 5.20a has perfect **pressure recovery**—that is, the pressure at the trailing edge is just equal to the pressure at the leading edge.

A fluid particle in the BL will never make it all the way to the back of the object on a shortest-distance basis. Because of viscous forces, it must come to a **halt** relative to the surface before it gets to the trailing edge. The actual point at which the BL has a relative speed of zero and **separates** from the surface

depends on (**a**) the ratio of pressure to viscous forces (given by the RN for the object, using its length in the fluid direction) and (**b**) the geometry of the object, which determines how rapidly the pressure changes occur. Because a turbulent BL **does** obtain more energy from the surface (Section B),

a turbulent BL normally separates later than a laminar BL.

When the BL does separate, it produces a region (not just a layer) of turbulent, time-varying fluid flow behind it; in this region, the low pressures pull departing fluid back toward the surface (horrors! that's not what we want), as is true for the cylinder of Fig. 5.20, and then there is a lot of pressure drag. The unstable, turbulent, vortex-filled region behind the separation point is known as the **wake**. If the wake is large, the average pressure on the front part is much greater than the average pressure on the rear, and the pressure drag is very large (dwarfing the viscous drag).

Consider in this light the actual fluid flow past an unstreamlined object at low, medium, and high RN (Fig. 5.21). At very low RN the viscous forces are comparable to the pressure forces and the fluid oozes around the object with negligible separation (Fig. 5.21a) so that pressure drag is negligible compared to

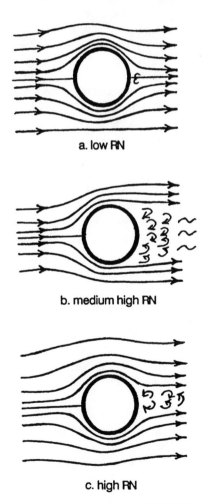

a. low RN

b. medium high RN

c. high RN

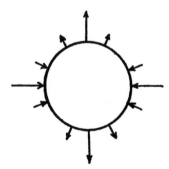

a.   THEORETICAL, INVISCID FLOW

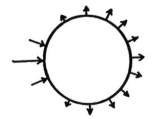

b.   MEASURED, RN = 1.6 x 10^5

Fig. 5.20 Pressure distributions around a long cylinder (air flow is from left to right)

Fig. 5.21 This is the actual nature of the fluid flow around a cylinder (very speed dependent).

viscous drag. An example of this kind of fluid motion would be a baseball sinking in a barrel of tar or a dust particle gently settling in still air.

At medium RN the BL is laminar and separates just about as soon as it hits the adverse pressure gradient going around the steep bend in the object; the result is a pressure drag that is **huge** compared to the viscous drag (Fig. 5.21b). An example of this kind of fluid flow is a smooth golf ball at normal fairway speeds.

At high RN, which encompasses most manned flight, the BL becomes turbulent before it gets to the thickest point; this allows the BL to **stay attached** much farther along on the object, resulting in much less pressure drag than for a laminar BL (Fig. 5.21c). Even though the skin friction drag increases with the commencement of turbulence in the BL, the pressure drag is so much less that the overall drag is reduced when this condition is first obtained. But even though the pressure drag has dropped dramatically now, the pressure drag is still far greater than the viscous drag for any unstreamlined object like this cylinder or wire.

Comparable flow conditions for a slightly streamlined object are most interesting and enlightening (Fig. 5.22). At very low RN the BL is almost completely attached and the pressure drag is negligible (Fig. 5.22a). The total drag for this "streamlined" object is much **greater** than for a cylinder of the same size (Fig. 5.21a) because the additional surface area has greatly increased the skin friction (viscous) drag. Therefore "streamlining" an object when the flow is already streamline does far more harm than good, because streamlining can only reduce pressure drag and it was already negligible! (This situation is **not** usually encountered when air is the fluid, even for model aircraft.)

At medium RN the BL tends both to stay laminar **and** to separate fairly early, despite the streamlining (Fig 5.22b). Sometimes a laminar BL will start to separate, transition to turbulent, and re-attach (forming what is called a "**laminar bubble**"); this typifies the air flow around many soaring aircraft. Sometimes on models it is desirable to artificially induce turbulence with some sort of roughening or with a sharp boundary using a "trip" wire.

At high RN the BL makes an early transition to turbulent flow and stays attached **much longer** (Fig. 5.22c), making the pressure drag significantly less than for the medium RN condition. The tendency is for the separation point to move farther and farther back as the RN increases. The result is that large, fast aircraft have an intrinsic efficiency advantage over smaller, slower aircraft.

Because of differing RN, a scaled aircraft is an entirely different animal than the original, as viewed by the air. Consequently, your ⅗ scale P-51 homebuilt will have very different flight characteristics from the World War II original. Very importantly, it also means that the properties measured in a wind tunnel at Reynolds numbers less than full-scale must be corrected to flight Reynolds numbers to be meaningful, and this extrapolation can be difficult to do correctly.

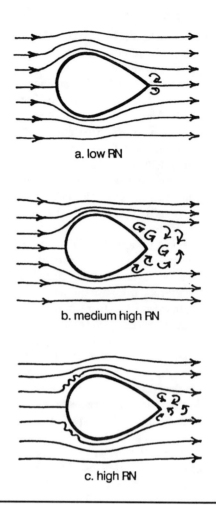

Fig. 5.22 The actual nature of the fluid flow around a (slightly) streamlined object

The dependence of aircraft properties on RN is known as **scale effect.** It is not just size or speed that is important but also the density and viscosity of the air, since they also appear in the definition of the RN. Some wind tunnels use variable density or cryogenic temperatures to try to bring the test RN closer to the flight RN, but the problem has been severe enough to cause some full-scale wind tunnels to be built, despite the enormous cost to construct them and the great gobs of power to run them.

At high Reynolds number (say, above 4 million), the flow around a streamlined shape is almost entirely attached and representative streamlines and the lift that is generated are readily calculated, even by a small personal computer. Reference 7 contains the Oshkosh Airfoil Program which uses the Joukowski transformation and five shape parameters to stretch a circle into an airfoil and then calculate streamlines and the pressure distribution around the airfoil, using inviscid potential flow theory. Because the viscosity of the air is ignored, neither the BL nor separation are modeled but the calculations should be pretty good for high speed airfoils when they are at a small angle relative to the oncoming air. Figure 5.23 presents a sample airfoil and calculated streamlines for three different angles, obtained from this computer program.

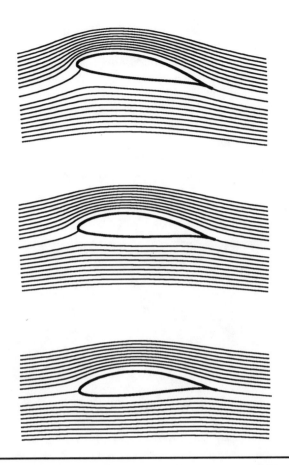

Fig. 5.23 Streamlines around a derived airfoil as calculated by the Oshkosh Airfoil Program

In a smoke tunnel, streamlines are made visible by inserting smoke into the air flow from vertically positioned, discrete sources located well in front of the airfoil. Picture 5.1, obtained in this fashion, shows the actual flow around a cambered airfoil at zero angle of attack. The Reynolds number is apparently quite low; there appears to be some separation near the trailing edge, especially on the upper surface.

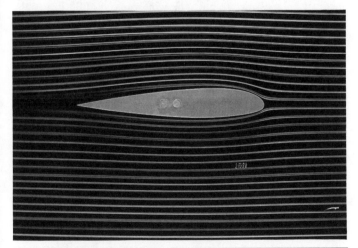

Pic. 5.1 Streamline flow around an airfoil in a smoke tunnel at 0° angle of attack (NACA photo courtesy of Robert Baals of NASA Langley)

Figure 5.24 sketches in exaggerated form the actual composition of the BL above an airfoil under typical cruise conditions. Notice the varying velocity profiles in the laminar and turbulent regions. Note also the reversed air flow in the wake. Some of the air there is moving **faster** than the aircraft! Now do you believe in pressure drag?

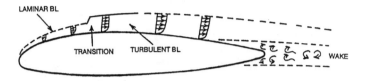

Fig. 5.24 An enlarged view of the boundary layer above a lifting airfoil

The velocity profile within the BL close to the point where the BL separates deserves additional enlargement (Fig. 5.25). The gradually thickening BL is outlined with the dashed line. Note how the velocity gradient at the surface decreases as the separation point (marked with an "S") is approached. Then, at the separation point, the velocity gradient is **zero** and there is no viscous drag! The bad news is that, downstream of that point, there is a region of reverse flow, and this is associated with a large pressure drag.

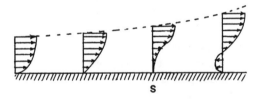

Fig. 5.25 Velocity profiles around a point where the boundary layer separates

The transition from laminar to turbulent flow in the BL occurs in a **range** of RN values, as noted earlier. An early transition is stimulated by (**a**) surface roughness or waviness comparable to the thickness of the BL, and by (**b**) an increasing pressure in the direction of fluid flow (i.e., an adverse pressure gradient). The first factor means that few lightplanes currently flying obtain very much laminar flow over their wings; manufacturing tolerances, intrinsic flexibility of the thin aluminum or fabric covering, and in-service accumulation of dirt and insect residue present obstacles like jagged mountain ranges to the friendly little fluid particles in the BL. The second factor has important ramifications for the design of new airfoils, as we shall see.

In cruising flight the wing and the fuselage are well streamlined with the air flow, and the viscous drag is much larger than the pressure drag. (Streamlining can be defined as the process of adjusting the shape of a body so that pressure drag is minimized for the flight RN.) Because viscous drag is much larger for a turbulent BL than for a laminar one, the cruising and top speed of an aircraft are maximized by obtaining as much laminar flow as possible without causing premature separation. (A simpler way is to reduce the aircraft size and therefore its skin area for a given load! This is what amateur aircraft builders are doing in great numbers! The penalty is higher landing speeds.)

It is much easier for a BL to stay laminar if it is speeding up (under the influence of a favorable, negative pressure gradient). As soon as the BL "rounds" the bend and has to begin its pressure recovery by slowing down, it will tend to transition to turbulence. (If it doesn't transition, or doesn't transition soon enough, it will separate prematurely.) This reasoning during the early part of World War II led to the development of **laminar** airfoils which are characterized by a gradually increasing thickness up to a maximum located about one-half way to the trailing edge (rather than one-quarter or so, for earlier designs; see Fig. 5.26). This really did work in the wind tunnel but it is not clear that much laminar flow was actually achieved on the aircraft that used it, because of roughness and rigidity problems. The famous P-51 Mustang fighter aircraft is considered to be the first airplane to utilize a laminar airfoil wing, and proved to be a good sales plane for the idea. Some of these early laminar airfoils have proved to be good high speed airfoils, though, and so have been used on many aircraft for this reason, rather than for their potential laminar properties!

However, an aircraft is **not** well streamlined during the climb, descent, or landing phases of flight. These flight conditions require the additional lift for a given speed that is obtained through an increased angle relative to the oncoming air (i.e., an increased angle of attack). The additional lift always comes with additional drag, but with excess power a slower and steeper climb and descent is possible. The drag increases with increasing angle of attack because the separation point moves closer and closer to the leading edge of the airfoil as the wing appears less and less streamlined to the air.

The pilot must be careful not to increase the angle of attack (the angle relative to the oncoming air, which is the angle relative to the horizon **only** in level flight) beyond a certain point while in flight or the aircraft will experience a **loss** in lift and a large increase in drag. This occurs at an angle of attack of perhaps 12° to 25° for most wings. What happens is that the separation point moves (sometimes very abruptly) close to the leading edge. The low average pressure above the wing becomes much closer to atmospheric pressure (so lift is lost) and

there is a great deal of pressure drag as the wake encompasses much of the region above and behind the airfoil. The unstable, turbulent wake makes the condition an unstable one for the whole wing. This is what is meant by the **stall** of an airfoil, or an airplane; the air in the BL really has stalled (stopped) and separated over much of the top surface of the wing. Much of the air immediately next to and above the airfoil is actually moving **upstream**!

Picture 5.2 shows the airfoil of Pic. 5.1 at an angle of attack of about 10°. Here the flow has separated over much of the upper surface but is well attached on the lower surface. Upper surface separation at such a relatively low angle of attack is again evidence of a low Reynolds number.

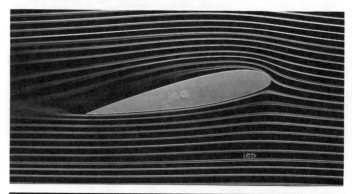

Pic. 5.2 Mostly separated (stalled) flow around the top surface of an airfoil in a smoke tunnel at 10° angle of attack (NACA photo courtesy of Robert Baals of NASA Langley)

The pattern of stall behavior for different parts of a wing on actual aircraft are often checked in flight by attaching short wool tufts to the wing; the wool tufts can be seen to be pointed upstream behind the separation point (see photos in Chap. 12). Later we'll look closely at the pressure distributions above an airfoil during these exciting aerodynamic events. Yet light aircraft normally are landed at a speed close to their stall speed because this permits the lowest possible landing speed; it is in fact just "plane" nice to have the wing run out of lift just as the wheels brush the ground.

A ball such as a baseball or golf ball can never hope to be streamlined, of course. If it is used in the particular range of speeds where the existence of a laminar BL promotes an early separation (Fig. 5.21b), there exists a great advantage in artificially inducing an early transition to turbulent flow in the BL. This is the function of the dimples on a golf ball. In fact a dimpled golf ball travels about five times farther than an equivalent smooth one. Golf ball manufacturers have been known to advertise for aerodynamicists to help them improve their dimples.

There is usually no reason to induce turbulence into the BL during streamlined cruising flight, for the BL is mostly attached (at least on the wing). Yet the principle of delayed separation for a turbulent BL is still true, and sometimes aircraft designers artificially induce turbulence to improve an aircraft's performance under unstreamlined flow conditions. More on this later.

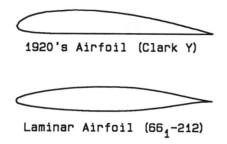

```
1920's Airfoil (Clark Y)



Laminar Airfoil (66₁-212)
```

Fig. 5.26 An old turbulent airfoil and a new laminar airfoil

The earliest aircraft used external wire bracing, as do some ultralights. Most biplanes, though, use streamline tubing, for reasons that Fig. 5.27 should make abundantly clear. The pressure drag produced by a small cylindrical section, only 1/8" in diameter, is so huge at even 100 kt that a good, smooth streamline shape can be **sixty** times thicker and still generate no more total drag!

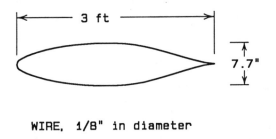

WIRE, 1/8" in diameter

Fig. 5.27 These objects have the same drag at 100 kt — streamlining pays big dividends!

The importance of BL **transition** and **separation** cannot be overemphasized. An understanding of these related phenomena makes it possible to understand qualitatively how an airfoil works, how new airfoils have improved on old ones, and how a wing will respond to varying flight conditions. The detailed mathematical prediction of transition, turbulence, and stall remain among the most difficult and challenging tasks facing theoretical aerodynamicists, however.

G. Lift: By Bernoulli or by Newton?

It is now clear that we have lift, abundant lift. But how are we to explain it to others? What fundamental principle of physics explains lift?

Bernoulli advocates tell us that the greater curvature/longer path of the upper surface of an airfoil causes the air to speed up so that it can meet up with the air that chooses to go the lower route around the airfoil; then, since the Bernoulli principle tells us that higher airspeeds are associated with lower pressures, the relatively higher pressures on the lower surface produce a net force up. Critics of this explanation point out that (a) the Bernoulli principle doesn't apply when energy is added to, or subtracted from, the air, (b) a symmetrical airfoil and an inverted, cambered airfoil are both capable of producing lift as long as a positive angle of attack is maintained and, most important of all, (c) the assumption that the top air "has" to get to the trailing edge at the same time as the bottom air surely requires justification beyond wishful thinking.

Newtonian proponents, on the other hand, tell us that the third law of motion, that every action (force) produces an equal and opposite (reaction) force, is the key. Thus lift is produced when, and only when, the wing is able to deflect the oncoming air. Further, some would add, this pushed-on air continues the chain by pushing down on the earth and getting an equal push up — so the earth is ultimately responsible for providing the lift.

Critics of these propositions point out that (a) except for holding the air in place with the gravitational force, the earth can't be a necessary part of the chain because lift is developed just as efficiently in a banked turn, even one with a 90° bank angle, and (b) an explanation for lift isn't very convincing if it isn't used in the experimental measurement of lift.

The author wishes to suggest that (a) both the Bernoulli principle and the Newtonian third law are valid, applicable, and useful in the understanding and measuring of lift and (b) it is foolish to talk about lift without talking also about the other component of the resultant aerodynamic force — drag. It is the ratio of lift-to-drag that measures the efficiency of an airfoil or wing and, ultimately, the feasibility of flight, because the lift force must be large enough to just equal the weight in level flight while the power required is just the product of the speed and the drag force, so a poor lift-to-drag ratio is associated with unreasonable power requirements.

Wings normally employ positively cambered airfoils because this shifts the minimum drag point to positive lift coefficients so that the lift-to-drag ratio in cruising flight is maximized (as well as usually yielding a greater maximum lift coefficient) — so positive camber is associated with **efficient** lift, not with the ability to produce lift. Even though the air within the boundary layer doesn't follow Bernoulli's principle because viscosity removes energy from the air, the actual airfoil pressures are largely determined by the flow and the pressures outside the boundary layer, where Bernoulli's theorem is valid; this is shown by the fact that airfoil theories that use this assumption are able to accurately predict the lift and drag characteristics of airfoils. And the join-up of upper surface air with lower surface air is a reasonable assumption, not black magic. Consider a symmetrical airfoil that is sufficiently well streamlined that the flow is attached all the way to the trailing edge. Then the upper surface air certainly joins the lower surface air. Isn't it reasonable to assume that the overall flow pattern will stay pretty much the same as the angle of attack is increased, at least up to a few degrees? In fact, this is just what experimental measurements tell us does happen.

Finally, the importance of pressures in airfoil theory and in the measurement of lift cannot be overemphasized. Airfoils are designed by modifying the geometry so as to obtain pressures that encourage maximum laminar flow or maximum lift or whatever the designer wishes to emphasize. Airfoils are tested in the wind tunnel for their lift properties by integrating the measured surface pressures. So Bernoulli's principle is a valid and useful principle (ultimately derived from Newton's laws of motion!) in the understanding and measurements of lift and drag.

Newton's third law, telling us that any object that exerts a force on another object will receive an equal and opposite reaction force on it, stands on a veritable mountain of corroborating evidence, all the way from baseball trajectories to space flight. So there can be no reasonable doubt that the aerodynamic force exerted by the air on an airplane is accompanied by an equal and opposite force of the airplane on the air. Lift is thus the component of the resultant aerodynamic force that is perpendicular to the flight path—independent of whether the aircraft is flying level or climbing or descending or in a steeply banked turn—and drag is the (rearward) component that is parallel to the flight direction. The location of the earth does not enter into this definition of lift and drag!

Sometimes the lift and drag forces on an airfoil or wing are measured in a wind tunnel with the use of force balances. These clearly rely on the validity of Newton's third law. More commonly, the lift force is determined from the measured pressure distribution while the drag is determined based on the momentum lost by the air in its passage past the airfoil—a direct use of the third law.

Section E noted that a simplistic picture of the aerodynamic force on a flat plate or airfoil, incorrectly assuming that only directly intercepted molecules of air are affected by the plate and that molecules are deflected like super balls (Fig. 5.13), yielded a discouraging estimate for the practicality of flight. The problem is not that lift is not produced but that it is produced so inefficiently. In fact, the ratio of lift to drag is predicted to be just equal to the inverse of the tangent (i.e., the cotangent) of the angle of attack so that either a great amount of speed is needed (at low angles of attack) or a tremendous amount of power is required (at high angles of attack) to generate lift equal to weight. Figure 5.28 contrasts the efficiency of a 12% thick, symmetrical, streamlined shape (airfoil), as determined experimentally, with the prediction of the

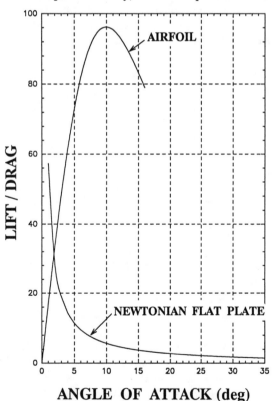

Fig. 5.28 Comparison of the efficiency of a real, symmetrical shape with the molecular model of Fig. 5.13 in the generation of lift

molecular model of Fig. 5.13 (Newtonian Molecular Model). The efficiency of the real airfoil reaches a maximum when a very usable amount of lift is being generated, but the efficiency of this fictional flat plate, with baseball-like approaching air molecules, plummets as the angle of attack is increased.

Few theories go entirely to waste. At supersonic speeds and at very high altitudes, where the flow ahead of the wing cannot be affected by the wing and where the molecules are so far separated that they act like independent particles, Fig. 5.13 usefully models the action.

Summary

Viscosity, the coefficient of viscosity (μ), and the kinematic viscosity (ν) were defined in Section A. It is the rate of change in fluid speed from one layer to another (the speed gradient) that indicates the effectiveness of the shear stresses within the fluid. For air, μ and ν vary significantly with changes in temperature (and therefore with altitude).

Section B introduced the Reynolds number as the dimensionless parameter that best predicts the nature of fluid flow and when a laminar BL will transition to a turbulent one. Other factors determining the location of this transition point are surface smoothness and waviness and intrinsic fluid turbulence. The turbulent BL has a much greater speed gradient close to the bounding surface and so has much more viscous drag than a laminar BL.

Fluid flow outside the BL was discussed in Section C through the concept of streamline flow and streamtubes. Within a streamtube, there must be a constant relationship between fluid speed, fluid density, and streamtube size because **no** fluid is lost along the way; this idea leads to the continuity equations for incompressible and compressible flow.

The Bernoulli equation gives the total energy density for a fluid in motion; this total energy density has a constant value for streamlines that don't contain a source of energy (conservation of energy), but the total can be composed of varying proportions of pressure energy and kinetic energy. The incompressible form of the Bernoulli equations was presented in Section D and applied to the fluid flow through a venturi tube. It became evident that the venturi tube could be used to measure fluid speeds or to obtain a pressure less than the static fluid pressure. Local dynamic pressure (q) and flight dynamic pressure (q_∞) were defined; dynamic pressure is just another word for kinetic energy density. The sum of static and dynamic pressure was given the name total pressure; the incompressible Bernoulli equation can be restated then as describing the constancy of total pressure within a streamtube.

The circulation theory of lift provided the first conceptual and analytical model that could predict the lift of a curved surface with reasonable accuracy; it predicted a significantly lowered pressure over the upper surface, thereby providing more of the lift force there than that due to the increased pressure on the lower surface, which is in accord with the experimental observations. As an inviscid theory, it predicted zero viscous and pressure drag for streamlined objects. The lift predicted by this theory is limited by its assumption of streamline flow over the airfoil surface; at low angles to the oncoming air this is a good assumption but it becomes increasing incorrect at large angles, as separation increases. Von Kármán was able to predict the periodic shedding of alternating vortices from the top and bottom of a cylinder held in a fluid at certain speeds; these von Kármán vortices serve as the driving oscillator in many experimental and natural occurrences of resonant behavior of wires or cylinders in air and other fluids. (You haven't lived a full aeronautical life until you've heard the "singing" of biplane wires.)

Air has an extremely small coefficient of viscosity but this viscosity is nonetheless responsible for the creation of a thin **BL** in which fluid speeds rapidly increase with increasing distance from the surface. This BL produces a viscous drag force on the surface but, particularly for an unstreamlined object, it can exercise an overwhelming influence on the large-scale fluid flow. This is because viscosity causes the fluid particles immediately next to the bounding surface to come to a halt before reaching the trailing edge, and this means that the BL will **separate** from the surface. The separation point is close to the trailing edge for a streamlined object pointing into the oncoming air but is much closer to the leading edge for an unstreamlined object or for a streamlined object at a large angle (about 12° to 25°) to the oncoming air. In the latter case there is a low pressure **wake** behind the object that produces very large values of pressure drag. The tendency to separate is stimulated by an adverse (positive) pressure gradient (increasing pressures).

Because a turbulent BL acquires more energy from the rest of the fluid through interlayer mixing, a turbulent BL always tends to separate farther from the leading edge than a laminar BL. Thus, the total drag of an object can be reduced by inducing turbulence into the BL **if** early BL separation and large pressure drag are present. Surface roughness and waviness and an adverse (positive) pressure gradient tend to stimulate an early transition of a laminar BL to a turbulent one. A **stalled** airfoil is one in which the separation point has moved so close to the leading edge that the high pressure wake covers much of the swept area behind the airfoil; much of the lift from the upper surface is lost and the pressure drag is very large.

The **Reynolds number** can be interpreted as a relative measure of the strengths of the pressure and viscous forces acting on a fluid particle; this explains the role of the RN as a **predictor** of the laminar-to-turbulent transition.

Bernoulli's principle and Newton's third law are both useful in understanding and measuring the generation of lift and drag.

Symbol Table (in order of introduction)

F	force
s	thickness of a thin layer of fluid
μ	coefficient of viscosity
ν	kinematic viscosity
BL	Boundary layer
RN	Reynolds number
x	a characteristic dimension (often the length) of a bounding surface in fluid flow
V_∞	fluid speed relative to a surface before it is affected by the surface
∞	as a subscript: refers to a fluid property before it is affected by a bounding surface
0	as a subscript: refers to SL standard value
A	cross-sectional area
V	fluid speed
Δt	an arbitrarily small time interval
q	dynamic pressure, often local dynamic pressure
q_∞	flight dynamic pressure (based on V_∞)
p_T	total pressure (= total energy density)
d	diameter within a tube, as for a venturi tube
R	resultant aerodynamic force; usually resolved into equivalent perpendicular components called lift and drag
L	lift force
D	drag force

Review Questions

1. Fluids, unlike solids, cannot sustain a static ___ force.

2. If the fluid speed is changing rapidly with separation distance from a certain bounding surface, that surface is experiencing a relatively (large, small) viscous drag.

3. The ___ of ___ is defined as the constant of proportionality between the shear stress on a fluid layer and the speed gradient within the layer.

4. This "constant" actually exhibits a distinct dependence on the ___ of the fluid, for air in motion.

5. This "constant," after being divided by the density of the fluid at the same point, is called the ___ ___.

6. The dependence of the Reynolds number on **fluid** properties is completely known if the ___ ___ of the fluid is known.

7. Disorder and efficient mixing between fluid layers are characteristics of ___ fluid flow.

8. An increase in fluid speed eventually causes a fluid to transition from ___ to ___ flow in the BL.

9. The best predictor of where or when this transition will occur is the local value of the ___ ___.

10. Two other factors that influence the location of such a transition are the ___ and ___ of the surface and the inherent ___ of the fluid.

11. The type of BL that always produces the least skin friction drag is a ___ one.

12. A BL exists for any flow of air around an object because air has a small but non-zero value for the fluid property called ___.

13. For a given flight speed, the BL over a wing tends to be slightly more laminar at high altitudes because the ___ ___ increases as the temperature decreases.

14. The lines that trace the path of a tiny particle imbedded in a steady fluid flow, or that represent the instantaneous directions of fluid velocities, are called ___.

15. The continuity equations are relationships between various fluid properties within a streamtube; they are derived by realizing that the ___ of the fluid within a streamtube doesn't change.

16. At low fluid speeds, air can perform as if it were incompressible because the varying pressures associated with intruding objects create compensating changes in fluid ___ **before** changes in fluid **density** occur.

17. The sum of all types of energy is constant for an object or for a fluid so long as the object or the fluid is a part of a system that is ___ .

18. The types of fluid energy that are important for air and aircraft (at low speeds) are ___ energy and ___ energy.

19. In incompressible fluid flow, the ___ pressure is the actual fluid pressure (and the total energy density) at a point where the fluid speed is zero; such a point is called a ___ point.

20. The ___ pressure is essentially constant within a venturi tube at low fluid speeds.

21. Assume that the narrowest diameter of a venturi tube is just one-half the entrance diameter. For incompressible flow, the continuity equation tells us that the fluid speed at the narrowest point will be ___ times the freestream speed.

22. The lift force on an airfoil is defined as the component of the resultant aerodynamic force that is perpendicular to the

    a. top surface of the airfoil

    b. bottom surface of the airfoil

    c. ground

    d. horizon

    e. oncoming air

23. The air flow over the bottom of an airfoil surface contributes (more, less) to the lift developed by the airfoil than does the air flow over the top surface, at least in normal cruise and climb conditions.

24. An airfoil produces lift when the pressure over the bottom surface of an airfoil is (greater, less) than the free airstream pressure or when the pressure over the top surface of the airfoil is (greater, less) than the free airstream value.

25. Inviscid flow theory predicts zero viscous drag for a non-spinning sphere in flight because it leads to the conclusion that the pressure on the rear half of the sphere is (greater than, less than, the same as) the pressure on the front half.

26. If a sphere in flight is spinning, inviscid flow theory predicts (zero, a small amount of, possibly a large amount of) lift and (zero, a small amount of, possibly a large amount of) drag.

27. A cylinder developing lift makes (more, less, the same) mass of fluid go over the top surface compared to zero-lift flow; this is evidenced by the movement of the ___ ___ to a point below the center line.

28. The ___ ___ condition is that the air travelling above the airfoil rejoins the air travelling below the airfoil at the trailing edge of the airfoil (assuming a sharp trailing edge); this condition permits the circulation theory to predict the magnitude of the lift produced by an airfoil.

29. This condition (previous question) is nearly true experimentally for an airfoil if

    a. the drag force is negligible

    b. no lift is being developed

    c. density changes are negligible

    d. it has a zero or small angle of attack

30. Von Kármán's alternating, periodic vortices behind an object produce spectacular effects when their frequency is equal to the ___ frequency of the object; this is the principle of ___.

31. The starting vortex formed by a wing just developing lift as it moves to the **right** would have a (clockwise, counterclockwise) rotation.

32. ___ forces on fluid particles dominate over ___ forces on fluid particles for almost all aircraft under actual flight conditions; this means that a transition to ___ flow can be expected somewhere before the trailing edge of an airfoil.

33. The driving force for a fluid particle is the ___ difference between its front and back; the retarding or restraining force is the ___ force.

34. ___ in the BL delays separation.

35. Separation produces a relatively (low, high) pressure wake behind an object; this results in a net horizontal pressure **on** the object that is in the (forward, backward) direction, and that can be quite large. This force is called the ___ drag force.

36. In Fig. 5.21b, the fact that air molecules are being dragged along with the cylinder is evidenced by the observation that the ___ behind the sphere are (less than, more than, the same as) the free airstream value.

37. Streamlining an object produces an **increased** total drag if the ___ drag was already less than the ___ drag force before streamlining.

38. True or False. Viscous-dominated fluid flow requires a high viscosity fluid. (Explain your answer and illustrate with an example.)

39. Can the viscous forces on an object ever be greater than the pressure forces on it? (Compare the friction force on sliding objects.) Why or why not?

40. The turbulent, unstable, vortex-filled region behind an airfoil that is stalled is called the ___.

41. If the drag force on an object in a fluid ever **decreases** as the fluid speed increases, you can be quite sure that the BL underwent a transition to ___ flow and that this reduced the size of the wake and the ___ part of the drag.

42. The wake tends to be (larger, smaller) for an airfoil pointing into the oncoming air (zero angle of attack) at **large RN** compared to the size of the wake at small RN.

43. Scale effect for a one-quarter size airfoil in a wind tunnel is negligible if the fluid speed is equal to the full-scale value and if the air ___ is ___ times greater than the full-scale value (and the Mach number is either the same or less than about 0.3!).

44. The surface properties of an airfoil that determine where the BL will transition to turbulence is the ___ and ___ of the airfoil.

45. The shape of an airfoil also influences the location of this transition point; if the transition hasn't already occurred at the maximum ___ of the airfoil, it will tend to occur there.

46. The "laminar" airfoils try to take advantage of this fact by using a shape that locates this point about ___ of the way back from the leading edge.

47. An adverse pressure gradient is one in which the BL fluid particles face an (increasing, decreasing) pressure as they proceed downstream.

48. Newspaper reporters (and therefore the general public) tend to think that a stalled airplane is one for which the engine has stopped, probably because that terminology is used with automobiles. What has actually "stalled" for an aircraft that is in a "stalled" condition?

49. An adverse pressure gradient tends to stimulate **both** the ___ and the ___ of the BL.

50. Increasing speed **decreases** the total drag on an object
    a. at sufficiently low RN
    b. at sufficiently high RN
    c. for a range of speeds at medium or high RN, only
    d. for a **range** of speeds at very low RN

51. What is the location of the **separation** point for an airfoil that is (fully) stalled?

52. The equal sign in Eq. 5.12 (incompressible Bernoulli equation) does not hold if there is an energy source or sink within the streamtube being described; specifically, if there is an energy sink between points #1 and #2 and point 2 is downstream of point 1, the equal sign in the equation should be replaced with a (greater than, less than) symbol.

53. When flying in rain, one can observe droplets of rain that remain in one spot on a wing strut or on the windshield. Explain how this can happen, even for flight speeds well over 100 kt.

54. The drag force that is present only when lift is being developed by a wing is called ___ drag, or sometimes drag-due-to-lift.

Problems

1. By what percent does the kinematic viscosity change in the range from SL to 30,000 ft in the StAt? Answer: 122%

2. What is the RN for the 4.95 ft wing of a Cessna 150 at standard SL conditions, traveling at 100 kt? Answer: 5.3×10^6

3. A river narrows from 80 ft to 65 ft as the depth increases by 15%. How much faster is the average speed in the narrower part of the river? Answer: 7%

4. What is the flight dynamic pressure for the Cessna of problem 2? Answer: 33.9 lb/ft^2

5. What is the total pressure for the Cessna of problem 2? Answer: 2150 lb/ft^2

6. A certain low-speed wind tunnel has a 10 ft × 10 ft square cross section that narrows to a 6 ft diameter circular test section. If the air speed in the square cross-section is 30 kt, what is the air speed in the test section? Answer: 106 kt

7. Often the air speed in a wind tunnel test section is determined by measuring the pressure difference between the test section and another part of the tunnel. Suppose that the wind tunnel in problem 6 has a measured difference of 12 lb/ft^2 between the given sections. What then is the airspeed in the test section, for SL standard air? (Hint: Solve Eq. 5.20 for V_2^2 and consider V_2 to be the speed in the test section.) Answer: 113 ft/s

8. The gyro horizon is a gyroscopic flight instrument that provides an artificial horizon reference for the pilot, thereby providing bank and pitch attitude information in clouds. The gyro horizon is operated by supplying an air pressure difference of at least 3.8" of mercury between the two connections to the instrument. This pressure difference draws air through the instrument at a sufficiently high speed to bring gyro rotational speed up to its rated value. The venturi tube used to obtain this pressure difference on old aircraft was mounted on the fuselage side and utilized an inlet diameter of about 3" and a throat diameter of about 1.25". What is the minimum (true) airspeed required to operate the gyro horizon under SL StAt conditions? at 8500 ft in the StAt? Answer: 50 kt, 56 kt

9. The turn and slip flight instrument requires a minimum pressure difference of 1.8" of mercury to operate properly. Answer the same questions as for the question #8, for this instrument. Answer: 34 kt, 39 kt

10. You've found an old airplane venturi tube like that described in the previous problem and have mounted it on the top of your roof with the hope that it will lift water out of your basement. What wind speed is needed to lift water 12 ft, under SL StAt conditions? Answer: 83 kt

11. A mountain pass can serve as nature's venturi tube. (The Student Center's plaza is KSU's venturi tube.) It has been suggested that an airplane's altimeter might have as much as a **thousand** ft error while flying through such a pass under high wind conditions. Suppose that the wind far away from a 9300 ft pass is blowing at 87 kt and the static air properties are standard for that altitude. Would the altimeter indicate a too-low or too-high altitude? Estimate the necessary speed inside the pass to give the thousand ft altimeter error. Partial answer: 172 kt

12. An aircraft is flying at 20,000 ft under StAt conditions. The flight speed is 150 kt but the maximum speed over the top of the wing is 200 kt. Estimate the maximum percentage drop in static air pressure over the wing. Answer: Only 3.2%

13. To eliminate scale effect, it is necessary to use a RN in the wind tunnel that is just equal to its value in flight. If the **size** of a test model is **less** than the full-scale article, this equalization must be accomplished by some combination of (**a**) increasing the air density to a value greater than its standard value or (**b**) increasing the wind tunnel speed over its flight value (which is usually too demanding from an energy standpoint) or (**c**) decreasing the coefficient of viscosity.

Suppose, for example, that we have a 1/12 scale model that we are able to test in a wind tunnel at the **same** speed as the expected flight speed under standard SL conditions. This means that we must use method (a) or (c) above, or both. Suppose, further, that we can cool the air in the tunnel to 35.11° F (see the StAt tables to see what viscosity corresponds to this temperature, since it depends essentially only on the temperature). How much do we have to **compress** the air in the tunnel at this temperature to provide the total compensation needed to keep the RN exactly the same and thereby eliminate scale effect (assuming low Mach numbers—see next chapter)? (i.e., what pressure do we need in the tunnel? Hint: It is easiest to work with ratios.) Express your answer in terms of the standard SL pressure. Answer: 11.0 atmospheres (so the reduction in viscosity **did** help out—the pressure would have had to have been 11.5 atmospheres if only method (a) had been used)

14. A new turboprop airliner is to cruise at 500 kt at 22,500 ft in the StAt. A test is to be carried out in a compressed-air wind tunnel in which the pressure can be increased up to 20 times the standard SL pressure while the temperature is maintained at 33.33° F by cooling. For a 1/24 scale model, what tunnel airspeed will provide the same type of air flow over the model as over the aircraft in flight, assuming compressibility effects are negligible in both cases? Answer: 307 kt (This is called **dynamic similarity.**)

15. What is the overall RN for the flat plate of Fig. 5.5, assuming SL StAt conditions? Answer: 9.54×10^5

References

1. Hunt, H. H., *Aerodynamics for Naval Aviators*, U.S. Navy, NAVAIR 00-80T-80, 1960 (Revised January, 1965). This is a basically qualitative discussion of low and high speed aerodynamics.

2. Hoerner, Sighard F., *Fluid-Dynamic Drag*, published by the author, 1965. A classic collection of drag data by the German aerodynamicist (Fieseler, Junkers, and Messerschmitt) who settled in the U.S.A. after World War II. The experimental data reported in this chapter were obtained from this source. This reference and a companion volume, *Fluid-Dynamic Lift*, are available from Hoerner Fluid Dynamics, 7528 Staunton Place NW, Albuquerque, NM 87120.

3. Talay, Theodore A., *Introduction to the Aerodynamics of Flight*, NASA SP-367 (Langley Field), NTIS N76-11043, 1975. A once-over-lightly, qualitative, readable treatment.

4. Von Kármán, Theodore, *Aerodynamics*, McGraw-Hill Book Company, Inc., 1954, 1963. Von Kármán was a Director of the Gugenheim Aeronautics Laboratory, a chairman of the Scientific Advisory Board of the USAF, and a chairman of the Advisory Group for Aeronautical Research and Development (AGARD) of the NATO alliance. Here he presents a brief, charming history of the growth in basic aerodynamic understanding up through early supersonic flight.

5. Von Kármán, Theodore, *The Wind and Beyond*, Little, Brown and Company, 1967. An aerodynamic autobiography of von Kármán, these are the personal recollections of a man who watched aerodynamics and aircraft mature over a period of fifty years. His account of Northrop's flying wing project is particularly intriguing.

6. Shapiro, Ascher H., *Shape and Flow—The Fluid Dynamics of Drag*, Doubleday & Company, Inc., published by Anchor Books, 1961. In pocket-book form, the narration and snapshots from the movie *The Fluid Dynamics of Drag* (following reference) are given; Professor Shapiro uses a low speed air source and a force balance in a discovery approach to drag phenomena at low and high RN. The development drags a little at times (pun intended) but the concepts of BL transition and pressure drag versus viscous drag do emerge nicely at the end. The fictitious "inertial force" approach is used.

7. Jones, R. T., *Modern Subsonic Aerodynamics*, Aircraft Designs, Inc., 1988. (A nice little introduction to various aspects of aerodynamics and airfoils and wings and stability, by the famous NACA/NASA aerodynamicist who is famous for his studies of swept wings; he also loves his Ercoupe. The BASIC language Oshkosh Airfoil Program is included so that any one with a personal computer can try his or her hand at airfoil design.)

8. Lugt, Hans J., *Vortex Flow in Nature and Technology*, John Wiley & Sons, 1983. (This is a beautiful description, with hundreds of figures and pictures, of rotating fluids—from airfoils to planetary atmospheres.)

9. Van Dyke, Milton, *An Album of Fluid Motion*, The Parabolic Press, 1982. (Two hundred and seventy-nine photographs of fluid flow, at high speeds and low speeds.)

10. Billah, K. Yusuf and Scanlan, Robert H., "Resonance, Tacoma Narrows bridge failure, and undergraduate physics textbooks," *Am. J. Phys.* Vol. 59, No. 2, February 1991. (The authors note that engineers currently believe that the bridge failure was caused by aerodynamically induced self-excitation or negative damping in a torsional degree of freedom, and should not be considered to be an example of resonance between the bridge and shed vortices. Some applied mathematicians, on the other hand, support a different explanation, so the case isn't closed yet.)

Films
(Available from Kent State University Film Services)
by Educational Services, Inc.,
Fluid Mechanics Series

1. *The Fluid Dynamics of Drag*, in four parts, 1960, 2 hr 10 min. *Some Curious Experiments*, *Fundamental Concepts*, *The Laws of Drag in Fluids of High and Low Viscosity*, and *How to Reduce Drag*. (C2711,C2712,C2713,C2714)

2. *Fundamentals of Boundary Layers*, 1968, 25 min. (C2715)

3. *Turbulence*, 1968, 28 minutes. (CC2729)

4. *Flow Visualization*, 1963, 30 minutes. (C2717)

Pic. 5.3 The far-aft location of the maximum thickness point on this P-51 Mustang's wing is clearly evident here.

Chapter 6

Transonic and Supersonic Fluid Flow

The diving characteristics of the P-51 are outstanding. Because of its clean-lined design, laminar-flow wing, exceptional aerodynamical characteristics, and small frontal area made possible by the single in-line engine, the P-51 outdives just about any plane built. . . . In making a high-speed dive the most important thing is to take it easy.

Since extremely high airplane speeds have been developed only in recent years, the phenomenon of compressibility is still pretty much of a mystery. . . . About all that is known for certain is this: Just as soon as an airplane approaches the speed of sound, it loses its efficiency. Compression waves or shock waves develop over the wings and other surfaces of the airplane. And the air, instead of following the contour of the airfoil, seems to split apart. It shoots off at a tangent on both the upper and lower surfaces. . . . The lift characteristics of the airplane are largely destroyed, and intense drag develops. The stability, control, and trim characteristics of the airplane are all affected. The tail buffets, or the controls stiffen, or the airplane develops uncontrollable pitching and porpoising, or uncontrollable rolling and yawing, or any combination of these effects. Each type of high-speed fighter plane has its own individual compressibility characteristics. If the speed of the airplane isn't checked and the pilot doesn't regain control of it, either the terrific vibrations of the shock wave cause structural failure or the airplane crashes while still in the compressibility dive.

In your P-51, the first effect of compressibility that you feel is a "nibbling" at the stick—the stick will occasionally jump slightly in your hand. If you don't check the airspeed, this will develop into a definite "walking" stick—the stick will "walk" back and forth and you won't be able to control it. At this stage the airplane is beginning to porpoise—that is, to pitch up and down in a violent rhythm like a porpoise. As the airplane accelerates further, the porpoising will become increasingly violent.

Once the airplane begins to porpoise, you won't be able to anticipate its porpoising movements by any counter-movements of the stick. Anything you do in this regard merely makes the situation worse. Or you may develop an aggravated case of reversibility—the control forces reverse . . . and you have to push forward on the stick in a dive to keep the airplane from pulling out too abruptly.

An airplane goes into compressibility before actually reaching the speed of sound. Some airplanes go into it when they reach 65% of the speed of sound; some when they reach 70% of the speed of sound. The percentage figure at which any particular airplane goes into compressibility is known technically as its critical Mach number. . . . The P-51 has one of the highest critical Mach numbers of any airplane now in combat. It can be dived to beyond 75% of the speed of sound before going into compressibility.

. . . It is possible to come out of compressibility safely if you don't go into it too far. The most important thing to remember about this is that while in compressibility you have virtually no control over your airplane. While in compressibility you can aggravate your situation, you can make it a lot worse. But outside of cutting off the power (if it isn't already off) and holding the stick as steady as possible, there's nothing you can do to help the situation. All you can do is ride it through until you decelerate enough and

lose altitude to the point where your speed is below the red line speed as given in the table. This usually means an uncontrolled dive of between 8,000 and 12,000 feet, depending . . . to a great deal upon the angle of dive in which you encountered compressibility.[1]

A. Compressible Fluid Flow

In this chapter we discuss the flow of a compressible fluid around a bounding surface at speeds up to and beyond the normal propagation speed (wave speed) for a pressure disturbance in that fluid.

You have probably dropped a pebble into a pond of still water and watched as a circular wave grew and spread out from the point of impact. Viscous forces in the water eventually returned the pond to its previous smoothness. Now if one very regularly touches the water with a pointed object, the waves will be periodically produced (Fig. 6.1a). The tip of the object produces a momentary depression in the surface of the water, pressure forces act to restore the fluid to its previous height, and these pressure forces cause the height disturbance to be propagated in all directions at a speed known as the **wave velocity**.

If the pointed object is moved above the surface of the water while still making periodic dips, each circular wave has a different center. Figure 6.1b shows the resulting wave pattern when the oscillating point is moved at a speed of just half the wave velocity.

If the pointed object is moved horizontally at a speed just **equal** to the wave speed, the pattern of Fig. 6.1c results. The point touches down each time at the edge of the wave created by the previous contact.

If the horizontal speed of the pointed object is **twice** the wave speed, the pattern of Fig. 6.1d results. The object touches the water after travelling a distance equal to twice the radius of the last wave it produced.

We can easily obtain a general expression for the half-vertex angle, θ (the Greek letter theta), which defines the **bow wave** which separates disturbed and undisturbed fluid for $V > V_{\mathrm{w}}$. Figure 6.2a shows the geometry when $V = 5\,V_{\mathrm{w}}$ and for a particular instant of time when the object has just reentered the water. We use the symbol Δt to represent the time interval between successive dips. (Mathematical notation: The Greek letter delta, Δ, is commonly used to denote a small change in the value of a variable, as here.) Then the disturbance just being produced is located a distance $V\Delta t$ **from** the previous disturbance center, and the previous point disturbance has grown to a circle of radius $V_{\mathrm{w}}\,\Delta t$. From the definition of the sine of an angle,

$$\sin\theta = \frac{V_{\mathrm{w}}\,\Delta t}{V\,\Delta t} = \frac{V_{\mathrm{w}}}{V} \tag{6.1}$$

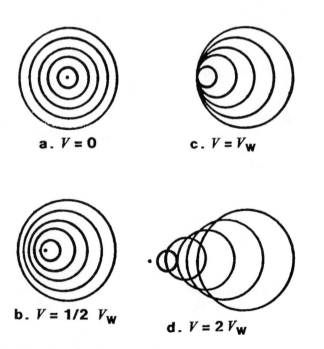

a. $V = 0$ c. $V = V_{\mathrm{w}}$

b. $V = 1/2\ V_{\mathrm{w}}$

d. $V = 2\,V_{\mathrm{w}}$

Fig. 6.1 A dipping, moving point in water (moving to the left), making waves

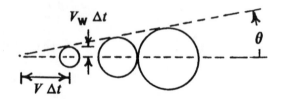

a. PERIODIC DISTURBANCE MOVED AT

$V = 5\,V_{\mathrm{w}}$

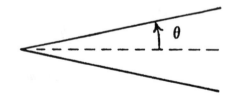

b. CONTINUOUS DISTURBANCE MOVED AT

$V = 5\,V_{\mathrm{w}}$

Fig. 6.2 A wave source moving to the left at greater than wave speed

[1]*Pilot Training Manual* (North American P-51D Mustang, August 1945).

If the pointed object is simply held in the water and moved at a speed $V = 5 V_W$, the pattern of Fig. 6.2b results.

Our discussion so far has been for two-dimensional waves on the surface of **water**, but the same principles apply to pressure disturbances in any fluid. If the fluid is air, the molecules are constantly gaining energy from collisions with the oncoming aircraft and are transmitting this energy to oncoming molecules of air when colliding with them. If the disturbing influence in air is a small, streamlined rod, the pattern of Fig. 6.2b would be pretty much a side or top view of the propagation of the pressure disturbance caused by the sharp tip of the rod as it pushed away the molecules. In three dimensions the pressure disturbances form a **cone**.

For any fluid, the vital parameter is the ratio of the object's speed to the wave speed. For air this ratio has been given the name **Mach number**, in honor of the Austrian physicist Ernst Mach (1838–1916).

$$M_\infty \equiv \frac{V_\infty}{V_A} \qquad (6.2)$$

FLIGHT MACH NUMBER

The wave speed in air is just the **speed of sound**, V_A, since V_A is the propagation speed for any pressure disturbance. The infinity subscripts are intended as a reminder that the reference point is free airstream. (Mathematical notation: This use of the "∞" subscript will be commonly employed in future pages.) In air the cone that separates the undisturbed air from the disturbed air is called the **Mach cone**; its half-vertex angle is the Mach angle, θ_M.

From Eqs. 6.1 and 6.2,

$$\sin \theta_M = \frac{1}{M_\infty} \qquad (6.3)$$

The Mach cone is the surface in space along which the oncoming air is experiencing an extremely rapid pressure rise due to the high speed motion of a point disturbance. But aircraft are hardly **point** sources of pressure disturbances! We address that complication shortly.

This section is necessary only because air **is** a compressible fluid. If air were incompressible, a pressure disturbance would have an **infinite** propagation speed because no one molecule could move without moving all the others. We have indicated already that density changes **are** negligible up to 250 kt or so, and this has allowed us to describe the operation of venturi tubes and the air flow around airfoils for this speed range. The treatment of air in motion in the previous chapter began generally enough with the derivation of the continuity equations, Eq. 5.7 for compressible streamtube flow and Eq. 5.8 for incompressible streamtube flow. The following section (Section 5D), though, presented only the incompressible form of the Bernoulli energy equation, and all of the subsequent discussions were also for incompressible fluids.

Here is the **Bernoulli equation** for **compressible** fluids, expressing again the constancy of the energy density for two points, 1 and 2, along a streamtube carrying a (compressible) fluid.

$$\left(\frac{\gamma}{\gamma-1}\right)\left(\frac{p_1}{\rho_1}\right) + \frac{1}{2}V_1^2 = \left(\frac{\gamma}{\gamma-1}\right)\left(\frac{p_2}{\rho_2}\right) + \frac{1}{2}V_2^2 \qquad (6.4)$$

BERNOULLI EQUATION FOR COMPRESSIBLE FLUID FLOW

As for the incompressible form, this relationship will not be valid for regions where energy is being transferred **into** or **out of** the fluid by an external energy source.

One revealing insight into the differences between compressible and incompressible fluid flow is obtained by using the compressible forms of the continuity and Bernoulli equations to study the venturi tube (Fig. 5.11) once again. After some messy algebra and some calculus, one finds that

Change in cross-sectional area with a change in speed

$$\propto \left(M_\infty^2 - 1\right) \qquad (6.5)$$

(Mathematical notation: The symbol "$\propto$" means "is proportional to.")

For $M_\infty < 1$, the quantity $[M_\infty^2 - 1]$ is **negative**, which means that if we wish to **increase** the speed of a fluid, the cross-sectional area of the tube must be decreased (which is reasonable enough). Thus we found that the greatest speed in the venturi tube occurred at the narrowest point, for incompressible flow.

For $M_\infty > 1$, however, the quantity $[M_\infty^2 - 1]$ is **positive**, meaning that supersonic speeds can be obtained in the venturi tube (or wind tunnel!) only where the cross-sectional area is **increasing**!

In a subsonic tunnel, the narrowest part is called the **throat** section and corresponds to the narrowest part in the venturi tube; it is in this throat section that test models and test airfoils are suspended. The air velocity in this test section at first increases as the power output of the electric motors driving the fans is increased. In accordance with the discussion above, though, the airspeed never will go supersonic in this throat section. When Mach 1 is (nearly) achieved, the tunnel is said to be "**choked**." What is happening is that the density of the air in the throat is increasing so as to just compensate for any additional mass flow provided by the propellers, rather than increasing the speed, when the tunnel is choked. Physically, we can see that additional air pressure before the throat cannot speed up the air flow beyond Mach 1 because pressure waves to do this cannot be transmitted into this region when the speed there becomes supersonic! With sufficiently powerful motors, however, the airspeed will become supersonic in the **diverging** part after the throat, and this area becomes the working section for a supersonic tunnel. Rocket nozzles are usually designed as convergent-divergent nozzles so that supersonic flow is obtained before the rocket gases leave the nozzle. Such a nozzle is often called a **de Laval nozzle** in honor of this man who first tried to patent the idea.

B. Compressible Flow over Airfoils

Next we examine the density changes that occur for two-dimensional air flow past a wing section. At a Mach number of 0.3, the true dynamic pressure $[p_T - p_\infty]$ is about 2% greater than $\frac{1}{2}\rho V_\infty^2$, and so compressibility effects are significant only above that Mach number. Looking at the V_A speeds in the StAt (Appendix B), note that this means that compressibility effects are significant for speeds greater than about **200 kt** at the standard **SL** temperature of 59° F, but compressibility effects are significant for speeds greater than only about **185 kt** at the standard **20,000 ft** temperature of –12° F.

The **pilot** doesn't notice compressibility effects (except perhaps indirectly) until much higher speeds than $M_\infty = 0.3$ are reached. The (local) air flow over the aircraft will reach Mach 1 **before** the flight Mach number, M_∞, reaches 1, because of the speeding up of the air flow over the curves of wing, fuselage, and tail. The flight Mach number at which **local** flow somewhere around the airplane first becomes supersonic is known as the aircraft's **critical Mach number**, M_{CR} (Fig. 6.3a). At flight speeds somewhat beyond M_{CR}, the pilot of an aircraft not designed for these speeds **will** notice compressibility effects because radical changes to stability and maneuverability begin to occur, as P-51 pilots were warned to expect. The value of M_{CR} for a particular aircraft depends strongly on the airfoil sections used and their thicknesses; current subsonic jet transports have a M_{CR} of about 0.75 to 0.85.

The pressure decreases as the air moves back from the leading edge and this pressure gradient accomplishes the task of speeding up the air to supersonic speeds; it is the push from behind that does the job, so to speak. But how is the air to know that a slow zone is dead ahead? Once the air passes the speed of sound there is **no** way for a higher pressure region to influence air flow before it, to gradually slow down the air flow. The result is a narrow zone, a **normal** shock wave, in which the air abruptly transitions from supersonic to subsonic. In this normal shock wave, the air pressure, the air density, and the air temperature all abruptly **increase**. The changes are so abrupt that they are said to be discontinuous.

The normal shock wave gets its name from the fact that it is at right angles (90°) to the direction of air flow, a direction of motion which isn't changed in passage through the shock. The fluid speed behind a normal shock wave is always **subsonic**. The shock wave extends from the outer part of the BL (within which the speed of the fluid at the surface is still zero) to the end of the region where the flow has become supersonic (Fig. 6.3b). The interaction of the shock wave with the critical BL is particularly significant in determining the reaction of the airfoil and the airplane to high speeds.

At some particular M_∞ greater than M_{CR}, the strong **adverse pressure gradient** (the increasing pressure) through the normal shock wave causes the BL to **separate**, and this is what really gets the pilot's attention. Recall that the BL separation which occurs at high angles of attack at low speeds is called the

"stall" of the wing or airfoil. Analogously, this shock-induced separation is often called the **shock stall**. The shock stall also produces lower pressures on the rear part of the airfoil so that pressure drag and total drag begin a sharp increase, just as for the low speed stall; the overall upper-surface pressure increases so that the lift of the airfoil is suddenly reduced, too, just as for the low speed stall. The Mach number at which this occurs is generally about 5% or 10% higher than M_{CR} and is known as the **drag divergence Mach number**, M_D.

Saunter out to peer carefully at the wing of a Lear business jet or Boeing 727 the next chance you get. You will find small metal tabs about an inch in height, located in front of the ailerons and inclined at a distinct angle to the direction of flight. These are the vortex generators that were mentioned in the first chapter. Now we can see that their purpose is to add vortices with rotational energy and these vortices **invigorate** the BL at speeds past M_{CR} so that shock-induced separation is significantly reduced. Vortex generators are often specifically placed in front of ailerons to prevent the rapid oscillations of the separated shock wave that appear to the pilot as **aileron buzz**. (Vortex generators are now used in a number of aircraft for low speed boundary layer control as well, as will be discussed later.) Incidentally, the early Learjets are also an example of a jet that has a maximum operating Mach number (0.81) that is greater than its M_D; its maximum range cruise Mach number is 0.70, however.

Whether a shock stall occurs or not, the control surfaces lose effectiveness at supersonic and near-supersonic speeds because they can no longer affect the air flow ahead of them. This is the reasoning that led to the development of the all-moving horizontal tail surface for supersonic aircraft (the slab tail or **stabilator**) as a replacement for the fixed horizontal stabilizer plus elevator. The use of a stabilator on low speed aircraft has generated mixed reviews, as will be discussed later.

At slightly higher speeds the normal shock wave on the top surface (where flow speeds are the greatest in level flight) will be joined by a normal shock on the lower surface, as a region of supersonic flow appears there also (Fig. 6.3c). Both shock waves then move toward the trailing edge as M_∞ approaches one (Fig. 6.3d). The supersonic region has grown a great deal and now nearly covers both surfaces. BL separation is considerably **reduced** from its value at M_D.

When the aircraft "goes" supersonic, the air flow away from the airfoil is all supersonic (Fig 6.3d). There is a region ahead of the wing where the air must go subsonic (and even to zero at the leading edge stagnation point); the BL remains subsonic because of the no-slip condition at the surface; and, finally, there is a small region behind the wing section which is subsonic. The front subsonic region borders the nose of the **bow wave**. (This bow wave is completely analogous to the familiar bow wave which forms in front of a boat that is moving at a speed greater than the water wave speed.) The bow wave may be a nearly normal shock wave if the leading edge is blunt, but the shock always bends back as an **oblique** shock wave away from the center of the wave. In an oblique shock wave the flow

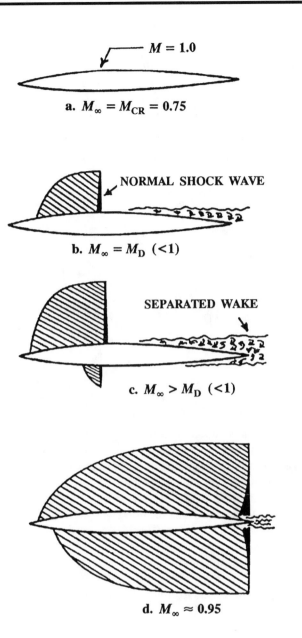

a. $M_\infty = M_{CR} = 0.75$

$M = 1.0$

NORMAL SHOCK WAVE

b. $M_\infty = M_D$ (<1)

SEPARATED WAKE

c. $M_\infty > M_D$ (<1)

d. $M_\infty \approx 0.95$

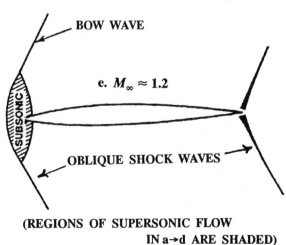

BOW WAVE

e. $M_\infty \approx 1.2$

SUBSONIC

OBLIQUE SHOCK WAVES

(REGIONS OF SUPERSONIC FLOW
IN a→d ARE SHADED)

Fig. 6.3 Compressible flow over a lifting airfoil at transonic to supersonic speeds

remains supersonic but changes its **direction** of flow. This same change in flow direction must occur at the trailing edge and so an oblique wave is also present there.

A thin airfoil reduces the amount that the air speeds up over the top surface compared to a thick airfoil. Therefore M_{CR} **increases** as the relative thickness of the airfoil **decreases**. M_{CR} and M_D can never reach one, of course, because some air flow speedup is necessary to obtain lift for level flight. A great deal of effort in recent years has gone into the design of airfoil shapes that increase M_D for a given airfoil thickness. These designs try to tailor the pressure distribution above the wing so that the local airspeed is more gradually increased; when the shock wave forms, it forms farther back from the leading edge and is weaker than for the airfoils previously used. These new airfoils have come to be called **supercritical** airfoils. (At one time there was a move afoot to use some other term than "supercritical airfoil," apparently because it was feared that passengers would be loathe to fly on a wing that was close to "critical." It appears, though, that the term is too well established to be displaced. NASA's Richard Whitcomb, well known for his work in this area, is credited with coming up with the term.)

An airfoil that has been used at supersonic speeds, especially in missiles that don't have to worry about landing, is the double-wedge airfoil. Figure 6.4a shows the zero-lift wave pattern at about Mach 1.2. The very sharp leading edge acts much like the point disturbance discussed at the beginning of this section and the oblique bow wave is **attached** to the leading edge. The supersonic flow has to speed up to follow the sharp middle point of the airfoil but it is able to do this without separation and without producing any sudden changes in air properties. This change in flow direction without the formation of a shock wave is called an **expansion wave**.

Both oblique waves bend farther back as the speed increases, as expected from the behavior of Mach waves, but then they approach a constant angle relative to the surface at high Mach numbers (Fig. 6.4b).

The distribution of the pressure forces for the double-wedge airfoil at $M_\infty = 1.2$ and for zero lift (pointing directly into the oncoming air) is shown in Fig. 6.4c. Notice that the pressures on the front half are greater than p_∞ and the pressures behind that are less. Evidently there is considerable pressure drag even when no lift is being developed and when no BL separation is occurring. This drag force is called **wave drag**. Wavemakers always have a bill to pay; shock waves represent fluid energy that has been gained through an interaction with the wing. On a local level, we can see that the force which does the work to deliver this fluid energy also tends to reduce the relative speed between the airfoil and the fluid, and so this is just another contributor to the total drag force on the wing. The strong shock waves produced by blunt, thick airfoils cause a great deal more wave drag than do the weak shock waves produced by sharp-edged, thin airfoils.

When this double-wedge airfoil is developing lift, it produces wave patterns like those sketched in Fig. 6.5. Notice that the resultant force of the air on the wedge acts at the **center**

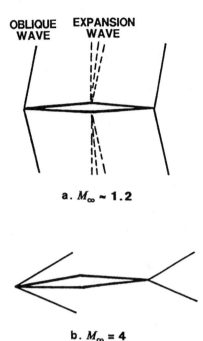

OBLIQUE   EXPANSION
WAVE   WAVE

a. $M_\infty \sim 1.2$

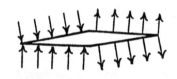

b. $M_\infty = 4$

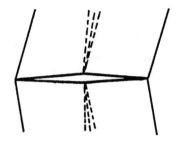

a. $M_\infty \sim 1.2$

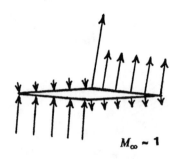

$M_\infty \sim 1$

Fig. 6.5 Shock patterns and pressure distribution on a double-wedge airfoil at supersonic speeds (developing lift)

c. PRESSURE FORCE DISTRIBUTION
FOR $M_\infty = 1.2$

Fig. 6.4 Shock patterns and pressure distribution on a double-wedge airfoil at supersonic speeds (no lift)

of the airfoil (just as for the early Newtonian theory) and that this resultant force is tilted back from the perpendicular to the direction of flight. Therefore there is a component of this resultant aerodynamic force that acts in the **rearward** direction. This is a drag force that is present only when lift is being developed by the airfoil, and this lift-induced drag force turns out to be present for all types of wings at all speeds. This is one more addition to our list of drag forces and is called **drag-due-to-lift** or **induced** drag. (The other current members of our drag list are viscous or skin friction drag, separation-induced pressure drag, and wave drag.) Induced drag won't be heard of again until Chapter 12 deals with wing theory.

C. The Heat Barrier

In compressible flow, the slowing of the fluid produces potentially large temperature increases. This is the **heat barrier** that has replaced the sound barrier as the primary obstacle to ever faster flight. The heated air raises the temperature of the

airframe to such high temperatures that aluminum loses some of its strength and more exotic materials such as titanium must be used for critical components, especially leading edges. We can obtain an interesting insight into this problem by applying the compressible form of the Bernoulli equation (Eq. 6.4) to air flow that is brought to a halt; the flow energy is transformed into heat energy at such a point. Consider point #1 to be the free airstream and point #2 to be a stagnation point ($V_2 = 0$). The stagnation point temperature is also referred to as the **total temperature**, T_T, in analogy with the definition of total pressure (Section 5D). We also substitute for (p/ρ) from the ideal gas law (Eq. 4.5). Then Eq. 6.4 becomes

$$\left(\frac{\gamma}{\gamma - 1}\right) k\, T_\infty + \frac{1}{2} V_\infty^2 = \left(\frac{\gamma}{\gamma - 1}\right) k T_T \qquad (6.6)$$

Solving for the stagnation temperature,

$$T_T = T_\infty + \frac{1}{2}\left(\frac{\gamma - 1}{\gamma\, k}\right)\left(V_\infty\right)^2 \qquad (6.7)$$

Substituting for V_∞ from the definition of the flight Mach number (Eq. 6.2) transforms this into

$$T_T = T_\infty + \frac{1}{2}\left(\frac{\gamma - 1}{\gamma\, k}\right)\left(M_\infty\, V_A\right)^2 \qquad (6.8)$$

A further substitution from Eq. 4.9 yields

$$T_T = T_\infty + \frac{1}{2}\left(\gamma - 1\right) M_\infty^2\, T_\infty \qquad (6.9)$$

For air the specific heat ratio, γ, is equal to 1.40, so this equation becomes

$$T_T = \left(1 + 0.2\, M_\infty^2\right) T_\infty \qquad (6.10)$$

where the temperatures are absolute temperatures.

The stagnation temperature represents an **upper limit** for the surface temperature of an aircraft in high speed flight.

EXAMPLE 6.1. What is the stagnation or total temperature for Mach 2 flight at 30,000 ft in the StAt?

Solution: From the StAt table, T_∞ = 411.9 R

from Eq. 6.10, $T_T = [1 + (0.2)(2)^2](411.9) = 741.4$ R

Thus the **increase** in air temperature is
$$\Delta T = (741.4 - 411.9) = \textbf{329.5° F!}$$

Certain terms have arisen to describe the various regimes of flight. **Transonic** flight[2] refers to flight speeds for which significant amounts of **both** subsonic and supersonic flow occur; conventionally, this speed regime is said to extend from Mach 0.8 to Mach 1.2, although it varies considerably for different aircraft. Transonic flow for a particular aircraft begins when M_{CR} is reached. It ends when the normal shock wave has moved well aft of the trailing edge and no longer affects BL flow. (However, a feisty pilot can easily induce a normal shock wave at much lower Mach numbers by engaging in flight maneuvers that require lift much greater than the weight — e.g., dogfighting. For civilian flying, the greater hazard when operating close to M_{CR} is air turbulence, especially clear air turbulence.)

Supersonic flight is said to exist from the end of the transonic speed regime to a rather arbitrary Mach 5, where **hypersonic** flight is said to begin. Hypersonic flight begins at the speeds where the problem of heating dominates aircraft design, and it includes the speeds where there may be more BL interaction problems as the oblique bow wave bends closer to the airfoil and fuselage. The "heat barrier" is particularly in evidence at hypersonic speeds because the total temperature varies as the **square** of the Mach number. Figure 6.6 summarizes these speed regime definitions.

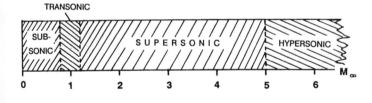

Fig. 6.6 How to describe your speeds to others

Summary

This chapter has dealt with the complications produced by the compressible nature of air. Compressibility produces differences from the predictions of the incompressible theory for

---

Pic. 6.1 The Concorde at the 1988 EAA Oshkosh, Wisconsin, convention

speeds of about 200 kt or greater, but the much larger and conceptually distinct effects are those that occur at speeds close to the fluid wave speed (which is the speed of sound if air is the fluid). Air that has speeded up to supersonic speed above a curved airfoil can receive no advance warning when it comes time to slow up to the overall subsonic speed. The result is a normal shock wave, a thin region perpendicular to the air flow direction in which the airspeed abruptly decreases and the static properties all increase in value. BL separation tends to be triggered by the large adverse pressure gradient in the shock wave so the pressure drag becomes large even for a streamlined airfoil pointing into the oncoming air; this is the **shock stall**. Artificially induced turbulence, such as via **vortex generators**, reduces separation and flow instability at the shock stall. At supersonic speeds a **bow wave**, composed of a normal shock wave directly in front of the leading edge and transitioning into oblique waves farther from the leading edge, forms in front of the airfoil; at the same time, expansion waves form where the flow changes direction around the airfoil and oblique waves form behind the airfoil. All of these waves obtain their energy from the airfoil (wing) and so their presence indicates a **wave drag** force that exists even when no lift is being developed. When lift is developed, the net aerodynamic force on the airfoil is always tilted back, indicating the presence of a (lift-)**induced** drag force.

Local and **flight Mach numbers** were defined; these predict the position and intensity of shock waves much as the Reynolds number predicts transition from laminar to turbulent fluid flow. Mach waves are oblique waves produced by a point disturbance traveling at speeds greater than the wave speed. A sharp-edged, thin airfoil at supersonic speeds produces a bow wave that is very much like a Mach wave.

The compressible forms of the continuity and Bernoulli equations predict correctly that supersonic speeds cannot be

---

[2]It appears that von Kármán and a colleague devised the word *transonic*; he reports that, for aesthetic reasons, he successfully argued for a single *s* in the word.

achieved in a converging streamtube (or in the throat of a venturi tube) but only in a diverging part that follows. The equations also predict a stagnation temperature that gives an idea of the maximum temperature on a wing or on an airframe at supersonic speeds.

M_{CR} is the flight Mach number at which the maximum local Mach number first reaches 1. M_D is the flight Mach number at which shock stall first produces a diverging (rapidly increasing) drag curve. The new **supercritical** airfoils succeed in pushing M_D closer to 1.

Symbol Table (in order of introduction)

V	speed of an object in a fluid
V_W	the wave speed in a given fluid
θ	half-vertex angle formed by a point object moving at $V > V_W$
M	local Mach number or flight Mach number
M_∞	flight Mach number
θ_M	Mach angle (half-vertex angle for a Mach wave)
M_{CR}	critical Mach number, when local fluid flow first becomes supersonic
M_D	drag divergence Mach number, where drag increases rapidly due to shock wave formation and the separation it causes
T_T	total temperature, the temperature at a stagnation point

Review Questions

1. A Mach wave and a Mach cone are formed when a point disturbance moves through the air at a speed ___.

2. It is because air is ___ that the wave speed is finite and a shock wave forms soon after the maximum local ___ ___ exceeds 1.

3. In a converging-diverging tube, the fluid won't exceed the pressure propagation (wave) speed in the converging part because the fluid ___ forces which otherwise can speed up the fluid are not able to affect the oncoming fluid; instead an increased pressure at the entrance to the tube produces an increase in the ___ of the fluid.

4. The critical Mach number for an airfoil is always (**a**) less than 1, (**b**) equal to 1, (**c**) greater than 1.

5. A ___ shock wave is a thin region in space in which the fluid abruptly (increases, decreases) its speed and in which the flow direction (changes, doesn't change).

6. The airspeed behind this kind of a shock wave is (**a**) always subsonic, (**b**) sometimes subsonic and sometimes supersonic, (**c**) always supersonic.

7. The **pilot** is most likely to first notice compressibility effects when the flight Mach number is equal to (**a**) 1, (**b**) M_{CR}, (**c**) M_D.

8. An expansion wave (which doesn't contain abrupt changes in fluid pressure or density or speed) is formed when fluid travelling at speeds (less than, equal to, greater than) the wave speed changes its **direction**.

9. The turbulent wake behind an airfoil is (greater, about the same, less) for $M_\infty = 1$ compared to $M_\infty = M_D$.

10. ___ ___ artificially induce turbulence into the BL and thereby reduce shock-induced (or, at low speeds, high angle-of-attack) separation.

11. This shock-induced separation is often called the ___ stall.

12. The wave in **front** of an airfoil that is the boundary between disturbed and undisturbed air at $M_\infty > 1$ is called the ___ wave.

13. In an ___ shock wave, the fluid changes direction and its speed decreases, but the speed doesn't ever decrease to less than the wave speed.

14. Airfoils designed to move the first appearance of a normal shock wave farther back on the airfoil are called ___ airfoils.

15. The result is that these airfoils (previous question), for a given thickness, increase ___ to a value closer to one.

16. An **attached** bow wave is characteristic of a (blunt, sharp) leading edge for an airfoil for M_∞ (less than, equal to, greater than) one.

17. The energy necessary to produce shock waves manifests itself as another type of drag on the wing, a drag called ___ drag.

18. Large-scale fluid energy is transformed into small-scale (molecular) ___ energy at a stagnation point; this is equivalent to stating that the stagnation temperature is always greater than the ambient temperature.

19. ___ flight involves a range of M_∞ from about 0.8 to about 1.2; ___ flight continues up to a M_∞ of about ___; ___ flight is flight at even higher speeds.

20. Attempts to break the world's speed record for propeller-driven aircraft are commonly made on hot days rather than on the cooler days on which the engine would be happier and probably deliver more power. Why?

Problems

1. True dynamic similarity (Problem 14 in Chapter 5) certainly includes identical Mach numbers, especially as Mach 1 is approached and exceeded. Compare the flight Mach numbers for the aircraft and the model of that problem to see if the air flow really will be identical. Answer: 0.82 versus 0.48 (no!)

2. Calculate the Mach angle for flight at the borderline between supersonic and hypersonic flight. Answer: 11.5°

3. What is the vertex angle for the bow wave formed by a sail boat traveling at 7 kt if the wave speed is 3 kt? Answer: 51°

4. Estimate the maximum temperature increase of the air for the Cessna 150 of Problems 5.2 and 5.4. Answer: 2.4° F

5. Estimate the maximum temperature **increase** of the air for a Boeing 727 which is cruising at an optimum long-range cruise speed of 402 kt at 28,000 ft in the StAt. Answer: 38.3° F

6. The standard temperature at the 60,000 ft cruise altitude of the English/French Concorde airliner is about –70° F. What is the total temperature for the Concorde's Mach 2.0 cruise speed? Answer: 242° F (Mach 2.2 is about the upper limit for an aluminum airplane and thus both the Concorde and the Russian SST have design cruise speeds around Mach 2. Actual maximum skin temperatures on the Concorde are reported as being about 260° F, at which temperature the fuselage lengthens by about 9 or 10 inches!)

7. How much faster must an aircraft fly at 25,300 ft if it is to achieve the same flight dynamic pressure achieved at SL, under StAt conditions? Answer: 50%

8. What is the flight Mach number for the Boeing of Problem 5?

9. Estimate the air temperature at the leading edge of the wing of an aircraft that is just barely hypersonic at 26,000 ft in the StAt. Answer: 2097° F!

References

Books

1. Nixon, David (editor), *Transonic Aerodynamics*, Progress in Astronautics and Aeronautics, Vol. 81, 1982.

2. Ower, E., and Nayler, J. L., *High Speed Flight*, Hutchinson & Co., 1958.

3. Hallion, Richard, *Supersonic Flight: The Story of the Bell X-1 and Douglas D-558*, The Macmillan Company, 1972.

4. Henshaw, J. T. (editor), *Supersonic Engineering*, William Heinemann Ltd., 1962.

5. Sedden J., and Goldsmith, E. L., *Intake Aerodynamics*, American Institute of Aeronautics and Astronautics, Inc., 1985.

6. Walker, Jearl, "The Amateur Scientist: Shock front phenomena and other oddities to entertain a bored airline passenger," *Scientific American*, September, 1988, pp. 132–135.

Films
Available from Kent State University Film Services

1. Educational Services, Inc., Fluid Mechanics Series
   Channel Flow of a Compressible Fluid, 1965, 29 min. Supersonic flow, normal shock waves, Mach waves, choking.
2. Shell Oil Company
   a. *Approaching the Speed of Sound*, CC4270, 27 min. (very nice flow visualization, good flight shots)
   b. *Transonic Flight*, BC4293, 20 min. (very nice flow visualization, good flight shots)
   c. *Beyond the Speed of Sound*, BC4280 (obsolete)

Pic. 6.2 The supersonic Rockwell International B-1B bomber with its wings swept back

Pic. 6.3 The B-1B with its wings in the forward position

Pic. 6.4 The nose gear is just coming up on this VariEze as it begins its climbout after takeoff.

Pic. 6.5 The Taylorcraft is a classic lightplane that has been in production for about 50 years; it uses the NACA 23012 airfoil.

Chapter 7

Airspeeds

My gloves are at work again, leveling the airplane at 33,000 feet. Throttle comes back under the left glove until the engine tachometer shows 94 percent rpm. The thumb of the right glove touches the trim button on the control stick once again, quickly, forward. The eyes flick from instrument to instrument, and all is in order. Fuel flow is 2,500 pounds per hour. Mach needle is resting over 0.8. . . . There is a circular computer in the clipboard strapped to my leg that tells me that the indicated airspeed of 265 knots is actually moving my airplane over the land between Abbeville and Laon at a speed of 465 knots.[1]

A. The Pitot Tube and Airspeed Determination

Ever since Chapter 1 we have spoken of an airplane's flight or true airspeed, V_∞, as if it were a simple matter to determine. It isn't. In this chapter we make true airspeed computations for subsonic and supersonic flight conditions.

Henri Pitot invented the **pitot tube** (pronounced peé toe) in 1732 for the purpose of measuring the speed at varying depths in a French river. By the second decade of powered flight, his device had become the standard airspeed instrument for aircraft. The aeronautical form of the pitot tube as a speed meter is sketched in Fig. 7.1. Notice that this airspeed indicator is nothing more than a differential manometer sensing the **difference** between the total air pressure and the static air pressure, $p_T - p_\infty$. The total pressure is the stagnation pressure existing within the pitot tube itself as it points into the oncoming air. The static pressure is measured by a second tube which

connects to openings that are perpendicular to the direction of the oncoming air. If the static source is a second tube that is attached to, or part of, the pitot tube, then the combination is known as a pitot-static tube, as shown in the figure. More often than not, though, the static pressure is obtained from small openings or ports on opposite sides of the fuselage. (These static ports need to be on both sides of a static tube or on both sides of the fuselage so as to minimize reading errors during asymmetrical – side slip – flight paths.)

So we see that the airspeed indicator no more measures airspeed than the altimeter measures altitude; they are **both** pressure-measuring devices and the pressures they measure are only partially determined by the property that labels them!

The precise relation of this differential pressure to the flight or true airspeed depends on the Mach number. Consider first incompressible flow $\left(M_\infty \leq 0.3\right)$ and the incompressible form of Bernoulli's equation (Eq. 5.12).

$$p_1 + \frac{1}{2}\rho\, V_1^2 = p_2 + \frac{1}{2}\rho\, V_2^2 \qquad (7.1)$$

Consider point #1 to be in the free airstream and point #2 to be a stagnation point ($V_2 = 0$, $p_2 = p_T$) within the pitot tube. Then

$$p_\infty + \frac{1}{2}\rho V_1^2 = p_T \qquad (7.2)$$

and

$$\boxed{\begin{array}{c} p_T - p_\infty = \text{total pressure} - \text{static pressure} \\ = \frac{1}{2}\rho V_1^2 \qquad \left(= q_\infty\right) \end{array}} \qquad (7.3)$$

PROPERTY MEASURED BY THE AIRSPEED INDICATOR

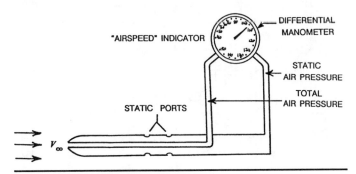

Fig. 7.1 Simplified geometry of a pitot-static tube and an airspeed indicator

---

[1]Richard Bach, *Stranger to the Ground* (New York: Harper & Row, 1963).

We see that the "airspeed indicator" at low speeds is measuring the **product** of the air density and the **square** of the true airspeed—i.e., the **dynamic pressure**. However, the indicator always has **speed** units marked on it. (If it were an honest instrument, it would have **pressure** units marked on it!) To use how this speed marking might be possible, we solve Eq. 7.3 for the flight speed and obtain

$$V_\infty = \sqrt{\frac{2(p_T - p_\infty)}{\rho}} \qquad (M_\infty \le 0.3) \qquad (7.4)$$

However, the airspeed indicator only measures the quantity $(p_T - p_\infty)$. Therefore it must be calibrated by **assuming** some value for the density of the air; sea level standard density, ρ_∞, is a natural choice. The **calibrated** airspeed is then

$$V_{CAL} = \sqrt{\frac{2(p_T - p_\infty)}{\rho_0}} \qquad (M_\infty \le 0.3) \qquad (7.5)$$

At higher Mach numbers, the compressible form of the Bernoulli equation must be used. (Also, an assumption must be made regarding the process by which the air is slowed to zero speed.) The result of such an analysis is

$$V_\infty = \sqrt{5}\, V_A \left(\left(\frac{p_T - p_\infty}{p_\infty} + 1 \right)^{2/7} - 1 \right)^{1/2} \quad (M_\infty \le 1) \quad (7.6)$$

in which we have assumed that $\gamma = 1.40$.

We see that the **indicated** flight speed at higher speeds depends on both the **speed of sound**, V_A, and the static pressure, p_∞, as well as depending on the quantity measured by the pitot-static tube, $(p_T - p_\infty)$. Therefore high speed airspeed indicators are calibrated by **assuming** sea level standard sound speed, V_{A0}, **and** sea level standard pressure, p_0.

$$V_{CAL} = \sqrt{5}\, V_{A0} \left(\left(\frac{p_T - p_\infty}{p_0} + 1 \right)^{2/7} - 1 \right)^{1/2} \quad (M_\infty \le 1) \quad (7.7)$$

The calculation of flight airspeed (**true** airspeed, V_∞) from a speed read from an airspeed indicator calibrated according to Eq. 7.7 is accomplished in two steps. The first step is to correct for compressibility and the second step is to correct for non-standard (non-sea level) air density. For incompressible flow conditions and non-standard air density, Eqs. 7.4 and 7.5 tell us that the second step correction can be made by multiplying the calibrated airspeed by the factor

$$\sqrt{\frac{\rho_0}{\rho_\infty}}$$. We make this a general second-step correction for all airspeeds and give the name **equivalent airspeed** to the airspeed that is corrected by this factor to yield the true or flight airspeed.

$$V_E \equiv V_\infty \sqrt{\frac{\rho_\infty}{\rho_0}} \qquad (7.8)$$

DEFINITION OF EQUIVALENT AIRSPEED

The word **equivalent** is appropriate here in the sense that it is the true airspeed under sea-level density conditions. Also, if one knows the equivalent airspeed of an aircraft, one also knows the flight dynamic pressure, **for**

$$q_\infty = \frac{1}{2}\rho V_\infty^2 = \frac{1}{2}\rho_0 V_E^2 \qquad (7.9)$$

where Eqs. 5.13 and 7.8 have been combined. This is a key relationship! Although V_∞ is the most important airspeed for navigational purposes, knowledge of the air density is also necessary before the (low speed) flight dynamic pressure can be determined. However q_∞ can be calculated just from knowledge of V_E since ρ_0 is a known constant (SL air density). **Therefore V_E is the most important speed *aerodynamically*.** (It is very important to note that, at low subsonic speeds, as is shown below, $V_E \cong V_{IND}$, the indicated airspeed.)

Now we can work backwards and determine that first-step correction, for compressibility. At low speeds the compressibility of the air is negligible and V_{CAL} is equal to V_E. At high speeds, though, the correct relationship is

$$V_E = V_{CAL} \sqrt{\frac{\rho_\infty}{\rho_0}} \left(\frac{\left(\frac{p_T}{p_\infty} \right)^{2/7} - 1}{\left(\frac{p_T}{p_0} \right)^{2/7} - 1} \right)^{1/2} \qquad (7.10)$$

Equation 7.10 never wins user-friendly prizes, so Fig. 7.2 presents this compressibility correction in graphical form. Note that the correction depends on both p_∞ (or, equivalently, on pressure altitude) and V_{CAL}. Note also that there is no correc-

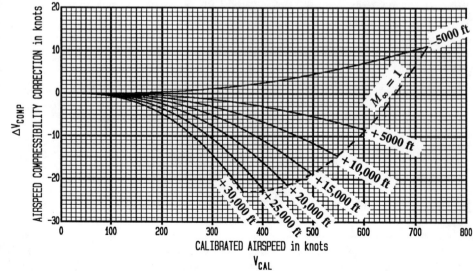

Fig. 7.2 Compressibility correction from Eq. 7.10: The equivalent airspeed is the calibrated airspeed plus the (positive or negative) correction.

tion if the pressure has its standard sea level value, as an inspection of Eq. 7.10 also reveals. Finally, note that the correction is negligible at low airspeeds, but that the real criterion is Mach number because even relatively low calibrated airspeeds require correction at high altitudes where the speed of sound is much less than its value at sea level. The dashed envelope marked $M_\infty = 1$ is the equivalent airspeed at which sonic flight speed is reached under the assumed sea-level density condition.

One last correction. The airspeed indicator isn't a perfect differential manometer; there will usually be some **instrument error** (i.e., an imperfect measurement of the differential pressure). But an even greater problem is the accurate measurement of the free airstream static pressure, p_∞, leading to what is called **position error**. A lot of engineering time has been spent trying to determine the best shape and best location for the static ports, but extreme flight conditions—especially low speeds—can result in significant errors. The magnitude of these errors is determined experimentally during the aircraft certification process by using a long pitot tube probe that is well away from the flow field of the airplane, either by using a location well under the aircraft or well in front of it. These corrections are then published in the flight manual for the aircraft.

The reading on the airspeed indicator is referred to as the **indicated airspeed**, V_{IND}, and the calibrated airspeed is determined by reference to this published correction table. Figure 7.3 gives typical instrument and position error correction information; it represents the published correction for the 1977 Cessna 182Q up to 160 kt, but is arbitrarily extrapolated to higher speeds for our instructional edification. Note that the landing configuration (just the addition of full flaps for this aircraft) disturbs the flow field around the static probe sufficiently that large errors are induced at speeds close to the minimum stall speed. Pilots who brag about how slowly their aircraft can fly ought to take a second look.

The flight or true airspeed is calculated from the indicated airspeed in this sequence:

1. $V_{CAL} = V_{IND}$ + correction for instrument and position errors
2. $V_E = V_{CAL}$ + compressibility correction
3. $V_\infty = V_E$ + density correction

(Note that the mnemonic **ICE T** is a handy aid in remembering this sequence.)

In equation form the steps in this computation can be summarized by

$$V_\infty = \left(V_{IND} + \Delta V_{POS} + \Delta V_{COMP}\right)\sqrt{\frac{\rho_0}{\rho_\infty}} \quad (7.11)$$

**CALCULATION OF THE TRUE AIRSPEED
FROM THE INDICATED AIRSPEED**

where ΔV_{POS} is the correction for instrument and position error and ΔV_{COMP} is the compressibility correction.

Note that Eq. 7.11 can be used with **any** desired unit of speed, so long as the unit is used consistently throughout, because the correction factor, the density ratio, is dimensionless.

EXAMPLE 7.1. A C-182Q is cruising along at a pressure altitude of 10,000 feet at an indicated airspeed of 120 kt. The outside air temperature is 65°F. What is the true airspeed of this aircraft?

Solution: From Fig. 7.3, $\Delta V_{POS} = -3$ kt, so

$$V_{CAL} = 120 \text{ kt} + (-3 \text{ kt}) = 117 \text{ kt}$$

From Fig. 7.2, $\Delta V_{COMP} = 0$, so $V_E = 117$ kt
From the StAt table and the ideal gas law (Eq. 4.5),

$$\rho_\infty = \frac{p}{kT} = \frac{1455.6}{(1716)(65+460)} = 0.001616 \frac{\text{sl}}{\text{ft}^3}$$

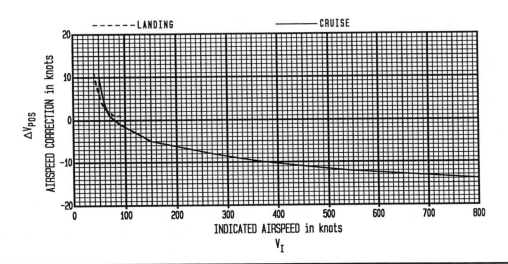

Fig. 7.3 An example of airspeed indicator errors due to instrument and position errors

From Eq. 7.8,

$$V_\infty = V_E \sqrt{\frac{\rho_0}{\rho_\infty}} = 117 \text{ kt} \sqrt{\frac{0.002377}{0.001616}} = \mathbf{142 \text{ kt}}$$

EXAMPLE 7.2. A Learjet is cruising at an indicated airspeed of 310 kt at a pressure altitude of 25,000 feet. The outside air temperature is 0°F. Compute the flight airspeed, assuming that Fig. 7.3 applies to this aircraft's airspeed indicator. Also determine the flight Mach number.

Solution:

$$\Delta V_{POS} = -9 \text{ kt} \text{ so } V_{CAL} = 310 \text{ kt} + (-9 \text{ kt}) = 301 \text{ kt}$$

$$\Delta V_{COMP} = -12 \text{ kt} \text{ so } V_E = 301 \text{ kt} + (-12 \text{ kt}) = 289 \text{ kt}$$

$$\rho_\infty = \frac{p}{kT} = \frac{786.3}{(1716)(0+460°)} = 0.0009961 \frac{\text{sl}}{\text{ft}^3}$$

$$V_\infty = 289 \text{ kt} \sqrt{\frac{0.002377}{0.0009961}} = \mathbf{446 \text{ kt}}$$

From Eq. 4.9,

$$V_A = \sqrt{\gamma k T_\infty} = \sqrt{(1.40)(1716)(460)} = 1051 \text{ ft/s}$$

Using the conversion factor from Sec. 5B,

$$V_A = (1051 \text{ ft/s}) \left(\frac{1 \text{ kt}}{1.689 \text{ ft/s}}\right) = \mathbf{622 \text{ kt}}$$

From Eq. 6.2,
$$M_\infty = \frac{V_\infty}{V_A} = \frac{446 \text{ kt}}{622 \text{ kt}} = \mathbf{0.72}$$

It is quite possible for the static source tube on an aircraft to be blocked, by insects or from the wax job of an attentive owner, for example. If this happens, the altimeter will indicate a fixed altitude as the aircraft climbs or descends while the airspeed indicator can read either too high or too low, as an inspection of Eq. 7.5 reveals. Pilots are told to break the glass face on one of their flight instruments under these circumstances **if** they don't have an on-board alternate static source; this breakage vents the static line for all the flight instruments to the cabin air. Most newer aircraft, however, have a valve that allows the pilot to switch the static pressure line to the cabin without the excitement associated with breaking glass. In either case, there is an additional position error associated with the use of cabin air as the source for the instrument static air pressure: Heater ducts may change the pressure at the point where the alternate pressure tube is vented; also, openings in the passenger compartment tend to produce a pressure in the cabin that is **less** than the true static by a venturi effect because the airspeed around the cabin is usually greater than the flight speed. The latter effect can overwhelm the former effect if the opening(s) become large enough, as for flight with open windows or with an open cockpit aircraft.

EXAMPLE 7.3. Assume that you are flying at 5000 ft in the StAt with a calibrated airspeed of 120 kt (202 ft/s). (a) If the static port becomes totally plugged at this altitude and you then descend 500 ft at a constant indicated airspeed, what is the actual calibrated airspeed at the lower altitude? (b) If you had instead climbed 500 ft at constant indicated airspeed after your static port was plugged, what is the actual calibrated airspeed at the lower altitude?

Solution: (a) The dynamic pressure at your 5000 ft cruising altitude was

$$q_\infty = \frac{1}{2}\rho_0 V_{CAL}^2 = \frac{1}{2}(0.002377)(202)^2 = 48.8 \text{ lb/ft}^2$$

while the pressure in the static port of the airspeed indicator (p_∞), was 1760.9 lb/ft^2 (from Appendix B) and the pressure in the pitot tube (the total pressure) must have been (Eq. 7.2)

$$p_T = p_\infty + q_\infty = 1760.9 + 48.8 = 1809.7 \text{ lb/ft}^2$$

Since the static pressure is assumed to stay the same and the calibrated airspeed hasn't changed, the total pressure in the pitot tube must also have stayed the same. The problem then is to see what calibrated airspeed would give the same total pressure if the static port were at its correct value (for 4500 ft in the StAt) of 1794.1 lb/ft^2. Thus

$$p_T = 1809.7 \text{ lb/ft}^2 = p_\infty + \frac{1}{2}\rho_0 V_{CAL}^2$$

$$= 1794.1 + \frac{1}{2}(0.002377) V_{CAL}^2$$

which can be solved by algebra to obtain $V_{CAL} = 114.6$ ft/s (68 kt). So you are travelling much slower than you would gather by looking at your airspeed indicator—a bad scene. What has happened is that the denser air at the lower altitude has compensated for the lower airspeed, showing how small the differential air pressures measured by the airspeed indicator really are.

(b) This time the static pressure should have decreased to 1728.2 lb/ft^2 (appropriate to 5500 in the StAt) rather than staying constant. Since the total pressure didn't change,

$$p_T = 1809.7 \text{ lb/ft}^2 = p_\infty + \frac{1}{2}\rho_0 V_{CAL}^2$$

$$= 1728.2 \text{ lb/ft}^2 + \frac{1}{2}(0.002377) V_{CAL}^2$$

which yields a value for V_{CAL} of 261.9 ft/s (155 kt) and now we are travelling much faster than our airspeed indicator tells us.

Figure 7.4 presents the position correction for a 1977 Cessna T210M when the alternate static source is selected **and** when windows are open. In this case, at least, the additional position errors tend to cause the pilot to make the approach for landing at a **higher** true airspeed than normally used, with an accompanying risk of overshoot; the indicated cruise airspeed would also be too high. The author has flown a small open cockpit biplane (experimental certification) that obtains static air from the cockpit and the airspeed indications appear to be significantly in error, in the direction suggested by Fig. 7.4. (Unfortunately, the advertisements for this aircraft appear to be using **uncorrected** stall and cruise speeds. *Caveat emptor!*)

An onboard computer chip **can** be used to obtain the true airspeed; knowledge of the outside air temperature and the static pressure allow one to calculate the ambient air density,

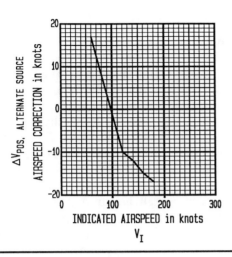

Fig. 7.4 An example of airspeed indicator errors using an alternate static source

as you did in Chap. 4, and this yields the true airspeed when combined with the flight dynamic pressure that is measured by the pitot tube (Eq. 7.4). However, at typical approach speeds, V_{IND} is a direct measure of V_E, and V_E is a direct measure of the flight dynamic pressure, and this is what determines the stalling speed and thus the proper approach speed for a particular aircraft. Therefore the V_{IND} and V_E speeds for landing approach or for takeoff do **not** depend on the density of the air and pilots can safely use the same approach speed to a SL density airport as they do to a 10,000 foot density altitude airport, for a given weight. This could not be done if the airspeed indicator always read true airspeed. This advantage for the critical approach and takeoff phases of flight far outweighs the disadvantage of having to calculate true cruise airspeeds from indicated values, particularly since radio equipment can provide the more important ground speed information directly.

Pilots are cautioned to remember, however, that their **flight** or **true** airspeeds for takeoff, approach, and landing are going to be considerably **higher** at a high density altitude airport compared to a low density altitude airport. A businessman-pilot once told this writer that he used a higher indicated airspeed for an approach to a high altitude airport because he had heard that the approach speeds would be higher in the "thin" air. Don't do it! The "airspeed" indicator automatically compensates for density variations as far as their aerodynamic influence is concerned. Pilots should trust and use their airspeed indicators in the same way at high density altitudes but expect their takeoff, approach, and landing speeds to **appear** faster and **be** faster.

EXAMPLE 7.4. The 1977 Cessna T210M has a recommended short-field approach speed of 71 kt indicated airspeed at its gross weight of 3800 pounds. (a) Determine the true airspeed for an approach to an airport under standard SL conditions and under standard 10,000 foot conditions. (b) Determine the true airspeed for these approaches if the alternate static source has been selected, the windows are open, and the pilot uses the

same V_{IND} of 71 kt. (Assume the corrections of Figs. 7.3 and 7.4 apply to this aircraft.)

Solution: (a) From Fig. 7.3, $V_{CAL} = 71$ kt $+ 1.5$ kt $\cong$ **73 kt**

and since $M_\infty < 0.3$, $V_E \cong V_{CAL} = 73$ kt

From Eq. 7.8, for SL density altitude,

$$V_\infty = V_E \sqrt{\frac{\rho_0}{\rho_\infty}} \cong V_E = \textbf{73 kt}$$

for 10,000 ft density altitude,

$$V_\infty = 73 \text{ kt} \sqrt{\frac{0.002377}{0.001756}} = \textbf{85 kt}$$

(b) From Fig. 7.4, $V_{CAL} = 71$ kt $+ 12$ kt $=$ **83 kt**

and since $M_\infty < 0.3$, $V_E \cong V_{CAL} =$ **83 kt**

From Eq. 7.8,

for SL density altitude, $V_\infty = V_E \sqrt{\frac{\rho_0}{\rho_\infty}} = V_E =$ **83 kt**

for 10,000 ft density altitude, $V_\infty = 83$ kt $\sqrt{\frac{0.002377}{0.001756}} =$ **97 kt**

B. The Machmeter

For subsonic jets the flight Mach number rather than the indicated airspeed provides the upper limiting speed for the cruise phase of flight. All such aircraft incorporate a Machmeter among their flight instruments. In addition, they must provide a speed warning device that jangles obnoxiously whenever the flight speed exceeds the maximum operating Mach number by 0.01. One version of the Boeing 727 trijet airliner, for example, provides this warning when $M_\infty = 0.902$ is reached.

When (if) we solve Eq. 7.6 for the flight Mach number, we obtain

$$M_\infty = \frac{V_\infty}{V_A} = \sqrt{5} \left(\left(\frac{p_T - p_\infty}{p_\infty} + 1 \right)^{2/7} - 1 \right)^{1/2} \quad (M_\infty \le 1) \quad (7.12)$$

and this is the calibration formula for a subsonic Machmeter. Note that both $(p_T - p_\infty)$ and p_∞ must be measured by the Machmeter. (Above Mach 1 a shock wave forms ahead of the probe and a different calibration must be used.)

A temperature probe pointing into the airstream will indicate a temperature close to the total temperature given by Eq. 6.10. This enables a flight crew to calculate the ambient temperature, T_∞, and thereby determine the true flight speed.

EXAMPLE 7.5. Determine the true airspeed for a jet aircraft whose Machmeter indicates $M_\infty = 2.0$ and whose total temperature thermometer indicates 200°F. Determine also the equivalent airspeed if the flight is being conducted at a density altitude of 30,000 feet.

Solution: From Eq. 6.10,

$$T_\infty = \frac{T_T}{1 + 0.2\,M_\infty^2} = \frac{200 + 460}{1 + 0.2(2)^2} = \frac{660}{1 + 0.8} = 367\ \text{R}$$

From Eq. 4.9 and the conversion factor of Sec. 5B,

$$V_A = \sqrt{\gamma\,k\,T_\infty} = \sqrt{(1.4)(1716)(367)} = 939\ \text{ft/s} = 556\ \text{kt}$$

Therefore $V_\infty = M_\infty V_A = (2.0)(556\ \text{kt}) = \mathbf{1112\ kt}$

From Eq. 7.8 and Appendix B,

$$V_E = V_\infty \sqrt{\frac{\rho_\infty}{\rho_0}} = 1112\ \text{kt}\ \sqrt{\frac{0.000891}{0.002377}} = \mathbf{681\ kt}$$

It is early in the winter and early in the night for a flight crew of three which prepares to ferry (deadhead) their Boeing 727-251 from JFK airport in New York to Buffalo. They are in the pre-takeoff checklist.

Second Officer: "Pitot heat." First Officer: "Off and on." The climbout is routine at an indicated airspeed of 305 kt and a rate of climb of 2,500 feet/minute. But, as the altitude increases above 16,000 feet, the indicated **airspeed** begins to **increase**. First Officer: "Do you realize we're going 340 kt and I'm climbing 5,000 feet a minute?" The Mach overspeed warning sounds as the aircraft reaches 23,000 feet; the indicated airspeed is 405 kt. Captain: "No, just pull her back, let her climb." At 24,800 feet: First Officer: "There's the Mach buffet, guess we'll have to pull it up." Captain: "Pull it up." Ten seconds later: "Mayday, mayday . . ." ". . . go ahead." "Roger, we're out of control, descending through 20,000 feet." Thirty-three seconds later: "We're descending through 12 (thousand feet), we're in a stall." [2]

The heaters for the pitot tubes had **not** been turned on. In the climb through cold clouds, ice had formed in the pitot system. The crew, ignoring the other flight instruments, believed the insistent but erroneous speed and Mach indications. What they interpreted as transonic shock buffet at 24,800 feet was instead the "stick-shaker" that automatically shakes the control wheel and tries to move it forward as the 727 approached the stall. After a wing stalled, the 727 rolled to the right and began a steep, nose-down spiral with a bank angle between 70° and 80°. Recognition of the true nature of the problem came too late for recovery.

"Mind set" and "attention fixation" are among the terms used by psychologists to explain why the crew did not look for other explanations for the increasing airspeed. Be alert for this problem when you face an aeronautical emergency!

Accident investigators found that a pitot tube blockage by ice on this aircraft could produce either erroneously high or erroneously low indicated airspeeds, depending on whether the tiny water drain hole past the pitot tube opening was also blocked by ice. The author has observed only a **decrease** (to zero) in the indicated airspeed as the failure mode due to pitot tube icing on light aircraft. Clearly, study of the construction and aerodynamics of airspeed indicators can be a healthy exercise for pilots.

[2] *Aviation Week & Space Technology*, Jan. 5, 1976, and Jan. 12, 1976.

C. Coffin Corner

As an aircraft climbs toward its service ceiling, the range of calibrated speeds available to it keeps shrinking. This is because the calibrated stall speed is not pressure dependent and stays constant while the maximum calibrated speed (in level flight) is reduced by the decrease in available power. The service ceiling is usually taken as the density altitude at which the maximum rate of climb has been reduced to 100 ft/min; at this point the cruise speed is only slightly greater than the maximum rate-of-climb speed and is relatively close to the stall speed.

A high speed aircraft has an additional problem. The maximum speed is limited not only by power (and the usual never-exceed indicated airspeed) but also by a maximum operating Mach number, M_{MO}, determined during its certification. If M_{MO} is exceeded, symptoms of the shock stall (Mach buffet of the stabilizers, aileron buzz, etc.) begin to appear.

Because an aircraft's Mach number is based on its true airspeed and the speed of sound (which decreases with increasing altitude as the temperature decreases), the true airspeed at which M_{MO} is reached decreases with increasing altitude in the troposphere. On the other hand, the true stall airspeed increases with increasing altitude because the calibrated stall airspeed is constant. The stall speed and M_{MO} approach each other as altitude is gained, eventually boxing the aircraft into a narrow range of speeds called the coffin corner. At these very high altitudes, severe turbulence can either slow the aircraft and cause a low speed stall or increase the airspeed and cause a shock stall upset.

EXAMPLE 7.6. A certain model of the Boeing 727 transport aircraft has a calibrated stall speed in the cruise configuration of 159 kt; it also has an M_{MO} of 0.902. What is the range of allowed true airspeeds at 30,000 ft in the StAt?

Solution: At 30,000 ft, $\rho_\infty = 0.000891$ sl/ft^3 (from Appendix B), so the equivalent stall speed (from Fig. 7.2) is 156 kt and the true stall speed (from Eq. 7.8) is

$$V_\infty = \sqrt{\frac{\rho_0}{\rho_\infty}}\,V_E = \sqrt{\frac{0.002377}{0.000891}} \times (156\ \text{kt}) = 255\ \text{kt}$$

The speed of sound at 30,000 ft in the StAt (from Appendix B) is 589.4 kt, so

$$V_{\infty,MO} = M_{MO} V_A = (0.902)(589.4\ \text{kt}) = 532\ \text{kt}$$

and, for these conditions, the aircraft has an available speed range of (532→255) or 277 kt.

Fig. 7.5 presents a graphical solution to this problem, right up to where the corner closes at 60,000 ft.

D. EFIS and the Glass Cockpit

Large multiengine, propeller-driven aircraft and jet transport aircraft are now designed with an integral Electronic Flight Instrument System (EFIS) that is based on a computer-

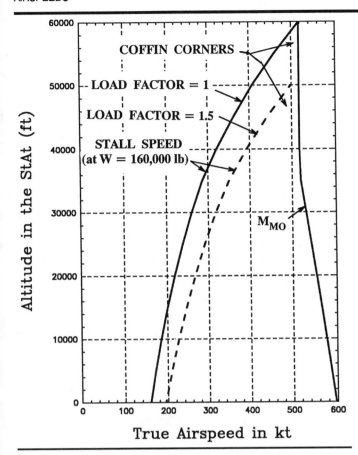

Fig. 7.5 The operating envelope at maximum gross weight for one model of a transport aircraft; loading the wings increases the stall speed and narrows the envelope (a 48° banked turn gives a load factor of 1.5)

controlled display using a glass-faced cathode ray tube (CRT), similar to the CRT used in standard graphics computer terminals. Flight, powerplant, and navigational information can all be called up to appear on the same display in such a system. CRTs require some depth in the instrument panel to provide room to accelerate, focus, and deflect electrons on their way toward the light-emitting phosphors on the face. Some day in the future it may be possible to replace these CRTs with cheaper, thinner, and less power-hungry displays based on liquid crystals or other technology.

With a computer running the show, true airspeed and ground speed can be computed and displayed when desired. The digital revolution that is fueling the continued growth of computer power promises inexpensive and much more accurate measurement of flight and engine parameters, but pilots have trouble obtaining rapid approximate values and trend information from digital displays. This writer has learned to appreciate the accuracy and ruggedness of an inexpensive, homebuilt digital tachometer in a certain homebuilt aircraft that delighted in destroying conventional galvanometer movements through vibration. Perhaps a display that provides circular bar graph information along with the more precise digital value is the ideal compromise. Because solid state pressure sensors that use a strain gauge element to provide a pressure-dependent resistance are now readily available and reasonably

priced, digital airspeed indicators that begin reading from 0 are also quite feasible.

The military has found it important to provide its pilots with vital flight and navigational information **while** the pilot is looking forward out the cockpit window. This is accomplished with the help of a **head-up display** (HUD), which projects and appears to superimpose the information on the cockpit windshield. Such capability would be greatly appreciated by all pilots, especially at the point when a transition must be made from a fully instrument approach through clouds to a visual landing. But it appears that most of us will use a HUD in a car before we use one in an aircraft.

Summary

The conventional airspeed indicator measures the difference between total pressure and static pressure. At low speeds this pressure difference equals the dynamic pressure and is therefore the most meaningful information regarding the aerodynamic forces acting on an airplane during the critical takeoff, approach, and landing phases of flight. The indicator is calibrated in terms of speed by assuming that air is incompressible and that it has sea level standard density. To obtain the actual airspeed of the aircraft from the indicated value, it is necessary to correct for installation and instrument error, for compressibility, and for non-standard air density. Installation or position error is primarily caused by an incorrect detection of static air pressure; this error can be quite large during very slow flight and when an alternate static source inside the cockpit is used. The equivalent airspeed is a direct measure of the flight dynamic pressure and thus is the most aerodynamically important of the various types of airspeeds.

A Machmeter provides information regarding the upper limiting cruise speed for high speed aircraft. Because the speed of sound depends on the outside air temperature, its measurement must be included.

Symbol Table (in order of introduction)

V_∞	the speed of an aircraft relative to undisturbed air (aka true airspeed and flight speed)
V_{CAL}	calibrated airspeed (the reading of the airspeed indicator when there is no instrument or position error)
V_E	equivalent airspeed (the airspeed that produces the same dynamic pressure at altitude as does a true airspeed of the same value under SL density conditions)
V_{IND}	indicated airspeed

Review Questions

1. What property of the air flow around an airplane does the "airspeed" indicator actually measure? How does it do this?

2. Under what cruising flight conditions will the airspeed indicator tend to indicate a speed **greater** than the flight speed?

3. An airplane flying at a pressure altitude of 10,000 feet encounters a total blockage of its static pressure line. Will the actual indicated airspeeds during the approach to a SL airport be higher or lower than they should be?

4. A bug makes its home in the static line of a certain aircraft while the aircraft is tied down for the night. The next morning the pilot notices that (based on the rate at which the altitude is changing) the plane isn't climbing as well as usual. Will the indicated airspeed tend to be greater or less than it should be during this climb?

5. If the small venting drain hole of a pitot tube becomes blocked by ice before the opening becomes plugged, the pressure in the total pressure line can stay constant at the value it had when blockage occurred. On the other hand, the pressure in the total pressure line will remain close to the static pressure if the drain hole remains open after the pitot tube opening becomes blocked. Which of these possibilities occurred in the 727 accident recounted earlier? Explain why.

6. If one aircraft is cruising at an equivalent speed of 240 kt at a density altitude of 10,000 feet and another aircraft is cruising at an equivalent airspeed of 240 kt at a density altitude of 20,000 feet, which aircraft has the greater true airspeed?

7. When will the equivalent airspeed be greater than the calibrated airspeed, and when will it be less, in general? (Hint: See Fig. 7.2.)

8. The approach to landing at a high density altitude airport should be made at the same ___ airspeed as for a SL airport; this will result in a higher ___ airspeed at the high density altitude airport.

9. Pilots tend to anticipate takeoff and pull back on the control wheel (but nowhere as much as Hollywood pilots!) when the ground is rushing by at what appears to be the right speed. Why is this such a bad habit? What **should** determine the point of rotation on takeoff?

10. What other information besides the flight Mach number is needed to obtain the true airspeed of an aircraft?

11. The greatest difference between indicated and true airspeeds will occur for pressure altitudes (higher, lower) than sea level and for temperatures (greater, less) than the standard value for that pressure altitude.

12. Suppose that a venturi tube is being used to provide the reduced pressure necessary to drive a gyroscopic flight instrument. Will it work as well at a high density altitude airport as it does at a SL density airport? Explain. (Hint: Refer to Prob. 5.8 and Eq. 5.21.)

13. There is only one obvious flight condition for which an airplane subject to the airspeed indicator corrections of Figs. 7.2 and 7.3 would have a true airspeed just equal to its indicated airspeed. What is that speed and what are the flight conditions?

14. What kind of a shock wave appears in front of a Machmeter probe at supersonic speeds? Explain.

Problems

Note: Assume that Figs. 7.2, 7.3, and 7.4 apply to the aircraft airspeed indicator for the purpose of solving the following problems.

1. Estimate the density of the air for the F-84F night flight by Richard Bach that introduced this chapter. (Use the 30,000 foot correction in Fig. 7.2.) Answer: 0.00067 sl/ft^3

2. What was the air temperature at Richard Bach's cruising altitude? (Use the flight Mach number.) Answer: –59°F

3. Figure 7.3 shows a +11 kt error for an indicated airspeed of 40 kt in the landing configuration. What percent error in the measurement of p_∞ does this represent? Answer: Only 0.2% (!)

4. Determine the true airspeed for an aircraft flying at a pressure altitude of 800 feet, an air temperature of –40°F, and an indicated airspeed of 120 kt. Answer: 108 kt

5. Determine the true airspeed for an aircraft flying at a pressure altitude of 3400 feet, an air temperature of 90°F, and an indicated airspeed of 100 kt. Answer: 107 kt

6. Determine the true airspeed for an aircraft flying at a pressure altitude of 25,000 feet, an air temperature of 0°F, and an indicated airspeed of 370 kt. Determine also the flight Mach number.
Answer: 530 kt, 0.85

7. Determine the true airspeed for a temperature of +80°F, a pressure altitude of –5000 feet, and an indicated airspeed of 700 kt. Determine also the flight Mach number. Answers: 651 kt, 0.964

8. The FAA suggests that pilots makes their approach at a speed 30% greater than the stall speed for each weight and flap configuration. This is the "reference speed" or target speed on final approach that is contained in flight manuals, especially for airliners. Suppose that you decide to **experimentally** determine what your approach speed should be, based on this "official" criterion. (On a new aircraft on its first flight, this might well appear to be a reasonable procedure to follow—but it isn't.) You find that your aircraft stalls at an indicated airspeed of 40 kt in the landing configuration. What **should** be your **indicated** airspeed on the approach, assuming the airspeed indicator errors given in this chapter? (Hint: Ask yourself what airspeed the FAA really is talking about.) Answer: about 63 kt

9. Make the determination of problem 8 for an indicated stall speed of 60 kt and an alternate static source with windows open.
Answer: about 101 kt

10. Determine the true airspeed for an aircraft whose Machmeter indicates 2.5 and whose total temperature thermometer reads 300°F. Answer: 1333 kt

11. When the needle of an airspeed indicator points to 100 kt under SL StAt conditions, what pressure value should it really be indicating? How about at 10,000 ft in the StAt? Answer: 33.9 lb/ft^2

12. Do manufacturers recommend using the 1.3 factor for short field landings? For example, the 1978 C-152 stalls at 43 kt (calibrated air speed) with the flaps down (power off) and the recommended approach speed for this configuration is 54 kt (indicated airspeed).

13. We'll see in Chapter 12 that the stall speed is proportional to the square root of the flying weight of an aircraft. What approach speed does this suggest that you should use for the C-152 of the previous problem if you are at 1400 lb rather than at the 1670 lb gross weight? (Start from the manufacturer's recommended approach speed at gross weight.) Answer: 44 kt indicated airspeed

14. For the conditions of Example 7.3, how fast would you actually have to be flying to obtain a calibrated airspeed of 120 kt at sea level? Answer: 345 kt

References

Andresen, Jack, *Fundamentals of Aircraft Flight and Engine Instruments*, Hayden Book Company, Inc., New York, 1969.

Dommasch, Daniel O., Sherby, Sydney S., and Connolly, Thomas F., *Airplane Aerodynamics*, Pitman Publishing Company, New York, 1967.

Gracey, William, *Measurement of Aircraft Speed and Altitude*, John Wiley & Sons, Inc., New York, 1981.

Hunt, F.L. and Stearns, H.O., *Aircraft Speed Instruments*, NACA Report #127, 1922.

Hilding, K., *Aircraft Speed Instruments*, NACA Report #420, 1932.

Chapter 8

Determining Airfoil Properties

Alone, off by himself between boat and shore, Jonathan Livingston Seagull was practicing. A hundred feet in the air, he lifted his beak, lowered his webbed feet, and strained to hold a steep-twisting camber through his wings. He slowed til the wind was barely a whisper in his face; the ocean stood still below him. His eyes were half closed in fierce concentration. He worked to force one single more degree of camber . . . then, abruptly, his wing-feathers ruffled, he stalled.[1]

A. Airfoil Geometry

In this chapter we examine in detail the aerodynamic forces and moments that result from the interaction of moving air with a wing section or *airfoil*. We will not be able to apply this knowledge immediately to the wings of actual aircraft because we make the initial assumption that there is no flow of air along the airfoil perpendicular to the section (i.e., toward or away from the wing tip on a real wing). This simplifying assumption of 2-D flow allows us to isolate the contributions of (**a**) properties of the air, and (**b**) the airfoil shape, on the aerodynamic forces and moments that are produced. In a later chapter we will see that the necessary corrections in going from an airfoil to a wing are fairly easily handled so long as the wing is relatively long and narrow.

The airfoil properties we shall discuss in this chapter are also the properties usually measured in wind tunnels, for the airfoil to be studied is extended into a "wing" that reaches from one tunnel wall to the other, so that (except for wall corrections) the flow really is two-dimensional. Holes (*orifices*) are drilled in this wing for pressure taps so that the pressures at various points can be measured. Then the lift and pitching moment of the airfoil are determined from the measured pressures while the drag is usually determined from the momentum loss of the air after it has passed the airfoil. Note that the principle of relativity underlies this wind tunnel technique; it is assumed that the same forces and moments exist whether the wing moves or the air moves, so long as the air flow conditions are identical.

It is quite feasible to obtain the lift force, the drag force, and the moment on a model wing through measurements of the external forces required to maintain equilibrium (a *force balance*), but then no clue is provided as to how the airfoil might be improved.

Figure 8.1 identifies the geometrical characteristics that are used to describe a particular airfoil. The **chord** is the distance from the leading edge of the airfoil to the trailing edge; the symbol for wing chord is c. The chord line is an imaginary line connecting the leading and trailing edges. The airfoil shape is normally defined by providing a discrete number of coordinates for various points, with the points closest together where the curvature is the greatest. All of these coordinates are given in terms of c so that the airfoil description is easily scaled up to any desired size.

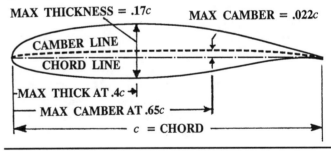

Fig. 8.1 Airfoil nomenclature, using the LS(1)-0417 airfoil as an example

The **mean camber line** (or just mean line) traces the points that lie just halfway between upper and lower surfaces, measuring perpendicular to the chord line. The **camber** of the airfoil (actually its maximum camber) is the maximum separation distance of the chord line and the mean line. Positive camber, with the mean line going above the chord line, is a clue that lift will be developed when the airfoil is pointing its chord directly into the oncoming air; such airfoils are used on most aircraft (except some aerobatic and some supersonic machines) because they produce the minimum drag at cruise speeds. The

[1]Reprinted with permission of Macmillan Publishing Company from *Jonathan Livingston Seagull* by Richard Bach. Copyright © 1970 by Richard D. Bach and Leslie Parrish-Bach.

mean line is closely related to the circulation produced by the airfoil; applying thin airfoil theory to the mean line is one way to analyze the performance of an airfoil.

The **maximum thickness** of the airfoil and the **location** of this point tell us a great deal about the intended use of the airfoil. Low speed aircraft typically use airfoils of $0.12c$ to $0.17c$ in thickness (12% to 17%) while supersonic aircraft may be down around $0.05c$. Ordinary airfoils reach their maximum thickness at about $0.3c$ while airfoils intended to support more laminar flow tend to have their maximum thickness closer to the mid-chord position.

The **sharpness** of the leading edge is very important because it strongly influences the angle at which the boundary layer separates from the nose region as the oncoming air approaches at increasingly larger angles relative to the chord line. The curvature is normally less **above** the leading edge than below because airfoils are primarily intended for use at positive angles relative to the air. The sharpness of the airfoil was specified on some early airfoils by indicating the **radius** of the circle that is the best fit when its center is placed on a line drawn from the leading edge and tangent to the mean line there. Newer airfoils are defined by successive iterations of a computer program rather than by purely geometric descriptions, and so the curious inspector of airfoils is left to his or her own devices in trying to quantify leading edge sharpness.

When airfoil coordinates are listed, the variable x is used to indicate the position component along the chord position in fractions of the chord length ($x = 0$ is the leading edge and $x = 1$ is the trailing edge) while the variable z gives the position component perpendicular to the chord position, also in terms of the chord length. It is assumed that a smooth curve is drawn through the listed points, and the listed points are normally much closer together near the leading edge. Some of the airfoils in the 1940s were completely defined by equations but most newer airfoils are defined just by their coordinates (and perhaps some rules for interpolation).

The aerodynamic forces acting on an airfoil (or a particular wing) depend **only** on the properties of the air and the angle between the chord and the oncoming air (the **angle** of attack). These forces do **not** depend on the angle of the chord relative to the **horizon**. (The horizon is important to a wing only to the extent that it defines the direction of the force of gravity.)

Appropriate to its importance, the **first** letter of the Greek alphabet (alpha, α) is used to symbolize the angle of attack of a wing or airfoil. The **nature** of the air flow over a given wing or airfoil is primarily determined by the current value of α. Specifically, the optimum approach speed, the optimum climb speed, and the stall speed all occur at weight-dependent speeds but at fixed values for α. It appears that an angle-of-attack instrument should be in all cockpits, but this is currently true only for military aircraft and some airline aircraft.

Figure 8.2 illustrates the fact that α depends **only** on the direction of the oncoming air relative to the chord line (or other fixed reference on the airfoil), for it is drawn for a constant angle of attack of 10°. (Note that, in no-wind conditions, the

direction of the relative wind for the airplane is just **opposite** to the direction of the **flight path vector**, for an outside observer.) In much of a pilot's flying experience, the angle of the attack is judged by looking out on the wing and comparing it with the horizon, but this reveals the angle of attack **only** if the plane is flying level (i.e., at constant altitude). In a climb the angle of attack is always **less** than the angle relative to the horizon; in a descent it is always **greater**. In fact, in the dangerous flat spin the horizon angle is close to zero but the angle of attack is perhaps 60° or more (i.e., well above the stalling angle) on both wings.

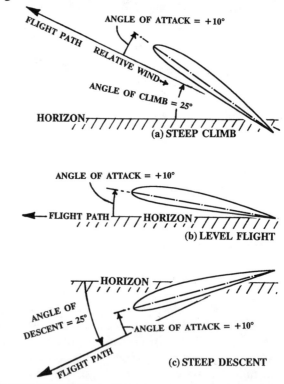

Fig. 8.2 Three different flight conditions—but all with the same angle of attack

One of the primary benefits of aerobatic proficiency to the pilot is the resultant ability to judge the angle of attack completely **independently** of the current location of the horizon. In an English bunt, for example, in which inverted flight is entered from level flight by pushing the nose forward through 180°, a too hasty push will generate a (negative angle of attack) stall and stall buffet even though the aircraft is pointing almost straight down at the ground! Whether through aerobatics or otherwise, all pilots must be permeated with the understanding that the angle of attack determines how their aircraft flies, and this angle is often **not** the same as the angle relative to the horizon. Often the angle of attack can be **felt** on the basis of control responsiveness and aural and visceral clues.

Any airfoil possessing a mean camber line that **coincides** with its chord line must have the same curvature and same absolute value of the z coordinate for every point on its chord line. Such an airfoil is reasonably enough designated as a **sym-**

metrical airfoil. Evidently a symmetrical airfoil has **zero camber** and, at $\alpha = 0°$, generates no lift (but still has some pressure drag and some viscous drag).

Symmetrical airfoils are commonly used for the horizontal and vertical tail surface on all types of aircraft. Supersonic aircraft also sometimes use them for their wings because they have good high speed properties and lift is generated mostly from high speeds rather than from high α; at approach speeds, though, flaps and other wing devices are used to give the wing the positive camber it needs to develop lift at low speeds. The top aerobatic aircraft use symmetrical airfoils to give symmetrical upright and inverted performance, a very important feature at the highest levels of competition.

B. The Pressure Coefficient

Uncounted millions of air molecules are attacking you from all sides — even now, I would guess. If just a small fraction of those attacking you from one side can be persuaded to cease and desist (as happens when there is a wind), it is possible for a large force to be generated. This is what airfoils are designed to do **efficiently**, and so we now turn to the way in which the varying pressures on an airfoil are scientifically described.

The best way to report the experimental measurement of pressures on an airfoil turns out to be through the **pressure coefficient**, C_P, which isolates the geometrical effects from the air-property effects. We will utilize simplified, fictitious pressure distributions on a flat plate to see how **all** the airfoil coefficients that describe an airfoil's properties can be obtained from a knowledge of these pressure distributions. In Section D, to make sure we do not lose sight of reality, we look at some actual pressure distributions on a flat plate.

Airfoil properties are measured in a wind tunnel by generating two-dimensional air flow (no span-wise component) over a wing that uses the airfoil to be studied as its cross-sectional shape. The wing has orifices that sample the local air pressure at many points on both surfaces along a line from the leading edge to the trailing edge. The pressure data are **not** reported directly because they are very dependent on the value of the freestream dynamic pressure, q_∞. Instead, the pressures are reported in terms of a **pressure coefficient**, C_P, which indicates the deviation of the local static pressure from the freestream static pressure as a fraction of the dynamic pressure. In symbols this definition of C_P can be written as

$$\text{local static pressure} = p \equiv p_\infty + C_P q_\infty. \quad (8.1)$$

but this equation can also be written, perhaps more informatively, as

$$\Delta p \equiv p - p_\infty = C_P q_\infty$$

or

$$C_P = \frac{\Delta p}{q_\infty} = \frac{p - p_\infty}{q_\infty} \quad (8.2)$$

DEFINITION OF THE PRESSURE COEFFICIENT

where $q_\infty = \frac{1}{2}\rho V_\infty^2$ is the incompressible flight dynamic pressure (Eq. 5.13). Because both p_∞ and q_∞ have units of pressure, C_P is non-dimensional, as a good little coefficient should be.

At low speeds the value of C_P has a simple interpretation in terms of the local airspeed. Applying Bernoulli's incompressible equation (Eq. 5.11) to the freestream and to a local point on an airfoil surface,

$$p_\infty + \frac{1}{2}\rho V_\infty^2 = p + \frac{1}{2}\rho V^2 \quad (8.3)$$

Rearranging,

$$p = p_\infty + \frac{1}{2}\rho\left(V_\infty^2 - V^2\right) = p_\infty + \frac{1}{2}\rho V_\infty^2\left\{1 - \frac{V^2}{V_\infty^2}\right\}$$

$$= p_\infty + q_\infty\left\{1 - \frac{V^2}{V_\infty^2}\right\} \quad (8.4)$$

Comparing Eqs. 8.2 and 8.4 shows us that, for incompressible flow,

$$C_P = 1 - \frac{V^2}{V_\infty^2} \quad (8.5)$$

From inspection of Eqs. 8.1 and 8.5, we can deduce four values or ranges for C_P that have obvious and important physical significance (Table 8.1).

(a) $C_P < 0$	implies	$V > V_\infty$	and	$p < p_\infty$
(b) $C_P = 0$	implies	$V = V_\infty$	and	$p = p_\infty$
(c) $0 < C_P < 1$	implies	$V < V_\infty$	and	$p > p_\infty$
(d) $C_P = 1$	implies	$V = 0$	and	$p = p_T$
Table 8.1				

When C_P is negative (**a**), the **local** airspeed is **greater** than the flight speed and the **local** pressure is **less** than ambient. Clearly this is what we need on the top surface of an airfoil because such a pressure, in conjunction with higher pressures on the bottom surface, will yield a lift force.

We can expect to find at least one point on the airfoil where the local airspeed is just **equal** to the flight speed (**b**). This will occur just beyond the leading edge where the flow, having been slowed by the approaching airfoil, speeds up to get around the thicker part behind the leading edge. This point is identified by a **zero** value for its pressure coefficient.

For maximum lift, we would like to encourage pressures **greater** than ambient on the bottom surface of the airfoil (**c**). This implies a **positive** value for C_P.

The **stagnation point** (**d**), where the local airspeed is **zero**, is identified by its possession of the maximum possible C_P of **1**. The pressure at a stagnation point also has its maximum possible value, $p = p_T = p_\infty + q_\infty$. We can guess that a stagnation point will be located very close to the leading edge at zero angle of attack and will move backwards along the lower surface as the angle of attack increases because the stagnation point is the dividing point between upper and lower surface air.

One more little item: Because we would like to see negative pressures on the top surface and positive pressures on the bottom, and because it hasn't seemed right to have the top surface data appear on the bottom part of a graph, the custom of plotting C_P with **negative** values plotted **above** the axis and positive values plotted below the axis has become well established.

C. Airfoil Coefficients from Airfoil Pressures

We are now in a position to understand how the lift and drag (and the lift and drag coefficients) are calculated from a knowledge of pressure coefficients.

EXAMPLE 8.1. Suppose that we have inserted a flat plate of chord c and span b (equal to the tunnel width) into an airstream at an angle of attack of 10°, and that the pressures and pressure coefficient distribution are as shown in Fig. 8.3. Note how the pressure distribution has been sketched, with an arrow pointing **away** from the surface representing a pressure **below** the ambient or static pressure and an arrow pointing **toward** the surface representing a pressure that is **greater** than ambient. The relative length of the arrows is chosen so as to indicate the relative strengths of the pressures. A pressure can produce a force at **right** angles to the surface, only, at any given point (Section A of Chap. 4) and so all the arrows are drawn perpendicular to the surface (a trivial task with a flat plate, fortunately). Determine the lift, drag, and moment coefficients, the latter about two different points.

Solution: The pressure distribution of Fig. 8.3 is clearly not physically possible; the pressures at low speeds must vary smoothly from one value to another without extremely sharp or discontinuous jumps and they would never be quite this constant over so much of a real airfoil. However, we start with this simplified version of the pressure envelope to make the method clear; we'll transition to real airfoils soon enough.

STEP 1. (Figs. 8.3 and 8.4) The force on each segment of this hypothetical airfoil is given by

$$F = (\Delta p) \times (\text{area}) = \left(C_P q_\infty\right) \times (\text{area})$$

by virtue of the definition of pressure (Eq. 4.3) and the pressure coefficient (Eq. 8.2).

Thus, because we are assuming a span of b and a chord of c for our flat plate airfoil, the forward 40% of the top surface has a force on it of

$$F_1 = \left(1.5\right) q_\infty \left(b \times 0.4c\right) = 0.60 \, q_\infty \, b \, c$$

while the rear 60% of the top surface has a force on it of

$$F_2 = \left(0.5\right) q_\infty \left(b \times 0.6c\right) = 0.30 \, q_\infty \, b \, c$$

and the bottom surface a force of

$$F_3 = \left(0.7\right) q_\infty \left(b \times c\right) = 0.70 \, q_\infty \, b \, c$$

Note that F_1 is shown to be acting at the $0.2\,c$ point because that is the **mid-point** of its constant-pressure region. The other two forces are similarly located.

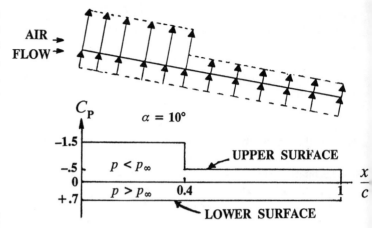

Fig. 8.3 Make-believe pressure distribution for a flat plate

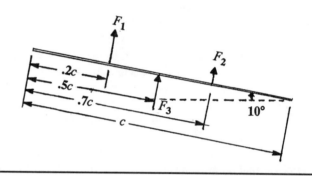

Fig. 8.4 Resolution of the pressure distribution into three equivalent forces

STEP 2. (Figs. 8.4 and 8.5) Find the resultant pressure-induced force on this airfoil, F_R, by vector addition.

$$\sum F \equiv F_R = F_1 + F_2 + F_3 = 1.60 \, q_\infty \, b \, c$$

In this case all the forces are in the same direction and so add arithmetically, but in general there would be some regions of positive pressure on the top surface and some negative pressures on the bottom, and these would have to be treated as resulting in negative ("pulling") forces, even for our simple flat plate.

If you have been geometrically alert during these steps, you now realize that the numerical coefficient (1.6) for the resultant force is just the **area** encompassed on the C_P diagram by the upper and lower surfaces! Therefore, from now on, you can just glance at a C_P graph and immediately know whether the airfoil is producing a little or a lot of aerodynamic force. Another useful conclusion is that a maximum aerodynamic force (including a maximum lift force!) requires as much of the lower surface as possible to be close to its **positive** maximum (+ 1) while the top surface should have the largest possible distribution of maximum **negative** pressures (caused by maximum possible speeded-up and yet attached air).

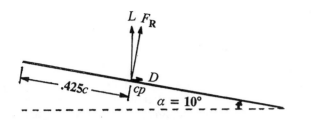

Fig. 8.5 Resolution of the pressure distribution into one equivalent force, with lift and drag components

STEP 3. (Figs. 8.4 and 8.5) Find the lift and drag forces on this fictitious airfoil.

$$L = F_R \cos \alpha = \left(1.60 \, q_\infty \, b \, c\right) \cos 10° = \mathbf{1.58} \, q_\infty \, b \, c$$
$$D = F_R \sin \alpha = \left(1.60 \, q_\infty \, b \, c\right) \sin 10° = \mathbf{0.28} \, q_\infty \, b \, c$$

STEP 4. (Figs. 8.4 and 8.5) Find the point where the equivalent force produced by the pressure distribution must be acting if it is to produce the **same** moment as well as the same force. This point is known as the **center of pressure** (cp) and, for actual airfoils, is strongly dependent on the angle of attack.

We can take moments about any point, but it is convenient here to use the leading edge.

$$\sum M_{LE} = -\left(0.2c\right)F_1 - \left(0.7c\right)F_2 - \left(0.5c\right)F_3$$
$$= -[(0.2c)(0.60) + (0.7c)(0.30) + (0.5c)(0.7)] \, q_\infty \, b \, c$$
$$= \mathbf{-0.68} \, q_\infty \, b \, c^2$$

Note that the moments about the leading edge are negative because we are using our previous definition of a positive moment as one that tends to pitch **up** the airplane nose.

The equivalent force, F_R, must act at a point $x = x_{CP}$ to produce the same moment, where we are still measuring from the leading edge:

$$M_{LE} = -\left(F_R\right)x_{CP} = -\left(1.60 \, q_\infty \, b \, c\right)x_{CP} = -0.68 \, q_\infty \, b \, c^2$$

Solving this equation for the location of the center of pressure yields

$$x_{CP} = \mathbf{0.425} \, c$$

Therefore the pressures around this "airfoil" produce lift and drag as if a **single** force of F_R was acting at a point 42.5% back from the leading edge.

STEP 5. Find the lift, drag, and moment **coefficients** due to the assumed pressure distribution.

The idea of an airfoil coefficient is to find a dimensionless number that expresses **completely** the effect of the **geometry** of the airfoil on the air in a fashion that is **independent** of the properties of the **air** and the **size** of the airfoil. We notice that the forces we've calculated depend on the properties of the air **only** through their dependence on q_∞ while they depend on the size of the airfoil only through their dependence on the **surface area**, S, of the airfoil, where $S = bc$ for a flat plate. Therefore we **define** the lift coefficient and the drag coefficients by

$$c_\ell \equiv \frac{L}{q_\infty \, S} \qquad (8.6)$$

DEFINITION OF AIRFOIL LIFT COEFFICIENT

$$c_d \equiv \frac{D}{q_\infty \, S} \qquad (8.7)$$

DEFINITION OF AIRFOIL DRAG COEFFICIENT

Similarly, the moment coefficient should depend only on the dynamic pressure, the surface area, **and** the chord c (because a moment is a force times a distance) and so we define

$$c_{m, x} \equiv \frac{M_X}{q_\infty \, S \, c} \qquad (8.8)$$

DEFINITION OF AIRFOIL MOMENT COEFFICIENT

where X is the point about which the moment is taken.

Therefore, for our fictitious flat plate airfoil, we find that

$$c_\ell = \frac{1.58 \, q_\infty \, b \, c}{q_\infty \, b \, c} = \mathbf{1.58}$$

$$c_d = \frac{0.28 \, q_\infty \, b \, c}{q_\infty \, b \, c} = \mathbf{0.28}$$

$$c_{m, LE} = \frac{-0.68 \, q_\infty \, b \, c^2}{q_\infty \, b \, c^2} = \mathbf{-0.68}$$

The moment coefficient about any other point can be obtained by calculating the moment about that point. For example, about the trailing edge,

$$M_{TE} = +\left(0.576c\right)\left(1.60 \, q_\infty \, b \, c\right) = 0.92 \, q_\infty \, b \, c^2$$

so $\quad c_{m,TE} = \dfrac{+0.92 \, q_\infty \, b \, c^2}{q_\infty \, S \, c} = \mathbf{+0.92}$

while the moment coefficient **about** the center of pressure is 0 because the moment arm, by definition, is zero there:

$$c_{m, CP} \equiv 0$$

D. A Real Flat Plate Airfoil

The analysis of the actual pressures around a flat plate is completely similar in principle. Figs. 8.6 and 8.7 depict the pressure distribution around a flat plate at two angles of attack and the resolution into equivalent single forces. When the pressures are not constant, as is true here, it is necessary to divide the pressure coefficient distribution into many strips in which the pressure is nearly constant. The force due to this strip is the average pressure times the area of the strip. The total force is obtained by summing the contribution of all the strips, as before. Similarly, the total moment is obtained by summing the products of the distance to a strip and its force contribution, for all the strips. The accuracy can be improved by making the

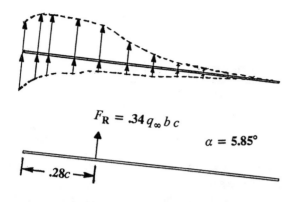

Fig. 8.6 The actual pressure distribution, and equivalent force, on a flat plate at low Reynolds number and a small angle of attack

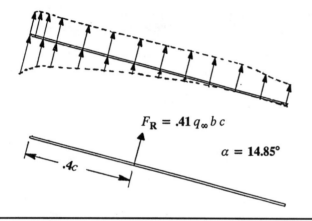

Fig. 8.7 The actual pressure distribution, and equivalent force, on a flat plate beyond the stalling angle

strips smaller and smaller, and it is relatively easy to obtain coefficient values that are as good as the experimental data.

We can glean some very useful information from examination of this actual flat plate airfoil, even though the data are for very low RN. First, for $\alpha = 5.85°$, note that the reduced pressure on the **upper** surface contributes about 70% **more** to the resultant force than does the positive pressure on the lower surface. The difference is even greater at $\alpha = 14.85°$. But notice the almost **constant** pressure over nearly the **entire** upper surface at this higher angle of attack. What could this mean? It could and does mean that the air flow has **separated** immediately behind that sharp leading edge, leaving the whole upper surface immersed in the turbulent wake. In general, whenever you see a region of (almost) constant pressure like this one on the rearmost part of an upper surface of an airfoil, even a good one, you can be pretty confident that the entire region is in the turbulent wake behind a separated boundary layer. This is a bonus from studying pressure distributions, although there are other ways of identifying the extent of separated flow in both wind tunnels and in flight.

Lift and drag are both proportional to q_∞ and S, so the ratio of lift to drag, (L/D), is **equal** to the ratio of the coefficients:

$$\frac{L}{D} = \frac{c_\ell}{c_d} \tag{8.9}$$

For our little flat plate we can calculate that $L/D = 9.75$ at $\alpha = 5.85°$ and $L/D = 3.8$ at $\alpha = 14.85°$. The L/D value is a good measure of the **efficiency** with which an airfoil is doing its job of pushing down on the air. A flat plate at $\alpha = 5.85°$ is **not** very good, but at $\alpha = 14.85°$ (good and stalled) it is a **disaster**. It does lift, even when stalled, but there is good reason to prefer curved and thicker airfoils for most aircraft (next chapter).

As you may have guessed already, L/D for the whole aircraft is also a common index of efficiency for the complete airplane.

E. A Fictitious Cambered Airfoil

So a flat plate airfoil isn't good for us, either structurally or aerodynamically. The next level of approximation to a real airfoil would be a wedge-shaped one—which would be structurally feasible and would develop lift at $\alpha = 0°$. Figures 8.8a and 8.8b present the geometry of our latest airfoil invention and the invented pressure distribution. The pressure on the rear end is assumed to be the same as on the rear half of the upper surface.

EXAMPLE 8.2. Obtain lift and drag and moment coefficients for the fictitious airfoil and pressures shown in Fig. 8.8.

Solution: With constant pressures on each segment, the equivalent force vector on each segment will again act at the

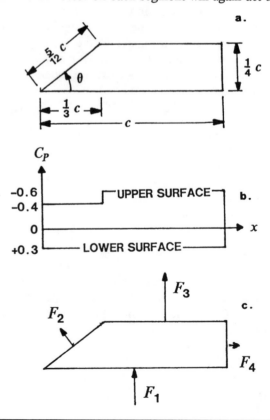

Fig. 8.8 A fictitious wedge airfoil, developing lift at zero angle of attack

mid-point of the segment. Assuming a span of b, the four forces are (Fig. 8.8c):

$$F_1 = (0.30)\, q_\infty (b \times c) = 0.30\, q_\infty\, b\, c$$

$$F_2 = (0.40)\, q_\infty \left(b \times \frac{5}{12} c\right) = 0.167\, q_\infty\, b\, c$$

$$F_3 = (0.60)\, q_\infty \left(b \times \frac{2}{3} c\right) = 0.40\, q_\infty\, b\, c$$

$$F_4 = (0.60)\, q_\infty \left(b \times \frac{1}{4} c\right) = 0.15\, q_\infty\, b\, c$$

Next we wish to find the lift and drag forces on this fictitious airfoil. All of the forces are either lift or drag except for force F_2 which has lift and **anti**-drag components (like the upper forward surface of real airfoils!). If the wedge angle is theta (θ), the lift component for F_2 is $L_2 = F_2 \cos \theta$ and the anti-drag component is $D_2 = F_2 \sin \theta$.

From Fig. 8.8a, $\sin \theta = \dfrac{c/4}{5/12\,c} = 0.6$

and $\quad \cos \theta = \dfrac{c/3}{5/12\,c} = 0.8,$ so

$$L_2 = \left(0.167\right)\left(0.8\right) q_\infty\, b\, c = 0.133\, q_\infty\, b\, c$$

and $\quad D_2 = \left(0.167\right)\left(0.6\right) q_\infty\, b\, c = 0.10\, q_\infty\, b\, c$

The overall lift and drag forces are then given by

$$L = F_1 + L_2 + F_3 = (0.30 + 0.133 + 0.40)\, q_\infty\, b\, c$$

$$= 0.833\, q_\infty\, b\, c$$

$$D = -D_2 + F_4 = \left(-0.10 + 0.15\right) q_\infty\, b\, c$$

$$= 0.050\, q_\infty\, b\, c$$

so $\quad c_\ell = 0.833 \quad$ and $\quad c_d = 0.050$

The pitching moment about the leading edge for our foolish wedge airfoil is

$$\sum M_{LE} = -\left(\frac{c}{2}\right) F_1 - \left(\frac{5c}{12}\right) F_2 - \left(\frac{2c}{3}\right) F_3 = \left(\frac{c}{8}\right) F_4$$

$$= -\left(\left(\frac{1}{2}\right) F_1 + \left(\frac{5}{24}\right) F_2 + \left(\frac{2}{3}\right) F_3 - \left(\frac{1}{8}\right) F_4\right) c$$

$$= -\left(\frac{1}{2}(0.30) + \frac{5}{24}(0.167) + \frac{2}{3}(0.40) - \frac{1}{8}(0.15)\right) q_\infty\, b\, c^2$$

$$= -0.433\, q_\infty\, b\, c^2$$

and therefore

$$c_{m,LE} = -0.433$$

F. Real Airfoils

For **real** airfoils, which typically are very curvaceous, it is necessary to divide the airfoil surface into many little strips and then determine the **slope** of the airfoil surface for each strip, and resolve the resultant force contribution into components along fixed directions (often perpendicular and parallel to the chord line), and finally add the components for the strips. At this point a computer program to do the work is a practical necessity.

First, an algebraic expression for a line that goes through the defining coordinates must be obtained. In practice, since no single function will go through all the points, a group of poly-

nomial functions that smoothly join together is usually calculated, using the principle of least squares to get the best possible fit. Then the slope of any point on the airfoil surface is easily calculated, using the first derivative of the function (differential calculus). Second, another set of functions that are a good fit to the experimental pressure distribution must be obtained. Finally, the summing is accomplished numerically.

The process just described is used by NASA and most other aeronautical groups to determine the lift and the moment coefficients for the airfoils that they test in a wind tunnel. The drag coefficient, in contrast, is usually obtained by measuring the momentum (or energy) loss of the air behind the airfoil. Data for this are obtained by a pressure-measuring device called a wake survey rake (because it looks like a garden rake, presumably). Fig. 8.9 presents top and side views of a wake rake used in recent NASA airfoil studies.

The reasons for using the wake rake to obtain the drag coefficient rather than using the pressure distribution, as we have done, are (**a**) the pressure forces give only the **pressure drag** and don't include the viscous drag contribution to the total drag and (**b**) it is felt that the small values of the drag coefficient for actual airfoils will make it hard to calculate the pressure coefficient accurately, especially because the drag calculation typically requires the subtraction of relatively large pressures. Thus the drag coefficient calculated from the pressure distribution is usually not even reported.

When the author calculated airfoil coefficients from published NASA pressure coefficients, it was with the eager anticipation of being able to estimate the fraction of the reported drag coefficient that was pressure-related and the fraction that was due to skin friction. Alas, the calculated pressure drag coefficients were greater than the reported drag coefficients (Fig. 10.35) even though there was very good agreement for the calculated lift and moment coefficients. The two NASA researchers who have been asked about this discrepancy agree that it is present but haven't been able to provide a satisfactory explanation. If nothing else, this points out that it is best to compare airfoils that have been tested in the **same** wind tunnel with the same analysis techniques. (And, if possible, with the same airfoil model!)

An older method for determining the lift, drag, and moment of an airfoil is to measure the actual force required to hold the airfoil stationary in the tunnel, as well as their points of application. Early aerodynamicists, including the Wright brothers, used this technique. However, the actual pressure distribution tells us a **great deal** about **how** the airfoil does its lifting and dragging and twisting, and this feature is very important for modern airfoil theory as well as for our purposes (next chapter).

Our introduction of the three airfoil coefficients has developed logically from the definition of the pressure coefficient and Bernoulli's energy equation. It might be worthwhile to note that the customary derivation of these coefficients proceeds instead from *dimensional analysis*. The idea of dimensional analysis is to algebraically determine the functions of the

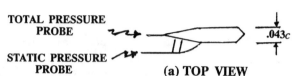

known variables that, when multiplied together, will yield a quantity having the right dimensions (e.g., force units for the lift equation). If, for example, we consider α, the ρ and the ν of the air, V_∞, and the dimensions of the airfoil x to be the significant parameters, we can (after a great deal of messy algebra) determine that

$$\text{Aerodynamic force} \propto \left(\frac{V}{V_A}\right)^? \left(\frac{\rho V x}{\mu}\right)^? (\alpha)^? (\rho V^2) (x^2)$$

(where "$\propto$" means "is proportional to.")

The first three terms in parentheses do not have intrinsic dimensions, and so **any** power (0.5, 1.0, 1.5, etc.) would keep the dimensions of the whole expression those of force, and that is the meaning of the superscript **?** over these factors. The fourth term in parentheses suggests a dependence on q_∞ and the last term can be interpreted as the area, S, of the airfoil, since x represents any dimension with the unit of length. (Note that α is a dimensionless parameter because it can always be defined as the ratio of two similar quantities, as it is when radian measure is used. The last two terms, proportional to the dynamic pressure and an area, are the ones that make this expression have units of force. Therefore we can guess that airfoil coefficients should also be expected to depend on the **Mach number** (the first term) and the **Reynolds number** (the second term) in some undefined fashion. Dimensional analysis is a powerful tool!

We certainly should have guessed that our airfoil coefficients would depend on M_∞ and RN, as well as on α, because we recognize that these three parameters are also important in determining the nature of the air flow around an airfoil. Sometimes the dependence of the coefficients on RN and M_∞ is very strong. You cannot hope to get the published performance from an airfoil unless you use it at a RN and a M_∞ consistent with its test environment. An added uncertainty is introduced by the fact that the typical wind tunnel generates considerably more small-scale turbulence than there is in atmospheric air and the effective RN of the tunnel may be considerably greater than the value that is calculated directly from the formula, which is often the value that is reported.

It would be a mistake to think that you'll never see a flat plate airfoil in action, for the little metal strips used for making vortex generators are dandy little flat plate airfoils! They must be pointed into the relative wind at such an angle as to produce suitable vortices but with minimum drag. Reference 2 describes the process by which optimum vortex generators were fitted to the top surface of the canard on the round-the-world-unrefueled Voyager to cure its rain problem. More of this in Chapter 11.

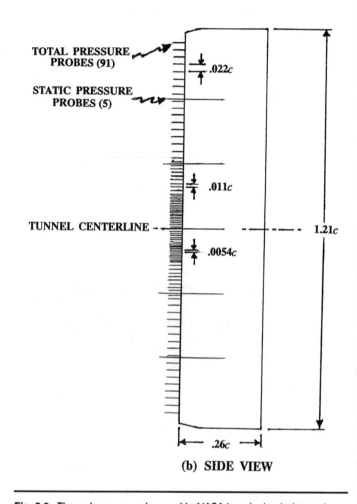

Fig. 8.9 The wake survey rake used in NASA Langley's wind tunnel

Summary

The chord, the thickness, the location of maximum thickness, the maximum camber and its location, and the leading edge radius are all key geometrical parameters of an airfoil or wing section. A symmetric airfoil is an airfoil with zero camber. The aerodynamic forces acting on a given airfoil depend only on properties of the air and the angle of attack of the airfoil, where the angle of attack is defined as the angle between the wing chord and the direction of undisturbed oncoming air. The

Pic. 8.1 Testing an airfoil in a two-dimensional wind tunnel (NACA photo courtesy of NASA Langley)

angle of attack is independent of the direction of the force of gravity.

The pressure coefficient at a given point on an airfoil is defined as the change in static pressure at that point divided by the dynamic pressure. The Bernoulli equation tells us that a positive pressure coefficient corresponds to slowed down local air flow, at least for incompressible flow. Experimental measurements of the pressures around an airfoil in wind tunnels are reported in terms of variations in the pressure coefficient. The area carved out on a C_P graph by the upper and lower surfaces is directly proportional to the aerodynamic force generated by the airfoil at that angle of attack. By converting the pressures into forces it is possible to calculate the lift force, the pressure drag force, and the pitching moment acting on the airfoil; in general, assuming fully chord-wise flow, one must consider the sum effect of the pressures at each small increment of arc in the chordwise direction along both upper and lower surfaces.

The pressure coefficient for each point on an airfoil depends only on the ratio of local air flow speed to overall air flow speed, at least for incompressible flow; therefore it leads to the existence of coefficients of lift, drag, and pitching moment that are uniquely descriptive of the airfoil for those flow conditions. Thus an airfoil's coefficients depend on the angle of attack, the Reynolds number, and the Mach number.

Symbol Table (in order of introduction)

c wing or airfoil chord (distance from the leading edge to the trailing edge)

x a location along the chord line of an airfoil, measured from the leading edge and expressed in fractions of the chord length

z a location perpendicular to the chord line of an airfoil, measured from the chord line and in fractions of the chord length ($+$ for the upper surface, $-$ for the lower surface)

α angle of attack: the angle of the chord line (or other fixed line on the airfoil) relative to the direction of distant, oncoming, undisturbed air (the relative wind)

C_P the pressure coefficient, which gives the local air pressure at a given point on the surface of an airfoil in terms of a fraction of the dynamic pressure

V local airspeed around an airfoil

b the width (span) of the airfoil being tested

F_R the resultant or equivalent force on an airfoil due to the distributed pressures on it

L the lift force on an airfoil

D the drag force on an airfoil

M the pitching moment on an airfoil, measured about some designated point

M_{LE} the pitching moment on an airfoil, measured about the leading edge

M_{TE} the pitching moment on an airfoil, measured about the trailing edge

cp the center of pressure, the point where the resultant aerodynamic force on an airfoil equivalently acts

$\dot{c}_\ell$ the airfoil lift coefficient

S the surface area of an airfoil (projected area, as in a shadowgraph — not the actual area)

c_d the airfoil drag coefficient

c_m the airfoil moment coefficient

Review Questions

1. What is the assumption regarding the experimental air flow conditions that is implicit in a report or discussion of **airfoil** properties as contrasted with **wing** properties? In what respect are these airfoil properties useful to us?

2. What is the name given to an airfoil possessing zero camber?

3. What is the significance of positive camber in an airfoil?

4. What airfoil property gives a clue as to whether a particular airfoil is intended to support substantial laminar flow?

5. Optimum climb, approach, and the stall can be determined for any aircraft under any operational conditions (various altitudes, power settings, weights, etc.) **if** the corresponding values for ___ are known for any **one** set of operating conditions. However, there remains some dependence on ___ and ___ which can be very important.

6. New observers are amazed at the steep angle of climbout of jet airliners and of overpowered aerobatic aircraft. What determines whether an observed climbout angle is really hazardous?

7. What symbols are commonly used to represent distances perpendicular and parallel, respectively, to the chord line on an airfoil?

8. Qualitatively, what is the angle of attack for an aircraft that is diving perpendicular to the ground under no-wind conditions? (Hint: Your answer depends on the type of airfoil on the wing; you may neglect fuselage and finite-wing effects.)

9. Where is the curvature of an airfoil most critical? Can you guess what operating conditions would tend to alter this curvature?

10. What kinds of aircraft can profitably use a symmetrical airfoil? Why?

11. The use of pressure **coefficients** to report the pressure distributions on a wind tunnel airfoil almost completely removes the dependence of the pressures on ___ ___. Remaining, however, are a dependence on ___ and ___.

12. What is the **maximum** possible value that is expected for any airfoil pressure coefficient? ___ What is such a point called? ___ ___

13. The lift force on an airfoil is defined as the component of the resultant force that is perpendicular to the ___ ___.

14. The drag force on an airfoil or wing is exactly parallel to the horizon **only** when the aircraft is in ___ ___.

15. What symbol is normally used to represent the **span** of an airfoil in a wind tunnel? ___

16. What is the convention for representing airfoil pressures on an outline of the airfoil?

17. The ___ is the point at which the resultant aerodynamic force effectively acts; therefore the pitching moment for pitching about an axis through this point is ___.

18. The part of the total aerodynamic drag on an airfoil that can be calculated from a knowledge of pressure coefficients is the ___ drag; the part that must be determined some other way is the ___ drag. Generally the sum of these two (which is the **total drag** for an **airfoil**), is obtained in a single measurement through the determination of the ___ loss of the air in the wind tunnel behind the airfoil, using a probe called a ___ ___.

19. It is possible to separate the measured aerodynamic forces on an airfoil into three contributing parts, for given air flow conditions: (**a**) the dynamic pressure, which contains the effect of the ___ of the air, (**b**) the ___ ___ of the airfoil, which contains the effect of the **size** of the airfoil, and (**c**) the ___, which contains the effect of the **shape** of the airfoil.

20. The pitching moment on an airfoil due to aerodynamic forces contains a fourth contributing part, namely the ___ of the airfoil.

21. One of these four contributing parts, the airfoil ___ , is constant only for constant air flow conditions. Therefore this contribution has an additional dependence on (**a**) ___, (**b**) ___, and (**c**) ___, because all these variables affect the nature of the air flow around the airfoil.

22. It is possible to derive the functional dependence of aerodynamic forces on the various properties of the air and the airfoil by requiring that the final equation have units of force. This approach to finding a suitable function is called ___ analysis.

23. How can a region of separated (stalled) air above an airfoil be recognized from its pressure distribution?

Problems

1. What pressure coefficient corresponds to a pressure of 2000 lb/ft^2 and a free airstream of 100 ft/s, under standard SL conditions? Answer: –9.78

2. For a free airstream speed of 100 ft/s, what local air speed corresponds to a pressure coefficient of +0.5, assuming standard SL conditions? Answer: 70.7 ft/s

3. For a free airstream speed of 100 ft/s, what local air speed corresponds to a pressure coefficient of –0.5, assuming standard SL conditions? Answer: 122 ft/s

4. Find c_ℓ, c_d, and $c_{m,LE}$ for the flat plate of Figs. 8.6 and 8.7. Partial answer: For Fig. 8.6, 0.34, 0.035, and –0.095, respectively

5. Suppose that a flat plate airfoil at an angle of attack of 13° has a pressure coefficient of –2.0 on the first 30% of its upper surface, –0.8 on the remaining 70%, and +1.0 on its lower surface. Sketch the pressure distribution diagram and, following the steps in the text and showing your work, determine the lift, drag, and moment coefficients (about the one-quarter chord point). Determine also the location of the *cp*. Partial answer: Lift coefficient is 2.10 and center of pressure is at 0.44*c*.

6. Repeat the calculation of problem 5 except assume that, for a 20° angle of attack, the pressure coefficient is –3.0 on the upper 20% of the upper surface, –0.4 on the remaining 80%, and +0.4 on the lower surface. Partial answer: The moment coefficient about the quarter chord point is –0.12.

7. Repeat the calculation of problem 6 except assume a **lower** surface pressure coefficient of –1.0 on the first 20% and +0.5 on the remaining 80%. Answer: –0.19

8. Suppose that a certain airfoil is wedge shaped, with a 30°–60°–90° triangle as its cross-section, and that the angle of attack is **zero**. If the upper surface has a pressure coefficient of –0.3, the lower surface +0.2, and the rear surface –0.4, find the coefficients and the *cp*. Answer: lift coefficient = 0.50, drag coefficient = 0.06, and moment coefficient about leading edge = –0.23; center of pressure is at 0.46 *c*.

9. Determine the moment coefficient about the quarter-chord point for the real flat plate airfoil at 5.85° angle of attack (Fig. 8.6).

10. Determine the moment coefficient about the quarter-chord point for the real airfoil flat plate airfoil at an angle of attack of 14.85° (Fig. 8.7). Answer: –0.061

11. Determine the **resultant** pressure force on the fictitious wedge airfoil of Fig. 8.8. Answer: 0.834 $q_\infty \, b \, c$

12. Determine the location of the center of pressure for the fictitious wedge airfoil. Answer: 0.52 *c*

13. Determine the moment coefficient about the trailing edge for the fictitious wedge airfoil. Answer: +0.40

14. Determine the moment coefficient about the quarter-chord point for the fictitious wedge airfoil. Answer: –0.225

15. A certain airfoil has a triangular shape with smallest angle pointing into the relative wind. The chord (base of the triangle) is c; the angle at the leading edge is 30°, the angle at the upper trailing edge is 60°, and the angle at the lower trailing edge is 90°. Suppose that, for $\alpha = 0°$, C_P on the upper surface is –0.3, on the lower surface is +0.2, and on the rear end is –0.4. What are the resultant forces on the lower surface, the rear surface, and the upper surface?
Answers: $+0.200\,q_\infty\,b\,c$, $0.231\,q_\infty\,b\,c$, and $0.347\,q_\infty\,b\,c$

16. For the airfoil of problem 15, what are the lift and drag components of the resultant force on the upper surface?
Answers: $0.301\,q_\infty\,b\,c$, $0.174\,q_\infty\,b\,c$

17. For the airfoil of problem 15, what are the lift and drag coefficients? Answers: 0.501, 0.057

18. For the airfoil of problem 15, what is the resultant pressure force? Answer: $0.504\,q_\infty\,b\,c$

19. For the airfoil of problem 15, what is the pitching moment about the leading edge due to (**a**) the lower surface, (**b**) the rear surface, and (**c**) the upper surface?
Answers: $-0.100\,q_\infty\,b\,c^2$, $0.067\,q_\infty\,b\,c^2$, $-0.200\,q_\infty\,b\,c^2$

20. For the airfoil of problem 15, what is the overall pitching moment coefficient about the leading edge? Answer: –0.233

21. For the airfoil of problem 15, where is the center of pressure located? Answer: $0.46\,c$

22. For the airfoil of problem 15, what is the pitching moment coefficient about the lower trailing edge? Answer: +0.27

23. Suppose that a flat plate airfoil at an angle of attack of 20° has a pressure coefficient of –0.5 on the entire upper surface, +0.7 on the forward 20% of the lower surface, and +0.3 on the remainder of the lower surface. Sketch the pressure distribution diagram and, following the steps in the text and showing your work, determine the lift, drag, and moment coefficients. The moment coefficient should be the one about the $0.25c$ point. Determine also the location of the cp. Partial answer: The drag coefficient is 0.301 and the requested moment coefficient is –0.188.

References

1. Fage, A., and Johansen, F. C., "On the Flow of Air Behind an Inclined Flat Pate of Infinite Span," R & M No. 1104, British A.R.C., 1927. The pressure data for the flat plate were obtained from this reference.

2. Bragg, M. B., and Gregorek, G. M., "Experimental Study of Airfoil Performance with Vortex Generators," *Journal of Aircraft*, Vol. 24, No. 5, May, 1987. (This is the story of how vortex generators—flat plates—solved the around-the-world Voyager's rain problem.)

Pic. 8.2 NACA/NASA aerodynamicist Robert T. Jones expounds on the theory of airfoils at an Oshkosh Experimental Aircraft Association Convention forum.

Pic. 8.3 Ultralights are becoming more airplane-like, as this Fisher Classic illustrates.

Pic. 8.4 The Mini Max, this one with a 2-cylinder, 1/2-VW engine, is a popular ultralight.

Chapter 9

Airfoil Coefficients

We were slipping through the air at 540 mph. I'd always liked the little XP-AZ5601-NG because of her simple controls and that Prandtl-Reynolds meter tucked away in the upper right corner of the panel. I checked over the gauges. Water, fuel, rpm, Carnot efficiency, groundspeed, enthalpy. All OK. Course 270°. Combustion efficiency normal at 23 percent. The good old turbojet was rumbling along as smoothly as always and Tony's teeth were barely clattering from the 17 buckets she'd thrown over Schenectady. Only a small stream of oil was leaking from the engine. This was the life.

I knew that the engine in my ship was good for more speed than we'd ever tried. The weather was so fair, the sky so blue, the air so smooth, I couldn't resist letting her out a little. I inched the throttle forward a notch. The regulator only hunted a trifle and everything was steady after five minutes or so. 590 mph. I pushed the throttle again. Only two nozzles clogged up. I pushed the small-slot cleaner. Open again. 640 mph. Smooth. The tailpipe was hardly buckled at all—there were still several square inches open on one side. My fingers were itching on the throttle and I pushed it again. She worked up to 660 mph., passing through the shaft critical without breaking a single window in the ship. It was getting warm in the cockpit so I gave the vertex refrigerator a little more air. Mach 0.9! I'd never been that fast before. I could see a little shocklet outside the port window so I adjusted the wing shape and it disappeared.

Tony was dozing now and I missed the smoke from his pipe. I couldn't resist letting the ship out another notch. In ten minutes flat we leveled off at Mach 0.95. Back in the combustion chamber the total pressure was falling like hell. This was the life! The Kármán indicator showed red but I didn't care. Tony's candle was still burning. I knew gamma was down but I didn't give a damn.

I was dizzy with the thrill. Just a little more! I put my hand on the throttle but just at that moment Tony stretched and his knee struck my arm. The throttle jumped up a full ten degrees! Crash! The little ship shuddered from stem to stern and Tony and I were thrown into the panel by the terrific deceleration. We seemed to have struck a solid brick wall! I could see the nose of the ship was crushed. I looked at the Mach meter and froze. 1.00! My God, I thought in a flash, we're on the peak! If I don't get her slowed down before she slips over, we'll be caught in the decreasing drag! I was too late. Mach 1.01! 1.02! 1.03! 1.04! 1.06! 1.09! 1.13! 1.18! I was desperate but Tony knew what to do. In a flash he threw the engine into reverse! Hot air rushed into the tailpipe, was compressed in the turbine, debusted in the chambers, expanded out the compressor. Kerosene began flowing into the tanks. The entropy meter swung full negative. Mach 1.20! 1.19! 1.18! 1.17! We were saved. She crept back, she inched back, as Tony and I prayed the flow divider wouldn't stick. 1.10! 1.08! 1.05! Crash! We had struck the other side of the wall! Trapped! Not enough negative thrust to break back through. As we cringed against the wall, the tail of the little ship crushed, Tony shouted, "Fire the JATO units!" But they were turned the wrong way! Tony thrust his arm out and swung them forward, the Mach lines streaming from his fingers. I fired them! The shock was stunning. We blacked out.

I came to as our gallant little ship, ragged from stem to stern, was just passing through Mach zero. I pulled Tony out and we slumped to the ground. The ship decelerated off to the east. A few seconds later we heard the crash as she hit the other wall.[1]

They never found a single screw. Tony took up basket weaving and I went to MIT.[1]

A. NASA's LS(1)-0417 Airfoil

The LS(1)-0417, described in a Technical Note in 1973, was the first new low speed airfoil to come out of the National Aeronautics and Space Administration (NASA) since the 1940s, when it was the National Advisory Committee on Aeronautics (NACA, often pronounced "Nahca," especially if one isn't in the presence of a former Nahca engineer). It was a spin-off from the newly available theories and computer programs that had been used to develop some successful new high speed airfoils. Initially known as the **GA(W)-1**, which stood for General Aviation (by **Whitcomb's** group) **#1**, its name was changed a year or two later to be more informative and more consistent with previous airfoil numbering schemes. Like its high speed relatives, it employs significant camber at the very rear of the lower surface. The intent is to increase the maximum lift at high angles of attack, but a side effect is that it becomes an "aft-loaded" airfoil with a relatively large pitching moment. Since a complete set of pressure distributions was included with its description, we can peer above and below and in front and in back of this airfoil and see how a modern airfoil does its job.

Figure 9.1 presents the pressure envelope for seven angles of attack, from $-7.99°$, well below the normal cruising angle, to $21.14°$, which is above the positive-angle stall. Let's look closely at these pressure distributions to see what we can learn. They are derived from pressure coefficient data taken at $M_\infty = 0.15$, $RN = 6.2 \times 10^6$, and with #80 roughness applied at $0.08c$ to ensure **turbulent** flow over most of the airfoil, as was expected to be the case in actual use.

Figure 9.1a is for $\alpha = -7.99°$. Note that there is some positive lift being developed by the rear half of the upper surface but the negative lift on the bottom surface overwhelms it.

Figure 9.1b is nearly the **zero-lift** angle of attack. To obtain zero lift from a positively-cambered airfoil, the angle of attack has to be negative. Did you notice where the *cp* is located for this angle of attack? The pressure distribution for the upper surface results in a resultant force coefficient of 0.61 at $0.49c$ which is nearly matched by a lower surface force coefficient of 0.66 at $0.18c$. However, the two resultant forces are separated by a large distance, $0.31c$, so the pitching moment is quite **large**. How can the small resultant force, 0.05, hope to generate such a large moment? It can do it only if it is located about 3.5 chord lengths in front of the airfoil nose! Clearly this is a little inconvenient. In general, whenever a cambered airfoil is producing exactly zero lift, it will still have a different pressure distribution on its upper and lower surface, resulting in a zero-lift pitching

moment. (Two equal forces that are **not** acting at the same point on a body produce a torque that tends to cause rotation, even though there is translational equilibrium. Such a combination is often called a *couple*.) So where is the *cp* when the resultant force is zero but the moment is not? The only way to get something from zero is to multiply by infinity, but placing the *cp* at infinity is likely to cramp our style even more. The consequence of all this is that the *cp* has fallen out of favor and has been largely replaced by the **aerodynamic center**, a concept discussed in the next section.

This airfoil is designed to spend most of its life at an angle of attack of about 0° (Fig. 9.1c)! (The "04" in the LS(1)-0417 designation means that the **minimum drag** occurs at a lift coefficient of 0.4, which is about where we are. This is also known as the **design lift coefficient**; and is often symbolized as c_{ℓ_i}.) The designers really were very successful in achieving their goal of uniform pressures over both surfaces for this angle of attack. The result is a very efficient airfoil, for its L/D at this angle is 67. (Compare this with the flat plate's 9.75 in the preceding chapter!) Note that the bottom surface is actually **hurting** the lift cause a bit because it has a less-than-ambient pressure on it as well. Very interesting.

Fig. 9.1d shows that increasing the lift by increasing the angle of attack to 8.02° yields more lift through much heavier loading (**forward thrust**) on the upper part of the nose section; the airfoil's efficiency, though, remains high at $L/D = 68$. At this angle of attack, the BL separated at about $0.9c$ on the upper surface, but this produced little effect on airfoil properties.

Doubling the angle of attack to 16.04° (Fig. 9.1e) accentuates this trend. Now the lower surface is helping out to the extent of 20% of the total lift, thanks to the overall above-ambient pressures there. L/D, however, has decreased to about 30. You know what that **constant pressure** region on the rear 40% of the upper surface means, don't you? Yep, the separation point is moving forward from the trailing edge and now the last 40% or so of the upper surface is immersed in the turbulent wake. No wonder the L/D is down!

Maximum lift from this airfoil is obtained in only three more degrees, at $\alpha = 19.06°$, and therefore this angle is defined as the **stalling** angle (Fig. 9.1f). Separation does not seem to be appreciably worse but the loading on the upper part of the nose section is terrific. Those pressure arrows are long and pointing partially forward, leading to the expressions "**leading edge suction**" or "**leading edge thrust**." If you load a wing too hard at too high a speed, this is where the crumbling usually begins.

Only two more degrees angle of attack and — **disaster** (Fig. 9.1g). Almost the entire upper surface is in the separated, turbulent wake, the lift is way down and the drag is way up (the

[1]C. D. Fulton, "Through the Sonic Wall," *Aviation Week & Space Technology*, December 12, 1949.

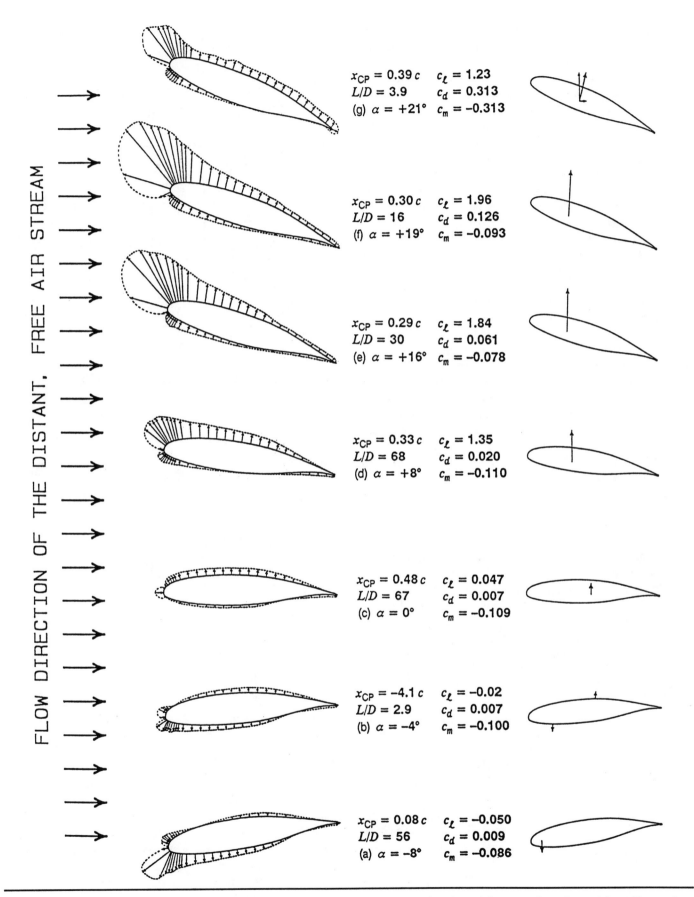

Fig. 9.1 LS(1)-0417 pressure distributions, lift, drag, and center-of-pressure location for angles of attack from negative to beyond the stalling angle

lift-to-drag ratio is only 3.9). The efficiency of our airfoil after the stall is almost exactly equal to the efficiency of our flat plate (Section 8D) at $\alpha = 14.85°$. If an aircraft has the power to overcome the drag, it is still possible to **fly** with this kind of lift and drag—except that the actual separation pattern is very unstable, causing the pattern to be different on the two wings and resulting in extremely violent rolling tendencies. At least the nose isn't hurting so much any more. Both wings may operate in this post-stall region during an upright spin. Light aircraft have no business trying to use angles of attack above the stalling angle in normal flight but, more and more, fighter aircraft are being designed to fly up to 60° or more in order to obtain maneuverability advantages in dog fights.

The stall exhibited by the LS(1)-0417 is a good type—the trailing edge type—in which separation begins at the trailing edge and gradually works its way forward. However, remember that this testing was done with an intentionally roughened airfoil to ensure a turbulent boundary layer beyond $0.08c$. If this airfoil is used at low RN with a smooth surface, laminar separation is possible and a wicked, abrupt stall might result.

Note too from Fig. 9.1 what has been happening to the stagnation point as the angle of attack increases from its zero-lift value. The flow dividing point evidently is moving from close to the leading edge to noticeably below the leading edge. This is why a little vane close to the leading edge can be used as an electrical switch to warn of an approaching stall by completing an electrical circuit through a stall horn or buzzer.

The distribution of the lift and drag along the chord, for both the design cruise configuration ($\alpha = 0°$) and the stall angle, are presented in Fig. 9.2. Note again the importance of the nose section at high angles and also the fact that the lower surface is actually pulling the plane down at cruise—but it ends up producing a steady 20% of the lift for α between about 10° and the stall angle. Detailed information (for this particular airfoil and at a relatively high RN) on the fractional contribution of the two airfoil surfaces to lift is presented in Fig. 9.3

At the stall, the nose section really is producing thrust or "negative drag," but it can't totally compensate for all that pressure drag on the rear 80% of the upper surface. The wide arrows on the drag distribution curves in Fig. 9.2 indicate where the net drag is zero, starting from the leading edge, so the actual drag is equal to the drag contribution of what is behind that point.

Figure 9.1 shows the center of pressure to be at mid-chord in cruise, then moving forward to nearly the quarter-chord point as the angle of attack is increased to the stall, then moving abruptly rearward at angles beyond the stall as the extreme pressures around the nose collapse. Figure 9.4 presents this center-of-pressure travel over the whole operating range.

The equilibrium of an aircraft in cruise or climb or descent requires that the moments add to zero. But the movement of the cp makes it difficult to perform this calculation because the moment arm for the lift on the wing keeps changing its value. The solution is to adopt a different viewpoint, that of the aerodynamic center.

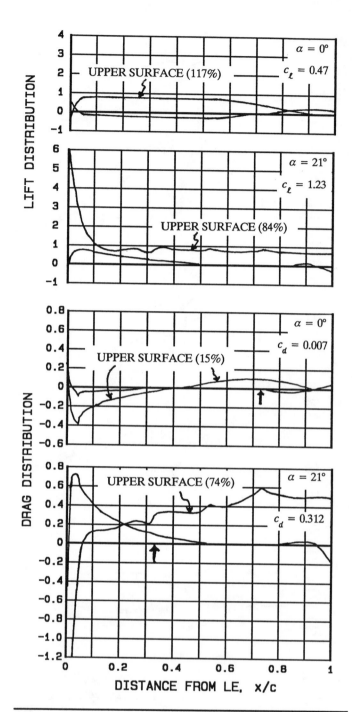

Fig. 9.2 The contribution of different parts of the LS(1)-0417 to its lift and drag, from the leading edge to the trailing edge

B. The Aerodynamic Center

We've been taking moments primarily about the leading edge, but we can choose another location if it is advantageous. Figure 9.5 presents the moment coefficient for moments about $0.20c$, $0.25c$, and $0.30c$, for the LS(1)-0417. Note that the moment coefficient increases with angle of attack for moments about $0.30c$ and decreases with angle of attack for moments about $0.20c$, but is nearly **constant** (so long as the airfoil isn't stalled) for moments about the quarter-chord point, $0.25c$. The

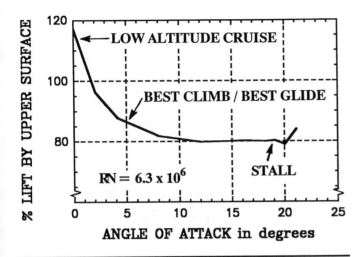

Fig. 9.3 Contribution of the upper surface to the net lift force for the LS(1)-0417

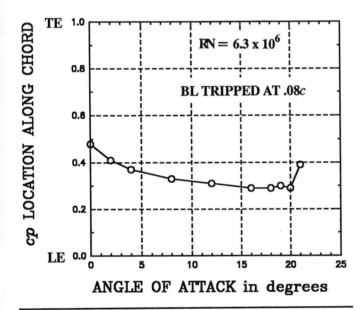

Fig. 9.4 Center-of-pressure travel for the LS(1)-0417

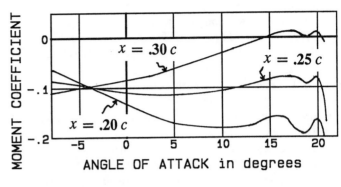

Fig. 9.5 Variation of the moment coefficient about three locations on the chord line for the LS(1)-0417

formed by the unequal position of the upper and lower surfaces forces when the net lift is zero, but the additional forces for other angles must act at the *ac* if the coefficient doesn't change. A logical consequence of this reasoning is that a symmetrical airfoil will have a moment coefficient that is nearly constant at **zero** when moments are taken about its aerodynamic center.

Because of its great usefulness in stability calculations and because it doesn't do anything foolish over the normal operating range of an airfoil, the aerodynamic center has pretty much replaced the center of pressure as a key location for airfoils.

C. Lift and Drag Curves

A lift curve is a smooth line drawn through the experimental c_ℓ versus α graph. Figure 9.6 presents the lift curve for NASA's LS(1)-0417. Note that the slope of the lift curve is nearly constant up to about 8° and then the curve starts to bend over. This is just the angle at which separation begins on the

point about which the moment coefficient is nearly constant is called the **aerodynamic center** (*ac*), and it is normally located at about 0.25*c* on every airfoil at subsonic speeds. The pitching moments generated by the wing are immensely important when calculating the stability of the airplane, and it is **so** much easier to work with a constant moment coefficient, that it has become the custom to always plot the moment coefficient about the 0.25*c* point.

What is the physical significance of the existence of an aerodynamic center? How can the moment **not** vary with angle of attack? (Actually, we're talking about the moment coefficient and not the moment itself here.) The answer is that the additional resultant force for all other angles besides the zero-lift angle must act **at** the aerodynamic center. The constant value for the moment coefficient is just the moment due to the couple

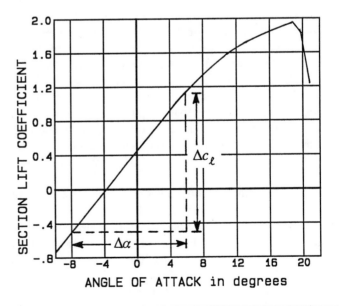

Fig. 9.6 Lift curve for the LS(1)-0417

upper surface! When separation is complete, the curve turns back down—but for some airfoils, especially at low RN, it actually recovers and begins to lift again. (Because this region is seldom of interest for civilian aircraft, it usually isn't tested—but see Fig. 17.11.) The nature of the maximum tells us whether the stall is a nice one or not; this one is. If separation starts at the leading edge, the lift curve will have a very sharp break and turn down very steeply. It is still possible to use such an airfoil by employing wing twist and stall strips and other aerodynamic crutches, but it seems smarter to start with an airfoil possessing inherently good stall characteristics, if at all possible.

The numerical value of the slope of the lift curve is of interest in some aerodynamic calculations. The slope is defined as the ratio of opposite side to adjacent side when a triangle is drawn with the linear part of the lift curve as the hypotenuse. Using the triangle marked on Fig. 9.6, we calculate the lift slope for the LS(1)-0417 to be

$$a_0 \equiv \text{slope of lift curve in linear region}$$

$$= \frac{\Delta c_\ell}{\Delta \alpha} = \frac{(c_\ell \text{ at } \alpha = 6°) - (c_\ell \text{ at } \alpha = -8°)}{6° - (-8°)}$$

$$= \frac{1.16 - (-0.51)}{14°} = 0.12 \text{ per degree.}$$

Converting this value to radians gives us 6.8/radian, which compares favorably with the general theoretical prediction of 2π per radian.

You can immediately deduce the nature and amount of the camber of an airfoil by inspection of its lift curve. This must be an airfoil with positive camber because it develops lift at $\alpha = 0°$. All symmetrical airfoils have lift curves which pass exactly through the origin (i.e., $L = 0$ at $\alpha = 0°$).

Because every aircraft operates at differing Reynolds number during cruise and approach and climb, airfoil designers try to present some indication of how their airfoil is affected by RN variations. Fig. 9.7 shows the variation in the LS(1)-0417's lift and moment curves for three different RNs over a narrow range of values. The trends are universal among airfoils in this RN range: The maximum lift is **less** at low RN because the BL is more laminar then and separates earlier. Laminar flow is good for minimum cruise drag but it is bad when separation time comes. One might be tempted to simply sketch in a lower curve for an even lower RN—but this is unlawful because the flow can become laminar enough that the stall is much more abrupt and occurs at a much lower α than expected from simple extrapolation.

To find out the optimum angle of attack for cruise, you need a plot of c_d versus c_ℓ (Fig. 9.8). Here we can see that low RN aircraft are penalized in drag as well as in maximum lift, and again the reason again is that there is more separation at the lower RN. Note that the minimum of the drag curve does occur at the advertised design lift coefficient.

For turbulent airfoils, at least, this drag curve typically has a nearly parabolic appearance [$c_d \propto c_\ell^2$], especially near the

minimum, and so this curve is often referred to as the **drag polar**. You are warned, however, that it looks much more like a bucket for some laminar airfoils, as Section B of Chapter 11 will demonstrate.

To confirm our suspicions about a symmetrical airfoil's properties vis à vis those of a positive-cambered one, take a look at Fig. 9.9. The symmetrical airfoil is a member of a group of 1940s laminar airfoils, but it has been artificially roughened

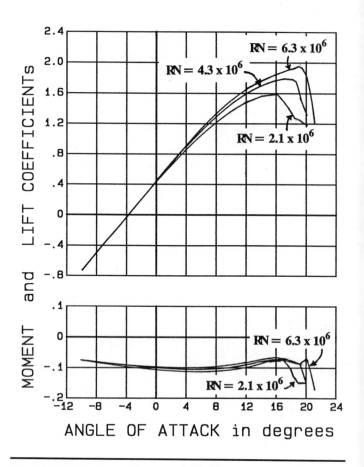

Fig. 9.7 Experimental dependence of the lift and moment coefficients for the LS(1)-0417 on Reynolds number

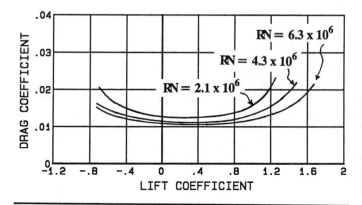

Fig. 9.8 Experimental dependence of the drag polar for the LS(1)-0417 on Reynolds number

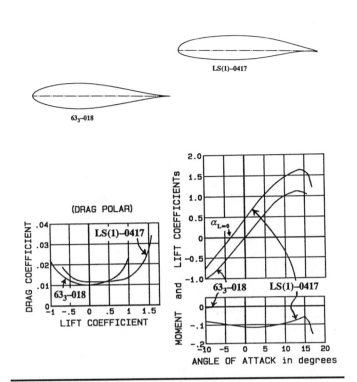

Fig. 9.9 Comparison of the lift, drag, and moment characteristics of cambered and symmetrical airfoils

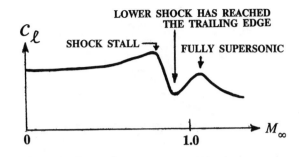

Fig. 9.10 Generalized variation of the lift coefficient through the transonic range (constant angle of attack)

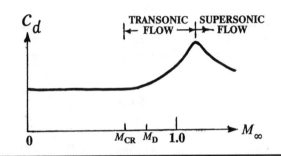

Fig. 9.11 Generalized variation of the drag coefficient through the transonic range (constant angle of attack)

to deliver the characteristics shown here. Note that the symmetrical airfoil does have symmetric lift and drag characteristics. It can provide just as much negative lift as positive lift. Even with artificial turbulence, the symmetrical airfoil has a lower minimum drag than the LS(1)-0417—but it occurs at an unusable zero value for c_ℓ. The LS(1)-0417 is much better at developing maximum lift. The moment coefficient of the symmetrical airfoil about its aerodynamic center is zero, as advertised.

D. Dependence on the Mach Number

The LS(1)-0417 is the first member of a low speed family of airfoils (yes, that is what the "LS" and "(1)" refer to) and isn't intended for use at high speeds. However, the designers do present airfoil characteristics for $M_\infty = 0.15, 0.20,$ and $0.28,$ and these data reveal only a slight loss in maximum lift coefficient as the effect of increasing Mach number.

A qualitative representation in general of the variation in c_ℓ with M_∞ for constant α for all airfoils over the entire subsonic→transonic→supersonic speed range is given in Fig. 9.10. (Compare the discussion of shock stall in Section B of Chapter 6.) At the shock stall, c_ℓ drops precipitously until the shock wave on the lower surface has reached the trailing edge, at which Mach number it begins to increase again and continues to increase until fully supersonic flow is attained. The maximum lift coefficient, however, is much less in the transonic range because of shock-induced flow separation.

In Fig. 9.11, a similarly qualitative curve for the dependence of the drag coefficient on the flight Mach number is given. Note especially the locations of M_{CR} (the critical Mach number) and M_D (the drag divergence Mach number), both of which were defined in Sec. 6B. Mixed subsonic-supersonic flow is responsible for the steep rise in c_d at M_D which abates only after the shock waves have moved to the trailing edge and supersonic flow prevails around the entire airfoil. Historically, this steep rise in the drag coefficient made it difficult to built a plane which had sufficient power to accelerate to supersonic speeds in level flight, even after the substantial controllability problems of transonic flight were mostly solved. As a result this drag rise was given the name "sound barrier."

Now it should be emphasized that the total drag of an airfoil is the product of its drag coefficient, the dynamic pressure, and the projected surface area. Even though the drag coefficient at constant angle of attack decreases as fully supersonic flow is reached, the total drag continues to **increase** with increasing Mach number because of the increase in dynamic pressure. This fact is illustrated in Fig. 9.12. A whimsical and imaginative disregard of Fig. 9.12 in favor of Fig. 9.11 is the "basis" for the delightful tale that began this chapter.

Pressure distribution graphs are helpful in understanding what is happening around an airfoil as the transonic and supersonic regimes are traversed. Because the pressure coefficient is directly related to the local air speed, there is a **critical pressure coefficient** associated with a local airspeed of Mach 1. When the upper surface pressure envelope reaches this critical pressure coefficient, we have reached M_{CR}. In Fig. 9.13a

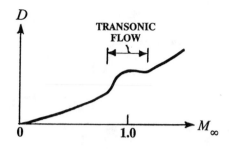

Fig. 9.12 The total drag force through the transonic range continues to increase

the flight Mach number is just subcritical and the minimum pressure coefficient is just short of the critical value. In Fig. 9.13b, we are in the transonic speed range and a shock wave has formed on the upper surface but not yet on the lower surface. In Fig. 9.13c, a shock wave on the lower surface has joined the one on the top. The center of pressure and the aerodynamic center have moved **aft**, the aerodynamic center to about **0.5c** from its typical subsonic value of about 0.25c, because the air well in front of the airfoil can no longer be affected by the airfoil. In Fig. 9.13d, the flow is completely supersonic and both shock waves have moved to the trailing edge.

Summary

At low angles of attack the pressures are often rather uniform over both surfaces on modern airfoils (good streamlining) but the pressures rapidly become more negative over the upper nose section as the angle of attack is increased toward the stall angle; after the stall occurs and as the angle of attack is further increased, the nose section pressure collapses to near atmospheric pressure, the lift decreases, and the pressure drag increases even faster.

For positive lift, the center of pressure tends to move forward as the angle of attack is increased to the stall; then it moves abruptly backward because of the breakdown of the highly negative pressures over the upper nose section. The aerodynamic center, defined as the point about which the pitching moment coefficient is nearly independent of angle of attack for an unstalled airfoil, is much more useful in the analysis of aircraft stability and control than is the center of pressure.

In cruise, the upper surface may have to overcome a negative lift generated on the lower surface. For greater, positive angles of attack, though, the subambient pressures on the upper surface produce about four times as much lift as the above-ambient pressures on the lower surface.

Important properties of the lift coefficient for a given airfoil include (a) the zero-lift angle of attack, (b), the stalling angle of attack (the angle for maximum lift), and (c) the maximum lift coefficient.

The drag polar of an airfoil is the variation of drag coefficient with lift coefficient; for many airfoils, it is a parabolic

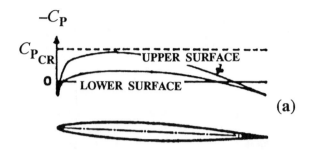

(a)

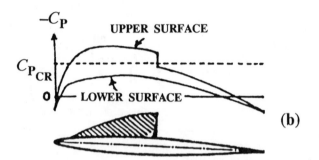

(b)

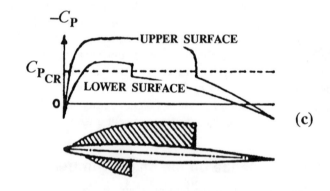

(c)

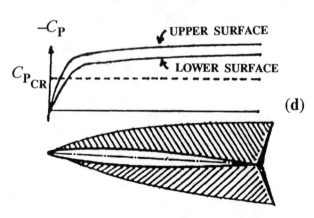

(d)

Fig. 9.13 The development of shock waves can be related to excursions of the pressure distribution through a critical pressure.

function with, by definition, a minimum at the design lift coefficient.

The stagnation point moves from about the leading edge to well below the leading edge as the angle of attack increases from its zero-lift value, and this is the basis for most stall warning devices.

The maximum lift coefficient tends to increase slightly as the upper surface air thins out at high speeds (at the onset of compressibility effects) but plummets when shock-induced separation occurs, and then partially recovers at supersonic speeds. The drag coefficient, meanwhile, steadily increases in transonic flight (= the "sound barrier"), abating only when fully supersonic flow prevails. The total drag, however, continuously increases with increasing speed.

The growth of shock waves with increasing Mach number can be profitably displayed on a pressure-distribution graph. A supersonic region is a region which has a pressure coefficient less than the critical pressure coefficient.

Symbol Table (in order of introduction)

ac aerodynamic center, the point on an airfoil where the moment coefficient is nearly constant (usually about 0.25c subsonically and about 0.50c supersonically)

a the lift slope, i.e., the slope of the lift curve

a_0 the lift slope in the linear region near $\alpha = 0°$

c_{ℓ_i} the design lift coefficient

Review Questions

1. How far and in what direction does the center of pressure move for the LS(1)-0417 airfoil as the angle of attack increases from (a) 0° to 8.02°, (b) 8.02° to 16.04°, and (c) 16.04° to 21.14°?

2. At the stall, the cp moves abruptly ___ relative to the leading edge because the highly negative pressures near the ___ have broken down.

3. The Piper Tomahawk and the Beechcraft Skipper both use the LS(1)-0417 airfoil. As you cruise along, what are you going to tell your passengers when they ask about how much of your current good fortune is due to the bottom surface of the wing?

4. Estimate how many times greater is the loading on the nose section of the LS(1)-0417 at its maximum compared to its value in cruise.

5. Do the pressures on the upper surface increase or decrease as the angle of attack is increased beyond the stall angle?

6. Because the location of the cp varies so much with ___ ___, the concept of an ___ ___ has largely replaced the concept of the cp, at least for stability and control analysis.

7. About what fraction of the lifting is done by the lower surface at angles of attack greater than 0°?

8. About what fraction of the pressure drag is attributed to the lower surface in cruise?

9. The aerodynamic center is defined as the point about which the unstalled airfoil has a constant value for its ___ coefficient. This constant value is zero for a ___ type of airfoil.

10. The lift coefficient attains a greater maximum value at large values for ___ because there is an earlier transition to turbulent flow and therefore a reduction in ___ flow at these large values.

11. The symbol used for the slope of the c_ℓ vs. α curve is ___, and this slope varies considerably at either end. In the region where the slope is constant, the symbol used is ___.

12. The dependence of the drag coefficient on the ___ coefficient is called the drag ___ because the dependence, for turbulent airfoils, is normally nearly parabolic.

13. The drag coefficient for the LS(1)-0417 airfoil, at any given angle of attack within the tested environment, is always less at (small, medium, large) values of RN.

14. This means that scaling down a given airplane, with no other changes, will tend to produce a (less, more) efficient machine.

15. The drag coefficient for the LS(1)-0417 attains its minimum value at an angle of attack of about ___°. This is intentional and is a direct result of it possessing positive ___.

16. The angle of attack that produces zero lift is always (negative, positive, zero) for a positively cambered airfoil and always (negative, positive, zero) for a symmetric airfoil.

17. For most positively-cambered airfoils, the moment coefficient about the aerodynamic center is (negative, zero, positive) while it is always (negative, zero, positive) for a symmetric airfoil.

18. As the flight Mach number increases for a given angle of attack, the lift coefficient abruptly drops at the ___ ___, but begins to recover somewhat when the lower shock wave _____.

19. As the flight Mach number increases toward one for a given angle of attack, the drag coefficient rises abruptly until _____ and then it begins to decrease.

20. M_{CR} occurs when the (minimum, maximum) value of the local pressure coefficient first reaches a value equal to the critical pressure coefficient for that freestream Mach number.

21. In the passage from subsonic to supersonic flight, the aerodynamic center for all airfoils moves from about ___ to about ___.

22. In cruising flight the upper surface shock wave forms (before, after) the lower surface shock wave.

23. The lift coefficient that the airfoil designer intends to be used in cruise is called the ___ lift coefficient.

24. What happens to the stagnation point as the angle of attack is increased from its zero-lift value?

Problems

1. The Beechcraft Skipper (a 2-place, side-by-side, low-wing trainer), which uses the LS(1)-0417, has a wing chord of 4.33 ft and, according to the manual, stalls at 56 mi/hr. At what RN is this airfoil operating at the stall, for flight at SL and at 10,000? Are these within the tested range? Answers: 2.26×10^6 and 2.06×10^6; just barely

2. Estimate the percent decrease in drag for an LS(1)-0417 airfoil as the RN increases from 2.1×10^6 to 6.3×10^6, at a cruise lift coefficient of 0.4. Answer: About 20% (a significant amount!)

3. The scale used in drawing the pressure vectors in Fig. 9.1 was about 0.27 cm per 1 in the pressure coefficient. Use this scale to estimate the maximum speed over the wing for an aircraft that is cruising along at 150 kt with a GA(W)-1 airfoil at an angle of attack of about 0°. (Cf. problem 5.12, in which you were given this information. Now you will see how it was determined.) Answer: About 200 kt

4. If the aircraft of problem 3 stalls at a true airspeed of 50 kt at some altitude, estimate the maximum speed around the wing for this condition. Answer: About 140 kt

5. If the drag polar of Fig. 9.8 is really a drag **polar**, the data should nicely fit an equation of the form

$$c_d = c_{d,min} + k\left(c_\ell - c_{\ell_i}\right)^2$$

where $c_{d,min}$ and c_{ℓ_i} can be obtained by careful inspection of the data. Do this for a RN of 6.3×10^6. Then obtain a value for k by substituting the last data point (at $c_\ell = 1.66$) into this equation and solving for k. Check the goodness of the fit by using your equation to estimate the value of the drag coefficient at a lift coefficient of 1.2 and comparing it with the experimental curve. Partial answer: The experimental value is about 87% of the calculated value. The correspondence should be better if the fit is restricted to lift coefficients farther below the stall.

6. If the airfoil of Fig. 9.1 is tested in a wind tunnel at an equivalent airspeed of 120 kt and with a chord of 2 ft and a span of 6 ft, what is the lift force on the airfoil at an angle of attack of 0°?
Answer: 275 lb

7. What is the total drag force (both pressure drag and viscous drag) on the airfoil for these same conditions? Answer: About 6 lb

8. What is the pitching moment about the leading edge for these same conditions? Answer: About –264 ft-lb

9. What is the lift force on the **upper** surface only, for these conditions? Answer: About 322 lb

10. Reading to the nearest 0.02 from Fig. 9.7, by what percent is the maximum lift coefficient for the highest Reynolds number greater than the maximum lift coefficient for the lowest Reynolds number?
Answer: About 20% — certainly a significant Reynolds number effect

11. By what percent is the drag coefficient for a lift coefficient of 1.2 and a Reynolds number of 2.1 million greater than the drag coefficient for the same lift coefficient at a Reynolds number of 6.3 million?
Answer: About 60% — a very large Reynolds number effect

References

McGhee, Robert J., and Beasley, William D., "Low Speed Aerodynamic Characteristics of a 17% Thick Airfoil Section Designed for General Aviation Applications," NASA-TN-D-7428, Langley Research Center, Hampton, VA, December, 1973.

Clancy, L. J., *Aerodynamics*, John Wiley & Sons, Inc., New York, 1975. (Available here is information on the role of the critical pressure coefficient and on wind tunnel methods.)

Abbott, I. H., and von Doenhoff, A. E., *Theory of Wing Sections*, Dover Publications, New York, 1949. (This is the basic reference for NACA airfoils up to about 1950.)

Dommasch, Daniel O., Sherby, Sydney S., and Connolly, Thomas F., *Airplane Aerodynamics*, Pitman Publishing Corporation, 1967. (Airfoil coefficients are derived using dimensional analysis in this text.)

Chapter 10

A Short History of Airfoils

It is not once nor twice but times without number that the same ideas make their appearance in the world.[1]

Dwarfs on the shoulders of giants see further than the giants themselves.[2]

Sherlock Holmes: "It is of the highest importance in the art of detection to be able to recognize out of a number of facts which are incidental and which are vital . . . I would call your attention to the curious incident of the dog in the night-time. The dog did nothing in the night-time. That was the curious incident."[3]

A. From the Birds to Early Airfoils

Observations of birds throughout mankind's history have reinforced the obvious fact that air **is** capable of supporting a denser-than-air object. Initial attempts to imitate the success of birds led to the *cul-de-sac* called the ornithopter (flapping-wing aircraft). But the wing sections of birds certainly provided an early clue that a **curved** shape was more efficient than an uncurved one in producing lift from the movement of air past a wing. Figures 10.1 and 10.2 show the approximate in-flight airfoil shapes of two birds (Airfoils #1 and #2, Refs. 2 and 3).

Fig. 10.1 Albatross wing section during gliding flight; about 13% thick. Wing span is as much as 7 ft; it can soar for hours.

Otto Lilienthal, a German aero enthusiast who was educated as an engineer, made scientific studies and measurements of bird flight which proved this point; he published his findings in *The Flight of Birds as a Basis for the Art of Flying* in 1889. Four years later he began a series of gliding experiments in which he suspended himself below a pair of wings and moved his body to

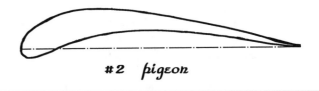

Fig. 10.2 Pigeon wing section, about 12% thick; the flight vehicle includes a small head, plump body, but long, tapered wings.

alter the *cg* location and thereby gain tenuous longitudinal and lateral control. He was imitating the gliding flight of birds, of course, and his apparatus became the inspiration for the thousands of hang gliders that have followed. In 1896, after some 2000 glides of up to one quarter mile in length, he was killed when his glider nosed into the ground from a height of about 50 feet, probably due to a turbulence-induced stall.

The Wright brothers in Dayton, Ohio, used the air pressure tables of Lilienthal and others to estimate the required wing area for their 1900 glider, but found that they were not getting as much lift as expected; therefore their next year's glider had a span increased by 5 feet (to 22 feet) and a chord increased by 2 feet (to 7 feet). Discouraged by the seeming inaccuracy of published information, they devised their own wind tunnel and began comparing various shapes and cambers for both wings and propeller sections. The result was the uniformly **thin** and only slightly cambered airfoils that we associate with the early successful flying machines. Throughout World War I, in fact, designers felt that airfoil efficiency demanded thicknesses of the order of only **6** to **8** percent of the chord. The Eiffel 36 (a French airfoil) and the R.A.F. 6 airfoil (a product of the Royal

---

[1]Aristotle, *On the Heavens.*
[2]Stella Didaceus, *Eximii verbi divini.*
[3]Sir Conan Doyle, *Silver Blaze.*

Aircraft Factory, England), appearing as Airfoils #3 and #4 (Figs. 10.3 and 10.4), illustrate this incorrect design assumption. Because the technology of the time could not produce a cantilever (internally supported) wing with the requisite strength, almost all of these early aircraft used a **biplane** configuration and relied on external wire bracing for strength. The R.A.F. 6 was more widely used and retained its popularity longer than its predecessors; it also could boast of its wind tunnel origin.

Fig. 10.3 Eiffel 36 airfoil from about 1912; about 7% thick. Used on the Curtiss Jenny (a WW I trainer and post-war barnstormer favorite).

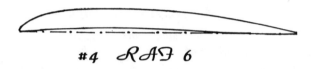

Fig. 10.4 R.A.F. 6 (MOD) airfoil from about 1912; about 7% thick. Used from 1912 to 1918 on many different aircraft; still used for prop blades.

The R.A.F. 15 (Airfoil #5, Fig. 10.5) succeeded the R.A.F. 6 and became probably the most commonly employed airfoil in both the U.S. and Great Britain from 1917 to 1925. In modified form it was used on the de Haviland DH-4 that the United States produced toward the end of World War I and which continued in service for many years. The designers of this airfoil thickened the airfoil at the probable locations of front and rear spars; they retained inward curvature on the bottom surface between these points, however, because of their understanding that this provided the greatest lift/drag ratio at cruise.

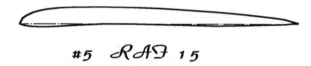

Fig. 10.5 R.A.F. 15 airfoil from about 1915; 6.4% thick. Used from 1915 to about 1925 on many types of aircraft.

Airfoil design, up to this time, had proceeded largely on the basis of intuition and trial and error. In the mid 1920s, though, the theories of Kutta and Joukowski (cf. Chapter 5) were combined in an analytical theory (using the assumption of inviscid flow) that yielded an infinite series of profiles (known as **Joukowski** profiles), all obtained by transformations from a circle. This process, now easily carried out on small personal computers, yields reasonable predictions for the lift, moment, and pressure distributions at low angles of attack but cannot calculate maximum lift coefficients or drag coefficients because it doesn't include boundary layer effects. (See the Oshkosh Airfoil Program, Ref. 7 in Chap. 5.) Some of these Joukowski

airfoils were quite thick and, with other thick airfoils, were considered a reasonable way to increase overall efficiency, because thick wings could now be made strong enough to eliminate the set-up time, maintenance, and high drag of external bracing wires. In the United States, Colonel V. E. Clark tested a number of airfoils in the Massachusetts Institute of Technology (M.I.T.) wind tunnel for the U.S. Army; his **Clark Y** airfoil (#6, Fig. 10.6) became probably the best-known airfoil in the world. It took Charles Lindbergh over the Atlantic to Paris in 1927, for example. (Clark began with the Clark A, so persistence paid off!) The bottom surface is flat over about 90% of the chord, making it easy to build.

Fig. 10.6 Clark Y airfoil from about 1925; 12% thick. This is the standard for comparison for low speed airfoils; also popular for prop blades. It carried Charles Lindbergh from New York to Paris in 1927.

Why did it take aircraft designers twenty years to realize that a gently cambered, thick airfoil like the Clark Y was far superior to the barely-curved flat plates they were using? It seems clear now that the answer lies in the fact that all their tests, from the Wright brothers on, had been made at **very low RN** and that they simply did not appreciate the importance of

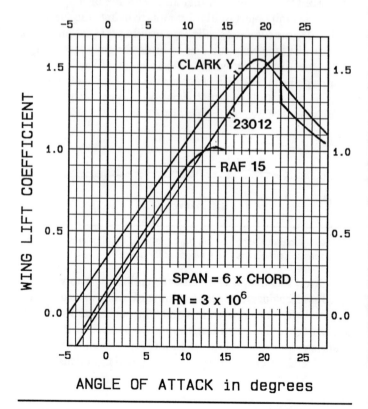

Fig. 10.7 Lift comparison for wings built up from the R.A.F. 15, the Clark Y, and the 23012 airfoils

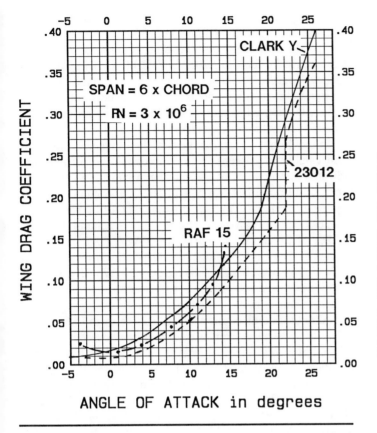

Fig. 10.8 Drag comparison for wings built up from the R.A.F. 15, the Clark Y, and the 23012 airfoils

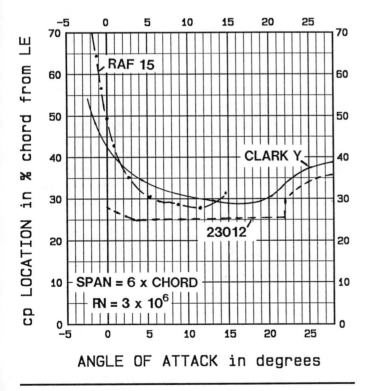

Fig. 10.10 Center-of-pressure travel comparison for wings built up from the R.A.F. 15, the Clark Y, and the 23012 airfoils

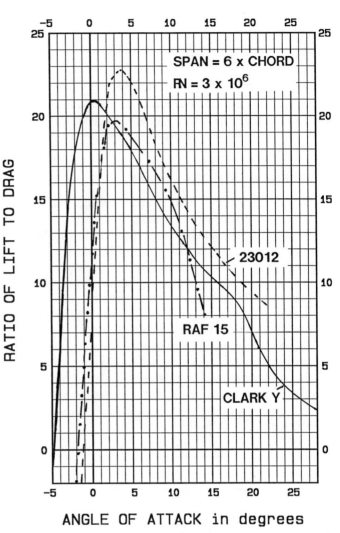

Fig. 10.9 Lift/drag comparison for wings built up from the R.A.F. 15, the Clark Y, and the 23012 airfoils

scale effect. In fact, a curved flat plate generates about twice as much maximum lift at RN= 40,000 as does the Clark Y, while also maintaining a far better L/D ratio throughout the narrow usable range of angles of attack. At a RN of about 3 million, however, the Clark Y generates 50% greater maximum lift than the R.A.F. 15 (Fig. 10.7). Note also its superior drag and *cp* properties (Figs. 10.8, 10.9, and 10.10).

In the U.S., the National Advisory Committee for Aeronautics (NACA) told the aeronautical community in its report #63 of 1930 that "It seems evident from these tests on models that wings may be designed with ample room for cantilever spars and have at the same time aerodynamic properties comparing favorably with the thin sections used now."

Airfoil #7 (Fig. 10.11), which is very similar to the Clark Y, is the airfoil that still provides good flight characteristics for thousands of Piper J-3 Cubs, Pacers, Tri-Pacers, and twin-engine Aztecs, at a ripe old age of 60 years!

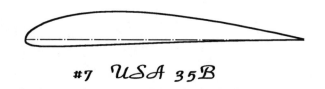

Fig. 10.11 U.S.A. 35B airfoil from about 1923; 12% thick. A modification of the Clark Y, it is used on many Piper singles and twins, from the Cub to the Aztec.

B. 1930s and 1940s Airfoils

NACA was created in 1915 and thereafter made many important contributions to aeronautical knowledge. Max Munk of NACA's Langley Field laboratory proposed a variable density wind tunnel in 1921 so that tests could be made at realistic values of the RN. (The viscosity of air isn't greatly affected by compression so the RN increases almost directly with increases in air density.) This new tunnel was instrumental in testing two new families of airfoils, as represented by five of their members (airfoils #8 through #12, Figs. 10.12 to 10.16).

As you have undoubtedly deduced already, there is no universal agreement regarding the naming or numbering of airfoils. All too often, airfoils are simply numbered consecutively by their loving developers, so that their designation gives no clue to their shape or probable application. Airfoils developed by NACA and NASA in the United States, though, have (al-

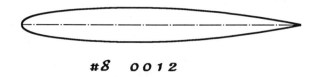

Fig. 10.12 NACA 0012 airfoil from about 1932; 12% thick. A symmetrical member of NACA's popular 4-digit series. Widely used for tail group airfoils and wind tunnel comparisons.

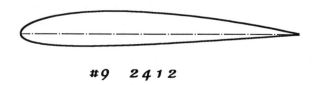

Fig. 10.13 NACA 2412 airfoil from about 1932; 12% thick. Used on Cessna singles (although some were modified with a leading edge cuff in 1970).

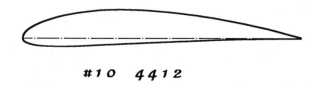

Fig. 10.14 NACA 4412 airfoil from about 1932; 12% thick. More camber yet. Used on the Aeronca L3, Bellanca Champions, and the Stinson Voyager.

most) always used the last two digits to indicate the **maximum thickness** of the airfoil, in percent of chord.

The NACA **4-digit** family resulted from a systematic study of **geometrical** effects in NACA's Variable Density Tunnel in Langley, Virginia. The mean camber line was formed by joining two parabolic sections and then the thickness distribution, camber, leading edge radius, and location of the maximum camber were systematically varied.

In the 4-digit series, (**a**) the first digit indicates the maximum camber in percent of the chord, (**b**) the second digit indicates the location of the maximum camber in tenths of chord, and (**c**) the last two digits give the maximum thickness in percent of chord. Thus the NACA 2412 airfoil has a maximum camber of 2% located at $0.4c$ and a maximum thickness of 12% of its chord. Similarly, the NACA 0012 airfoil is a symmetrical airfoil (no camber) and has a thickness of 12% of its chord. Because these airfoils are defined through geometric equations, it is easily possible to generate intermediate or related members in the 4-digit family.

As noted in their captions (Figs. 10.12 through 10.14), these 4-digit airfoils have been used on a great number of light aircraft. Some early World War II military aircraft also used members of this popular family; for example, the Boeing B-17 bomber used a 0018 section for the wing (tapering to a 0010 section at the tip) and the Curtiss P-36 and P-40 fighters used 2215 sections for the wing (tapering to a 2209 section at the tip). Symmetrical sections of this family are still widely used for tail members; for example, Cessna tends to use the 0009 section (tapering to the 0006.2 section) for their light aircraft; they also used the 0012 and the 0010 sections (both tapering to 0008 sections) on one model of their Citation business jet. Piper used 0010 sections for the tail members of its new Tomahawk trainer. NACA's 0012 occupies a special place in the airfoil world because it has been tested in over 40 different wind tunnels and has become a "touchstone" airfoil by which different wind tunnels can be compared.

From the tests of the 4-digit series, it was concluded that the **maximum lift coefficient** increased as the position of maximum camber was moved either forward or backward of the $0.5c$ point. Moving it backward produced unacceptably large pitching moments, and so another airfoil family (the **5-digit** series) was generated with a more **forward maximum camber location** and different mean lines (but all still geometrically defined). Airfoils #11 and #12 (Figs. 10.15 and 10.16) are representative members of this series, which has found considerable acceptance among designers of subsonic, high speed aircraft, particularly because of their low pitching moments (which reduce the size of the tail required for stability and control). The Beechcraft Bonanza, for example, uses a 23016.5 section (tapering to a 23012) for its wing. Large transport aircraft such as the Douglas DC-4 and DC-6 used a 23016 section (tapering to a 23012).

The first three digits in the 5-digit series indicate the type of mean line used and the last two digits indicate the maximum airfoil thickness in percent of chord. An airfoil with a 230 mean

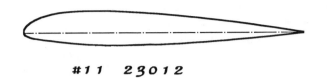

Fig. 10.15 NACA 23012 airfoil from about 1935; 12% thick. Used on the Taylorcraft and the Helio Stallion and on wingtips.

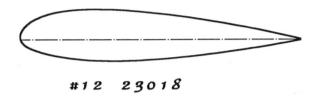

Fig. 10.16 NACA 23018 airfoil from about 1935; 18% thick. Used on the Beech Queen Air and B99 and the Cessna 310.

line has a maximum camber of 1.8% of the chord. The second and third digits give the location of the maximum camber point as a number which is just **twice** the distance from the leading edge in percent of chord. The design lift coefficient is three-halves of the first digit of the mean line designation. Thus the popular 23012 has a design lift coefficient of 0.3, a maximum camber of 1.8% located at $0.15c$, and a maximum thickness of 12%. The unusual 43013 airfoil, used on the Ercoupe, has very noticeable camber on its lower surface, close to the leading edge; its maximum camber is 3.7% of its chord.

The NACA series has been very widely used in foreign countries, in part because there is such a large family from which to choose and because they are so well documented. Reference 4 is the primary reference for all of the NACA airfoils. Airfoil #13 (Fig. 10.17) is a minimal representative from the many airfoils designed in countries other than the U.S. during this general period of time. By the early 1930s it was evident to a number of researchers that it was possible to maintain laminar flow over a substantial fraction of an airfoil **if** the surface was sufficiently smooth and free of waves and if the air had only the large-scale turbulence common to the atmosphere and not the small-scale turbulence common to wind tunnels. It was also known that a negative pressure gradient (for which the pressure is decreasing) pushed the laminar-to-turbulent transition RN for the boundary layer to a higher value (and therefore prolonged laminar flow) while an adverse, positive pressure gradient had the opposite effect.

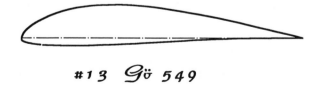

Fig. 10.17 Göttingen 549 airfoil from about 1930; 14% thick. A popular glider airfoil, expecially in the 1930s.

This theoretical knowledge and the existence of a new low turbulence wind tunnel resulted in another airfoil series, designed specifically to enhance the fraction of laminar flow, just as World War II began in Europe. The first result from these studies was the NACA 1 series. The members of this group that have proved useful for propeller blade sections are those having the minimum pressure point located at $0.6c$; because their designations begin with the number 16, this is often called the 16 series. An example of a member of this series is the NACA 16-212. The Clark Y airfoil and this series currently dominate the low speed propeller blade market. The Clark Y may be as good but the 16 series has some structural advantages. The efficiency of these sections is 80% or better up to 500 miles/hour but the efficiency deteriorates rapidly if the tip speed exceeds M = 0.91. Hamilton Standard has used the 16 section for their propellers; in 1974 Cessna announced that its McCauley props would have their Clark Y airfoil modified so as to effect a 2% to 2.5% efficiency improvement.

The best of these laminar flow airfoils for wings are the so-called 6 series (which reasonably enough followed series 1 through 5 into the wind tunnel). They all have the minimum pressure point located close to mid chord. Airfoil #14 (Fig. 10.18) is a happy member of this tribe that has found favor with Piper for its post-war designs; we shall see later that the hoped-for laminar flow probably does not materialize in practice on these particular aircraft but the greater thickness of this profile (for fuel, for example) and the fact that its greatest thickness occurs at a convenient location for the main spar (and thence the main spar at a good place relative to occupants' legs) has made it a reasonable choice.

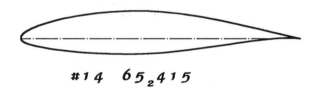

Fig. 10.18 NACA 65$_2$-415 laminar flow airfoil from about 1944; 15% thick. Used on the Piper Cherokee, Warrior, and Navajo.

Within the 6 series, **(a)** the first digit is a series indicator, **(b)** the second digit gives the location of the minimum pressure point in tenths of the chord, **(c)** the third digit indicates the extent of the low drag range if laminar flow is successfully obtained (this is the famous **laminar drag bucket**), **(d)** the fourth digit is the design lift coefficient in tenths (basically, design lift coefficient here means the center of the low drag range, where cruising flight would normally take place), and **(e)** the last two digits once again give the maximum thickness in percent of the chord. When the third digit is moved down to become a subscript, it denotes an improved version. Thus airfoil #14 (Fig. 10.18, the NACA 65$_2$-415) is an improved member having a minimum pressure (at cruise) at $0.4c$ for a design lift coefficient of 0.4; it also has a theoretical low drag range extending from $c_\ell = (0.4 - 0.2) = 0.2$ to $c_\ell = (0.4 + 0.2) = 0.6$ and a thickness of 0.15c. Six-series airfoils have appeared on many airplanes designed since World War II. Cessna used one on its Cardinal, Ted Smith choose one for his Aerostar 600, Beechcraft used one on its Sport 150 Model B19 (Musketeer), Mitsubishi choose one for its MU-2, Mooney used a 63$_2$-215 tapering to 64$_1$-412 on its never-produced M22 Mustang, and the Learjet Model 24c uses one.

The 6 series has enjoyed considerable popularity with aircraft designers even though most wings aren't manufactured with sufficient rigidity and smoothness to obtain the low drag benefits of laminar flow. With their rearward location for the minimum pressure, the normal shock wave that forms at high speeds will form rather far back on the wing, giving thin 6-series airfoils a high critical Mach number. Fighter aircraft from the P-51 Mustang and later have used them. (The P-51 was developed very quickly as an export aircraft for a besieged England at the beginning of World War II; the designers used a 66-216, tapering to a 66-211, and it worked out well.) Among jet fighters, the McDonnell RF-101C Voodoo used a NACA 65A007(mod), the Republic XF103 used a NACA 65A003, and the XF107 used a NACA 66A005(mod).

Designers and builders of soaring aircraft (gliders) **have** developed techniques that permit wings of considerable smoothness to be built, and gliders have enjoyed the benefits (and problems) of laminar flow for many decades. Airfoil #15 (Fig. 10.19) seems to have been the most popular 6-series airfoil for gliders, but many different ones have been used; the airfoil illustrated is used on the U.S. built Schweizer 2-33, for example. With a laminar airfoil, there isn't much drag penalty associated with the extra thickness of an 18% section, and a thick section is very much needed structurally if long, narrow wings are to be built.

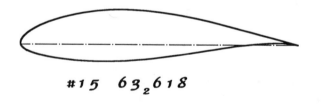

#15   63$_2$618

Fig. 10.19 NACA 63$_3$-618 laminar flow airfoil from about 1944; 18% thick. A popular glider airfoil in the 1960s.

C. 1960s to the Present

And then, for nearly 30 years, NACA and its successor, NASA, seemed to be resting from their airfoil labors. Wing shapes, the new jet engines with their highly demanding inlets, the problems of supersonic flight, and finally the new frontier of space, occupied the attention of the aerodynamicists. By the late 1960s, though, airflow theories that included boundary layer effects were matched with computers that could make the millions of required calculations in a reasonable amount of time, and out popped—new airfoils! One of the early airfoil designers in this new era was Professor Franz Xaver Wortmann

of the Institute for Aerodynamics and Gas Dynamics of the University of Stuttgart, Germany. Almost all of the high performance gliders of the past two decades use one of his laminar flow airfoils; #16 (Fig. 10.20) is one of his special successes. A rear flap, which normally occupies the last 17% of the chord on the whole wing, is intended as an in-flight camber changing device and thus is called a **cruise flap**; it is deflected slightly upwards for efficient dashing to the next cloud and the next thermal, but slow speed circling and the gaining of altitude in a thermal is accomplished with the flap straight or deflected partially downward.

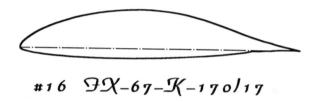

#16   FX-67-K-170/17

Fig. 10.20 Wortmann FX-67-K-170/17 from about 1967; 17% thick. A very popular glider airfoil in the 1970s (uses a cruise flap).

The numbering of the Wortmann airfoils is not completely uniform but, in general, (**a**) they start with Professor Wortmann's first two initials, (**b**) the first two digits indicate the year of origin of the airfoil, (**c**) a "K" designates a flapped airfoil, (**d**) the next three digits give the airfoil thickness in percent of the chord times 10, and (**e**) the number following a slash gives the flap chord in percent of the airfoil chord. Thus airfoil #16 (Fig. 10.20), the FX-67-K-170/17, dates from 1967, has a maximum thickness of 17.0% of the chord, and has a 0.17c cruise flap. These FX airfoils are designed to maintain laminar flow (and very low drag) over a wider range of lift coefficients than the earlier NACA 6 series (Refs. 5 and 6). Some Wortmann airfoils are also good low Reynolds number airfoils. Dr. Wortmann died in 1985.

Airfoil #17 (Fig. 10.21), the GU 25-5(11)8, easily the most bulbous of the lot at 20% thickness, was one of a series of airfoils designed by T. Nonweiler of the Department of Aeronautics and Fluid Mechanics of the University of Glasgow, Scotland (Report No. 6801). The airfoil's anticipated use apparently was for a man-powered aircraft, but NASA aeronautical engineer-turned entrepreneur Burt Rutan switched from a GA(W)-1 [LS(1)-0417] airfoil on the canard (front wing) of his VariEze (Figs. 1.4 and 1.6) in about 1975 when he experimentally deter-

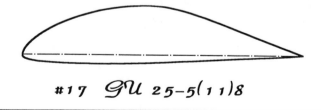

#17   GU 25-5(11)8

Fig. 10.21 University of Glasgow GU 25-5(11)8 airfoil from about 1968; 20% thick. Gives high lift at low RN; popular for canards but can give trouble when wet or when it isn't very smooth.

mined that the poor low RN performance of the GA(W)-1 was severely limiting low speed performance of the aircraft. The GU airfoil, with a 50% greater maximum lift coefficient at a RN of 500,000 than the GA(W)-1, changed a nose-down stall of the canard at 60 kt to a gentle nodding (oscillating) stall at 52 kt (Ref. 7). Other new canard designs have also used this airfoil, at least until 1983, when it became clear that, on some designs, the **lift** (as well as the drag) of the airfoil was dangerously sensitive to a loss of laminar flow due to variations in construction, an accumulation of bugs, or rain (Refs. 8, 24, and 25), causing BL separation at mid-chord in **cruising** flight. This behavior presents a special problem for canard-type aircraft because the *cg* is quite far from the canard, making the moment arm large. Any loss in lift immediately translates into a very noticeable trim change, perhaps even one too large to be adequately countered by the elevator.

Work on new airfoils that would push the drag divergence Mach number closer to 1 began in about the late 1950s—with the British and French. In 1962 H. H. Pearcy of the National Physics Laboratory (NPL) in England described what are called "peaky" airfoils (because of the sharp peak in the pressure coefficient at the leading edge) that featured relatively flat upper surfaces and positive camber toward the trailing edge; derivatives of these Pearcy sections were used on the Boeing 747, the Douglas DC-10, and the A-300 airliners (Ref. 9). In the United States, Richard T. Whitcomb (already famous for the Area Rule) was deeply involved in the design of airfoils for the proposed U.S. supersonic airliner (which was cancelled), but he quit in disgust when he couldn't achieve efficiencies similar to those achieved at subsonic speeds. Shortly thereafter, though, he began experimenting with various subsonic airfoil shapes, beginning with a slotted 6-series section. Guided by Sherlockean intuition, he apparently followed the simplest and oldest of all experimental techniques—cut and try—and developed airfoils with a nearly flat upper surface and significant aft camber (now commonly referred to as "aft-loaded" airfoils since a significant amount of the lift is developed by the rear one-third of the airfoil). Meanwhile, computer programs that could adequately describe subsonic and supersonic flow were being developed and they led to airfoils that looked very much like Whitcomb's; this was fine evidence that the theory worked.

Airfoil #18 (Fig. 10.22) is a slightly modified version of Whitcomb's first supercritical airfoil; the coordinates of this airfoil became generally available only many years after the 1974 design date (Ref. 10). Whitcomb himself is credited with suggesting the "supercritical" appellation (Ref. 11). Supercriti-

cal wing sections were tested on the Navy F-8, the T-2C, and the F-111 (military jet aircraft) and it was soon apparent that these new sections really were a significant improvement; the larger leading edge radius compared with earlier high speed airfoils gave them superior low speed and maneuvering capabilities. As an alternative to higher cruising speeds, it was found that a thicker—and therefore significantly lighter—wing could be made that would have as good high speed properties as previous thin sections (Refs. 11 and 12).

By 1983 it had become clear that Whitcomb's work had sparked a real resurgence in high speed airfoil research. Today, no new business jet or transport aircraft can hope to succeed unless it can boast of possessing a supercritical wing. Among the aircraft in this category are the Cessna Citation 3, the Canadair, the Grumman Gulfstream III, the Lear Star 600, the Falcon 50, the Westwind 2, the North American Sabreliner Models 65 and 80, as well as the Boeing 767 airliner. Whitcomb applied for a patent on his supercritical airfoil design in 1971, and the patent was granted in 1976, but the principle of the supercritical airfoil is well known and the patent doesn't appear to be enforceable (Ref. 13).

Fig. 10.23 shows how the new supercritical airfoils have improved on older transonic airfoils by tailoring their pressure distributions to produce a weaker shock wave that is farther back on the airfoil, for a given airfoil thickness and flight Mach number. Note the aft-loading (Ref. 35).

NASA soon realized that the air flow theory/computer alliance that produced supercritical airfoils could also be used to produce low speed airfoils with a desired pressure variation.

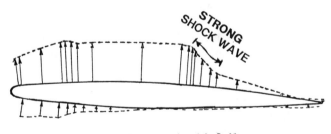

(a) 1960s Transonic Airfoil

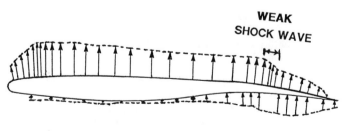

(b) 1970s Supercritical Transonic Airfoil

#18 *Whitcomb*

Fig. 10.22 Whitcomb supercritical airfoil from about 1974; 11% thick. Designed to raise the drag divergence Mach number for a given thickness.

Fig. 10.23 Pressure distribution and shock wave development on old and new transonic airfoils.

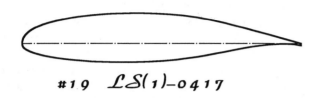

Fig. 10.24 NASA LS(1)-0417 or GA(W)-1 airfoil from about 1974; 17% thick. The first member of a new tribe of airfoils for General Aviation applications.

The first result of this effort was the 1974 GA(W)-1 airfoil, whose properties were discussed in the previous chapter (airfoil #19, Fig. 10.24). (Recall that GA stands for "General Aviation," W stands for "Whitcomb," and 1 is a series number.)

The NASA GA(W)-1 was clearly superior in maximum lift coefficient to the earlier NACA airfoils, especially at high RN (6 million) and with "standard roughness"; under these conditions its wind tunnel maximum lift coefficient was about 2.0 while the 2412 has about 1.2, the 23012 has about 1.2, and the 64₂-415 has about 1.1. However, the GA(W)-1's minimum drag coefficient was about "10 counts" greater under the same conditions, at 0.0110 versus 0.0100, and its moment coefficient was nearly twice as great as the 2412's and eight times as great as the 23012's. (Incidentally, "standard roughness" is obtained by applying number 60 grit to both surfaces of the airfoil from the leading edge back to 8% of the chord.)

A year later, in 1975, the other shoe dropped in the form of the GA(W)-2, which was obtained by linearly scaling down the thickness distribution of the GA(W)-1 to a 13% thick section while retaining the same mean camber line. In return for the structural disadvantage of a thinner section, the GA(W)-2 offered a 0.1 increase in maximum lift coefficient and a usable decrease in drag coefficient of about 20 counts – but still with the same large pitching moment. The authors noted (Ref. 14) that

> In practical general aviation application, no "laminar bucket" would be expected because boundary-layer transition usually occurs near the leading edge of the airfoils, a result of roughness of construction or insect remains gathered in flight. . . . The addition of a roughness strip (BL "trip") at .075c resulted in essentially full chord turbulent flow which was confirmed by oil-flow techniques.

(This oil-flow technique, which involves spreading a very thin oil over the upper leading edge and observing how smoothly it is carried backward on the airfoil model, is a common wind tunnel technique for directly determining the transition point and the separation point.)

In 1976 NASA came back with another new airfoil in the series as well as a new and more informative labelling scheme. Hereafter the GA(W)-1 was to be the LS(1)-0417 and the GA(W)-2 was to be the LS(1)-0413, where "LS" stands for Low Speed, "1" stands for first series, "04" indicates that the design lift coefficient is 0.4, and "13" or "17" reveals the thickness in percent of chord. With a complete airfoil family, it was now possible to directly compare the effect of varying thickness

while keeping the mean camber line and the thickness distribution constant. The lift coefficient at zero angle of attack should be determined just by the camber but the investigators found that the very thick 21% section had a slightly smaller value than the others; they attributed this to an effective camber decrease caused by boundary layer thickening over the greater length of the upper surface. This also resulted in a more sluggish boundary layer at the thickest point for the 21% airfoil and a correspondingly greater tendency to separate; therefore the maximum lift coefficient was significantly reduced and there was a strong dependence of the value on RN. At an RN of 4 million the maximum lift/drag ratios were 100, 80, and 60 for the 13, 17, and 21% airfoils respectively (Ref. 16).

In 1977, right on the yearly schedule, a modification to the LS(1)-0413 was reported. The upper surface was modified by adding material from 0.025c to 0.4c and then removing material to the trailing edge; the lower surface was modified by adding material from 0.5c to the trailing edge. The intent was to reduce the magnitude of the adverse pressure gradient on the upper surface at typical climb lift coefficients and thereby produce less separation and a greater lift coefficient. The wind tunnel showed that the expected decrease in the adverse pressure gradient had been realized. The end result was a better lift/drag ratio for the modified airfoil but only at the lowest RN (2 million); at higher RN the more turbulent boundary layer could more easily overcome the greater adverse pressure gradient of the unmodified airfoil. (Airfoil #20, Fig. 10.25, and Ref. 17)

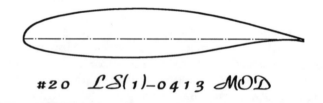

Fig. 10.25 NASA LS(1)-0413 MOD airfoil from about 1979; 13% thick. A modified member of NASA's low speed airfoil series. Used on the speedy Glasair experimental/homebuilt aircraft.

Modifications to the LS(1)-0417 and the LS(1)-0421 as well as a new medium speed airfoil series were announced in 1978. The LS(1)-0421 was modified by reducing its aft adverse pressure gradient; the result was a large increase in lift/drag ratio at a (climb) lift coefficient of 1.0 as well as a welcome reduction in the pitching moment coefficient. The LS(1)-0417 Mod was used successfully for the canard of the Q-200, a 2-place adaption of the Quickie design, when the airfoil contamination problem of the GU airfoil became apparent.

The new medium speed airfoils, designated MS(1)-0313 and MS(1)-0317, had design (drag rise) Mach numbers of 0.72 and 0.68, respectively, at a design RN of 14 million. These medium speed airfoils were designed to fill the gap between the supercritical and the low speed airfoils; they would be applicable to medium size business aircraft.

In the MS series, the good maximum lift and gentle stalling characteristics of the LS series were to be retained while the drag divergence Mach number was to be increased. Wind tunnel measurements again provided confirmations of the theory. There was a slight reduction in the maximum lift coefficient for the MS(1)-0313 versus the LS(1)-0413 but also a reduction in the pitching moment coefficient (because the new sections were becoming less aft-loaded!). The overall performance of the MS(1)-0317, on the other hand, was superior in all respects to the LS(1)-0417; the drag was uniformly less, there was a smoother decrease of maximum lift with a decrease in RN, there was a reduced sensitivity to roughness, and the pitching moment coefficient was also reduced. Inspect Figs. 10.26, 10.27, and 10.28 for a graphical comparison of these recent NASA families and previous NACA families. (Refs. 18, 19, 20, and 21).

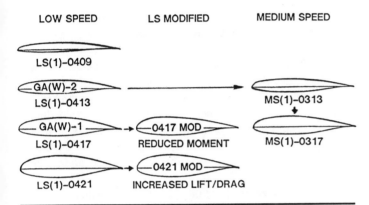

Fig. 10.26 Members of NASA's LS and MS airfoil families

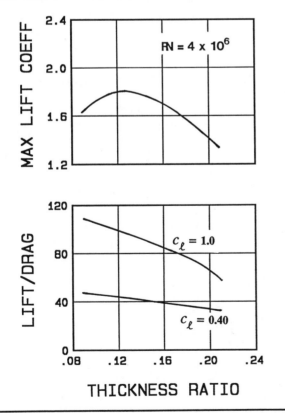

Fig. 10.27 Effect of thickness on the maximum lift coefficient and the maximum L/D for NASA's LS airfoils

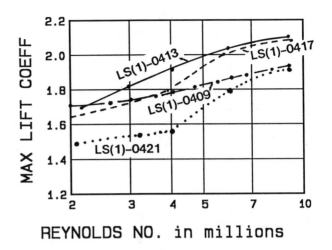

Fig. 10.28 Effect of Reynolds number on maximum lift coefficient, NASA LS airfoils

Note from Figs. 10.27 and 10.29 that the maximum lift coefficient obtained from either the new or the old families of airfoils seems to peak at an airfoil thickness of about 13%, at least for these intermediate values of the RN. But then note the almost linear decrease in lift/drag ratio with increasing thickness, which favors the thinnest possible airfoils. Then Fig. 10.28 tells us that the thinnest airfoil also experiences the least loss in maximum lift coefficient as the RN is decreased; in fact, at the lowest RN the 9% section has a greater value than all the others (which is where this chapter began!). NASA is now extending its low speed airfoil tests down to a RN of 1 million or lower.

The relatively small profile changes in the LS(1)-0417 MOD airfoil produce surprisingly large changes in the design pressure distribution, according to Fig. 10.30. Clearly the modified airfoil is less aft-loaded, for the moment coefficient is reduced by 28%. We also see from Fig. 10.31 that the modified airfoil has 5% more attached flow on the upper surface at a lift coefficient of 1.0 because of the smaller adverse pressure

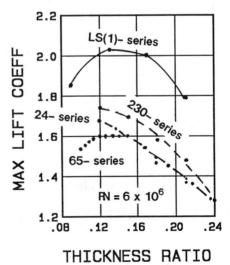

Fig. 10.29 Maximum lift coefficient for old NACA and new NASA airfoils at high Reynolds number

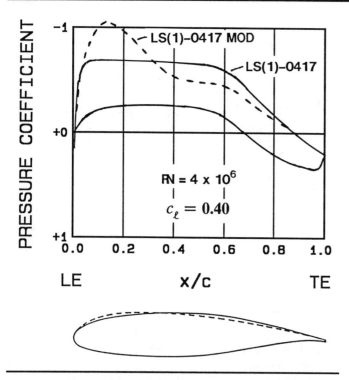

Fig. 10.30 Effect of a small profile change on the pressure distribution at the design lift coefficient, for two NASA LS airfoils. The modified airfoil has a moment coefficient only 72% that of the original.

gradient (as seen from the slope of the pressure coefficient versus x coordinate graph).

We see in Fig. 10.32 that the thinner of the two MS profiles has a clear advantage at high speeds. We can conclude that, at these intermediate values of the RN, a 13% or thinner section should be chosen if it allows a sufficiently strong wing to be built, within given weight and cost constraints.

The next airfoil in this introductory parade is a new laminar airfoil, the NLF(1)-0215F (#21 and Fig. 10.33) — another good example of history repeating itself, for by 1981 it was clear that laminar flow was present in large bunches on airplanes with

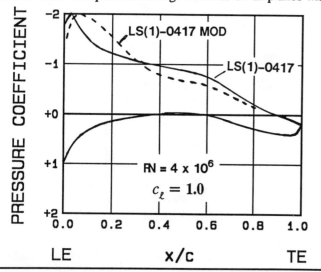

Fig. 10.31 Pressure distributions for an LS airfoil and an LS MOD airfoil at a larger lift coefficient. The BL separates at 0.92c for the original, at 0.97c for the modified airfoil.

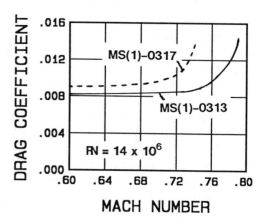

Fig. 10.32 The calculated drag rise for NASA's MS airfoils

smooth fiberglass wings, in contradiction to what NASA had said when they first introduced the turbulent GA(W) airfoils. Perhaps the turning point for powered aircraft was the rediscovery of the sublimation technique for experimentally determining the amount of laminar flow on wings, **in flight**, by Bruce Holmes of NASA. Until these experiments, many designers were not convinced of the possibility for extensive laminar flow in practice on powered aircraft.

With a renewed belief that laminar flow was achievable, it was clearly worthwhile to try to improve on the old 6-series airfoils. The "NLF" in the designation of this airfoil emphasizes that the laminar flow is to be obtained **naturally** through a careful tailoring of the pressure distribution and **not** by suction, blowing, or cooling in the boundary layer. Actually, sailplanes have used composite construction for decades and have obtained significant fractions of laminar flow over their wings, but it is the next generation of powered aircraft (taunted by the success of hundreds of experimental/homebuilt aircraft) that will be turning to this construction. The Voyager couldn't have made it around the world unrefueled without laminar flow and the new Beech Starship is counting on significant laminar flow on its smooth, composite wings.

The NASA author of the report on this NLF airfoil is a bright-eyed sailplane enthusiast. He stated in 1981 that

> The goal of the present research on NLF airfoils at Langley Research Center is to combine the high maximum lift of the NASA low speed airfoils with the low drag characteristics of the NACA 6-series airfoils.

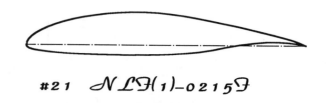

Fig. 10.33 NASA NLF(1)-0215F airfoil of 1981; 15% thick, with a cruise flap. A recent laminar flow airfoil, it is used on the Lancair experimental/homebuilt aircraft.

His wind tunnel supports his claim for success, although the pitching moment coefficient is even more negative than it was for the original GA(W)-1 (Ref. 22). The "F" at the end of the designation means that it uses a cruise flap, like some of the Wortmann airfoils. The whole wing should be flapped and the ailerons should move up and down together to some extent as well, to optimize the advantages of this variable-camber design feature.

Another NLF airfoil, the NLF(1)-0414F, appeared in 1984. A little later it was fitted to a modified 1985 Cessna T-210 (Ref. 34) and tested in NASA Langley's 30×60-ft full-scale wind tunnel. In the cruise configuration, laminar flow appeared over the first 70% or so of both top and bottom wing surfaces. It was estimated that the laminar flow would give an additional 10% more cruise speed at the same power or up to 29% more range at lower power settings. With an untwisted wing, it was necessary to add a drooped outer panel on the wing, along with an inner drooped section or a vortex generator, to obtain satisfactory directional stability at the stall (i.e., no tendency to roll off into a steep bank), but most of the low drag benefits were retained.

There is considerable interest in extending the benefits of laminar flow to large aircraft, including jet transports, which fly at very high RN. A 1985 airfoil, the HSNLF(1)-0213, where "HS" stands for High Speed, illustrates this interest (Ref. 31). The HSNLF(1)-0213 is designed for a lift coefficient of 0.20 at a flight Mach number of 0.70 while maintaining 50% laminar flow on the top surface and 70% on the lower surface.

Supercritical airfoil designers have also continued their development work since the 1974 invention. For example, the SC(3)-0712 appeared in 1985, the SC(2)-0012 in 1987, and the SC(2)-0714 in 1988 (Refs. 27 through 29).

Gliders, ultralights, and remotely-piloted vehicles (RPVs) operate at relatively slow speeds and with relatively small wing chords—i.e., at low Reynolds numbers. RPVs are of considerable interest for their reconnaissance potential, especially to the U. S. Navy. Therefore there is considerable activity toward the design of airfoils that can operate efficiently at Reynolds numbers below 1 million. The problem at low RN is that there is **too** much laminar flow—a laminar BL tends to separate if it doesn't transition to a turbulent BL before it reaches a region of increasing pressure. NASA's LRN(1)-1007 airfoil (Airfoil #22, Fig. 10.34), for example, is designed to provide attractive lift/drag ratios at Reynolds numbers of 200,000 and below.

Evidently airfoil designers are now able to compute an airfoil shape for any reasonable pressure distribution but they

must still proceed from an understanding of the relationship between various pressure distributions and final airfoil properties and they must be careful to obtain good performance over a range of angles of attack and RN. There are plenty of pitfalls that must be avoided on the trip from the wind tunnel to the airplane. For example, the GA(W)-1 and GA(W)-2 have been used on a number of aircraft but there have been complaints that the large pitching moment of these airfoils (due to their aft loading) **(a)** leads to a greater trim drag, **(b)** adversely affects control surfaces in that area so that they tend to "float" to an uncommanded location if there is slack in the control system and cause a heavy aileron feel, and **(c)** produces so much fuselage-wing interference drag that the anticipated benefits from using these new sections may not be realized. The new sections, because of their undercamber, are also harder to manufacture with traditional sheet metal technology (Ref. 23, p. 285).

It was noted in the previous chapter that, in airfoil tests, NASA obtains its lift and moment coefficients from an integration of measured surface pressures but obtains the drag coefficient from a wake rake which measures the momentum deficit of the air flow in the wake. The reasoning is that the lift is typically a cooperative effort for both the upper and lower surfaces but the drag is the difference between the very much smaller contributions of the nose and rear section, respectively, and so might be difficult to determine accurately from the pressures. Additionally, the drag calculated from surface pressures is just the **pressure** drag and does not include the skin friction (viscous) drag. Contrariwise, it seemed probable that using the surface pressures to obtain the pressure drag might give some very interesting information regarding the fraction of total airfoil drag that was caused by unavoidable viscous drag. Viscous drag is all that remains if we can achieve perfect streamlining (in two-dimensional flow) at the low angles of attack used in cruising flight, after all. And, sure enough, at **low** angles of attack the calculated pressure drag coefficients for the LS(1)-0417 were less than the reported drag coefficient and the difference was very nearly constant. To the authors surprise, though, the integrated pressures yielded pressure drag coefficients that were **greater** than the reported drag coefficients at all lift coefficients **above** about 1.1 (Fig. 10.35). Even though the wake rake is not expected to measure accurate drag values after BL separation occurs, the amount of the discrepancy at intermediate lift coefficients has not been explained to the author's satisfaction.

Table 10.1 presents a summary of the significance of various airfoil designations.

We have presented an airfoil history almost entirely in terms of NACA and NASA airfoils, but this is only because these airfoils are in the public domain and are well documented, as well as being tested in wind tunnels. However, essentially every large aircraft company in the world now develops their own proprietary airfoils to suit their own needs and government research laboratories in other countries are active as well. Too,

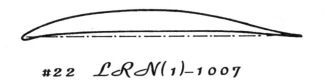

Fig. 10.34 NASA LRN(1)-1007 airfoil of 1984; 7% thick; low Reynolds number (100,000 to 150,000) airfoil

Airfoil Family	Example	Significance of Characters in the Designation
NACA 4-digit $\approx$ 1932	2412	1st digit (2) is the maximum camber in percent of chord (0.02c) 2nd digit (4) is the location of the maximum camber (0.4c) 3rd and 4th digits (12) are the maximum thickness (0.12c)

NACA 5-digit $\approx$ 1935	23015	1st digit (2) indicates the approximate amount of camber (0.02c)
		    (2 means 1.1% for the 210 series, 1.5% for the 220 series,
		    1.8% for the 230 series, and 2.3% for the 250 series)
		2nd and 3rd digits (30) indicate the location of maximum camber
		    —just twice the distance in percent of chord (0.15c)
		4th and 5th digits (15) are the maximum thickness (0.15c)

NACA 6 series $\approx$ 1944	63_2-215	1st digit (6) is the series designator 2nd digit (3) is the location of the minimum pressure (0.3c) Subscript (2) indicates the $\pm$ extension of the low drag range (from $c_\ell = c_{\ell_i} - 0.2$ to $c_\ell = c_{\ell_i} + 0.2$) 3rd digit (2) indicates the design lift coefficient ($c_{\ell_i} = 0.2$) 4th and 5th digits (15) are the maximum thickness (0.15c)

NASA (1973 to present)	MS(1)-0313	1st two letters (MS) indicate the intended application
		    HSNLF = High Speed Natural Laminar FLow
		    LRN = Low Reynolds Number
		    LS = Low Speed
		    MS = Medium Speed
		    NLF = Natural Laminar Flow
		    SC = Supercritical
		Number in parentheses (1) is the series designator
		Next two digits (03) give the design lift coefficient ($c_{\ell_i} = 0.3$)
		Last two digits (13) give the thickness in percent of chord (0.13c)

Wortmann	FX-67-K-170/17	1st two letters (FX) indicate the designer (F. X. Wortmann) Next two digits (67) indicate year of design (1967) A "K" indicates a flapped airfoil The next three digits (170) indicate the thickness (0.170c) Two digits after the slash (/17) indicates a cruise flap (0.17c)
Eppler	E591	Richard Eppler's 591st airfoil design

Table 10.1 Decoding of Airfoil Designations

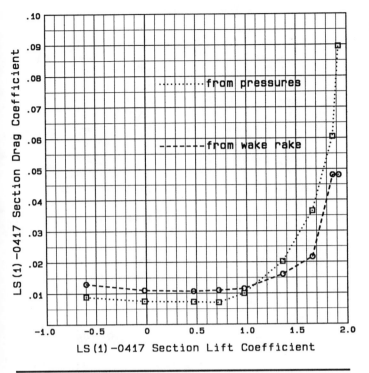

Fig. 10.35 Comparison of drag coefficients calculated from pressure distribution with drag coefficients calculated from wake rake measurements

computers can now spit out new (unproven) airfoils at the rate of hundreds per day!

Summary

A compact history of airfoil development has been presented, proceeding from the very thin sections of the earliest aircraft to the efficient Clark Y and then to the newest computer-designed airfoils. The decoding of airfoil numbers was presented. For the new NASA low speed turbulent airfoils, at common flight Reynolds numbers, the maximum lift coefficient is greatest for a wing of about 13% thickness. New airfoils, designed for maximum natural laminar flow or good efficiency at low Reynolds numbers, have been added to the early low speed and supercritical airfoils.

Review Questions

1. From its designation, what can you deduce about the origin and properties and date of a 23024 airfoil?

2. From its designation, what can you deduce about the origin and properties and date of a 2418 airfoil?

3. From its designation, what can you deduce about the origin and properties and date of a 63_1-212 airfoil?

4. From its designation, what can you deduce about the origin and properties and date of a FX 62-K-131/17 airfoil?

5. From its designation, what can you deduce about the origin and properties of a hypothetical MS(2)-0211 airfoil?

6. Are the LS and MS airfoil families considered to be laminar or turbulent flow airfoils?

7. It appears that World War I airplane designers used very thin airfoils because they were not aware of ___ effect.

8. The 5-digit NACA family evolved from a desire to increase the maximum ___ coefficient by moving the point of maximum ___ closer to the leading edge while keeping the ____ ____ small.

9. The NACA 16-series airfoils are still very much used for ___.

10. The NACA family that has a "laminar drag bucket" when smooth is the ___ series.

11. A member of the new NASA family of low drag airfoils is identified by the first three letters of its designation, which are ___. (What is the significance of the first letter?)

12. The unacceptable aspect of the GU 25-5(11)-8 airfoil, as it was used on some aircraft designs, is the apparent dependence of its ___ on the maintenance of laminar flow.

13. What do the letters "LS" in front of a NASA airfoil designation mean?

14. The airfoils that have essentially taken over the glider market are those developed by ___.

15. The designer most closely associated with supercritical airfoil design in the U.S. is ___.

16. Two distinguishing characteristics of supercritical airfoils are (a) the ___ of most of the upper surface and (b) the ___ that exists very near the trailing edge on the lower surface.

17. The greatest problem with using the GA(W)-1 appears to be the large value for its ___ coefficient.

18. "Counts" refers to increments of ___ in the ___ coefficient.

19. Transition and separation are often determined in the wind tunnel by spreading ___ on the airfoil.

20. Different members of the LS family were obtained by linearly scaling the ___ of the original member.

21. Within the LS family, the member giving the largest maximum lift coefficient, over the tested range of RN, is the one with a thickness of ___.

22. All the NASA airfoils show a decrease in their maximum lift coefficient as the RN is ___, but the effect is least pronounced for the ___ members.

23. Maximum lift/drag ratio within the LS family is obtained by the ___ member.

24. The best reason for building a 17% wing rather than a thinner one is ___.

25. The supercritical, LS, and MS airfoils were designed by specifying a desired ___ ___ and letting the computer program go from there (within some reasonable constraints).

26. The airfoil on most of the Cessna singles is a NACA ___, sometimes modified with a leading edge cuff (C-172).

27. The airfoils that pre-date the 4-digit series and that are still used for propeller sections are the ___ and the ___.

28. The prime intention in the design of the new NASA laminar airfoil was to obtain better ___ than the 6 series offers.

29. The LS family was tested in the RN range from about ___ to ___ million but now NASA is testing down to about ___ million.

30. What is the design lift coefficient for the 6-series airfoil (65_2-415) used on the Piper Warrior?

31. From their designations, what can you say about the likely difference between the aerodynamic performance of a 23012 airfoil and a 2412 airfoil?

32. Under favorable conditions of smoothness and rigidity, would you expect the 63-216 or the 66-216 to have the greater amount of laminar flow?

33. What do the letters "HSNLF" and "LRN" and "SC" indicate when used as part of the designation for a NASA airfoil?

34. Jane's *All-The-World Aircraft* is a yearly encyclopedia of aircraft, the back copies of which are to be found in many libraries. The airfoil sections used by various aircraft are indicated fairly often, especially for civilian aircraft and old U.S. military designs. Obtain the airfoils used by two powered aircraft and one sailplane from this reference series. Indicate the volume number (or year) and the page where you find the reference.

Problems

1. What are the stalling angles of the three airfoils/wings described in Fig. 10.7? Which airfoil has the gentlest stall and which has the most abrupt? How do these two properties correlate with their ages?

2. Which airfoil in Fig. 10.8 has the smallest drag coefficient and at what lift coefficient is it obtained?

3. If *L/D* is considered to be a measure of airfoil efficiency, how do the efficiencies of the three airfoils of Fig. 10.9 compare with their ages?

4. At what lift coefficient does the 23012 obtain its greatest efficiency (Fig. 10.9). Answer: About ████ .40

5. Based on *cp* travel, which of the three airfoils in Fig. 10.10 is most similar to the LS(1)-0417 described in Fig. 9.4?

6. At cruise, about where on the chord does the maximum local airspeed occur for the LS(1)-0417 Mod airfoil (Fig. 10.30)? For an aircraft travelling at 100 kt, what is this speed? Estimate the critical Mach number for this airfoil at this angle of attack, under SL conditions. Partial answer: About 460 kt

7. Same question as the previous, except for a climb lift coefficient of 1.0 (Fig. 10.31). Partial answer: About 380 kt

8. Estimate the wing area required if a 1600 lb airplane is to cruise at 100 kt under SL conditions, at zero angle of attack and with a Clark Y airfoil (Fig. 10.7). For a span that is six times the chord, what is the chord and span of a rectangular wing with this area?
Partial answer: *b* = 26.6 ft.

9. Same question as the previous except with the 23012 airfoil and at an angle of attack of 5°.

10. What is the drag force for the wing of question 8?
Answer: About 68 lb

11. What is the drag force for the wing of question 9?
Answer: About 68 lb

12. Estimate the percentage of total drag that is unavoidable viscous drag for the airfoil of Fig. 10.35. If the viscous drag is constant, estimate the actual total drag coefficient at a lift coefficient of 1.5 and compare it with the reported value. Partial answer: Viscous drag is about 32% of the total drag at $\alpha = 0°$.

References

1. Mackay, Alan L., editor, *The Harvest of a Quiet Eye*, Crane, Russak and Company, Inc., N.Y., 1977. (A selection of scientific quotations.)

2. McMasters, John H., and Henderson, Michael L., "Low-Speed Single-Element Airfoil Synthesis," Science and Technology of Low Speed and Motorless Flight, NASA Conference Publication 2085, 1979. (Available from NTIS as N79-23890.)

3. Magnan, A., "Bird Flight and Airplane Flight," NASA TM-75777, 1981. (Translation of *Le vol des oiseaux et le vol des avions*, Services Technique de L'Aeronautique, Paris, France, Technical Bulletin No.

74, June, 1931.) (Available from NTIS as N81-14962. Theory of bird flight, especially gliding and soaring types; applications to aircraft; lift/drag ratio for birds and aircraft; stresses in maneuvers.)

4. Abbott, Ira H., von Doenhoff, Albert E., and Stivers, Louis S., "Summary of Airfoil Data," NACA Report No. 824, 1945. (Also available in Abbott, Ira H., and von Doenhoff, Albert E., *Theory of Wing Sections*, Dover Publications, Inc., 1949.) (This is the basic reference to the 4-digit, 5-digit, and 6-series NACA airfoils.)

5. Althaus, Dieter and Wortmann, Franz Xaver, *Stuttgarter Profilkatalog I*, Friedr. Vieweg and Sohn Verlagsgesellschaft mbH, Braunschweig, 1981. (Here is a catalog of mostly Wortmann laminar airfoils that date from 1962 to the date of publication and that cover the RN range of 0.2 million to 6 million. Lift and drag for these airfoils was primarily determined from the pressure distribution on the tunnel walls and in the wake area. For comparison, tunnel data for NACA airfoils 4415, 23012, 63₃-618, 64₁-012, and 64₂-415 at RN between 0.7 million and 3 million are presented.)

6. Althaus, Dieter, *Profilpolaren für der Modellflug*, Neckar-Verlag Vs-Villingen, 1980. (This is a German collection of low RN airfoils, primarily for model aircraft. Profiles tested include Wortmann, Eppler, NACA, Gottingen, Clark Y, and Sukolov, at RN between about 60,000 and 200,000.)

7. Downie, Don and Downie, Julia, *Complete Guide to Rutan Homebuilt Aircraft*, Tab Books Inc., 1981.

8. Kelling, F. H., "Experimental Investigation of a High-Lift Low Drag Airfoil," Report No. 6802, University of Glasgow, Department of Aeronautics and Fluid Mechanics, 1968. (GU 25-5(11)8 airfoil report.)

9. Pearcey, H. H., "The Aerodynamic Design of Section Shapes for Swept Wings," Vol. 3, *Advances in the Aeronautical Sciences*, Pergamon Press, 1962.

10. Spaid, Frank W., Dahlin, John A., Roos, Frederick W., and Stivers, Louis S., Jr., "An Experimental Study of Transonic Flow About a Supercritical Airfoil," NASA TM-81336-Suppl, 1983. (Available from NTIS as N83-33846. Tests on two slightly modified versions of the original NASA Whitcomb supercritical airfoil section are described.)

11. Becker, J.V., *The High Speed Frontier*, NASA SP-445(01), 1980. (Available from NTIS as N81-15969. Discussion of the history of our attempts to get to and beyond Mach 1, especially with regard to airfoils on propellers and on wings.)

12. Whitcomb, R.T., "Review of NASA Supercritical Airfoils," Paper No. 74-10 presented at 9th Congress of the ICAS, Haifa, Israel, Aug. 1974. (Available from AIAA as A74-4132.)

13. Covault, Craig, "NASA to Enforce Airfoil Patent Rights," *Aviation Week & Space Technology*, April 24, 1978, pp. 24–25.

14. McGhee, Robert J., and Beasley, William D., "Low Speed Aerodynamic Characteristics of a 17-Percent-Thick Airfoil Section Designed for General Aviation Applications," NASA TN-D-7428, 1974. (Available from NTIS as N74-11821. First report on the LS(1)-0417—when it was called the GA(W)-1.)

15. McGhee, Robert J., Beasley, William D., and Somers, Dan M., "Low Speed Aerodynamic Characteristics of a 13-Percent-Thick Airfoil Section Designed for General Aviation Applications," NASA TM-X-72697, 1975. (Available from NTIS as N77-23049. First report on the LS(1)-0413—when it was called the GA(W)-2.)

16. McGhee, Robert J., and Beasley, William D., "Effects of Thickness on the Aerodynamic Characteristics of an Initial Low Speed Family of Airfoils for General Aviation Applications," NASA TM-X-72843, 1976. (Available from NTIS as N79-130001.) (Designation

change for the GA(W) series; report on the new "fatty" member, the LS(1)-0421; corrections to earlier measurements.)

17. McGhee, Robert J., and Beasley, William D., "Low Speed Wind Tunnel Results for a Modified 13 Percent Thick Airfoil," NASA TM-X-74018, 1977. (Available from NTIS as N79-24960. Report on the LS(1)-0413 Mod airfoil.)

18. McGhee, Robert J., Beasley, William D., and Whitcomb, Richard T., "NASA Low- and Medium-Speed Airfoil Development," Advanced Technology Airfoil Research, Volume II. (Proceedings of a conference held at Langley Research Center, Hampton, Virginia, on March 7–9, 1978.) (Available from NTIS as N80-21283. Modified versions of the LS(1)-0417 and LS(1)-0413; new MS(1)-0317, and MS(1)-0313.)

19. McGhee, Robert J., and Beasley, William D., "Low-Speed Aerodynamic Characteristics of a 13 Percent Thick Medium Speed Airfoil Designed for General Aviation Applications," NASA TP-1498, 1979. (Available from NTIS as N81-12015. Basic reference to the MS(1)-0313.)

20. McGhee, Robert J., and Beasley, William D., "Wind Tunnel Results for a Modified 17 Percent Thick Low Speed Airfoil Section," NASA TP-1919, 1981. (Available from NTIS as N82-11033. LS(1)-0417 Mod airfoil; tested from RN of 1 to 12 million.)

21. McGhee, Robert J., and Beasley, William D., "Low Speed Aerodynamic Characteristics of a 17-Percent-Thick Medium Speed Airfoil Designed for General Aviation Applications," NASA TP-1786, 1980. (Available from NTIS as N83-16290. MS(1)-0317 compared with MS(1)-0313.)

22. Somers, Dan M., "Design and Experimental Results for a Flapped Natural-Laminar-Flow Airfoil for General Aviation Applications," NASA TP-1865, 1981. (Available from NTIS as N81-30386. Report on the NLF(1)-0215F airfoil.)

23. Kohlman, David L, Matsuyama, Garey T., Hawley, Kevin E., and Meredith, Paul T., "A Feasibility Study for Advanced Technology Integration for General Aviation," Kansas University, 1980. (Available from NTIS as N81-15974. An exhaustive survey of possible advanced technology gains through (a) changes in research emphasis, (b) new airfoils, (c) design for crashworthiness, (d) new control systems, (e) new display and information technologies, (f) reduced noise, (g) more efficient propulsion, (h) composite construction, and (i) canard designs.)

24. Dwiggins, Don, "Dangerous When Wet," Homebuilt Aircraft, March, 1983, and April, 1983.

25. Bragg, M. B., and Gregorek, G. M., "Experimental Study of Airfoil Performance with Vortex Generators," Journal of Aircraft, Vol. 24, No. 5, May 1987. (Discusses the recent history of problems with laminar flow airfoils on canard-type aircraft and the development of vortex generators on the around-the-world Voyager to solve its problem.)

26. Somers, Dan M., "Design and Experimental Results for a Natural-Laminar-Flow Airfoil for General Aviation Applications," NASA TP-1861, 1981. (Available from NTIS as N81-24022. This paper describes properties of the first NLF airfoil, the NLF(1)-0416.)

27. Mineck, Raymond E., and Lawing, Pierce L., "High Reynolds Number Tests of the NASA SC(2)-0012 Airfoil in the Langley 0.3-Meter Transonic Cryogenic Tunnel," NASA TM-89102, 1987. (Available from NTIS as N87-23594.)

28. Ferris, James C., McGhee, Robert J., and Barnwell, Richard W., "Low-Speed Wind-Tunnel Results for Symmetrical NASA LS(1)-0013 Airfoil," NASA TM-4003, 1987. (Available from NTIS as N87-26033.)

29. Johnson, William G., Jr., Hill, Acquilla S., and Eichmann, Otto, "High Reynolds Number Tests of a NASA SC(3)-0712(B) Airfoil in the Langley 0.3-Meter Transonic Cryogenic Tunnel," NASA TM-86371, 1985. (Available from NTIS as N85-29926.)

30. McGhee, Robert J., Viken, Jeffrey K., Pfenninger, Werner, Beasley, William D., and Harvey, William D., "Experimental Results from a Flapped Natural-Laminar-Flow Airfoil with High Lift/Drag Ratio," NASA TM-85788, May, 1984. (Available from NTIS as N86-26286. This paper presents the NLF(1)-0414F airfoil, with up to 70% laminar flow on top and bottom surfaces.)

31. Sewall, William G., McGhee, Robert J., Viken, Jeffrey K., Waggoner, Edgar G., Walker, Betty S., and Millard, Betty F., "Wind Tunnel Results for a High-Speed, Natural Laminar-Flow Airfoil Designed for General Aviation Aircraft," NASA TM-87602, November, 1985. (Available from NTIS as N88-14078. This is the HSNLF(1)-0213 airfoil, designed for a lift coefficient of 0.20 at a Mach number of 0.70 and a Reynolds number of 11 million.)

32. Jenkins, Renaldo V., Hill, Acquilla S., and Ray, Edward J., "Aerodynamic Performance and Pressure Distributions for a NASA SC(2)-0714 Airfoil Tested in the Langley 0.3-Meter Transonic Cryogenic Tunnel," NASA TM-4044, 1988. (Available from NTIS as N88-24580.)

33. Fisher, S. S., and Abbitt, J.D., "A Smoke-Wire Study of Low Reynolds Number flow over a NASA LRN(1)-1007 Airfoil Section," Proceedings of the International Conference on Aerodynamics at Low Reynolds Numbers, Vol. I., Paper #5, 1986. (Available from AIAA.)

34. Murri, Daniel G., and Jordan, Frank L., Jr., "Wind-Tunnel Investigation of a Full-Scale General Aviation Airplane Equipped With an Advanced Natural Laminar Flow Wing," NASA TP-2772, 1987. (Available from NTIS as N88-10009. This is the test of a modified Cessna T-210 with the NL(1)-0414F airfoil.)

35. Whitford, Ray, Design for Air Combat, Jane's Publishing Inc., 1987.

36. Vincenti, Walter G., "Design and Growth of Knowledge: The Davis Wing and the Problem of Airfoil Design, 1908–1945," Chapter 2 in What Engineers Know and How They Know It, The Johns Hopkins University Press, 1990. (This is a fascinating description of the design considerations that led to the one-time use of an airfoil that quickly returned to obscurity after being chosen for the World War II B-24 bomber. The author believes that the apparent superiority of the Davis over competing NACA airfoils was due to its greater support for laminar flow, and that this superiority didn't extend beyond the wind tunnel.)

Document sources
National Technical Information Service (NTIS)
5285 Port Royal Road
Springfield, VA 22161

Technical Information Service, AIAA
750 Third Avenue
New York, New York 10017

NACA/NASA document key
CP = Conference Proceedings
CR = Contractor Report
TM = Technical Memorandum
TN = Technical Note, TR = Technical Report
TT = Technical Translation
WR = Wartime Report (NACA)

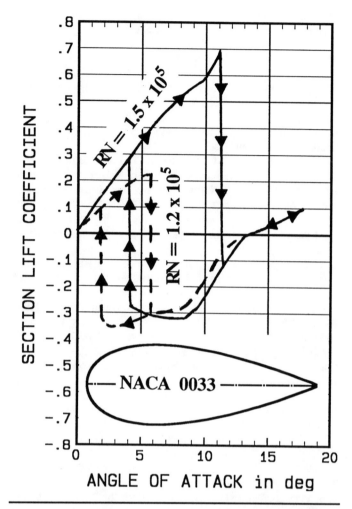

Fig. 10.36 What an exciting aerobatic wing this bulbous NACA airfoil would make! At low Reynolds number it undergoes a very early leading edge separation of the laminar BL on its upper surface, resulting in negative lift with a positive angle of attack—but eventually positive lift is restored. When the angle of attack is reduced, there is a long hysteretic delay before the BL is able to reattach. Note the tremendous importance of Reynolds number here. See Section 11F and Fig. 11.13 for additional information. (Based on wind tunnel measurements reported in Ref. 6.)

Pic. 10.1 This Quickie canard displays the many vortex generators which are required to prevent mid-chord BL separation in cruise when bugs or rain contaminate its GU 25-5(11)8 airfoil.

Chapter 11

Airfoils Compared; Airfoil Problems

Some canard aircraft have trouble with rain cutting lift on the canard by disturbing what the engineers call laminar flow. We had to know whether this one would behave this way—and better sooner than later. . . . Just as we started picking up the first few drops of rain, the airplane started to pitch down. I corrected with back stick—more elevator—but it kept coming down. The stick forces became very heavy, and as I kept coming back with the stick, I had a horrible feeling in the pit of my stomach that the airplane was coming down and I couldn't stop it. . . . I pushed the power up. No matter what I did with the stick, we were going down pretty steeply, and I looked around for some way to get out of the rain. But the only thing to do was press on through, looking at the ground and at the edge of the shower and hoping we got through the shower before we got to the ground. . . . Jeana got the parachutes ready, stowed the loose items, and I went through the emergency procedures, but there was no way I was going to bail out. I kept fighting it, fighting with such concentration and belief that I could save the airplane that I'm sure, if we hadn't come out of the shower, I would have stayed with the airplane all the way to the ground. . . . But we just did make it through, the canard started to dry off, and we turned and flew around the shower and landed at Mojave.[1]

At the design point of an angle of attack of 3 deg, the lift (coefficient) drops from 0.9 to 0.55 when the boundary layer is tripped . . . the vortex generators return the lift within approximately 0.1 of the natural transition level.[2]

A. Effects of Camber

The NACA 1412, 2412, and 4412 sections present an interesting opportunity to study the effects of changing camber while retaining the same maximum thickness and the same location for the maximum camber point. (The 1412 is the nearly symmetric section that Bellanca chose for its Decathlon—in place of the 4412 used on Champs and Citabrias—to improve its aerobatic capabilities, especially in inverted maneuvers.) Figures 11.1 and 11.2 present comparable data for the three sections at medium RN. Inspection of these figures leads to the conclusions that increasing camber (a) increases the lift at zero angle of attack and (b) moves the drag minimum to a higher value of lift coefficient, without much affecting its value. In other words, the design lift coefficient increases with increasing camber.

We also note that the rather severe "standard roughness" reduces the maximum lift for these airfoils by only about 10% but increases the minimum drag by about 60%! Note also the strange kink in the lift curve of the 2412 at this RN. The kink is caused by a boundary layer that is just on the brink of becoming turbulent when it separates; it reattaches after separation when it does become turbulent and so the lift continues to increase for a while; this kinky behavior probably wouldn't be found in production-quality wings. Reference 4 in Chap. 10 shows that the kink disappears into a smoothly rounded (lower) maximum at lower RN (for which reattachment doesn't occur) and into a sharper (but still rounded) and higher maximum at higher RN for which the transition occurs before real separation begins.

The maximum lift coefficient is not very sensitive to camber, at least after a moderate amount of camber is attained (as in the 2412), contrary to what one might guess. Thickness, on the other hand, is a geometrical property to which the maximum lift coefficient is sensitive, for most airfoil families, and the coefficient usually peaks for thicknesses around 12% or 13%.

---

[1] Dick Rutan, *Voyager* (New York: Knopf, 1987).
[2] M. B. Bragg and G. M. Gregorek, "Experimental Study of Airfoil Performance with Vortex Generators," *Journal of Aircraft*, Vol. 24, No. 5 (1987).

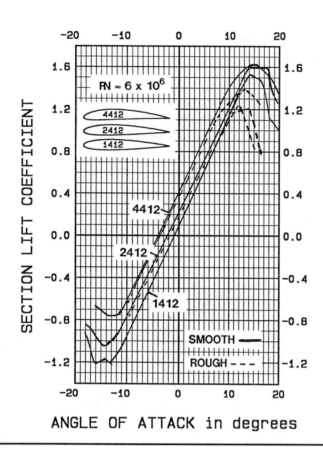

Fig. 11.1 Effect of camber and roughness on the lift of three NACA 4-digit airfoils

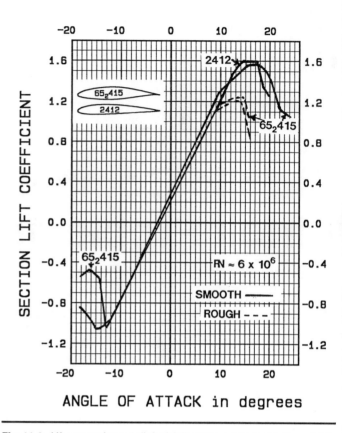

Fig. 11.3 Lift comparison: 4-digit airfoil versus a 6-series laminar airfoil

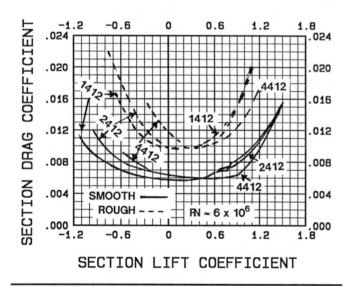

Fig. 11.2 Effect of camber and roughness on the drag of three NACA 4-digit airfoils

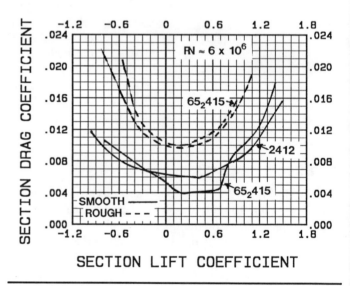

Fig. 11.4 Drag comparison: 4-digit airfoil versus a 6-series laminar airfoil

B. Comparison of a Popular 4-Digit Airfoil with a Popular 6-Series Airfoil

It is also good clean fun to compare the 2412 section with the 65_2-415 section since they are the basis for wing sections used on rival Cessna and Piper low speed aircraft, respectively.

Figures 11.3 and 11.4 present section characteristics for a RN appropriate to cruise. Note that the lift characteristics for the two sections are very similar; both yield about the same maximum lift at about the same angles of attack, both in the smooth and in the rough condition. However, with respect to drag the laminar 65_2-415 airfoil is much better within its laminar "drag bucket" **when** it is smooth and rigid; however, it is uniformly

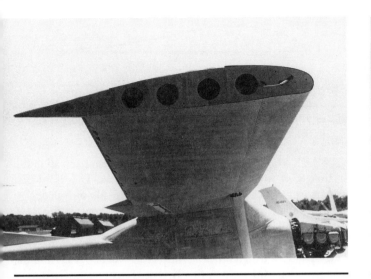

Pic. 11.1 This Cessna shows off its airfoil when the wingtip is off for painting; note the leading edge cuff.

Pic. 11.2 This Cherokee's 6-series airfoil is revealed when the wingtip is off, too.

poorer than the 2412 when it is rough and has a predominantly turbulent BL. Notice that the drag bucket really **is** centered on the design lift coefficient of 0.4. Outside the bucket the drag quickly becomes worse than for the 2412 because the laminar flow becomes turbulent on one or the other surface and the drag climbs out of bottom of the bucket. This laminar airfoil does have the structural advantage of being 3% thicker; this results in a lighter structure for equivalent strength and provides extra room for a retracting landing gear and for fuel, etc. Probably no production aircraft obtain significant amounts of laminar flow over their wings in flight because of manufacturing and service-related roughness, as well as a lack of contour stiffness.

The sudden rise in drag outside the drag bucket suggests a potential for trouble. If a pilot should try to force his aircraft off the ground at a very low speed, he will be operating in this region and the aircraft may take off but refuse to climb out of ground effect (Ref. 1). (A wing is more efficient, producing less drag for its lift, when it is very close to the ground, as will be discussed in the next chapter.) This problem is particularly likely to occur when the density altitude is very high because then the takeoff run will be much longer than usual and climb performance is at its minimum.

C. Searching for Maximum Lift from a Monoelement Airfoil

The problem of designing an airfoil for the maximum possible lift coefficient, **without** the complexity or expense of flaps or other lift-enhancing devices, has attracted the attention of numerous airfoil designers. Aside from their considerable intrinsic theoretical interest, such airfoils are suitable for the simplest recreational aircraft and man-powered aircraft.

We have seen that the new NASA airfoils develop a maximum lift coefficient of 1.7 (at a RN of 2 million) to about 2.0 (at a RN of 6 million). These are general purpose airfoils, intended to operate efficiently over a wide range of lift coefficients. Robert Liebeck of McDonnell Douglas Aircraft startled the

aeronautical world in 1970 when he presented a series of hump-backed airfoils that were designed to produce maximum lift coefficients greater than 2.5! Liebeck utilized the 1959 theory of B. S. Stratford (Ref. 2), who calculated the shape for the adverse pressure gradient region that would allow the upper surface pressure to **recover** to the ambient value in the **minimum** distance.

So maximum lift should be generated by (**a**) obtaining a maximum positive pressure on the lower surface, (**b**) quickly accelerating the upper surface air around the nose to a maximum speed and therefore a minimum pressure, (**c**) maintaining this most-negative pressure as long as possible, and (**d**) then decelerating the flow with a Stratford pressure distribution (Fig. 11.5). Clearly, though, we can't hope to make the entire lower surface a stagnation point (!) and we can't expect instantaneous acceleration of the upper surface air, but we can (we are told) hope for something like Fig. 11.6a. In fact, that is the theoretical pressure distribution for the airfoil of Fig. 11.6b! It

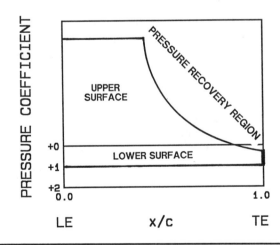

Fig. 11.5 Pressure distribution for maximum lift (the Stratford distribution)

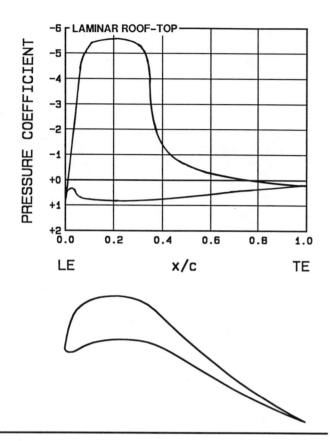

Fig. 11.6 An airfoil designed for maximum lift, with its pressure distribution (Liebeck)

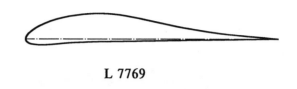

Fig. 11.7 An airfoil for a human-powered aircraft (Lissaman L 7769)

near the leading edge — and it was impossible to keep all of the flexible 85-ft wing at the same angle of attack. Therefore Lissaman went back to his computer and to his airfoil design program and came up with the L 7769 airfoil shown in Fig. 11.7. This airfoil was designed for low drag over a "good" range of angles of attack, for a low and constant pitching moment coefficient, and for a high lift/drag ratio at a RN of about 1.4 million. The resulting wing won the first Kremer Prize (on the Gossamer Condor) by flying a figure 8 course on August 23, 1977, and then won a second Kremer Prize (on the Gossamer Albatross) by flying across the English Channel on June 12, 1979 (Ref. 6), all with a single ½ hp human engine.

D. Laminar Flow True Believers

By the 1940s it was well known that laminar flow over a wing could dramatically decrease the drag, a fact that was proven both in the wind tunnel and in flight tests. However it was also found that typical sheet metal construction, even with flush riveting, was not smooth enough or rigid enough or wave-free enough to obtain significant laminar flow in practice. And even if laminar flow could be maintained as far as the run-up pad, it was felt that insect strikes on climbout would soon destroy most of the laminar flow before much cruising could be done. So the conventional wisdom had it that laminar flow wings and fuselages were not worth pursuing.

By the 1960s the matter was viewed quite differently in sailplane circles. The highest performing sailplanes were factory-made with an extremely smooth fiberglass surface — and the significant loss of performance from an encounter with rain or bugs made it very clear that much of their good performance was due to laminar flow.

Burt Rutan's first popular design, the VariEze, was offered to interested builders along with a new contruction technique: epoxy-saturated fiberglass cloth over a rigid foam core. Previous homebuilt designs had been generally based on all-wood construction (as in the first couple of decades of flight), welded steel-tube fuselages with fabric cover and wooden wings (typical construction through the 1930s), or sheet aluminum semi-monocoque construction (as in most manufactured aircraft since the 1940s). All these methods required special tools, special skills, and so much construction time that few projects were completed and then typically over a decade or so. Fiberglass and foam construction promised to require no special tools (just some wooden jigs), a very short learning curve, freedom to make any desired shape, potentially smooth and rigid surfaces, and less builder time — and it has largely

has a predicted maximum lift coefficient of 2.54 at a drag coefficient of 0.013 at a RN of 5 million (Ref. 3).

This airfoil looks like it could be difficult to build both light and strong, but in 1973 and 1978 Liebeck presented another series of airfoils that were more practical and these have been put to use on aircraft, on racing cars, and on windmills! (Refs. 4, 5, and 6) Several of his airfoils have been patented. Some of his recent work has been with low RN airfoils.

There is another very practical problem with these hump-backers. If you are flying with a Stratford pressure distribution on the top surface of your wing and you decide to increase the angle of attack just a trifle, the boundary layer may go "boom" — and you will experience a "hard" stall with sudden total separation. Also, the performance of the airfoil may be very poor away from its maximum lift design point. There is no longer any doubt, though, that these airfoils really can produce astonishing maximum lift coefficients. Because of the shape of the upper surface pressure distribution curve, Liebeck-type airfoils are sometimes referred to as "rooftop" airfoils.

When Paul MacCready, already well known in sailplane circles, was preparing an assault on the lucrative Kremer Prize for the first man-powered aircraft to fly a specified figure 8 course, he assigned the task of airfoil design to a friend, Peter Lissaman. The first airfoil they tried had a single surface — like the earliest human fliers. It turned out to have too much sensitivity to changes in angle of attack. Either above or below the design angle of attack, regions of separated flow would develop

delivered on its promises. As a result, many new homebuilders surfaced overnight and now there are hundreds of fiberglass and foam airplanes slipping through the skies.

For those who aren't familiar with this construction technique, a brief description follows. A typical built-up composite fuselage begins with slabs of a special semi-rigid foam, perhaps 2" or more in thickness. One side of the foam is carved out as desired to form either an inside side or the inside bottom of the fuselage and is overlaid with successive layers of woven fiberglass cloth that are saturated with either a special epoxy or a vinylester resin. The fiberglass strands must be pulled straight and their orientation carefully chosen so as to coincide with the expected load paths. The glass strands in the fiberglass cloth are extremely strong (for their weight) in tension but have no resistance to compressive stresses. The epoxy or resin is used to attach the cloth to the core and to fittings as well as providing some compressive strength. The bottom and side slabs are then joined by using additional layers of cloth around the corners. A top, also finished on the inside, can then be attached to part of the fuselage. Then the outside foam can be carved into a rectangular cross-section with smoothly rounded corners and the outside of the fuselage is given two or more layers of cloth, all carefully oriented, to form a hard, strong surface. The weave of the cloth is filled with a plastic filler and then sanded to glass smoothness.

To build the wing with this technique, the first step is to cut out airfoil templates (from plywood, for example); these are then tacked onto opposite ends of a rectangular piece of relatively dense styrofoam. Any change in airfoil thickness or any desired wing sweep or wing twist is easily obtained by making different templates or by skewing their relative locations. Next a taut wire is heated electrically to such a temperature that it will slice through the foam like a hot knife through soft butter. The hot wire is moved around the outside of the templates in a carefully coordinated manner by two people and a section of a wing is produced in a matter of a few minutes. The various sections are then epoxied together to form a complete foam wing. On small wings, adequate strength may be obtained from just covering the foam with fiberglass cloth and properly saturating it with epoxy or resin, but it is common practice on larger wings to cut the foam core at about the 0.3c location and install the fiberglass equivalent of a spar. Fig. 11.8 shows a typical sequence of operations, in simplified form. The trailing edge of the wing must be sealed with a glass-to-glass bond rather than relying on adhesion to the foam. Many builders have had allergic reactions to volatile ingredients in the epoxy or filler, so appropriate precautions must be taken. In addition to being very smooth, the front half of the surface must satisfy a waviness criterion that states that no dips or peaks greater than 0.002" can exist over any 2"×2" region, if extensive laminar flow is to be expected.

The technique displayed in Fig. 11.8 is an inside-to-out fabrication method. Disadvantages include the requirement for good foam cores to begin with (to avoid excess fill and accompanying weight) and the tremendous amount of work in

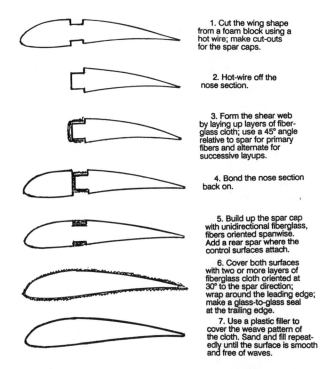

1. Cut the wing shape from a foam block using a hot wire; make cut-outs for the spar caps.

2. Hot-wire off the nose section.

3. Form the shear web by laying up layers of fiberglass cloth; use a 45° angle relative to spar for primary fibers and alternate for successive layups.

4. Bond the nose section back on.

5. Build up the spar cap with unidirectional fiberglass, fibers oriented spanwise. Add a rear spar where the control surfaces attach.

6. Cover both surfaces with two or more layers of fiberglass cloth oriented at 30° to the spar direction; wrap around the leading edge; make a glass-to-glass seal at the trailing edge.

7. Use a plastic filler to cover the weave pattern of the cloth. Sand and fill repeatedly until the surface is smooth and free of waves.

Steps in Building a Composite Wing with Spars, using Fiberglass cloth and Epoxy: Typical hand lay-up method fo homebuilt and prototype aircraft.

Fig. 11.8 Hand lay-up, moldless construction method for building a composite wing

filling and sanding (filling and sanding, filling and sanding) to get the preferred glass-like finish. For this reason, the most popular composite designs for homebuilt construction now offer premolded parts which, having been formed in a female mold, possess a naturally smooth exterior surface. There is still plenty for the homebuilder to do, though, and few aircraft are completed in less than a thousand, or even a couple of thousand, hours. Airplanes built with these room-temperature-cured adhesives must be painted white to keep them from getting so hot that the adhesive loses some of its strength while sitting on a hot ramp. The paint is also needed to protect the foam and epoxy from ultraviolet rays.

Once a sufficiently smooth wing and fuselage have been built, the biggest operational problem is to keep the surface smooth and free of insect remains. As early as the 1940s, an aircraft was flown with paper covering the leading edge of the wing; after some altitude had been gained and most of the bugs left behind, the paper was pulled away from inside the cockpit. Clearly this isn't a practical procedure for everyday flying! Instead, those with laminar wings simply plan to clean them every time they land on a warm summer day, and then try to minimize the amount of time flown in buggy air. NASA has investigated the possibility of pumping a continuous stream of fluid through a porous leading edge to prevent bugs from sticking to the wing during the low altitude part of a climb. F. X. Wortmann has suggested using a resilient leading edge that will "bounce off" the insects. Independent airfoil designer John Roncz has even designed airfoils that attempt to minimize "successful" insect encounters; after the Voyager landed, he was observed carefully counting the number of successes (only 69 on the entire airframe!). Besides insects, it is known that laminar flow is destroyed by rain or by ice crystals above 40,000 feet or even by the moisture in the vicinity of clouds. For safety, the critical requirement is that the **lift** of the wing must **not** depend on the attainment or retention of laminar flow, and this has just recently been written into the FAA certification requirements (Ref. 80). The only penalty for a loss of laminar flow should be a slight loss in cruising speed with no effect on stability or control. Unfortunately, this has not been true for many of the homebuilt, canard-type aircraft that use airfoil #17 in Fig. 10.21, the GU 25-5(11)8, for their canard.

Pilots of the VariEze soon began to report experiencing a trim change when flying through rain; usually it was a nose-down trim change but one VariEze apparently reacted with a pitch-up. In any case, the problem was only annoying, especially when flight was in and out of rain, because there was little effect on cruising or landing speeds. Fig. 11.9 shows the trim change as it was measured in a test in NASA's full-scale wind tunnel (Ref. 78).

The slightly larger and longer range Long-EZ, on the other hand, seems to show a stronger effect and always pitches down. Although there may not be enough trim authority to maintain hands-off flight, cruise and landing are not much affected generally. Clearly the rain is causing some BL separation but apparently the slot in front of the elevator prevents any wholesale separation.

The Quickie (Fig. 1.8 and Pic. 11.3), a very small (4 feet tall, 17 feet long, 300 pound empty weight) single-seat aircraft that also uses the GU airfoil for its front wing, is a different story. Most – perhaps all – Quickies show a strong and immediate reaction to even a few drops of rain; the boundary layer separates near mid-chord on the upper surface and half or more of the available elevator travel may be required to maintain level flight. The minimum flight speed is correspondingly increased by 10 kt or more so that a normal (tail-down, bounce-free) landing is not possible. The GU airfoil question is clouded by the fact that aircraft designers typically give templates only for cutting out the foam cores and not for the final profile; small variations from machine to machine are therefore inevitable because of individual variations in cloth application and in the filling and smoothing process. Fortunately, the Quickie's problems appear to be completely solved when a row of dozens of tiny vortex generators are placed span-wise at about mid-chord. The Quickie's big brother, the 2-place Q-200, uses the LS(1)-0417 MOD airfoil and apparently does not suffer from a rain problem. And John Roncz has designed a new airfoil for the Long-EZ that eliminates the pitch-trim problem that it had. The Voyager aircraft began with perhaps the most devastating rain effect of all (Fig. 11.10) but was very satisfactorily tamed by the addition of 210 vortex generators to the canard (0.4" high, 1.8" long, at a 20° angle of attack) at the 0.45c point. Finally, the new Beech Starship required vortex generators at a late stage in its certification to solve a high altitude canard separation problem.

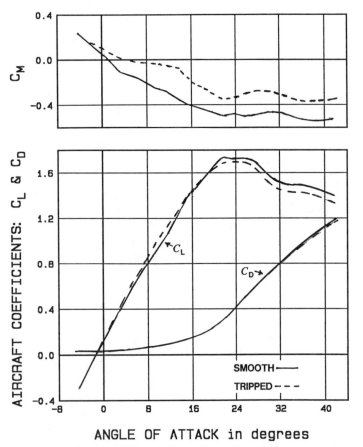

Fig. 11.9 Effect of artificial tripping of the boundary layer on the canard, for the VariEze. (Trimmed flight at 5.5° becomes untrimmed flight with a pitch-down moment coefficient of –0.14 when transition occurs.)

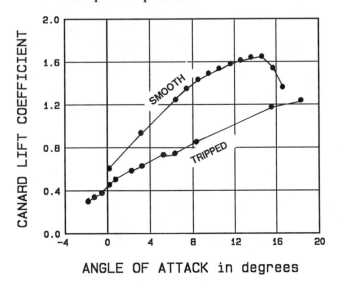

Fig. 11.10 Effect of tripping on the lift of the original Voyager canard (Ref. 76)

Pic. 11.3 A single-seat Quickie in flight (author is pilot, photographer is T. Brumbaugh)

Pic. 11.4 Sublimating chemical test of the VariEze's GU canard in the Langley 30 by 60 ft wind tunnel. Transition is at about 55% of chord, as shown by the darker area where the chemical has sublimated away because of the energy transfer encouraged by a turbulent BL. Note the forward regions of locally turbulent flow where the BL has been tripped by roughness. (Photo courtesy of NASA Langley.)

Why should rain or bugs cause premature BL separation on the GU airfoil? If an early transition to a turbulent BL occurs, it would seem that this more energetic BL should be less likely to separate. The answer appears to be that the tendency to separate is also strongly dependent on the thickness of the BL. Rain or insects or a paint stripe at the leading edge causes the BL to be as much as four times thicker when it reaches mid-chord on the 20% thick GU, and this thick BL simply can't withstand the severe adverse pressure gradient that it meets as it tries to round the bend, even at low angles of attack. The vortex generators take in energetic air from just outside the BL and generate a tiny rotating mass of air that tends to hug the surface and entrain the BL.

When the Rutan Aircraft Factory decided to try to find an airfoil for the Long-EZ canard that would not undergo a trim change, it enlisted the services of John Roncz. The problem with testing his computer-generated airfoils, though, was that it almost never rains in Mojave! Then it was discovered that a ¼" strip, torn from a roll of (heat) duct tape and applied to the 0.05c point on the wing with the ragged edge to the rear, duplicated the rain effect very nicely, and this led to the new Roncz canard that seems to solve the trim problem although the maximum lift coefficient is still slightly reduced by rain. This duct tape tripping method is now used both in the wind tunnel and in flight tests.

Airfoil designers emphasize the importance of tailoring the pressure distribution around the airfoil in such a way as to obtain laminar flow while retaining good stall characteristics. After an initial acceleration region, the boundary layer must be teased and eased into turbulence by a transition region consisting of a mildly adverse pressure gradient region, and then the **pressure recovery** can be effected expeditiously with a stronger adverse pressure gradient region. It is estimated that present-day sailplanes obtain up to about 60% laminar flow on the upper surface and close to 100% laminar flow on the lower

surface of their wings. Until the late 1970s, NASA felt that laminar flow for powered aircraft was not practical. But Bruce Holmes of NASA Langley rediscovered and exploited the **sublimation** technique for in-flight experimental determination of the extent of laminar flow over an aircraft; he has found that powered aircraft with composite construction are achieving much more laminar flow than was previously believed possible and now NASA is busily cranking out laminar flow airfoil designs (previous chapter). The sublimation technique involves the application of a thin coating of a chemical such as acenaphtene which will sublime from the solid state directly to the vapor state; the sublimation occurs much faster under a turbulent boundary layer than under a laminar one. The triggering effect of insect hits and construction irregularities is readily discerned with this technique. Aircraft that have been tested with the sublimation technique include the VariEze, the Long-EZ, the Quickie-like Rutan biplane racer, the Bellanca Skyrocket II, and a modified Cessna 210.

Holmes' more recent technique for the mapping out of in-flight laminar flow is through the use of painted-on, shear-sensitive liquid crystals. (Liquid crystals have the fluidity of liquids but some of the molecular ordering of solids.) The liquid crystal coating shows different colors in the laminar and turbulent BL regions. The advantages of this technique are the rapid response time and the ability to measure many angles of attack in one flight; disadvantages include the sensitivity to local temperature variations due to fuel cells or spars beneath the skin and the necessity to record each flight condition as it is flown (Ref. 77).

Wichita (General Aviation manufacturers) and airline transport manufacturers are listening—and doing similar research on their own. Inspect Figures 11.2, 11.4, and 11.11 to

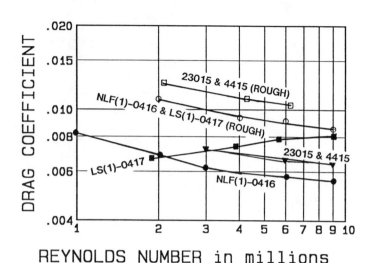

Fig. 11.11 Drag coefficients for various airfoils at a lift coefficient of 0.4 and over a range of Reynolds numbers (Ref. 26 of Chapter 10)

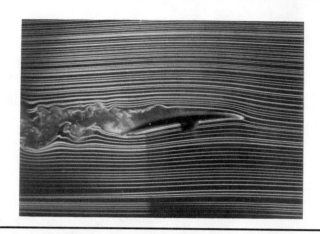

Pic. 11.5 This NASA LRN-1007 airfoil, designed by S. M. Mangalam and W. Pfenninger, shows significant separation at RN = 40,000 and $\alpha = 8°$.

Pic. 11.6 Separation is dramatically reduced when the air flow is acoustically excited (398 Hz). See Ref. 86. (Photos courtesy of S. M. Mangalam, Analytical Services & Materials, Inc.)

judge for yourself whether laminar flow is worth the bother (Refs. 7, 8, and 9.) There **is** hope that transport category aircraft will soon share in the delights of laminar flow in spite of the unfavorably large RN at which they operate. Special problems for them also include the generally deleterious effect of wing sweep and the difficulty of maintaining leading edge smoothness when then there are leading edge high lift devices as well (Ref. 10, 11, and 12).

E.   Special Problems at Low Reynolds Number

Some insects such as the dragonfly and the common housefly operate in the RN range from 1000 to 10,000. The dragonfly uses a sawtooth single surface airfoil and the housefly uses hairs projecting perpendicular to the surface of its wing to turbulate the air and delay laminar boundary layer separation. For indoor airplane models in this range, a single surface curved plate is the best airfoil. For model gliders in the RN range of 10,000 to 30,000, the lift coefficient doesn't exceed about 0.5 and fully laminar flow is the rule rather than the exception.

In the RN range from 30,000 to 70,000, drag is reduced and lift increased if the boundary layer is **tripped** (by, for example, a wire stretched in front of the leading edge). A proposed Mars airplane would fly in this range, as would remotely-piloted vehicles (RPVs) at very high altitudes. The most versatile tripping device for this and the next range is a **loudspeaker** generating sound waves in the 50 Hz to 1000 Hz range! (See Pic. 11.6.)

Bats and radio control models share the RN range from 70,000 to 200,000. It is easy to get extensive laminar flow in this range but there can be problems with separation of the laminar boundary layer; artificial BL tripping is required for thick airfoils. Large soaring birds and many ultralights operate in the RN range from 200,000 to 700,000. Laminar separation is a problem but there is often enough chordal length for the boundary layer to transition and then reattach with a "laminar bubble" in between.

The range from 700,000 to 3 million has been extensively studied. Sailplanes and single-engine aircraft operate mostly in this range. The possibility for laminar separation still exists (in fact, it actually occurs for the thin 6-series airfoils in this range: Ref. 14). The LS(1)-0417 produces about 14% more maximum lift and less drag if vortex generators are installed at the 0.60c point when the Reynolds number based on chord is 2.5 million. Some gliders have tried "pneumatic turbulators" which direct impact air to tiny orifices located where the laminar separation bubble begins. These pneumatic turbulators are doing somewhat the same job as a BL trip strip but are more versatile and can be used over a wider RN range.

Figure 11.12a presents the Eppler 61 airfoil, which is designed for low RN use; specifically, this airfoil has been suggested for use in a Mars airplane (which would be at very low RN because of the very low atmospheric density). The E 61 has a maximum lift coefficient of about 1.3 down to a RN of 40,000. (By now, you probably aren't surprised by the slim shape of this section.) (Ref. 6 of Chap. 10 and Ref. 16)

Figures 11.12b and 11.12c present two likely candidates for your next ultralight design. The Eppler 591 comes with Dan Somer's recommendation; it offers a theoretical maximum lift coefficient of 2.0 at a RN of 700,000, a soft (type 1) stall (next

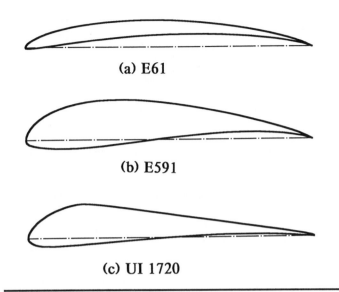

(a) E61

(b) E591

(c) UI 1720

Fig. 11.12 Three low Reynolds number airfoils: (a) the Eppler 61 for a Reynolds number of 100,000, (b) the Eppler 591 for ultralights, and (c) the University of Illinois UI-1720 for ultralights

–0.22. Alternately, the University of Illinois UI-1720 (Refs. 17 and 18) promises a lift coefficient of 1.96 at a relatively low RN of 1.8 million. If you anticipate operating at an RN of about 150,000 and want 100% laminar flow and don't mind a maximum lift coefficient of 0.24 at an angle of attack of 1.3 degrees, be sure to check out Ref. 19.

The classic source for model aircraft airfoils is Schmitz (Ref. 20). A recent book by Simons is an excellent modern guide to both model airplane aerodynamics and airfoils (Ref. 21). See also the recent Althaus airfoil catalog (Ref. 6 of Chap. 10).

F. Types of Stalls

Airfoil stall is usually considered to consist of three primary types: (1) **Trailing edge** stall occurs when the separation begins at the trailing edge and progresses steadily toward the leading edge as the angle of incidence is increased. This is the desirable type of stall; the lift curve continues to climb until the separation point nears the leading edge. (Compare Figs. 9.5 and 9.1c through 9.1f for the LS(1)-0417.) This is the normal type of stall for relatively thick airfoils with well rounded noses at medium to high RN; the lift coefficient is parabolically rounded about the maximum value. (2) **Leading edge** stall occurs when a laminar bubble forms near the leading edge. If the RN is large enough, the boundary layer will transition and reattach; the effect then is rather similar to increased thickness and camber near the leading edge. As the angle of attack is increased, the bubble moves closer to the leading edge and then very suddenly bursts, and separation is complete. There is an extremely sharp drop in the lift curve when the bubble bursts, which presents severe controllability problems for the pilot if a whole wing is involved. The stall for the 23012 shown in Fig. 10.1 is of this type. After the stall, the airfoil drops down to the thin airfoil lift curve (see next type) and then begins to increase in lift once again as

the angle of attack is further increased. (3) **Thin airfoil stall** occurs when a laminar bubble forms at the leading edge and then grows large enough to cover the whole airfoil surface. The resulting lift curve has a highly rounded maximum (Ref. 15).

Perhaps all common airfoils have a critical RN at which they transition between the type 1 and type 2 stall. There is then a RN for which there exists **hysteresis** in the lift curve: Increasing the angle of attack brings on separation when the bubble bursts, but when the angle is subsequently reduced the bubble does not re-form until the angle is considerably less.

Figure 11.13 presents experimental lift curves for the Wortmann FX 63-137 airfoil at three different Reynolds numbers, illustrating all three types of stall (Ref. 81). At a RN of 250,000, the lift curve has a nicely rounded maximum as the stalling angle is reached because, in this case, the boundary layer has separated as a laminar BL, transitioned, reattached as a turbulent BL, and finally separated again as a turbulent BL. At a RN of 100,000, the separation bubble lengthens with increasing angle of attack until it reaches the trailing edge as a long bubble and then it experiences a very abrupt loss of lift as the bubble bursts, a classic leading edge type of stall. At a RN of 70,000, the laminar BL separates as the angle of attack increases and there is insufficient chord length to provide any significant reattached flow; the maximum lift is very low but at least it occurs at a high angle of attack and is the least abrupt

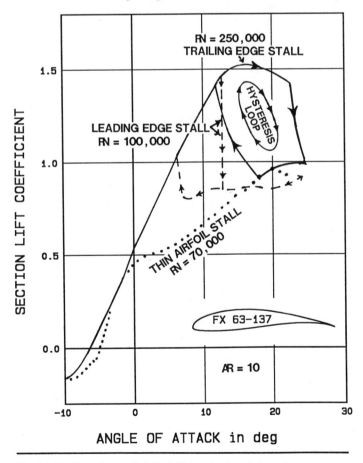

Fig. 11.13 Three types of stalls (with two hysteresis loops)

of all, a typical thin airfoil type of stall. This is called "subcritical" flow.

The story can get even wilder for some airfoils at certain Reynolds numbers. Hysteresis can also occur at **low** angles of attack when the flow transitions to supercritical flow as the angle increases, to form a mid-chord bubble; at decreasing angles the closed bubble is resistant against bursting to a lower angle. An airfoil can show **both** types of hysteresis loops at the same time. With hysteresis in the lift curve, you can bet there is also hysteresis in the drag curve; the drag may be the same at two very different lift coefficients. Incidentally, the word "hysteresis" is most commonly used in science to describe what happens when a substance such as iron is magnetized; the magnetism remains even after a magnetizing field is entirely removed. You undergo a hysteresis loop yourself when you enter a classroom or watch TV; you can never be quite the same when you leave the room or when the influence of the TV is removed.

Fig. 11.14 depicts what is happening around and within a laminar bubble in the BL.

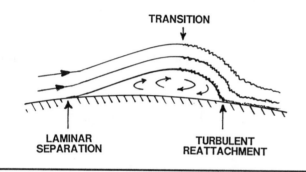

Fig. 11.14 The laminar separation bubble

G. Effects of Frost, Snow, Ice, and Rain

Federal Aviation Regulations include the statement that "no pilot may take off an airplane that has snow or ice adhering to the wings or stabilizing or control surfaces; or any frost adhering to the wings, or stabilizing or control surfaces, unless that frost has been polished to make it smooth." (FAR 91.209, Large and Turbine-Powered Multiengine Aircraft)

On January 13, 1982, an Air Florida Boeing 737 crashed into a bridge shortly after takeoff from Washington National Airport under snowstorm conditions. Because of clearance delays, there was a span of 49 minutes between the time when de-icing procedures were terminated and the takeoff roll was initiated. The National Transportation Safety Board (NTSB) later noted that

> The most significant effect of even a small amount of snow or ice on the wing surface is the influence on the smooth flow of air over the surface contour. Changes in the contour shape and roughness of the surface will cause the airflow to begin to separate from the wing at a lower angle of attack than normal and cause a reduction in lift which will normally be developed by a wing at a given angle of attack

and a given airspeed. Both the maximum lift which can be developed and the angle of attack at which it will be developed will be reduced significantly. Since the total lift developed depends upon both airspeed and angle of attack, an aircraft having snow or ice on the wings will be maintaining a higher than normal angle of attack at a given airspeed, or conversely must maintain a higher airspeed at a given angle of attack, in order to produce the lift required to support the aircraft weight. Stall buffet and stall will be encountered at a higher than normal airspeed. . . . Aside from altering the lift-producing properties of the wing surfaces, the most significant detrimental effect of snow or ice contamination on performance is the increase in the aircraft's total drag, that is, the force which resists the aircraft's forward motion through the air. . . . If snow or ice is present outboard on the wing, the lift distribution along the wing span will be changed so that the aircraft will be out of trim during takeoff. As it approaches takeoff airspeed and during initial rotation, the forward movement of the center of lift will cause the aircraft to pitch noseup. If the flightcrew failed to, or was unable to, counter the pitching moment of the aircraft with sufficient forward control column force, the aircraft could become airborne at an excessively high pitch attitude. The aircraft would not accelerate and it would retain a high angle of attack and high drag. . . . Any swept wing aircraft is vulnerable to such flight characteristics if takeoff is attempted with the outboard portions of the wings contaminated with snow or ice.

However, the NTSB report concluded that the B-737 still could have made a successful takeoff except for the fact that the crew failed to apply engine anti-icing, which caused erroneous thrust indications and resulted in the throttles not being advanced as far as they should have been (Ref. 22).

A 1970 study in England on the 0012 airfoil suggested that the effects of frost on wings is not very significant unless the frost covers the initial 10% of the wing and then it can cause up to a 33% decrease in the maximum lift and up to a 100% increase in drag (Ref. 23). These conclusions presumably do not apply to all airfoils or to all RN.

There are two basic types of ice that form on aircraft: **glaze ice** and **rime ice**. Glaze (sharp, smooth, clear) ice forms on an airplane when the outside air temperature is within about 10° F of freezing, while rime (rough, textured) ice forms at still lower temperatures. Both types of ice accumulate when an aircraft is flying in a cloud for which there exists supercooled water droplets. At temperatures close to 0° C, the droplets are very large and splatter over the cold wing before they freeze into glaze ice; at much colder temperatures, the drops are very small and individually freeze into rime ice. The rate of accumulation of glaze ice can be astonishing high under the proper circumstances. Figure 11.15 shows the typical shapes of these two types of ice under cruising flight conditions (Ref. 24). With such radical changes in airfoil shape, especially in the critical leading edge region, it is not surprising that airfoil performance is greatly degraded, as shown in Figs. 11.16 and 11.17. Note that glaze ice does about twice as much damage as rime ice.

Another recent study, this one in the NASA Lewis Research Center's Ice Tunnel, provides interesting data on the time dependence of ice formation (Ref. 25). Figure 11.18 presents the shapes versus time for both glaze and rime icing

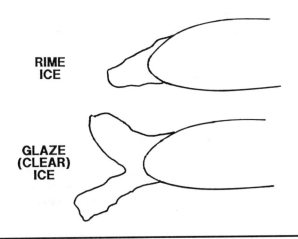

Fig. 11.15 The two general types of airfoil ice under **cruise** conditions

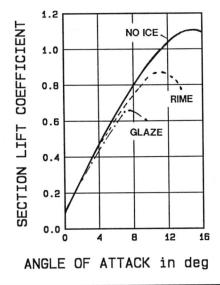

Fig. 11.16 Effect of ice on the lift of the 65A215 airfoil

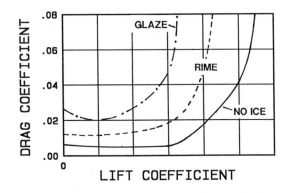

Fig. 11.17 Effect of ice on the drag polar of the 65A215 airfoil

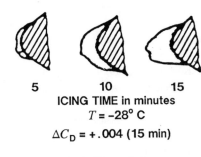

ICING TIME in minutes
$T = -28°$ C

$\Delta C_D = +.004$ (15 min)

(a) Rime Icing

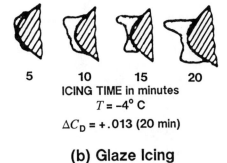

ICING TIME in minutes
$T = -4°$ C

$\Delta C_D = +.013$ (20 min)

(b) Glaze Icing

Fig. 11.18 Growth pattern for rime and glaze (clear) ice, **cruise** conditions

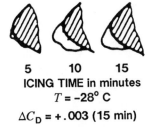

ICING TIME in minutes
$T = -28°$ C

$\Delta C_D = +.003$ (15 min)

(a) Rime Icing

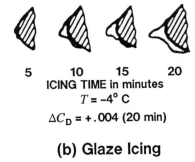

ICING TIME in minutes
$T = -4°$ C

$\Delta C_D = +.004$ (20 min)

(b) Glaze Icing

Fig. 11.19 Growth pattern for rime and glaze (clear) ice, **climb** conditions

under cruise conditions; Figure 11.19 presents similar information for climb conditions. This study also used a 6-series airfoil.

Researchers are now busily trying to correlate predictions of the new airfoil computational programs with the experimental measurements of icing effects with the hope that icing effects for the new airfoils can be accurately predicted without the time and expense of individual ice tunnel tests.

Reference 26 is a selected bibliography of pre-1981 NACA and NASA publications relating to the icing problem. As you might expect, the radical recontouring of the leading edge with ice formation can dramatically alter the landing characteristics of an aircraft. Generally speaking, the pilot should plan on a higher approach speed than normal and "fly" the aircraft onto the ground well above the normal landing speed because the stall could occur at a much lower angle of attack than normal and could be accompanied by an uncommanded roll.

It would not be too surprising if rain, especially heavy rain, could alter the flow pattern over a wing. Some recent transport accidents and some recent predictions suggest that rain effects could even be as large as wind shear effects—an unexpected conclusion. (See Refs. 27, 28, and 29.)

H. Airfoil Whimsy and Vortex Lift

What chance is there that you will accidentally discover a new airfoil that is superior to all that have preceded it? And if you do, what should your next step be? These were the questions that presented themselves in 1968 to one Richard Kline, a New York art director for an advertising agency, when one of the paper airplanes he was folding for his son turned out to have exceptional stability and long range. The airfoil on this "discovery" paper airplane was completely flat on top but wedge shaped with a step on the bottom (Fig. 11.20). With the help of a pilot friend, Floyd Fogleman, and $14K, the new airfoil was granted U.S. Patent Number 3,706,430 in 1972. *Time* magazine and CBS's "60 Minutes" were among those who jumped on the "little guy out-foxes the big guys" bandwagon (Ref. 30). It was claimed that the new Kline-Fogleman airfoil had demonstrated its superiority on a paper airplane that flew nearly the entire length of a football field, and that it resisted stalling even beyond an angle of 19°.

The success of this new airfoil in later years has been truly modest; it has apparently only been used on model airplanes. Some tests have been made on it in smoke and wind tunnels, however (Refs. 31 and 32). They indicated that the sharp leading edge was causing separation when an angle of 7° was exceeded but that a vortex was formed which caused the flow to reattach immediately **aft** of the vortex (Fig. 11.21). As the angle of attack was further increased, the size of the leading edge vortex increased and the reattachment point gradually moved toward the trailing edge. By about 25°, the flow had totally separated.

This "**vortex lift**" effect actually occurs and is used on many high speed aircraft today (Refs. 33 and 34; Figs. 12.21 to 12.24). On the F-16, the strake generates a vortex at high angles of attack, and this vortex spreads over the main wing and helps

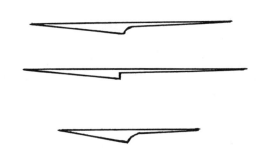

Fig. 11.20 General shapes of the original Kline-Fogleman airfoils

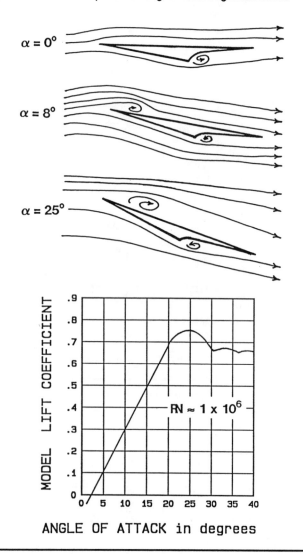

Fig. 11.21 Flow patterns and lift curve for a wing using the Kline-Fogleman airfoil (the model had a 55° leading edge sweep and an aspect ratio of 1.455; Ref. 31)

keep the flow attached to much higher angles than otherwise. A delta wing airplane such as the supersonic Concorde airliner obtains considerable lift in this fashion, too. The effect might be thought of as equivalent to placing a vortex generator at the leading edge; the vortex helps keep the BL attached and the circulation (lift) maintained. For low speed, high RN applications, though, a sharp leading edge is **not** an efficient way to generate lift; we gave up on flat plates a long time ago.

Anyway, the Kline-Fogleman airfoil's claim to be resistant to stall appears to be valid, thanks to these leading edge vortices; the airfoil also maintained its lift more than usual beyond the angle for maximum lift (like a thin airfoil stall) and it did prove to be stable (unlike conventional cambered airfoils). However, at the tested RN of about 1 million, the maximum lift coefficient was only about 0.75 and the maximum lift/drag ratio never even reached 3.0! Clearly the airfoil was very inefficient under the tested conditions, in addition to its obvious structural problems.

More recently, the inventors have written a little book on model airplanes (Ref. 75) in which they present their side of the story; the famous step here appears with conventional (rounded nose, thick) airfoils and sometimes on the upper surface as well.

Vortex generators can be thought of as turned-over flat plate airfoils, oriented at an angle of attack such that a maximum vortex is generated but below the angle at which the vortex bursts. The NACA flush inlet, a diverging channel cut into the body of a fuselage for cooling air for the engine or for people, has perpendicular walls with sharp corners to generate a vortex on each side (Fig. 11.22) that greatly helps in encouraging the boundary layer air to come in. Most Long-EZ's use a large NACA flush inlet on the bottom of the fuselage, rather than a projecting scoop such as the P-51 used, to obtain carburetor and cooling air. Many production and homebuilt aircraft use the NACA scoop as a low drag inlet for cabin air. Incidentally, the initial application was to jet engine inlets and so the NACA report was classified for many years (Ref. 79). It didn't work out in that application because, located on the side of the fuselage, it wasn't effective over all angles of attack.

I.   The Future Is Here: The Airfoil Store

In the past, aircraft designers have had to page through airfoil catalogs in search of the best airfoil for their needs. Now the catalogs are thicker than ever but, for the discriminating designer, a made-to-order airfoil is much tastier. A custom airfoil is just what the new computer programs are quite capable of generating; you can get maximum lift or maximum lift/drag ratio at the RN of your choice (within reason). The programs for low speed airfoils, most perhaps related to Richard Eppler's work, are now readily available but considerable practice and insight (and a reasonably powerful desk computer) are needed if they are to be used effectively (Refs. 35, 36, 37, and 84).

John Roncz has achieved unique status as an airfoil designer and aerodynamics popularizer among homebuilders. Unlike his counterparts in universities or industrial or government research labs, with one exception (Ref. 83) he has not presented learned papers at professional engineering conferences or written research articles for engineering journals. While he has developed many airfoils for Burt Rutan (for the powered-glider Solitaire, the around-the-world Voyager, the Long-EZ canard, the Catbird), he has retained his independent status. The Beech Starship (Chap. 1) uses 5 different Roncz airfoils. By 1984 he could report that 23 of his airfoils were flying on eight different types of airplanes. The unlimited-class Pond racer, designed by Burt Rutan and to be test-flown by brother Dick Rutan (Pic. 11.7), is in the development stage as this is written; it uses Roncz airfoils for its flying surfaces and its propellers. For this airfoil designer, every new aircraft design seems to require one dozen or more custom airfoils! Roncz is also one of the most popular speakers at the forums held at the week-long Oshkosh EAA Convention (Pic. 11.8). His talks have ranged from introductory aerodynamics and airfoil design to a hilarious analysis of the probable flight performance of his pet dragon.

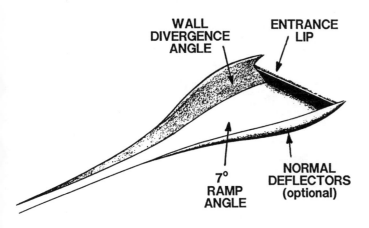

Fig. 11.22 The NACA low drag, flush inlet ("NACA scoop")

Pic. 11.7 Test pilot Dick Rutan (left) and designer Burt Rutan (right) with a model of the Pond Racer (Oshkosh 1988); prop and wing airfoils by Roncz

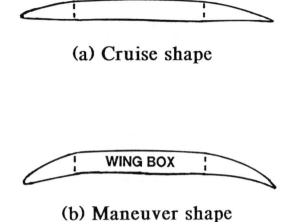

(a) Cruise shape

WING BOX

(b) Maneuver shape

Fig. 11.23 The USAF "mission-adaptive" wing — a modern version of the Wrights' warpable wing of 1903

The U.S. Air Force is working on what it calls a "mission-adaptive" wing for which the camber and wing twist are computer-controlled while in flight (Ref. 38 and Fig. 11.23). It is possible to get much higher maneuvering performance if the camber of the leading and trailing sections of the wing can be **smoothly** increased during maneuvering flight, rather than relying on hinged flap surfaces. In a sense it is a return to the Wright brothers' wing-warping technique — but the speeds involved are *noticeably* greater.

It has been suggested that research on airfoils is now reaching a point of diminishing returns and that it is time for fundamental research to turn to more sophisticated programs that can predict multi-element properties (for example, airfoils with one or more flaps or spoilers) and whole-airplane properties.

Summary

The camber of an airfoil was shown to control the magnitude of the design lift coefficient: a symmetrical airfoil has 0 for its design lift coefficient.

The earliest laminar airfoils had a narrow low drag "bucket" and tended to be less efficient than other airfoils when operating outside that bucket. The new composite construction methods allow the promises of laminar flow to be fulfilled and the new laminar airfoils aim to increase the maximum lift and broaden the laminar bucket, while minimizing any dependence of lift on laminar flow. A single-element airfoil can yield an astonishingly large maximum lift coefficient (the Liebeck airfoils, for example) if it is optimized for one angle or for a narrow operating range by use of the Stratford pressure distribution theory. The boundary layer becomes too laminar at low RN and it therefore becomes increasingly difficult to prevent laminar separation at low angles of attack from occurring (stall types 2 and 3) as the RN is reduced. Yet these low RN regions are of considerable interest for such "aircraft" as remotely-piloted vehicles and Mars explorers.

Pic. 11.8 Airfoil designer John Roncz, a forum speaker at the Experimental Aircraft Association's Oshkosh 1988 Convention

The three general types of airfoil stall, trailing edge, leading edge, and thin airfoil, were discussed and compared. It was noted that the forming and bursting of a laminar separation bubble was important to the understanding of low RN airfoil characteristics.

Quantitative estimates of the airfoil degradation to be expected from frost, ice, or heavy rain were presented. The drag coefficient can triple and the maximum lift coefficient can be reduced by a factor of 2.

The principle of vortex lift was presented in the context of a "miracle" airfoil that has quietly faded away. It was noted that computer programs that can produce airfoils tailored to almost any special application are now readily available.

Review Questions

1. The proper camber for an airfoil is determined by the value of the ___ coefficient required for cruising flight.

2. An airfoil jumps out of its drag bucket when the boundary layer ___.

3. The amount of lift developed at an angle of attack of 0° is determined by the ___ of the airfoil.

4. The 6-series airfoil on the Cherokee appears to be better than the 4-digit airfoil on the Cessnas only when it is ___.

5. The steepness of the sides of the drag bucket could lead to trouble if a pilot tries to take off at a ___ lift coefficient.

6. The maximum lift coefficient of the LS series, at the most favorable RN, is about ___; the maximum lift coefficient of some of the Liebeck sections is about ___.

7. The Stratford distribution is a pressure distribution for the ___ gradient region on the upper surface of an airfoil that will permit pressure recovery in the minimum ___.

8. The Lissaman L 7769 airfoil, apparently based on some of Liebeck's work, was used on a successful ___ aircraft.

9. A "___" airfoil is one that has a very constant pressure coefficient over most of the front half of its upper surface.

10. A wing made from composite materials (such as foam and fiberglass) can achieve considerable laminar flow because such a wing can be both ___ and ___, as well as sufficiently rigid.

11. The original Quickie design resulted in an aircraft in which laminar flow over the canard is lost when flying in ___; the result is a great loss in ___ and a great increase in ___.

12. NASA has investigated the possibility of using a ___ and a porous ___ ___ to deal with the bug problem for laminar airfoils.

13. To obtain good stall characteristics on a laminar airfoil, there must exist a carefully tailored region around mid-chord which induces the boundary layer to ___.

14. Present day sailplanes can achieve up to about ___ % laminar flow on the upper surface of their wings and up to about ___ % on the lower surface.

15. A rediscovered experimental technique for determining the extent of laminar flow on a wing is the use of a ___ substance.

16. For indoor airplane model gliders, the best airfoil is a ___ ___.

17. At low RN, the lift and drag characteristics can often be improved by artificially inducing transition; this is called "___" the boundary layer.

18. The most desirable type of stall for an airfoil is the ___ - ___ type. This lead to a ___ maximum for the ___ coefficient.

19. Leading edge stall is associated with the formation of a laminar "___" at the leading edge; the stall occurs very abruptly when this ___.

20. Many airfoils have a critical value of RN at which the maximum lift coefficient depends strongly on whether the ___ is increasing or decreasing (hysteresis).

21. The worst type of ice, from both a lift and a drag standpoint, is ___ ice.

22. A swept-wing aircraft is susceptible to ___ and ___ instability when it initiates a takeoff with snow or ice on its wings.

23. An airfoil with a maximum lift coefficient of 1.8 might be degraded to 1 with only about ___ if it is carrying a heavy load of ice.

24. The Kline-Fogleman airfoil had good ___ characteristics but very low values for its ___ and ___.

25. "___ lift" occurs when a leading edge of a wing or strake causes the flow to start to circulate and then reattach.

26. Which airfoil would you expect to stall at the higher angle of attack: an LS(1)-0413 or an LS(1)-0013? (Hint: Sketch their probable lift curves, assuming nearly equal maximum lift coefficients.)

27. An airfoil which has a different lift coefficient (and a different drag coefficient) when the angle of attack is increasing than when it is decreasing is said to exhibit ___ in its section characteristics. This is very common for airfoils that are operating at ___ .

Problems

1. Suppose you have designed a wonderfully cambered airfoil in which, at zero angle of attack, the air is not moving at all on the bottom surface but is moving uniformly at flight speed on the flat upper surface. What fraction does each surface contribute to the generated lift? What is the lift coefficient?
Hint: Refer to Secs. B and C in Chap. 8. Answer: 1.0

2. For a flat-bottomed airfoil at zero angle of attack, what must be the average speed on the upper surface if the lower surface has uniform flow at flight speed and the lift coefficient is 2.5?
Answer: 87% faster than the flight speed

3. How many drag counts better is the NLF(1)-0416 than the 23015 if both are smooth and operating at a lift coefficient of 0.4 and a RN of 3 million?

References

1. Trammell, Archie, "Laminar Flow," *Flying*, August 1968, pp. 54–59. (Nice popular description of the design and structural advantages of the 6-series laminar section over earlier airfoils, plus an interesting discussion of their possible disadvantages for takeoffs and landings.)

2. Stratford, B. S., "The Prediction of Separation of the Turbulent Boundary Layer," *Journal of Fluid Mechanics*, Vol. 5, 1959. (Presentation of the *Stratford distribution*, the upper surface pressure distribution that should place an airfoil on the brink of separation.)

3. Liebeck, R. H. and Ormsbee, A. I., "Optimization of Airfoils for Maximum Lift," *Journal of Aircraft*, Vol. 7, No. 5, pp. 409–415, Sept.–Oct. 1970. (Using the Stratford pressure recovery criteria for the aft part, an airfoil yielding a maximum lift coefficient of 2.54 at an RN of 5 million is described.)

4. Liebeck, Robert H., McDonnell Douglas Corporation, "A Class of Airfoils Designed for High Lift in Incompressible Flow," *Journal of Aircraft*, Vol. 10, No. 10, Oct. 1973, pp. 610–617. (Two humped airfoils which yielded experimental maximum lift coefficients of 1.8 at an RN of 3 million and 2.15 at an RN of 1 million just before catastrophic separation occurred are described.)

5. Liebeck, Robert H., "Design of Subsonic Airfoils for High Lift," *Journal of Aircraft*, Vol. 15, No. 9, September, 1978, pp. 547–561. (A series of high lift airfoils designed for high-altitude, long-endurance aircraft, for sailplanes, for propellers and windmills and fans, and for racing car wings are presented.)

6. Burke, J. D., AeroVironment Inc., "The Gossamer Condor and Albatross: A Case Study in Aircraft Design," Report No. AV-R-80/540, June 1980. (Available from AIAA. This is a technical history of the privately-financed aircraft development program that terminated in winning the Kremer Prize for man-powered aircraft over a figure 8 course and then the second Kremer Prize for the first man-powered aircraft to cross the English Channel.)

7. Main-Smith, J. D., *Chemical Solids as Diffusible Coating Films for Visual Indication of Boundary Layer Transition in Air and Water,* ARC British R and M No. 2755, February, 1950.

8. Owen, P. R. and Omero, A. O., "Evaporation From the Surface of a Body in an Airstream," ARC British R and M No. 2875, September, 1951. (Available from NTIS as N78-78495. Sublimation technique for detecting transition.)

9. Holmes, Bruce J., Coy, Paul F., Yip, Long P., Brown, Philip W., and Obara, Clifford J., "Natural Laminar Flow Data from Full Scale Flight and Wind Tunnel Experiments, Eighth Annual AIAA General Aviation Technology Test, Wichita, Kansas, 1981. (Sublimation detection of transition on various composite aircraft.)

10. "Advanced Aerodynamics and Active Controls," NASA-CP-2172, 1980. (Available from NTIS as N81-19001. Possibilities for natural laminar flow on large and high speed aircraft.)

11. Quast, Armin, "Laminar Airfoils for Transport Aircraft," European Space Agency Technical Translation ESA-TT-680, 1981. (Available from NTIS as N82-18190.)

12. Korner, H. and Redeker, G., "Recent Airfoil Development at DFVLR," Aircraft Systems and Technology Conference, 1982. (Available from AIAA as A82-40986. Discussion of new laminar airfoils for

transport, sailplanes, propellers, helicopter rotors, and combat aircraft.)

13. "Laminar Flow Control Systems Advance," *Aviation Week & Space Technology*, November 12, 1979, p. 197. (Plans for a hybrid supercritical-laminar airfoil for long-range transports.)

14. Carmichael, B. H., "Low Reynolds Number Airfoil Survey," NASA-CR-165803, Vol. 1, 1981. (Available from NTIS as N82-14059. Useful survey of airfoils for RN from one thousand to one billion.)

15. McCullough, G. B. and Gault, D. E., "Examples of Three Representative Types of Airfoil-Section Stall at Low Speed," NACA TN 2502, September, 1951.

16. Clark, Victor C., Kerem, Abraham and Lewis, Richard, "A Mars Airplane," *Aeronautics and Astronautics*, January, 1979, pp. 42–54. (Proposal for a battery- or hydrazine-engine-driven airplane, resembling a competition glider, that could make a high resolution survey of Mars.)

17. Sivier, Kenneth R., Orsmbee, Allen I., and Awker, Randal W., "Low Speed Aerodynamic Characteristics of a 13.1-Percent-Thick, High-Lift Airfoil," NASA-CR-153937, April, 1974. (Available from NTIS as N77-28069. Experimental study of the UI-1720 airfoil which yielded a maximum lift coefficient of nearly 2.0 at a RN of 1.75 million. Coordinates are given.)

18. Ormsbee, Allen I., "Low Speed Airfoil Study," Final Report, NASA-CR-153914, July 1977. (Report on the UI-1720 which had a maximum lift coefficient of 1.96 along with a drag coefficient of .028 at an RN of 1.8 million.)

19. Sapuppo, J. and Archer, R. D., "Fully Laminar Flow Airfoil Sections" *Journal of Aircraft*, Vol. 19, No. 5, May 1982, pp. 406–409. (Using the Stratford criteria, two airfoils that have laminar flow over the whole of both surfaces at RN up to 166,000 are described; unfortunately, the maximum lift coefficient of 0.24 occurs at an angle of attack of only 1.3°.)

20. Schmitz, F. W., "Aerodynamik der Flugmodells" (Aerodynamics of the Model Airplane), 1942. (1976 English translation of Part I on Airfoil Measurements is available from NTIS as N70-39001.)

21. Simons, Martin, *Model Aircraft Aerodynamics,* Argus Books Limited, 1978. (Recommended introduction to low RN aerodynamics; quite a few suitable airfoils, including Eppler and Wortmann sections, are presented.)

22. *Aviation Week & Space Technology*, Sept. 27, Oct. 4, Oct. 18, Oct. 25, Nov. 1, and Nov. 22, 1982. (Report on the B-737 accident at Washington National in which snow and ice on the wings was a contributing factor.)

23. Gregory, N. and O'Reilly, C. L., "Low-Speed Aerodynamic Characteristics of NACA 0012 Aerofoil Section, including the Effects of Upper-Surface Roughness Simulating Hoar Frost," R. and M. No. 3726, 1973. (Available from NTIS as N74-17709.)

24. Bragg, M. B., "Rime Ice Accretion and its Effect on Airfoil Performance," NASA-CR-165599, 1982. (Available from NTIS as N82-24166. Study of the effect on lift and drag coefficients of glaze and rime icing on the 65A215 Airfoil. Many references.)

25. Shaw, Robert J., Sotos, Roy G., and Solano, Frank R., "An Experimental Study of Airfoil Icing Characteristics," NASA-TM-82790, 1982. (Available from NTIS as N82-17083. Study of the effect of rime and glaze ice on the cruise and climb performance of a 63_2-A415 airfoil in the NASA Lewis Icing Tunnel.)

26. Reinmann, J. J., "Selected Bibliography of NACA-NASA Aircraft Icing Publications," NASA-TM-81651, 1981. (Available from NTIS as N82-11053.)

27. Luers, J. and Haines, P., "Heavy Rain Influence on Airplane Accidents," *Journal of Aircraft*, Vol. 20, No. 2, February 1983, pp. 187–191. (Suggests that heavy rain may be as serious a problem as wind shear; a number of transport aircraft accidents occurring in heavy rain are described.)

28. Haines, P. and Luers, J., "Aerodynamic Penalties of Heavy Rain on Landing Aircraft," *Journal of Aircraft*, Vol. 20, No. 2, February 1983, pp. 111–119. (Theoretical estimates of the lift and drag penalties due to very heavy rain.)

29. Lee, Otto W. K., "Preliminary Investigation of Effects of Heavy Rain on the Performance of Aircraft," NASA-TM-832782, 1982. (Available from NTIS as N82-20145. Attempt to verify predicted B-737 performance deterioration in rain.)

30. "The Paper-Plane Caper," *Time*, April 2, 1973, pp. 51–52. (Popular promotion of the strange Kline-Fogleman airfoil.)

31. McFarland, Charles, "An Experimental Investigation of the Aerodynamics of the Kline-Fogleman Airfoil," Wichita State University, Aeronautical Engineering Department Report No. AR76-5, 1976. (Available from NTIS as N77-16988. Used two- and three-dimensional smoke tunnel testing and force measurements at a RN of about one million.)

32. DeLaurier, D. and Harris, J. M., "An Experimental Investigation of the Aerodynamic Characteristics of Stepped-Wedge Airfoils at Low Speed," AIAA Paper 74-1015. (Available from AIAA as A74-42037. Kline-Fogleman airfoil again, but this time at an RN of about 2×10^4 and including tests on a glider supplied by Kline. The measured maximum lift coefficients and lift/drag ratios were all inferior to those measured for a thin, flat plate airfoil, which is itself inferior to a thin curved plate. However, the authors suggest that the stability of the airfoil during a high energy hand launch might account for some of its reported successes on a paper glider.)

33. Polhamus, E. C., "Predictions of Vortex-Lift Characteristics by a Leading-Edge Suction Analogy," *AIAA Journal*, Vol. 8, No. 4, April 1971, p. 193.

34. Wentz, W. H. Jr. and Kohlman, D. L., "Vortex Breakdown on Slender Sharp-Edged Wings," *AIAA Journal*, Vol. 8, No. 3, March 1971, pp. 156–161.

35. Eppler, R., "Direct Calculation of Airfoils from Pressures Distribution," NASA-TT-F-15417, March 1974. (Translation of an article published in German in 1957; available from NTIS as N74-19641. Early paper on the Eppler airfoil program.)

36. Eppler, Richard and Somers, Dan M., "A Computer Program for the Design and Analysis of Low-Speed Airfoils," NASA-TM-80210, 1980. (Available from NTIS as N80-29254.) Eppler, Richard and Somers, Dan M., "Supplement to: A Computer Program for the Design and Analysis of Low-Speed Airfoils," NASA-TM-81862, 1980. (Available from NTIS as N81-13921. All you need to design your own airfoils.)

37. Thomson, William G., "Design of High Lift Airfoils with a Stratford Distribution by the Eppler Method," University of Illinois Technical Report AAE 75-5, June, 1975. (NASA-CR-153913. Available from NTIS as N77-27108. General description of the Eppler inverse conformal mapping technique which yields the coordinates of an airfoil from a given velocity (or pressure) distribution. Various high lift airfoils obtained by use of this computer program are described.) Thomas, William G., "Program Manual for the Eppler Airfoil Inversion Program," University of Illinois Technical Report AAE 75-4, May 1975. (NASA-CR-153926. Available from NTIS as N77-28068. Summary of the theory and of the program, with a listing of the FORTRAN program.)

38. "Versatile Fighter Capabilities Emerge," *Aviation Week & Space Technology*, January 29, 1979, pp. 59–66. "Air Force Technical Plans Aimed at Next 20 Years," *Aviation Week & Space Technology*, January 29, 1979, pp. 174–178. (The mission-adaptive wing with programmed variable camber and twist is predicted.)

39. McMasters, J. H. and Henderson, M. L., "Some Recent Applications of High-Lift Computational Methods at Boeing," *Journal of Aircraft*, Vol. 20, No.1, January 1983, pp. 27–33. (A computational analysis design methodology for multi-element high lift wing/body combinations is described along with three applications to transport design.)

40. Dwiggins, D., *On Silent Wings*, Grosset and Dunlap, New York, 1970. (Sailplane technology but including the airfoils of a buzzard, the Wright Flyer of 1903, the Lilienthal Hang Glider No. 6 of 1893, a goose, the albatross, and various Goettingen sections.)

41. Chen, M. K. and McMasters, J.H., "From Paleoaeronautics to Altostratus—A Technical History of Soaring," AIAA Aircraft Systems and Technology Conference, August 11–13, 1981, Dayton, OH. (AIAA Paper No. AIAA-81-1611. A compact history of soaring, from the birds to the present time, including a summary of the evolution of sailplane airfoils. Good bibliography.)

42. Riegels, F., *Aerofoil Sections*, translated by Randall, D. G., Butterworths, London, 1961. (A hard-to-find but useful compendium of airfoils up to about 1950; a good reference for Goettingen profiles.)

43. Jeracki, Robert J. and Mitchell, Glenn A., "Low and High Speed Propellers for General Aviation—Performance Potential and Recent Wind Tunnel Test Results," SAE Paper 811090, October 1981. (A good survey of previous propeller technology as well as an outline for advanced technology improvements such as new airfoil sections, composite construction, sweep, integrated blade/spinner and prop/nacelle, and proplets.)

44. Wentz, W. H. Jr. and Ostowari, C., "Summary of High-Lift and Control Surface Research on NASA General Aviation Airfoils," SAE Paper No. 810629, April, 1981. (A good reference for the investigations that have been made on the design and testing of control surfaces and flaps for the new NASA airfoils.)

45. Keiter, Ira D., McCauley Accessory Div., Cessna Aircraft Co., "Impact of Advanced Propeller Technology on Aircraft/Mission Characteristics of Several General Aviation Aircraft," SAE Paper No. 810584, April, 1981. (Describes current and advanced technology propellers with anticipated cruise performance gains.)

46. Poole, R. J. D. and Teeling, P., "Airfoils for Light Transport Aircraft," SAE Paper No. 810576, April, 1981. (This paper illustrates the fact that the commercial aircraft builders—here de Havilland—are doing airfoil research, as well as NASA, but they don't normally publish coordinates for their airfoils!)

47. McMasters, J. H., Nordvik, R. H., Henderson, M. L., and Sandvig, J. H., "Two Airfoil Sections designed for Low Reynolds Number," *Technical Soaring*, Vol. VI, No. 4, June, 1981. (Soaring pilots have led the way for many airfoil innovations; here one section for use as a 29% thick strut for wind tunnels and another section for an ultralight glider are presented.)

48. Somers, Dan M., "Experimental and Theoretical Low-Speed Aerodynamic Characteristics of a Wortmann Airfoil as Manufactured on Fiberglass Sailplane," NASA-TN-D-8324, February 1977. (Available as NTIS N77-18049. Interesting comparison of an FX glider section in the Langley low turbulence pressure tunnel with the University of Stuttgart tunnel results.)

49. Smith, A. M. O., "High-Lift Aerodynamics," 37th Wright Brothers Lecture, *Journal of Aircraft*, Vol. 12, No. 6, June 1975, pp. 501–530. (A nice review of theoretical work up through the Liebeck airfoils by the chief designer for the Douglas B-26 and other aircraft.)

50. Wortmann, F. X., "Drag Reduction for Gliders," NASA-TM-75293, May 1978. (Translation of 1965 paper. Available from NTIS as N78-24116. General discussion of the importance of, and means for achieving, extensive laminar flow over the whole of a glider; some critical roughness heights are given; the aural method of distinguishing between a laminar and a turbulent boundary layer are described.)

51. Eppler, Richard, "Turbulent Airfoils for General Aviation," *Journal of Aircraft*, Vol. 15, No. 2, February 1978, pp. 93–99. (In contrast with recent NASA airfoils, Eppler here presents four simple sections—Numbers 1210,1211,1212, and 1213—with low moment coefficients, low trailing edge loading, and high flap-down maximum lift at a high RN of 16 million.)

52. Nagamatsu, H. T. and Cuche, D. E., "Low Reynolds Number Aerodynamic Characteristics of Low-Drag NACA 63-208 Airfoil," *Journal of Aircraft*, Vol. 18, No. 10, October 1981, pp. 833–837. (These investigators found a nearly constant maximum lift coefficient but a strong dependence of the pitching moment coefficient on angle of attack at RN between 20,000 and 500,000.)

53. Yip, Long P. and Coy, Paul F., "Wind-tunnel Investigation of a Full-Scale Canard-Configured General Aviation Aircraft," ICAS Paper Number 82-6.8.2, August, 1982. (Description of wind tunnel tests on a VariEze.)

54. Storer, John H., "Bird Aerodynamics," *Scientific American*, Vol. 186, pp. 24–29, 1952. (Discusses how birds fly and their wingtip devices. States that a duck hawk, when pressed by an airplane, can zip along at 175 to 180 miles per hour!)

55. Storer, John H., "The Flight of Birds: Analyzed Through Slow-Motion Photography," Craybrook Institute of Science, Bull. #28, 1948.

56. Carnish, Joseph J. III, "The Boundary Layer," *Scientific American*, August, 1954, pp. 72–77. (A nice popular-level article.)

57. Hazen, David C. and Lehnert, Rudolf F., "Low Speed Flight," *Scientific American*, April 1956, pp. 46–51. (How the boundary layer can be controlled with various types of flaps to decrease landing speeds; also an air pump circulation-control system in a Cessna 170 is shown!)

58. Weis-Fogh, Torkel, "The Flight of Locusts," *Scientific American*, March 1956, pp. 116–124. (The forces required for flight were measured on this tandem-winged insect!)

59. Weis-Fogh, Torkel, "Unusual Mechanisms for the Generation of Lift in Flying Animals," *Scientific American*, Nov. 1975, pp. 81–87; 148. (Hovering can only be understood through theories of nonsteady aerodynamics; the mechanisms are best described as clap-fling and flip.)

60. Pennycuick, C. J., "Mechanics of Flight," (Chapter 1 in *Avian Biology*, Edited by Donald S. Farner, Academic Press, 1975. Power requirements for birds; gliding and soaring flight.)

61. Cone, Clarence D., "The Soaring Flight of Birds," *Scientific American*, April, 1962, pp. 130–140. (Use of thermals for soaring; slotted wingtips on soaring birds.)

62. Smith, Roger L., "Closed-Form Equations for the Lift, Drag, and Pitching-Moment Coefficients of Airfoil Sections in Subsonic Flow," NASA-TM-78492, August, 1978. (Available from NTIS as N78-29068. FORTRAN program with sample calculations for angles of attack through 360°.)

63. Pierpont, P. Kenneth, "Bringing Wings of Change," *Aeronautics and Astronautics*, October, 1975, pp. 20–27. (History, up to 1975, of NASA's airfoil research programs.)

64. Freeman, Hugh B., "Comparison of Full-Scale Propellers Having R.A.F. 6 and Clark Y Airfoil Sections," NACA Report No. 378, 1931. (Early propeller tests.)

65. Miley, S. J., "On the Design of Airfoils for Low Reynolds Numbers," International Symposium on the Technology and Science of Low Speed and Motorless Flight, September 1974. (Available from AIAA as A74-42038.)

66. NACA Technical Report Numbers 93, 124, 182, 244, 186, and 315. (Tabulation of airfoil coordinates and airfoil measurements made in wind tunnels around the world, to about 1930; more than 800 sections are listed; good source for early airfoils.)

67. Hicks, Raymond M., Mendoza, Joel P., and Bandettini, Angelo, "Effects of Forward Contour Modification on the Aerodynamic Characteristics of the NACA 64_1212 Airfoil Section," NASA-TM-X-3293, 1975; Hicks, Raymond M. and Schairer, Edward T., "Effects of Upper Surface Modifications on the Aerodynamic Characteristics of the NACA 63_2-215 Airfoil Section," NASA-TM-78503, 1978. (Available from NTIS as N79-14024. Here's how to renovate your old 6-series airfoil and make it into a new technology version.)

68. Advanced Technology Airfoil Research, Vol. 1, NASA-CP-2045-Pt-1 (Available from NTIS as N79-20030. Papers by Eppler, Somers, Whitcomb, and many others. Advanced Technology Airfoil Research, Vol. 2, NASA-CP-2046. (Available from NTIS as N80-21283)

69. The Science and Technology of Low Speed and Motorless Flight, Part 1, NASA-CP-2085, 1979. (Available from NTIS as N79-23889. Many good papers in a broad range of related areas.)

70. Carmichael, B. H., "Summary of Past Experience in Natural Laminar Flow and Experimental Program for Resilient Leading Edge," NASA-CR-152276. (Available from NTIS as N79-26024.)

71. Bragg, M. B., Zaguli, R.J., and Gregorek, G. M., "Wind Tunnel Evaluation of Airfoil Performance Using Simulated Ice Shapes," NASA-CR-167960, 1982. (Available from NTIS as N83-15265. Lots of experimental pressure coefficient curves for simulated ice shapes; the measured effects were not as large as expected.)

72. Holmes, B. J., and Obara, C. J., "Observations and Implications of Natural Laminar Flow on Practical Airplane Surfaces," *Journal of Aircraft*, Vol. 20, No. 12, December, 1983, pp. 993–1006. (A valuable survey paper on the current state of the art in the achievement of natural laminar flow.)

73. Cox, Jack, "Voyager...The Adventure Begins," *Sport Aviation*, July, 1984, p. 23. "Rutan Prepares Two-Place Voyager for Global Flight," *Aviation Week & Space Technology*, June 11, 1984, p. 19. "Lightweight Rutan Voyager Makes Its First Test Flights," *Aviation Week & Space Technology*, July 2, 1984, p. 28.

74. Yeager, Jeana and Rutan, Dick, with Phil Patton, *Voyager*, Alfred A. Knopf, New York, 1987.

75. Kline, Richard, *The Ultimate Paper Airplane*, Simon & Schuster, New York, 1985.

76. Bragg, M. B. and Gregorek, G. M., "Experimental Study of Airfoil Performance with Vortex Generators," *Journal of Aircraft*, Vol. 24, No. 5, May, 1987, pp. 305–309. (Discusses the recent history of problems with laminar flow airfoils on canard-type aircraft and the development of vortex generators on the around-the-world Voyager to solve its problem.)

77. Holmes, Bruce J., Gall, Peter D., Croom, Cynthia C., Manuel, Gregory S., and Kelliher, Warren C., "A New Method for Laminar Boundary Layer Transition Visualization in Flight — Color Changes in Liquid Crystal Coatings," NASA TM-87666, January 1986. (Available from NTIS as N86-21518.)

78. Holmes, B. J., Obara, Clifford J., and Yip, Long P., "Natural Laminar Flow Experiments on Modern Airplane Surfaces," NASA TP-2256, 1984. (Available from NTIS as N84-26660. Tests of eight different aircraft, including homebuilts and production aircraft, to determine the amount of laminar flow they could sustain.)

79. Frick, Charles W., Davis, Wallace F., Randall, Lauros M., and Mossman, Emmet A., "An Experimental Investigation of NACA Submerged-Duct Entrances," NACA ACR No. 5I20, October, 1945. (Tests of the NACA inlet.)

80. Verstynen, Harry A. and Sexton, Bobby, "Certification of Natural Laminar Flow Technology," SAE Technical Paper No. 871848, October, 1987. (Available from AIAA as A88-30809.)

81. Marchman, J. F., III and Sumantran, V., "Control Surface Effects on the Low Reynolds Number Behavior of the Wortmann FX 63-137," Paper #11, Proceedings, Aerodynamics at Low Reynolds Numbers, London, October 1986. (Experimental measurements for Reynolds numbers between 70,000 and 300,000 are reported.)

82. Noland, David, "Wing-Man," *Air & Space*, Dec. 1990/Jan. 1991 (the John Roncz story).

83. Roncz, John G., "Propeller Development for the Rutan Voyager," SAE Paper #891034, April 11–13, 1989.

84. Eppler, Richard, *Airfoil Design and Data*, Springer-Verlag, 1990. (An essential reference for the airfoil enthusiast. Eppler here presents theory and practice for the most recent version of his airfoil design computer program (in FORTRAN), available from the University of Stuttgart. Among other things, he argues that the Stratford pressure distribution is not the one yielding maximum lift. Data for over one hundred airfoils, grouped according to intended application, are also presented.)

85. Smith, Stephen C., "Use of Shear-Sensitive Liquid Crystals for Surface Flow Visualization," *Journal of Aircraft*, Vol. 29, No. 2, March–April, 1992, pp. 289–293.

86. Zaman, K. B. M. Q., Bar-Sever, A., and Mangalam, S. M., "Effect of acoustic excitation on the flow over a low-*Re* airfoil," *J. Fluid Mech.*, Vol. 182, 1987, pp. 127–148.

Document sources
National Technical Information Service (NTIS)
5285 Port Royal Road
Springfield, VA 22161

Technical Information Service
AIAA
750 Third Avenue
New York, New York 10017

NACA/NASA document key
ACR = Advance Confidential Report
CP = Conference Proceedings
CR = Contractor Report
TM = Technical Memorandum
TN = Technical Note
TR = Technical Report
TT = Technical Translation
WR = Wartime Report (NACA)

Chapter 12

Properties of Wings

We don't fly airfoils.[1]

As he sank down low in the water, a strange hollow voice sounded within him. There's no way around it. I am a seagull. I am limited by my nature. If I were meant to learn so much about flying, I'd have charts for brains. If I were meant to fly at speed, I'd have a falcon's short wings, and live on mice instead of fish. I must forget this foolishness. I must fly home to the Flock and be content as I am, as a poor limited seagull.

He felt better for his decision to be just another one of the Flock. . . . And it was pretty, just to stop thinking, and fly through the dark, toward the lights above the beach.

Get down! Seagulls never fly in the dark! If you were meant to fly in the dark, you'd have the eyes of an owl! You'd have charts for brains! You'd have a falcon's short wings!

There in the night, a hundred feet in the air, Jonathan Livingston Seagull — blinked. His pain, his resolutions, vanished.

Short wings. A falcon's short wings!

That's the answer! What a fool I've been! All I need is a tiny little wing, all I need is to fold most of my wings and fly on just the tips alone! Short wings![2]

A. The Origin of Induced Drag

In the wind tunnel, it is usually possible to make the air move almost entirely in the chord-wise direction, although there may be some real problems near and beyond the stalling angle of attack. However, the flow of air over actual wings in flight will certainly not be solely in the chord-wise direction **when** lift is being developed because then the high pressure air on the lower surface tends to move around the wingtip to the lower pressure regions existing above much of the upper surface, and the resultant flow will have a significant span-wise component over much of the wing. (When no lift is being developed, the average pressures on the lower and upper surfaces must be the same and there is then little excuse for such undesirable behavior.)

The extent to which the air flow departs from two-dimensionality clearly depends on both the amount of lift being developed and the geometry of the wing. A long, narrow (untwisted) wing can be expected to have nearly chord-wise flow over much of its surface but a short, broad (untwisted) wing would not. Thus the degree of two-dimensionality is some

function of the ratio of the span to the chord (b/c) for a **rectangular** wing. It turns out, however, that we can account for most of the geometric effects for all different shapes of wings by defining what is known as aspect ratio:

$$\text{AR} \equiv \frac{b^2}{S} \qquad (12.1)$$

ASPECT RATIO OF A WING

(Just as with Reynolds number, one often sees this variable denoted by a single letter [A] or by both letters [AR]; it seems preferable to me to write both letters again but to cosy them up to show that a single variable is meant.) In this definition, as before, b is the wing span and S is the projected wing surface area. Further, if we make the reasonable definition that the average chord of a wing, $\bar{c}$, is the value that yields the wing surface area when it is multiplied by the span, then $S = b\bar{c}$ and we see that $\text{AR} = (b/\bar{c})$ where $\bar{c}$ is sometimes called the mean geometric chord (**MGC**) or the standard mean chord (**SMC**).

It is good geometry fun to calculate the aspect ratio for a given wing shape. For example, Fig. 12.1 shows approximately the wing planform of the Lockheed F-104, a Mach 2.2 interceptor that first flew in 1954. We can calculate S by dividing up the

[1]Ira H. Abbott, *Advanced Technology Airfoil Research*, 1978.
[2]Reprinted with permission of Macmillan Publishing Company from *Jonathan Livingston Seagull*, by Richard Bach. Copyright © 1970 by Richard D. Bach and Leslie Parrish-Bach.

planform into rectangles and triangles and using the appropriate area formula for each. Thus

$$S = \text{Areas I} + \text{Areas II} + \text{Areas III} + \text{Area IV}$$
$$= 2{\times}0.5(3.75\text{ ft})(6.95\text{ ft}) + 2{\times}(4.95\text{ ft})(6.95\text{ ft})$$
$$+ 2{\times}0.5(1.5\text{ ft})(6.95\text{ ft})$$
$$+ (3.75\text{ ft} + 4.95\text{ ft} + 1.5\text{ ft})(8.0\text{ ft})$$
$$= 187\text{ ft}^2$$
$$\text{then } AR = \frac{b^2}{S} = \frac{(6.95\text{ ft}+8.0\text{ ft}+6.95\text{ ft})^2}{187\text{ ft}^2}$$
$$= 2.6 \quad \text{(very low!)}$$

Note that aspect ratio is a dimensionless number.

The simplest treatment of the flow about a wing of finite span is to assume that the circulation which causes lift along the wing (the bound vortex) is simply turned 90° aft at the wingtips to become wingtip vortices (Fig. 12.2). However, the actual situation is more like that shown in Figs. 12.3 and 12.4. In humid air these vortices sometimes can be seen because the local reductions in pressure can cause condensation trails; also, aerobatic aircraft at air shows sometimes use smoke generators to display the vortices in a vivid fashion. But normally these

vortices are invisible and are capable of causing uncomfortable turbulence or even upset for a following aircraft. Although the viscosity of the air eventually dissipates them, they can persist for a matter of **minutes** after the aircraft making them has departed. The problem is particularly severe around airports when an aircraft (transport or smaller) is behind a **heavier** aircraft; a wing vortex from the heavier aircraft can generate a greater rolling moment than that provided by the flight controls. Also, the generated vortices tend to be strongest when an aircraft is in the landing or takeoff configurations (for reasons that will shortly become apparent).

The circulation around a wing causes the air to turn up before it tilts down to meet the wing. Thus there is upwash ahead of a wing and downwash behind it. If lift is being developed, though, there must be a **net** downwash because lift is generated as a reaction force to the downward force of the wing on the air.

Air molecules that gain vorticity and participate in the general upwash outside the wingtips are not doing the job that we really want them to do; this inefficiency exacts a penalty that must be paid, and we account for it by considering the wingtip vortices to be the cause of an **induced drag** that adds to the pressure drag and the viscous drag (i.e., the **parasite** drag) of a two-dimensional wing.

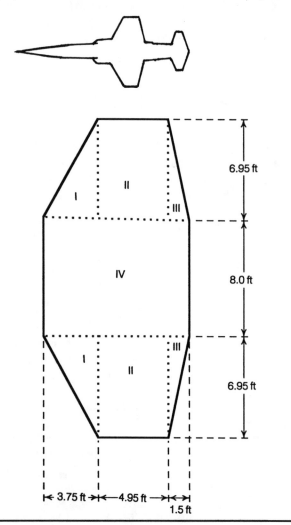

Fig. 12.1 Aspect ratio calculation for the F-104

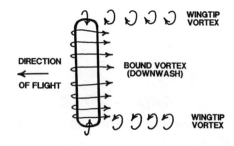

Fig. 12.2 Simplified picture of wingtip vortex generation by a lifting wing

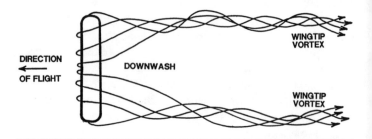

Fig. 12.3 More accurate representation of the flow pattern behind a lifting wing

Fig. 12.4 Directions of air flow immediately behind a lifting wing

There are at least **three** viewpoints that can be utilized in trying to understand the origin of induced **drag**. One is to recognize that the pressures on the wing must be altered by the span-wise flow field and the trailing vortices in such a way that lift is reduced. A second viewpoint is from energy conservation, namely, that we will have to pay extra for the rotational energy that we have given the air when only a change in direction was desired. (For this reason, British aerodynamicists commonly refer to this drag as **vortex drag** rather than induced [by lift] drag.) Third, we can see that the circulation-induced upwash-downwash has produced an aerodynamic angle of attack that is different than the geometrical angle (Fig. 12.5).

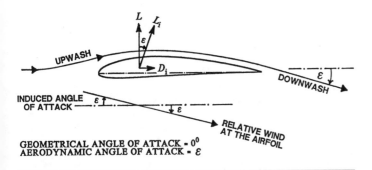

Fig. 12.5 The downwash angle and the induced drag due to lift for a wing of finite span

This third viewpoint allows us to calculate the magnitude of this induced drag. We can consider the resultant force vector (which, to first order, is equal to the lift vector) to be tilted backward by an amount just equal to the net downwash. Aerodynamic theory tells us that the downwash angle (denoted by the small Greek letter epsilon [ε]) depends on the way lift is distributed along the span of the wing. If this lift varies elliptically from a maximum in the center of the wing to zero at the tips, then the downwash angle is at its smallest possible value and is given by

$$\varepsilon = \frac{C_L}{\pi \, \text{AR}} \tag{12.2}$$

DOWNWASH ANGLE FOR AN ELLIPTICAL LIFT DISTRIBUTION

where C_L is the lift coefficient for the finite, **three**-dimensional wing and is defined by

$$L = C_L q_\infty S \tag{12.3}$$

DEFINITION OF THE THREE-DIMENSIONAL LIFT COEFFICIENT

Note the use of capital letters to designate three-dimensional coefficients.

For other types of span-wise lift distribution, it is customary to define a span efficiency factor e which is then less than 1 for these distributions.

$$\varepsilon = \frac{C_L}{\pi \, e \, \text{AR}} \tag{12.4}$$

DOWNWASH ANGLE IN GENERAL

The least efficient wings are therefore those possessing an e significantly less than 1; correspondingly, such wings produce the greater downwash angles for a given lift coefficient.

The product of e and AR is often called the **effective aspect ratio**,

Effective Aspect Ratio $\equiv e$AR

DEFINITION OF EFFECTIVE ASPECT RATIO

and this is a particularly useful term because the combination appears so often.

From Figure 12.5, the induced drag is given approximately by

$$D_i \approx L \sin \varepsilon \tag{12.5}$$

but, since ε is normally a small angle, we can replace $\sin \varepsilon$ by ε (in radians). Therefore

$$D_i \approx L\varepsilon = \left(C_L q_\infty S\right)\left(\frac{C_L}{\pi \, e \, \text{AR}}\right) = \left(\frac{C_L^2}{\pi \, e \, \text{AR}}\right)\left(q_\infty S\right) \tag{12.6}$$

INDUCED DRAG

It is convenient now to consider the first term in Eq. 12.6 as an induced drag coefficient, C_{Di}, and henceforth calculate induced drag on the basis of flight dynamic pressure and wing area, just as we do for other lift and drag forces.

$$C_{Di} \equiv \frac{D_i}{q_\infty S} = \frac{C_L^2}{\pi \, e \, \text{AR}} \tag{12.7}$$

INDUCED DRAG COEFFICIENT

In **level, unaccelerated** flight, the total lift equals the weight of the aircraft. Therefore $L = W = C_L q_\infty S$ and we can substitute for C_L in Eq. 12.6 to obtain

$$D_i = \frac{W^2}{\pi \, e \, \text{AR} \, q_\infty^2 S^2}\left(q_\infty S\right) = \frac{W^2}{\pi \, e \, \text{AR} \, q_\infty S}$$

Substituting then from Eq. 12.1 for the AR,

$$D_i = \frac{W^2}{\pi \, e \, b^2 \, q_\infty} \tag{12.8}$$

INDUCED DRAG FORCE IN LEVEL FLIGHT

This equation yields the important information that the induced drag (and the strength of the wingtip vortices!) depends **(a)** directly on the **square** of the weight of the aircraft, **(b)** inversely on the efficiency of the wing planform used, and **(c)** inversely on the dynamic pressure. Therefore wingtip vortices are strongest when they are being produced by heavy aircraft at their minimum speeds (which occur during approach, landing, and takeoff).

B. Coefficient Transformation to Three Dimensions

The total drag on a finite wing, by itself, is the sum of the pressure drag, the viscous drag, and the induced drag. Therefore the total drag coefficient is the sum of the airfoil drag coefficient (c_d of previous chapters) and our new induced drag coefficient, and we can write

$$C_D = c_d + C_{Di} = c_d + \frac{C_L^2}{\pi \, e \, AR} \qquad (12.9)$$

DRAG COEFFICIENT FOR A FINITE WING

This is the equation to use when you need to calculate a three-dimensional drag coefficient from wind tunnel section data. And because pressure drag and viscous drag for a wing are often lumped together under the label *profile drag*, the first term is sometimes called the **profile** drag coefficient to contrast it with the induced drag of the second term.

How about the geometrical angle of attack required to obtain a given lift coefficient from a given airfoil on a real wing? At zero lift ($C_L = 0$), there will be no downwash ($\varepsilon = 0$) and therefore the angle of attack for **zero** lift is the same for the airfoil as for the real wing. But when lift is generated, the aerodynamic angle of attack is less than the geometrical angle. Since we can't easily measure the aerodynamic angle of attack, it is convenient to adjust the slope of the lift curve so that we can calculate directly from it the greater geometrical angle of attack required for a real wing to generate a given lift coefficient compared with the same airfoil in two-dimensional flow.

At medium and high aspect ratios (greater than 5 or 6), the **maximum** lift coefficient for the wing and the airfoil are about the same; for low aspect ratios the maximum lift coefficient is reduced. This general behavior is summarized in Figure 12.6. Note that the geometrical angle of attack for zero lift ($C_L = 0$, $\varepsilon = 0$) is the only value that is unchanged.

The amount that the geometrical angle exceeds the aerodynamic angle for a given lift is just the downwash angle, so $\alpha_{3D} = \alpha_{2D} + \varepsilon$ and, substituting for ε, we obtain

$$\alpha_{3D} = \alpha_{2D} + \frac{C_L}{\pi \, e' \, AR} \qquad (12.10)$$

(angles in **radians**)

$$\alpha_{3D} = \alpha_{2D} + \frac{57.3 \, C_L}{\pi \, e' \, AR} \qquad (12.11)$$

(angles in **degrees**)

**CHANGE IN ANGLE OF ATTACK
WITH CHANGE IN ASPECT RATIO**

where we have added a prime to the efficiency factor because it may be somewhat different than the value for the induced drag. Incidentally, efficiency factors are generally in the range from about 0.7 to about 0.95. The 57.3 in Eq. 12.11 converts radians to degrees because 2π radians equals 360°.

Equation 12.11 and Fig. 12.6 tell us that an aircraft with a low aspect ratio wing is going to have a very large angle of attack when it is landing if it is using its maximum lift coefficient; to some degree this is a real limitation on how small the aspect ratio can be since a long, heavy landing gear will probably be necessary. Hopefully, you'll get a chance to work a problem or two that will reinforce this idea.

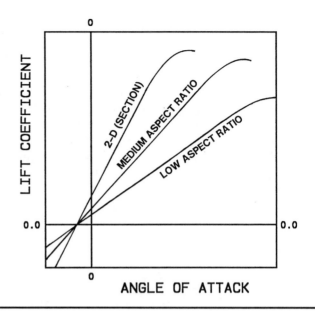

Fig. 12.6 Effect of aspect ratio on the lift curve

EXAMPLE 12.1. Suppose now that we wish to use an LS(1)-0417 [GA(W)-1] airfoil on a wing of aspect ratio 6.0 and that we estimate both span efficiency factors to be equal to 0.8. (**a**) What (geometrical) angle of attack is required to obtain the design lift coefficient of 0.4 for a Reynolds number of 6.3 million? (**b**) How does the induced drag compare with the profile drag for this condition?

Solution: (a) From Fig. 9.7, $C_L \approx c_\ell = 0.4$ at $\alpha_{2D} = -0.3°$. Then, from Eq. 12.11,

$$\alpha_{3D} = -0.3 + \frac{(57.3)(0.4)}{\pi(0.8)(6.0)} = \mathbf{1.2°}$$

(b) From Fig. 9.8, $c_d = 0.0107$ at $c_\ell = 0.4$.

From Eq. 12.7, $C_{Di} = \dfrac{0.4^2}{\pi \left(0.8\right)\left(6\right)} = 0.0106$

so $\qquad \dfrac{C_{Di}}{c_d} = \dfrac{0.0106}{0.0107} \approx 1$

and, for this wing all by itself, the induced drag is just about **equal** to the profile drag.

Suppose we wish to change the aspect ratio of a wing without changing the airfoil. Equations 12.9 and 12.11 then lead directly to these transformation equations:

$$C_{D2} = C_{D1} + \frac{C_L^2}{\pi}\left(\frac{1}{e_2 AR_2} - \frac{1}{e_1 AR_1}\right) \qquad (12.12)$$

$$\alpha_2 = \alpha_1 + \frac{57.3 \, C_L}{\pi}\left(\frac{1}{e_2' AR_2} - \frac{1}{e_1' AR_1}\right) \qquad (12.13)$$

(angles in **degrees**)
Effects of Changing the Wing on an Aircraft

EXAMPLE 12.2. As an example of the use of these equations, suppose that we plan to exchange our wing of the last example for a better one that has an efficiency factor of 0.9 and an aspect ratio of 7.0. What is the new drag coefficient and geometrical angle of attack at the design lift coefficient of 0.4 for this new wing?

Solution: From Eq. 12.9, $C_D = c_d + C_{Di}$

$$= 0.0107 + 0.0106 = 0.0213.$$

Then from Eq. 12.12,

$$C_{D2} = 0.0213 + \frac{0.4^2}{\pi}\left(\frac{1}{(0.9)(7.0)} - \frac{1}{(0.8)(6.0)}\right)$$

$$= (0.0213 + (0.0509)(0.15873 - 0.20833)$$

$$= \mathbf{0.0188}$$

so that the more efficient wing has reduced the total drag coefficient of the wing to 88% of its previous value ($0.0188/0.0213 = 0.88$).

From Eq. 12.13, the angle of attack required to obtain the design lift coefficient has been reduced to

$$\alpha_2 = 1.2 + \frac{(57.3)(0.4)}{\pi}\left(\frac{1}{(0.9)(7.0)} - \frac{1}{(0.8)(6.0)}\right)$$

$$= 1.2° + (-0.4°) = 0.8°$$

We need to remind ourselves at this point that, not only do we not fly airfoils, we also don't normally fly wings all by themselves. The effect of the fuselage, propeller slipstream, and tail members is to greatly increase the drag, to modify the efficiency factors, and to make some (usually small) changes in the lift (through a lifting fuselage or through loads on the horizontal tail members).

C. Ground Effect

We are now in a position to understand how an airplane can take off before it is able to climb, how it can float along when it is supposed to be landing, and how it can fly a greater distance when flown very close to the earth's surface. What happens is that the air flow over the wing, and especially around the wingtips, is altered by the presence of the ground. Specifically, the upwash and downwash angles are reduced. The result is an effective increase in aspect ratio and a resulting reduction in total drag.

Theoretical estimates of the influence of the ground on landing aircraft date from Prandtl's *Wing Theory* in 1919. The method of images is commonly employed: The ground can be replaced by an identical aircraft flying upside down an equal distance below the ground, generating upwash, because this satisfies the boundary condition that there be no vertical component to the air flow velocity at the ground. A recent treatment (Ref. 43) that includes a dependence on the wing efficiency factor, e, is as follows:

$$\frac{\text{Induced drag in ground effect}}{\text{Induced drag out of ground effect}} = 1 - \frac{2e}{\pi^2}\ln\left[1 + \left(\frac{\pi b}{8h}\right)^2\right]$$

$$(12.14)$$

where h is the height above the ground, b is the wing span, and "ln" is the natural logarithm (the *ln x* key on most scientific calculators). Table 12.1 provides a tabulation of ground effect, calculated from this equation, for a wing efficiency of 0.9.

$\left(\dfrac{h}{b}\right)$	Effective Increase in AR	Decrease in Induced Drag
0.50	10%	9%
0.45	12%	10%
0.40	14%	12%
0.35	17%	15%
0.30	22%	18%
0.25	29%	23%
0.20	40%	29%
0.15	60%	38%
0.10	104%	51%
0.05	308%	75%
Table 12.1 Estimated Magnitude of Ground Effect		

You can see that a really large effect requires snuggling very close to the ground, and this degree of propinquity is physically possible only for low wing aircraft. Even they can not expect to gain a real advantage in lift/drag during cruising flight, and thereby in range, unless the ground is flat enough to permit flying at very low altitude, and this in turn virtually rules out every environment except smooth water.

EXAMPLE 12.3. Suppose that we have the second, more efficient wing of the previous section down to about 3 feet above the ground. For a typical wing span for a lightplane of 30 feet, this means $\left(\dfrac{h}{b}\right) = 0.1$ and (from the table) we can estimate an increase in effective aspect ratio from 6.3 to

Effective AR (in ground effect) $= (6.3)(1 + 1.04) = 12.85$

and a corresponding decrease in the **induced** drag coefficient (at $C_L = 0.4$) from $(0.0188 - 0.0107) = 0.0081$ to

$$C_{Di} = (0.0081)(1.0 - 0.49) = 0.0041$$

so that the **total** drag coefficient (for the wing alone) is

$$C_D = 0.0107 + 0.0041 = 0.0148$$

which is a $\dfrac{(0.0188 - 0.0148)}{0.0188}(100) = 21\%$ decrease.

Small, single-seat homebuilt aircraft show some startling behavior in ground effect because their wings can be a foot or less from the ground when landing. The Quickie is a good example; the machine tends to plop onto the runway in a displeasing, embarrassing, and disgraceful fashion unless it is flared within inches of the runway. The maximum angle of attack very near the ground is very noticeably less than that even a few feet above the runway. Also, the author has flown a tiny biplane which possesses a schizoid lower wing. If allowed to take off at a relatively low speed, it would apparently jump out of ground effect only to encounter a pitch-down moment that sent it back into ground effect—and an uncommanded, rapid pitch oscillation ensued until additional speed had been gained. The solution was to use additional forward control stick (nose-down elevator deflection) during the second half of the takeoff roll and hold the machine on the ground just a little longer.

Ground effect is where you find it, of course. For Dick Rutan, assaying first flight in the Voyager,

> I was about a hundred feet up before I dared to look out the window—and saw that shadow, that crazy long network of shapes, spread out across the runway. "Holy bananas! Look at that shadow! What have we done?" I said, and when Jeana in the chase plane heard that, knowing I was still in ground effect, she reminded me to pay attention. The airplane could still turn around and bite me.

but Voyager had a wing span of 110.8 ft! (Ref. 37.)

D. The Making of an Efficient Wing

Aerodynamically efficient wings are those that combine an efficiency factor (*e*) close to 1 with a large aspect ratio. Equation 12.8 tells us that the drag due to lift for an airplane depends geometrically only on the efficiency factor (i.e., the planform) and the square of the wing span—and **NOT** on the aspect ratio. Therefore induced drag is minimized by using a tapered wing with the longest possible span. Because $\left(\frac{\pi b^2}{4}\right)$ is the "cylinder" of air intercepted by the wing from a head-on viewpoint, we can interpret this as meaning that the induced drag gets smaller and smaller as this cylinder of "intercepted" air gets larger and larger. A biplane will have even less induced drag than a monoplane of the same wing span because the "intercepted" air is that between a semicircular arc below the lower wing and a semicircular arc above the upper wing.

The other drag term, the parasite or **profile** drag (i.e., the sum of the viscous and pressure drag) is quite a different story! This drag depends directly on the surface area (*S*) of the wing. Therefore you can reduce this part of the total drag by using the smallest possible chord for the wing as well as the shortest possible wing span. Therefore, to reduce the total drag of an aircraft, we should increase the (span)² and decrease the surface area—i.e., **increase the aspect ratio!** This is why large aspect ratios are used. So don't let anyone tell you that a high aspect ratio wing is designed to reduce the induced drag; it is only the span that does that.

Specifically, a high aspect ratio wing minimizes the total drag under conditions where induced drag and parasite drag are roughly equal. At the highest speeds, at SL, many airplanes will have a higher top speed if their wings are clipped because this reduces the parasite drag more than it increases the in-

duced drag. We'll make a sample calculation for a specific airplane in Chapter 14.

With regard to the efficiency factor, *e*, as was pointed out in Section A, simple aerodynamic theory predicts that only an elliptical lift distribution will yield the maximum possible value of 1. This suggests using something close to an elliptical wing planform but such a wing is difficult to make (using conventional metal construction techniques, at least) because every wing rib on a side is unique. It is in fact generally better structurally and as good aerodynamically to use a wing whose shape approximates an ellipse through combining segments that individually have straight trailing edges.

The early Piper Cherokee 2-, 4-, and 6-seat aircraft all used a rectangular wing planform; it was easy and inexpensive to build but wasn't aesthetically pleasing and had rather sluggish roll response. The new wing that Piper uses on the Warrior and the Cadet and other Cherokee derivatives corrects both of these deficiencies. A tapered wing has a clear structural advantage, too, for it places more of the lift close to the fuselage attach points (so the wing bending moment is decreased).

Wings are commonly given a twist to obtain a closer approximation to an elliptical lift distribution and to promote good stalling characteristics (next section) and this considerably alters the lift distribution, and muddies the picture regarding the best planform shape. In sum, the best wing is probably one that provides the most reasonable compromise between the often conflicting demands of high efficiency factor *e*, cost and weight to build, and handling/control properties. The shape of such a wing will certainly be tapered to some extent but there is no good reason (other than an aesthetic one) to require a true elliptical planform. (The Supermarine Spitfire fighter aircraft which was effectively used by England in World War II is one of the few aircraft boasting an elliptical planform. Modern designs by Claude Piel, such as the 2-seat Emeraude, also use it.)

Wingtip design has been an aeronautical battleground for decades. The successful aeronautical warrior creates a tip shape that hinders the air flow around the tip of the wing. In other words, the designer is trying to keep the axis of the wing vortex as close to (or beyond) the tip of the wing as possible. A sharp tip (as viewed from behind) tends to keep the vortex close to the tip rather than letting it roll up and move inboard on the upper surface. Also, as viewed from above, the span should increase from the leading edge to the trailing edge; a tip that curves back toward the fuselage (as on a Cub wing) just encourages the vortex to move toward the fuselage, too. Presumably the design of the tip is most critical for a relatively untapered wing, where considerable lift is being developed near the tip.

One of the best-known wingtip designs is the so-called **Hoerner** tip, which has an estimated efficiency (on a rectangular wing) of 0.80; another recommended tip treatment is one which rakes the tip outward from the trailing edge (about 20° for an aspect ratio of 6), and this is said to have an efficiency factor of 0.82 (Figures 12.7a and 12.7b; Ref. 3).

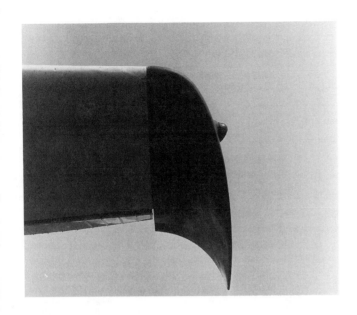

Pic. 12.1 Aftermarket wingtip treatment on a Cessna

Pic. 12.2 Burt Rutan's Catbird shows off its sheared wingtips.

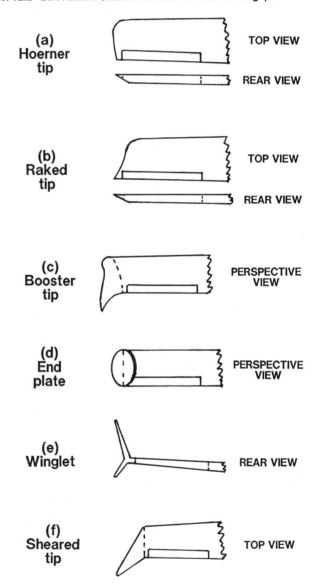

Fig. 12.7 Various wingtip treatments, all designed to increase the effective aspect ratio of a wing

Another rather popular wingtip treatment is the "booster" tip, which combines a raked tip with significant downwards curvature (Figure 12.7c).

End plates are a very old idea for increasing the effective aspect ratio of a wing (Figure 12.7d). In order to be effective, they should be as tall as they are wide. Their disadvantages are that they are directionally destabilizing (because they are ahead of the *cg*), they add viscous drag, and—they are *ugly*.

The newest wingtip design is the "sheared" tip in which both the leading and trailing edges sweep backward and meet at the tip (Fig. 12.7f); the leading edge should sweep at about a 60° angle and the wing should taper to about 25% of the chord at the tip. This shark-like, bird-like tip apparently is not only more efficient than others, both at cruise and at high angles of attack, it also appears to help retain aileron control down to the lowest speed with a smaller surface. This wingtip is used on both the wings and horizontal stabilizing surfaces for recent Rutan designs such as the high efficiency, 5-place, turbo-charged Model 81 Catbird (Pic. 12.2), on the jet-powered Model 143J, and on the new unlimited class Pond Racer (Pic. 11.7) as well as the German Dornier Do-328 short-haul transport. If it proves successful, expect to see many other aircraft with this distinctive wingtip treatment (Refs. 41 and 42). Perhaps the next step will be to go even more bird-like with a fully crescent planform (Ref. 52)

Winglets are another recent treatment for wingtips that were suggested and developed by NASA's Richard Whitcomb (of Area Rule and supercritical wing fame) in about 1975 (Ref. 3). Figure 12.7e shows a front view; see also Figures 1.4 and 1.9. Winglets can be thought of as devices for partially "unwinding" the wingtip vortex before it leaves the wingtip, thereby generating a forward force from a reduction in the vortex flow. It is now clear that winglets can produce the same beneficial reduction in induced drag as wingtip extensions but with a smaller bend-

Aircraft	W (lb)	b (ft)	S (ft^2)	$\left(\dfrac{W}{S}\right)$ (lb/ft^2)	$\left(\dfrac{W}{b}\right)$ (lb/ft)	AR
Curtiss JN4 "Jenny" (WW I)	2130	43.8	353	6	49	5.4
Gee Bee Model "Z" racer (1931)	2380	25.5	75	32	93	8.7
Cessna 152 (trainer)	1670	33.3	160	10	50	6.9
Piper Cherokee (4-place, straight wing)	2400	30.0	160	15	80	5.6
Piper Warrior (4-place, tapered wing)	2440	35.0	170	14	70	7.2
Piper PA-46-310P Malibu	4100	43.0	175	23	95	10.6
Cessna T210 (turbocharged)	3800	36.8	175	22	103	7.7
Fournier RF9 (powered glider)	1543	55.8	194	8	28	16.0
Lancair 320 (experimental/homebuilt)	1685	23.5	76	22	72	7.3
Glasair III (experimental/homebuilt)	2400	23.3	81	30	103	6.7
Light Aircraft (twin engine)						
Beechcraft King-Air C90A	9050	50.3	294	31	180	8.6
Cessna 441	9850	49.3	254	39	200	9.6
Gates Learjet 55	19,500	43.8	265	74	446	7.2
Mitsubishi MU-2 Marquise	11,575	39.2	178	65	295	8.6
Voyager (around the world unrefueled)	9695 (t-o) 2699 (land)	110.8	424 (total)	23	88	33.8 (wing) 18.1 (canard)
Piaggio Avanti	10,510	45.4	170	62	231	12.1
Beechcraft Starship	14,000	54.0	281	50	259	10.4
Sailplanes						
Bryan HP-18A	920	49.2	115	8	19	21.0
Schweizer SGS 2-33A (2-place)	1040	51.0	219	5	20	11.9
Schweizer 1-36	710	46.2	141	5	15	15.1
Standard Libelle 201B (15 meter class)	772	49.2	106	7	16	22.8
Military Aircraft						
Sopwith Camel (WW I fighter)	1482	28	231	6	53	6.8
Republic P-47D (WW II fighter)	15,000	40.8	300	50	368	5.5
Supermarine Spitfire (WW II fighter)	7875	40.2	242	33	196	6.7
North American P51-D (WW II fighter)	11,600	37.3	233	50	311	6.0
Boeing B-17 (WW II bomber)	48,720	103.8	1420	34	469	7.6
Boeing B-25 (WW II bomer)	33,500	67.6	610	55	496	7.5
Lockheed P-80C (Korean fighter jet)	15,336	39.9	238	64	384	6.7
McDonnell F-4E (Viet Nam fighter)	60,630	38.4	527	115	1579	2.8
Grumman F-14 (variable-sweep fighter)	58,715	38.2→64.1	565	104	916	2.6→7.3
McDonnell/Douglas F-15C (USAF fighter)	44,630	42.8	608	73	1043	3.0
General Dynamics F-16D (USAF fighter)	24,500	31.0	300	82	790	3.0
Transport Aircraft						
Douglas DC-3 (1930s)	25,200	95.0	987	26	265	9.1
Grumman IV (winglets: eAR = 6.9)	69,700	77.8	950	73	896	6.4
Boeing 737-300	124,500	94.8	1135	110	1314	7.9
Concorde	400,000	83.9	3856	104	4768	1.8
Boeing 757	126,880	124.8	1994	64	1016	7.8
Boeing 767-200T	194,900	156.1	3050	54	1249	8.0

Table 12.2 Wing Specifications for Various Aircraft

ing moment at the wing root. The Rutan VariEze was the first aircraft to use winglets, probably, and they provide here the additional benefits of a rearward vertical surface for the rudders. Grumman Gulfstream and Lear business jets are now using winglets. More recently, they are being employed on several jet transports (Ref. 4). A side benefit of the winglet, which may in itself justify its use, is the effective increase in wing span that it provides without the parking and storage problems of actual extended span.

Another innovative approach to improving span efficiency has been NASA's demonstration that "vortex turbines" at the wingtips can recover 21% of the energy lost to induced drag, in cruise. These turbines might prove useful for powering boundary layer control (BLC) or cabin pressurization systems, for example (Ref. 49).

Besides the efficiency factor, which we have seen is closely related to the type of planform and the wingtip design, the efficiency of a wing is even more strongly influenced by the wing's aspect ratio. The choice of aspect ratio revolves around tradeoffs among cost, weight, strength, maneuverability, and the relative importance of induced drag for the intended use of the aircraft. Some fighter aircraft have used very low aspect ratios; they have gained top speed and structural advantages in return for poor low speed characteristics. Table 12.2 shows that modern jet fighters have aspect ratios of about 3 (although the variable-sweep F-14 can increase that to about 7 for landing). Most light aircraft use aspect ratios in the range of 6 to 8.5, with the faster and larger aircraft tending toward the high end of the range. Note from the table that the Gee Bee racer had a respectable value for its aspect ratio even though it had an extremely small wing for its weight; this was necessary because of the large lift coefficients required during the steep pylon turns and the resulting importance of induced drag at that time. Modern pylon racers are rediscovering larger aspect ratios because modern structural materials make it possible without exacting too great a weight penalty.

With the rapid rise in fuel costs in the 1970s, designers of business and transport aircraft also began rediscovering high aspect ratios. These are also the aircraft that are trying on winglets for size, which can often be fitted to an old wing much more cheaply than designing a new wing from scratch.

Sailplanes are in a class all by themselves! Their very high aspect ratio wings are proof of their concern with maximum L/D for gliding but also of the importance of low induced drag during the tight, steep turns they use to keep within the narrow confines of a narrow, invisible shaft of heated, rising air above reflective patches of ground ("thermals" to sailplaners).

EXAMPLE 12.4. Let us look at some numbers to help us see what gains can be made by changing a wing. A Cessna 152 has a maximum listed speed under sea level StAt conditions of 110 kt with a 110 horsepower engine at its design gross weight of 1670 pounds. If we assume a propeller efficiency of 75%, we can determine with the methods of a later chapter that the total

Pic. 12.3 Seagulls use the proprietary "gull" wing planform.

drag force under these conditions is 244 lbs. Then we calculate the dynamic pressure as

$$q_\infty = \frac{1}{2}\left(0.002377\right)\left(110\,\text{kt} \times \frac{1.689\,\text{ft/s}}{\text{kt}}\right)^2 = 41.0\,\text{lb/ft}^2$$

and, if we assume an efficiency factor of 0.8 and obtain the aspect ratio from Table 12.2, Eq. 12.8 tells us that

$$D_i = \frac{W^2}{\pi\,e\,AR\;q_\infty\,S} = \frac{(1670\,\text{lb})^2}{\pi(0.8)(6.9)(41.0)(160)} = 24.5\,\text{lb}$$

and therefore we see that the induced drag is only about

$$\frac{24.5}{244}\left(100\right) = \mathbf{10\%}$$

of the total (complete airplane) drag under these conditions. We can generalize this result by noting that the efficiency of a wing is least important in the determination of maximum SL speed.

At the minimum flight speed of 43 kt, though, the dynamic pressure is down to 6.26 lb/ft², the lift coefficient is 1.67, and the induced drag force is 160 lb (using the same equations). The induced drag is **very** important whenever the lift coefficients are large, as in steep turns or in short-field takeoffs and landing.

At 12,000 feet in the StAt, the Cessna 152 has a maximum listed true airspeed of 100 kt. Under these conditions the total drag can be calculated as 167 lb and the induced drag as 43 lb, so that the latter is now 26% of the total. This should help you appreciate the reason for the large wing span of the U2 (Lockheed's ultra-high altitude reconnaissance aircraft). Thus induced drag is also **very important** at altitudes approaching an aircraft's service ceiling.

EXAMPLE 12.5. Suppose we simply try **extending** the Cessna 152 wings some 4 ft on each side and thereby also increasing the wing area to 200 ft^2. This would result in a new aspect ratio of $\dfrac{(43.3 \text{ ft})^2}{200 \text{ ft}^2} = 9.37$. The new drag force under SL StAt conditions and 110 kt would be composed of 14 lb of induced drag force and 274 lb of parasite drag.

We've decreased the induced drag with our stratagem, all right, but the total drag increased significantly because of the increase in wing surface area; therefore we have reduced the maximum speed of the aircraft while improving its low speed, banking, climb, and altitude performance. It would probably have been better to have built a new wing with the same surface area but with greater aspect ratio (i.e., a skinnier one).

It is generally stated that a wing that is mounted in the middle of the fuselage (a **mid-wing** configuration) will have its efficiency least affected by the fuselage (and this is why racers are often built this way). Next best is the high wing and then the low wing. The low wing is worst because it has the critical upper surface right where the fuselage can most easily affect the flow over it. (Perhaps the most compelling reason to put a low wing on an aircraft is to provide easily-utilized storage room for retractable landing gear.) Careful filleting of the wing-fuselage juncture is highly recommended; a fairing should be used to change the 90° juncture in the horizontal plane into a well-rounded transition.

E. Stalling Speeds

Now that we are using the three-dimensional lift coefficient, we can make estimates of the stalling speeds of aircraft based on the maximum lift coefficients of their wings. We assume that this speed is approached in straight-and-level flight by reducing the power to about idle and then gradually increasing the angle of attack at just such a rate that level flight is maintained until the stall occurs. Then there is no acceleration in the vertical plane and the upward lift force must just equal the downward weight force. (This is called a "1-g stall.") Therefore we can write

$$L = C_{L, \max} q_\infty S = C_{L, \max} \left(\frac{1}{2}\rho V_{\infty, \text{STALL}}^2\right) S = W$$

and solving this for the true stalling speed yields

$$V_{\infty, \text{STALL}} = \sqrt{\frac{2W}{\rho S C_{L, \max}}} \qquad (12.15)$$

TRUE STALLING AIRSPEED

The definition of equivalent airspeed (Eq. 7.8) allows us to rewrite this as

$$V_{E, \text{STALL}} = \sqrt{\frac{2W}{\rho_0 S C_{L, \max}}} \qquad (12.16)$$

EQUIVALENT STALLING AIRSPEED

and this gives us the valuable information that the equivalent stalling airspeed for a given aircraft depends only on weight and not on any properties of the air; therefore the calibrated and the indicated stall speeds for any given aircraft must also be independent of air temperature, air pressure, and air density—but not the weight of the aircraft.

We should not expect the maximum lift coefficient of Eq. 12.15 to be quite as large as the wing's maximum lift coefficient because of the download on the tail (for the more common tail-last configuration) during this kind of flight. Percent-wise, this should be a fairly small effect—perhaps around 10% in cruising flight. (Numerically, it can be impressively large for heavy aircraft; for instance, the download on the tail of a Boeing 747 airliner while landing has been estimated at 40 to 60 thousand pounds! [Ref. 5])

EXAMPLE 12.6. Let's try a sample calculation with the Cessna 152. It has a listed calibrated stall speed of 48 kt (flaps up). Under SL StAt conditions and using the mean wing chord of $\bar{c} = \dfrac{S}{b} = \dfrac{160 \text{ ft}^2}{33.3 \text{ ft}} = 4.80$ ft, this implies an operative RN of about

$$\frac{\left(48 \text{ kt} \times \dfrac{1.689 \text{ ft/s}}{\text{kt}}\right)(4.80 \text{ ft})}{0.00015723 \text{ ft}^2/\text{s}} = 2.5 \text{ million},$$

using Eq. 5.4.

The maximum lift coefficient for the 2412 airfoil at this RN under smooth conditions is about 1.5 (from the original NACA report). Therefore we can calculate a stall speed for the Cessna 152, using Eq. 12.16, of

$$V_{\text{CAL, STALL}} \approx V_{E, \text{STALL}} = \sqrt{\frac{(2)(1670)}{(0.002377)(160)(1.5)}}$$

$$= 76.5 \text{ ft/s} = \mathbf{45 \text{ kt}}$$

Our estimate is very close to the book value of 48 kt; the tail download would be expected to reduce the effective maximum lift coefficient and to raise the stalling speed a little, as it apparently does.

Equation 12.16 tells us that a slow landing speed can be obtained by either having a very **low** wing loading $\left(\dfrac{W}{S}\right)$ or by increasing the maximum lift coefficient, since landings are normally made at speeds close to the power-off stalling speed (at least for aircraft that have good stalling characteristics). Ultralights must have very large wings for their weight in order to get their stall speeds down to the maximum value of 24 kt that is currently permitted to "aircraft" in this category. But a large wing area produces a great drag penalty at high speeds and therefore fast and heavy aircraft try to keep their landing speeds manageable by artificially increasing the maximum lift coefficient above that given by a bare airfoil when in the landing configuration, and this is the subject of the next section.

F. High Lift Wing Devices and Other Tricks

We saw in Sec. D that a tapered wing is more efficient aerodynamically and structurally, in addition to having better roll characteristics. However, a simple tapered wing (except for the rare forward-swept planform) does **not** have satisfactory stalling characteristics, because of span-wise flow and because the tapered regions operate at lower Reynolds number, for which the maximum lift coefficient is reduced. Figure 12.8 shows that the stall begins near the wingtip for these tapered planforms and then works its way toward the fuselage as the angle of attack is increased. This means that the ailerons will stall first and that effective rolling control will be **lost** before the whole wing is stalled. And, because of the dynamic nature of the stall, it will normally occur on one wing first and send the aircraft into a rolling, nose-down turn from which a spin entry is quite likely.

Rather, we want the stall to begin near the fuselage so that the onset of separation will buffet the tail, thereby providing a natural aerodynamic warning to the pilot that a full wing stall is not far away, while the ailerons are still effective. There are many adequate tricks to accomplish this without returning to the rectangular wing.

One way to keep at least the ailerons from stalling is to cut a **slot** in the wing in the region ahead of the aileron (Chap. 1 and Fig. 12.9). This slot changes the effective shape of the leading edge and shortens the distance to the aileron in which separation can take place. The Northrup Flying Wing, some

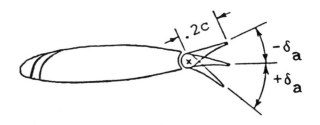

Fig. 12.9 A 20% aileron with a slot ahead of it, designed to keep the aileron effective through wing stall

Stinson aircraft, and the Globe Swift (see Pic. 1.1 in Chap. 1) are examples of airplanes that have used this technique.

The two tricks that are used most often to provide good stalling characteristics for low and medium speed aircraft are (a) geometrical or aerodynamic twist and (b) stall strips. When a wing is geometrically twisted so that the tips operate at a lower angle of attack than the roots, the wing is said to have "washout." On most light aircraft this twist is very apparent. A similar, if smaller, effect can be obtained by aerodynamic means—changing the airfoil section or the thickness toward the tip, for example. Composite wings lend themselves to this recommended approach, if the airfoil designer is up to the task.

Stall strips are sharp-edged strips of triangular cross-section that are attached to the leading edge of the wings of many aircraft. Most commonly they appear inboard so that their sharp edge will promote early separation, and vortices, in the part of the wing just in front of the tail surfaces, but engineers many times have used extensive cut-and-try techniques to get a combination of stall strips and strip placement that works. The abrupt ends of the stall strips can be expected to operate as vortex generators at high angles of attack. It does seem a pity to throw away a nicely rounded leading edge in this fashion, but good stall behavior is not optional for production aircraft. A twisted wing may be rather inefficient, too, because the tips may

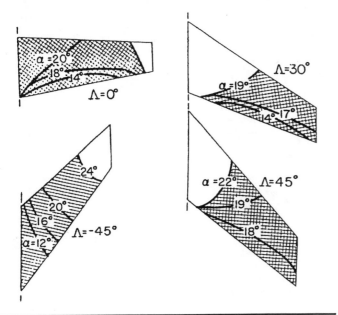

Fig. 12.8 Stall patterns for untwisted wings with various planforms (Ref. 39)

Pic. 12.4 Stall strip on the leading edge of the wing of a T-tailed Piper Tomahawk

be doing more dragging than lifting at cruise speeds; also, the twist can cause the wing to tend to twist some more at high dive speeds and this can easily get out of hand.

Simple swept wings, whether they sweep forward or rearward, have a potentially dangerous pitch-up tendency at the stall. On a rearward-swept wing, the tips stall first and so the center of pressure moves forward; on a forward-swept wing the root section stalls first so again the center of pressure moves forward at the stall (refer to Figure 12.8 again, perhaps). The big advantage of forward sweep over rearward sweep in that roll control is still very much available at the stall. If you're piloting a fighter aircraft, you'll appreciate that.

High speed aircraft typically need an extensive array of high lift devices just to obtaining reasonable landing speeds and therefore they will place the more efficient devices on the **outboard** sections to prevent tip stalling. (But remember the pitch-up problem for a swept-wing aircraft in takeoff configuration when it is carrying a load of snow or ice — Sec. F of Chapter 11.) Rows of vortex generators, one each for the wings and the stabilizers, have become popular aftermarket add-ons for light twin-engine aircraft, sometimes providing dramatically improved low speed controllability.

Some jet fighters, such as the F-5, F-15 and F-18, use a "snag" or sawtooth (a discontinuous increase in chord) in a wing or stabilizer leading edge to produce a vortex that solves buffeting or other separation problems. Vortex generators are also a common and simple band-aid for them. So are "fences," the 6" or taller thin flat plates that protrude vertically upward and parallel to the longitudinal axis of the aircraft; their purpose is to hinder the boundary layer in its desire to move toward the tips and stimulate tip stalling (Pic. 12.5).

Fences are commonly employed in short-takeoff-and-landing (STOL) conversions of light aircraft in which roll control must be extended to lower speeds and swept-wing aircraft which are particularly troubled by span-wise flow at high lift coefficients.

The **vortilon** is a newer type of fence which is used on aircraft such as the DC-9, the Long-EZ (see Pic. 1.3 in Chapter 1), and the new BAe 125-10000 business jet. Vortilons are small, airfoil-shaped plates that jut forward from the lower leading edge of the wing. These provide the flow benefits belatedly discovered for the engine pylons on the wings of jet aircraft when the engines were moved to the rear fuselage. Vortilons shed vortices over the upper surface at high angles of attack but present negligible drag at cruise speeds; therefore they are replacing fences in many applications.

An interesting example of the low speed use of sweepback is the upper wing on most aerobatic biplanes. The upper wing does most of the lifting on a biplane because only its lower surface has to interact with the other wing; if the aircraft is abruptly stalled while simultaneously yawing it with the rudder, a swept upper wing will stall abruptly and produce a crisp "snap" roll. (Just as in a spin, both wings are stalled but the outer wing is producing more lift and chasing the inner wing.) When the Great Lakes biplane of the 1920s was being designed, it began with two straight wings. But when flight tests showed that the center of pressure of the wings was too far forward, the quickest and easiest solution was to give the top wing some sweepback. The incidental result was to make the Great Lakes probably the best aerobatic aircraft in the United States for three decades or so. Canard aircraft such as the Long-EZ and the Starship, with rear-mounted pusher propellers, use sweep to move the aerodynamic center rearward to compensate for the heavy engine(s) in the rear of the fuselage.

Early versions of the Cessna Cardinal possessed the unsuspected capability of stalling the stabilator, causing a sudden pitch-down, when the aircraft was severely "side-slipped" (by use of opposite rudder and aileron). Cessna's solution was to cut a reverse-direction slot into the stabilator so that it couldn't be stalled. Look for the slot the next time you meet a Cardinal.

Pic. 12.5 The de Havilland Twin Otter uses a wing fence to block span-wise flow near the wingtip at high lift coefficients.

Pic. 12.6 This Christen Eagle shows off its swept upper wing in a knife-edge flyby.

cause of boundary layer separation over the fuselage and in investigating wing-fuselage interference problems.

Pictures 12.9 to 12.12 (courtesy of Ben MacKenzie, Norton Company) were taken during the certification process for a radome pod to be mounted on the right wing of the Beechcraft Bonanza. Picture 12.9 shows the air flow during a flap-down approach to a stall; notice that just the inboard section is showing separation but (surprise!) the flow over the aileron has

Pic. 12.7 Slots in a Cardinal's stabilator

(The general rule for either a conventional or canard configuration is that the **rearmost** horizontal surface must never be allowed to stall, to prevent an uncontrolled pitch-up or pitch-down.)

How do manufacturers and homebuilders determine the separation characteristics of their new flying machines? **Wool tufts** taped to the wing and fuselage are an old trick that works every time. As the angle of attack is increased, the tufts will start a wild oscillation at the point of separation and will turn around and point upstream in thoroughly separated regions. This tuft technique is also very useful in determining the amount and

Pic. 12.9 A partially stalled wing, flaps down; note the flow over the aileron.

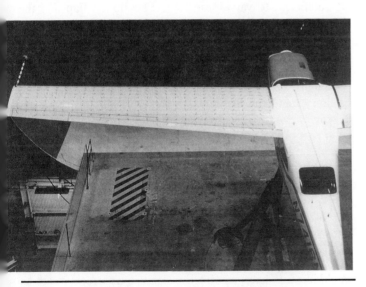

Pic. 12.8 A modified Cessna 210 with an NLF(1)-0414F airfoil is here being tested in NASA Langley's full-scale wind tunnel with woof tufts to reveal the pattern of the air flow at an angle of attack of 20° (photo courtesy of NASA Langley).

Pic. 12.10 A fully stalled wing

a strong inboard component. Picture 12.10 shows the fully stalled wing—look at all the air rushing forward!

Picture 12.11 shows the attached flow on the radome pod and on the wing during cruising flight. Picture 12.12 shows a flaps-up stall with the radome pod on the wing; note the separation inboard but attached flow outboard, maintaining aileron effectiveness.

Pic. 12.11 Cruising flight with the radome pod on the right wing

Pic. 12.12 Flaps-up stall with radome pod on the right wing

Figure 12.10 shows the effect of plus and minus deflection of an aileron on the lift curve of a basic airfoil section. The curves are **not** symmetrical about the zero deflection curve because of the camber of the airfoil; therefore upward deflection is seen to be more effective than downward. However, there do not appear to be any significant "dead spots" where control response is extremely non-linear. Note that these characteristics are given for a specific air gap between the aileron and the wing. Any flow of air through the aileron-wing gap reduces the effectiveness of the aileron and greatly increases the drag. Sailplanes and various new go-fast machines seal the gaps on all their control surfaces, as do aerobatic aircraft, for mostly different reasons.

Most aircraft use some form of high lift, high drag wing devices to lower their landing speeds and steepen their glide angle (to assist in efficiently clearing obstacles close to the

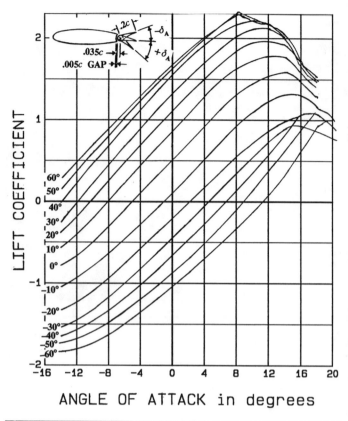

Fig. 12.10 Effect of deflecting a 20% aileron on the lift curve of the LS(1)-0417 (Ref. 9)

runway). Figure 12.11 through 12.17 give some important details on high lift devices. Note from Figure 12.11 that **all** the various types of flaps move the lift curve to the left (so that a given angle of attack gives more lift) and increase the maximum lift coefficient to varying degrees. The slotted flap and the Fowler flap are seen to be the most effective in increasing the lift coefficient, but all of them are good at adding drag. The Fowler flap gains an extra advantage by effectively increasing the wing area.

Figure 12.12 shows that both leading edge devices and suction **extend** the lift curve to higher angles of attack (by delaying separation) without causing any significant shift in the main part of the curve. The high angles of attack required to obtain the large lift coefficients is a distinct disadvantage on landing because it makes it harder for the pilot to see where the ground is and may require very long landing gear to avoid a tail strike. So slots or slats are usually used in conjunction with flaps; the leftward shift of the lift curve provided by flaps allows a high lift coefficient to be obtained without requiring such an extreme pitch attitude.

If a wing has a trailing edge type stall (Fig. 11.13), as we would like it to have, an effective flap will transform it into an abrupt, leading edge type stall. This is where the leading edge devices come into play to extend the lift curve to still higher angles of attack and maximum lift coefficients (Ref. 53).

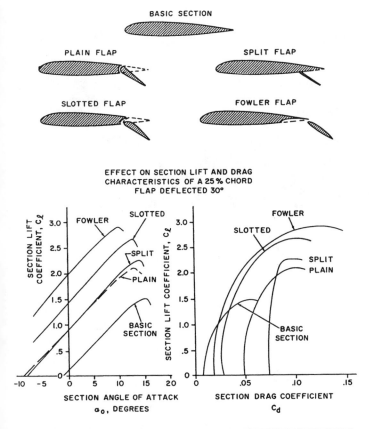

EFFECT ON SECTION LIFT AND DRAG
CHARACTERISTICS OF A 25% CHORD
FLAP DEFLECTED 30°

Fig. 12.11 Effect on the lift curve of various types of 25% flaps, deflected 30° (Ref. 9)

Pic. 12.13 An AT-6 (WW II Advanced Trainer) in landing configuration, flaps and gear down

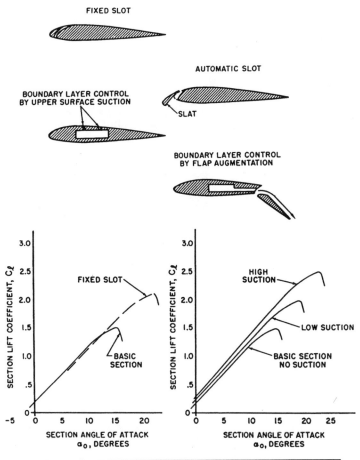

Fig. 12.12 Effect of slots and suction on the lift curve of a typical wing (Ref. 9)

Jet aircraft can use some of their excess power to blow air over their flaps. These "blown" flaps have become very common on jet fighter aircraft ever since the Lockheed F-104 Starfighter jet of the 1950s started using them. Another idea that has been proven to work extremely well on STOL (short-takeoff-and-landing) transports is to mount the jet engine ahead of and just above the centerline of the wing so that the exhaust swishes over the top of the wing at all times. This is called **upper surface blowing** (USB) and can result in maximum lift coefficients up to an amazing **10** or so. The key problem that must be solved is how to ensure that the aircraft remains controllable if one engine fails, because of the loss of thrust and decreased lift and increased drag on the side of the wing where it happens.

When flaps are deployed, an increase in the nose-**down** pitching moment occurs. In low wing aircraft, this and the extra drag produce a nose-down pitching moment which can be quite disconcerting. In high wing aircraft, though, the altered downwash from the flaps and the extra drag above the center of gravity typically produces a significant pitch-**up** tendency.

Figure 12.13 depicts with some detail the very serious high lift devices employed by Boeing on its 727 airliner. It is quite interesting to watch such a wing when it is deploying these surfaces! (The uninitiated may be heard to remark cautiously

about the way the wing is growing rapidly larger.) Can you guess why the slat is used on the outboard part and the leading edge flap on the inboard part? (Hint: Refer to Figures 12.8, 12.11, and 12.12.)

The effectiveness of these elaborate high lift devices is readily calculated. At an aircraft weight of 150,000 lb, the B-727 has a listed stall speed of 153 kt (equivalent airspeed) with no

Pic. 12.14 Takeoff flap configuration on a Boeing 727

Pic. 12.15 Slotted flaps on a Cessna

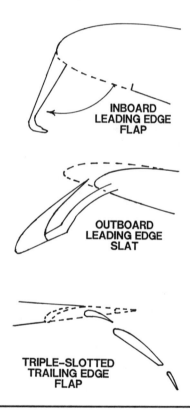

INBOARD
LEADING EDGE
FLAP

OUTBOARD
LEADING EDGE
SLAT

TRIPLE-SLOTTED
TRAILING EDGE
FLAP

Fig. 12.13 High lift devices employed by the Boeing 727 transport aircraft

flap or slat deployment, but this is decreased to 101 kt with full deployment. The corresponding maximum lift coefficients, using Table 12.2 and Eq. 12.15, are **1.9** and **4.4**!

There are some dangers intrinsic to the complexity of these devices; sometimes Murphy's Law appears to be extra effective in the air. On the night of April 4, 1979, for instance, a B-727 made an unscheduled descent from 39,000 ft to about 5,000 ft of altitude in 63 seconds. En route to this low altitude and during recovery to 11,000 ft, the aircraft achieved a maximum Mach number of 0.96, flight path angles from 55° above the

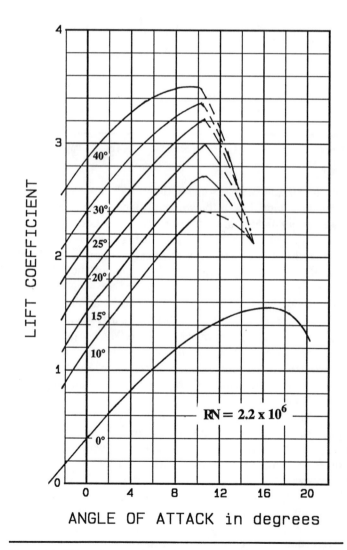

Fig. 12.14 Effect of a 29% slotted flap on the lift curve of an LS(1)-0417 airfoil (Ref. 38)

horizon to 90° below the horizon, and a maximum load factor of **6.0** (i.e., a lift of 6.0 times the aircraft's weight). The National Transportation Safety Board concluded that the captain had deployed some trailing edge flap while cruising at a Mach number of 0.816, apparently in an attempt to increase performance. In any case, an unwanted extension of the leading edge slats apparently followed the flap extension, and when retraction was attempted the Number 7 slat (the next-to-the-end slat on the right wing) failed to retract under the heavier-than-design air loads and because of a pre-existing misalignment. This resulted shortly thereafter in the aircraft rolling uncontrollably to the right. The aircraft continued in this uncontrollable roll and in nearly continuous Mach buffet until the slat was torn from the wing. A safe landing was then made (without the benefit of any high lift devices). (Ref. 6)

The next four figures present recent measurements of the effectiveness of flaps on one of the new-generation airfoils, the LS(1)-0417. Note first that the no-flap lift curve is consistent with the RN = 2.1 million curve of Fig. 9.5; the relatively low RN for these measurements results in premature trailing edge separation and a relatively poor maximum lift coefficient, but this is a realistic RN for many relatively slow aircraft during landing. Observe that even a 10° deployment of this slotted flap increases the maximum lift by over **50%** while reducing the stall angle by about **35%**. Full 40° flap deployment increases the maximum lift coefficient by a very impressive **125%**. To gain an insight into how this slotted flap accomplishes such good things, a study of the pressure distributions is certainly in order.

First we watch the pressures as the flap is progressively lowered and moved backward while keeping the angle of attack constant at 0° (Fig. 12.15). Note that the first 10° of flap nearly triples both the lift and the pitching moment (much more nose-down) but increases the drag by less than a factor of two. By energizing the upper surface BL, the slot has powerfully increased the circulation, correspondingly moving the stagnation point well below the leading edge. The flap cove itself contributes some nice, efficient lift.

Continued flap extension generates an astonishing increase in leading edge thrust, but the forces on the flap itself have an increasingly rearward tilt, causing the drag to increase rapidly also. Even though BL separation on the rear part of the main airfoil is signaled by the constant pressure there, the flap surface remains unstalled throughout, thanks to the slot. At this angle of attack, we see that maximum flap extension increases the lift, the drag, and the pitching moment by seven or eight times!

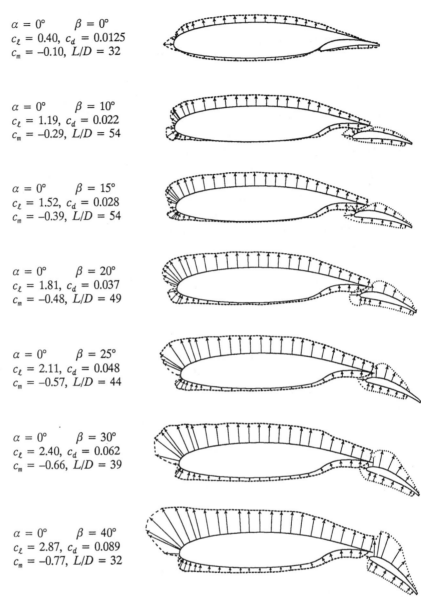

$\alpha = 0°$ $\beta = 0°$
$c_\ell = 0.40$, $c_d = 0.0125$
$c_m = -0.10$, $L/D = 32$

$\alpha = 0°$ $\beta = 10°$
$c_\ell = 1.19$, $c_d = 0.022$
$c_m = -0.29$, $L/D = 54$

$\alpha = 0°$ $\beta = 15°$
$c_\ell = 1.52$, $c_d = 0.028$
$c_m = -0.39$, $L/D = 54$

$\alpha = 0°$ $\beta = 20°$
$c_\ell = 1.81$, $c_d = 0.037$
$c_m = -0.48$, $L/D = 49$

$\alpha = 0°$ $\beta = 25°$
$c_\ell = 2.11$, $c_d = 0.048$
$c_m = -0.57$, $L/D = 44$

$\alpha = 0°$ $\beta = 30°$
$c_\ell = 2.40$, $c_d = 0.062$
$c_m = -0.66$, $L/D = 39$

$\alpha = 0°$ $\beta = 40°$
$c_\ell = 2.87$, $c_d = 0.089$
$c_m = -0.77$, $L/D = 32$

Fig. 12.15 Changing pressure distribution on the LS(1)-0417 airfoil at 0° angle of attack as a slotted flap is lowered (based on data in Ref. 38)

If we were flying this airfoil and approaching to land, the normal procedure might be to extend the flaps to 20° and hold an angle of attack close to 0° until it came time to break the glide in the landing flare. Figure 12.16 shows us what this slotted flap would do for us under these circumstances. For example, we would achieve the maximum available lift coefficient by milking the angle of attack only **10** more degrees as we "feel" for the ground. During this flare maneuver the lift coefficient increases by about 60%, the drag by 140%, and the pitching moment by almost nothing. The large negative pitching moment and the reduced elevator effectiveness close to the ground would both fight our attempt to bring the nose up so as to land on the main gear and not the nosewheel (for a tricycle gear aircraft) or execute a bounce-free 3-point landing (for a tailwheel-type aircraft). Usually this problem is just what determines the forward-most location for the center of gravity of the aircraft. The most astonishing aspect of these 20° pressure distributions is the enormous increase in the local airspeed (and decrease in pressure) over the leading edge. These leading edge pressures are about **eight** times greater (more negative) than the normal pressures above the airfoil in cruise, for a given airspeed.

Figure 12.17 gives us the opportunity to watch the changing pressures as we carry a 40° flap extension to and **beyond** the stall angle of about 10°. Note that the maximum lift occurs **after** the upper surface flow has already started to break down. After the stall, the leading edge thrust doesn't collapse anywhere as much as it does on the unflapped airfoil (Fig. 9.1g), but most of the upper surface does have separated flow over it. Interestingly, the pitching moment becomes a little less nose-down just before the stall because the pressures on the flap are decreasing while leading edge pressures are at their maxima.

It would be nice to be able to use **spoilers** for roll control so that the whole trailing edge of the wing would become available for flaps. Then we could hope for maximum whole-aircraft lift coefficients like the 3.5 airfoil coefficient shown in Fig. 12.14. It has proven difficult, though, to design spoilers which provide good roll response all the way down through the wing stall. There may even be an initial control reversal when a spoiler is first deployed because the boundary layer may momentarily stay attached and generate more lift before it separates and the lift is reduced.

G. High Speed Wings: Sweep, Vortex Lift, and Area Rule

In 1935 Professor Adolf Busemann of Germany suggested at a conference on supersonic flight that a swept wing would have a higher critical Mach number (lower wave drag for a given Mach number) than a similar straight wing. Little attention was paid to his theory at the time, though, probably because such speeds seemed unattainable. However, at the end of World War II it was discovered that Germany had verified the theory in their wind tunnels and had applied the idea to their jet fighters. In the United States, NACA's Robert T. (R. T.) Jones independently discovered the idea while working on thin airfoil theory. Soon thereafter North American decided they could handle the known low speed problems and gambled on using the new idea on their F-86 Sabre jet—and won big. (The Sabre replaced the straight-wing F-84 in the Korean War and regained air superiority for the United Nation forces.)

The critical Mach number for a wing is increased by sweepback because the air that actually travels along the wing chord line has a lower speed than the flight airspeed, and it is this component of the flight speed that determines the airfoil's critical Mach number. This sug-

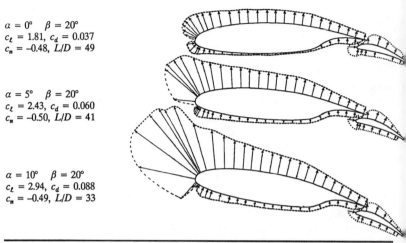

$\alpha = 0°$ $\beta = 20°$
$c_\ell = 1.81,\ c_d = 0.037$
$c_m = -0.48,\ L/D = 49$

$\alpha = 5°$ $\beta = 20°$
$c_\ell = 2.43,\ c_d = 0.060$
$c_m = -0.50,\ L/D = 41$

$\alpha = 10°$ $\beta = 20°$
$c_\ell = 2.94,\ c_d = 0.088$
$c_m = -0.49,\ L/D = 33$

Fig. 12.16 Changing pressure distriubtion on the LS(1)-0417 airfoil as the angle of attack is increased, for a flap deflection of 20° (based on data in Ref. 38)

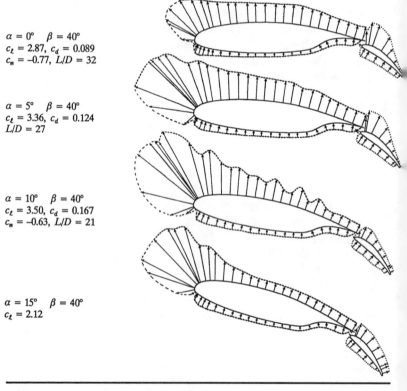

$\alpha = 0°$ $\beta = 40°$
$c_\ell = 2.87,\ c_d = 0.089$
$c_m = -0.77,\ L/D = 32$

$\alpha = 5°$ $\beta = 40°$
$c_\ell = 3.36,\ c_d = 0.124$
$L/D = 27$

$\alpha = 10°$ $\beta = 40°$
$c_\ell = 3.50,\ c_d = 0.167$
$c_m = -0.63,\ L/D = 21$

$\alpha = 15°$ $\beta = 40°$
$c_\ell = 2.12$

Fig. 12.17 Changing pressure distribution on the LS(1)-0417 airfoil as the angle of attack is increased, for a flap deflection of 40° (based on data in Ref. 38)

gests that a wing with a sweepback angle of Λ (capital Greek letter lambda) will have its airfoil Mach number, $M_{CR,\ airfoil}$ increased to $\dfrac{M_{CR,\ airfoil}}{\cos(\Lambda)}$ if both are in two-dimensional flow; for a real wing, however, the benefit is less and the following equation is recommended:

$$M_{CR,\ wing} \approx \frac{M_{CR,\ airfoil}}{\cos\left(\dfrac{\Lambda}{2}\right)} \qquad (12.17)$$

Pic. 12.16 The Rallye is one of the few light aircraft sporting automatic leading edge slats.

There are ongoing attempts to gain the high speed advantages of sweepback without suffering the normal rear sweepback penalties of increased induced drag and reduced maximum lift coefficient (from the low aspect ratio) and the pitch-up stalling tendencies. One of these attempts is the variable-sweep aircraft of which the F-111 and F-14 are current examples. The disadvantage of a variable-sweep aircraft is the very heavy wing fittings required to get the proper type of sweep because the wing loads are then highly concentrated.

A relatively recent brainchild of R. T. Jones is the oblique- or swing-wing aircraft, in which one wingtip swings forward and the other swings rearward. A simple one-piece wing, pivoted at a single point at the fuselage is much easier, cheaper, and lighter to build than dual sweepable wings. The immediate question is whether the flight characteristics will be satisfactory in the fully swept configuration. Wind tunnel tests were sufficiently encouraging that NASA contracted with Rutan's Scaled Composites company to design and build a subsonic proof-of-concept aircraft using the same moldless method of building sandwich all-composite aircraft that are featured in his Quickie, VariEze, and Long-EZ designs. The result, the AD-1, is depicted in Figure 12.19. A full-size, full-speed aircraft using this technology is now under development.

Another recent development is the realization that the new composite technology makes a forward-swept wing feasible for the first time. The problem with forward sweep is that in-flight aerodynamic forces tend to increase the angle of attack of the tips so that they produce even more lift — and twist off (**static divergence**). The requisite torsional stiffness to prevent this has made the concept unusable except for a few examples of mild forward sweep. However, composite wings can now be built that are strong and stiff in just the directions needed (this is called "aeroelastic tailoring") so that the weight penalty is small. Figure 12.20 shows a planform view of the NASA/USAF Grumman X-29 aircraft which is currently being flight tested.

Pic. 12.17 Jet aircraft show off their swept planform during a flyby.

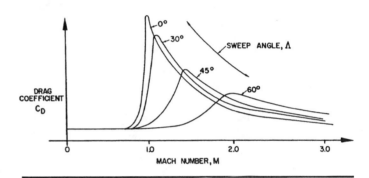

Fig. 12.18 Benefits of wing sweep for transonic and supersonic aircraft

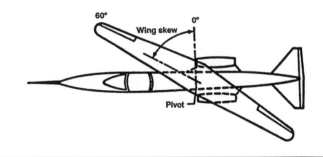

Fig. 12.19 The NASA/Ames AD-1 oblique-wing research aircraft

It uses a boron and graphite/epoxy composite wing. Its forward-mounted, close-coupled canard is expected to reduce wave drag at supersonic speeds. Like the F-16, it is a **fly-by-wire** aircraft. (That is, there is no mechanical or hydraulic link between the pilot's control stick and the flight controls. The connection instead is through electrical wires with a computer in the middle, and the pilot is but a single voting member of the committee that decides what the control surfaces will be told to do.)

The X-29A is designed to be statically and dynamically unstable with the triplexed computers providing the stability needed by human pilots. The benefits expected from this configuration include better stall properties (no tip stall, lower stall speeds), exceptional stability at high angles of attack (including being spin-proof), and higher maneuverability. (Because the aerodynamic center always moves rearward about 25% of the chord in the transition from subsonic to supersonic flight, a typical supersonic aircraft becomes far **too** stable and resists maneuvering flight after going supersonic.) Flight tests on the X-29A have fully confirmed the expected benefits of forward sweep.

Rearward sweepback does add lateral stability (stability in roll so that an aircraft will tend to roll back to level after one wing is deflected by air turbulence), but forward sweep reduces it. This is another reason not to use forward sweep on a slow speed aircraft—but it will look pretty jazzy when it happens anyway (see Ref. 37 in Chap. 19, for example). There are some possibly good reasons to use forward sweep on a horizontal stabilizer, though. It does place the surface out in the undisturbed air ahead of the vertical surface and there may be some bending-moment advantages.

Sweepback is desirable even at supersonic speeds if the sweepback angle is greater than the Mach angle (Eq. 6.3), for then the component of the airspeed along the wing chord is subsonic and a subsonic airfoil can be used.

At speeds greater than about Mach 1.5, an unswept rectangular wing of very small aspect ratio has less wave drag than a swept wing. This is the idea behind the stub wings on the Lockheed F-104 (Fig. 12.1).

Vortex lift is ideally a three-dimensional experience, an interaction between the air moving over a forebody or strake and the main wing, rather than just the two-dimensional affair discussed in the previous chapter. It was discovered in the 1950s that a slender delta-wing planform with a very **sharp** leading edge, could obtain greater maximum lift coefficients than with a rounded leading edge because twin vortices would roll up over the wing. The vortices not only help maintain attached flow but produce even lower pressures than ordinary attached flow ("potential lift") could produce, and this extra lift is very properly called vortex lift (Fig. 12.22). The Concorde was specifically designed to take maximum advantage of vortex lift while landing. Vortex lift is not **efficient** lift, but for aircraft with an excess of power on landing that is not a problem. The reason vortex lift is not efficient lift is because the lift is obtained through reduced pressures on a surface that is tilted significantly backwards (since the angle of attack is large), so the force vector is also pointing backward. Efficient lift is produced by camber because then the minimum pressures are produced over the leading edge and the resulting force vector points **forward** (leading edge thrust).

In the 1960s, this vortex lift concept was successfully transferred to fighter aircraft such as the F-16. There, vortex lift on the strake moves the overall center of pressure forward, reducing trim drag. At high angles of attack the forebody vortices

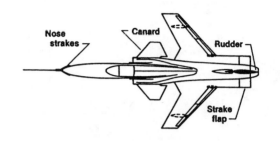

Fig. 12.20 The X-29A forward-swept proof-of-concept aircraft

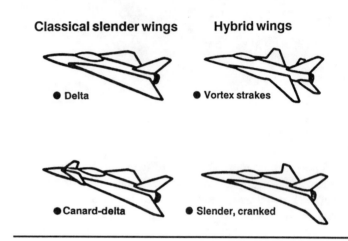

Fig. 12.21 Some design applications of vortex lift (Ref. 45)

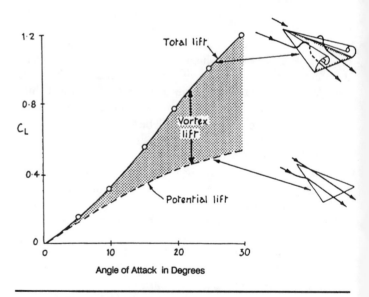

Fig. 12.22 Vortex lift on a highly swept delta wing aircraft with a sharp leading edge, at high angles of attack (Reprinted from *Design for Air Combat* [Ref. 40] by courtesy of Jane's Information Group)

generate vortex lift on the primary wing panels as well as reducing separation there; roll control is enhanced and buffeting is reduced (Fig. 12.23).

To better control vortex lift and make it more efficient, various forms of vortex flaps are being investigated (Fig. 12.24).

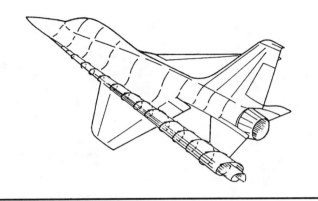

Fig. 12.23 Controlled vortex lift generated by forebody strakes on the General Dynamics F-16 (Reprinted from *Design for Air Combat* [Ref. 40] by courtesy of Jane's Information Group)

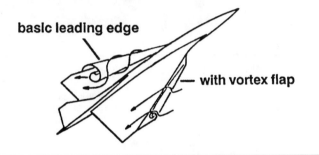

Fig. 12.24 Controlling vortex lift with a vortex flap (Ref. 45)

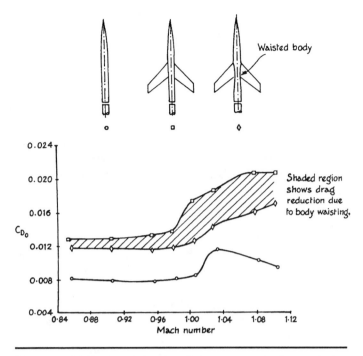

Fig. 12.25 Application of the Area Rule to transonic drag reduction (Reprinted from *Design for Air Combat* [Ref. 40] by courtesy of Jane's Information Group)

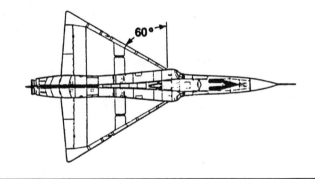

Fig. 12.26 The F-106 (developed from the F-102) has an obvious area-ruled fuselage.

Another important wing design principle for high speed aircraft is **conical camber,** in which the wing is cambered progressively from the root to the tip ahead of an imaginary line drawn between the leading edge at the wing root to the trailing edge at the wingtip. The result is an increase in drag coefficient at small values for the lift coefficient but a decrease in the drag coefficient at the lift coefficients needed for efficient and effective maneuvering. This design principle has been used for many supersonic aircraft, starting with the F-102, for various reasons. Designers of the modern F-15 fighter aircraft chose to use a conical camber wing of aspect ratio 3, with no leading edge devices and only a plain trailing edge flap, because the next best wing, one with an aspect ratio of 2.5 and variable camber provided by leading and trailing edge flaps, was significantly heavier, more expensive to build and maintain, and produced more induced drag overall (Ref. 40).

Richard Whitcomb is credited with the discovery of the **Area Rule,** the idea that transonic drag will be minimized if the whole-aircraft cross-sectional area increases smoothly in a pass from the nose to the tail. This can be thought of as transonic streamlining. It is important for transonic efficiency but not supersonic efficiency. The first production F-102s were unable to obtain supersonic flight in level flight until they were modified according to this rule (Fig. 12.26). Application of the rule requires a narrowing in the fuselage at the point where the wings are attached, a widening immediately thereafter, and

another narrowing at the tail, resulting in what has been called the "coke-bottle" look.

The idea behind the Area Rule is that a "body of revolution" (one with perfect symmetry about the longitudinal axis) has the minimum theoretical drag, so when it is necessary to add appendages such as wings and tail feathers to an aircraft fuselage, the fuselage should be indented there so that overall the air is not pushed to the side any more than it would be for a body of revolution. There is also a supersonic Area Rule which deals with cross-sectional areas that are directly related to the Mach angle.

Some of the newest go-fast experimental/homebuilt aircraft are using fuselages that taper in rather abruptly after the cabin; these are often called **pressure-recovery** fuselages because the idea is to return the air pressure to its static value before it get too close to the end of the surface, and thereby reduce separation and drag, just as for the pressure recovery region on the top surface of an airfoil. Some propeller spinners (for pusher aircraft) are trying this too.

H. Applications to Propellers

Propellers can be thought of as rotating wings and this suggests that composite materials, proplets (as in winglets), sweep, and the new supercritical airfoils should also be tried on propellers in search of both more efficiency and a higher usable maximum speed. By the 1950s, NACA propeller research, like airfoil research, had been abandoned in favor of research on high speed wings and jet engines. Although propellers were more efficient than jet engines at Mach 0.6, airline transports prefer to cruise at Mach 0.8 or higher, and jet fuel was cheap enough that fuel was only about 25% of the direct operating cost. This changed dramatically and abruptly in 1973 with the Middle East oil embargo; suddenly fuel became **50%** of the direct operating cost for airline transports.

Propellers were losing efficiency at aircraft speeds greater than about Mach 0.6 because of **compressibility** losses (i.e., wave drag). Perhaps thin, highly swept blades could be made from the new materials so that efficiency could be maintained to much higher speeds? Theoretical and experimental tests since 1975 have confirmed that a 30% fuel saving over the latest generation of turbofan-powered aircraft is possible, with no more noise or vibration for the passengers. The highest efficiency measured so far for these "**propfans**" is 79.3% at Mach 0.8 for a model tested in the NASA Lewis wind tunnels (Ref. 50 and Fig. 12.27).

Holding back the introduction of propfan-powered transport aircraft into airline service is the cost of certifying and building new-design aircraft, ones with sufficiently low noise levels and ones that will appeal to jet-familiar passengers, especially since fuel prices declined and then stabilized in the 1980s. However, McDonnell Douglas has worked on an MD90 series design and Boeing has worked on a 7J7 design which use propfan propulsion. Propfans no doubt merely await the next big fuel crunch.

A propeller produces thrust by pushing air backwards and relying on the action/reaction principle to receive an equal push forward. The air is always pushed a little sideways as well as backward, though, and this non-axial rotational component is called **swirl**. Swirl, which represents an energy loss, can be reduced by placing a second set of **counter-rotating** blades behind the first set, as in General Electric's Unducted Fan (UDF™) propfan design. It also may be possible to recover most of the efficiency lost to swirl in a simpler way by using non-rotating vanes on the engine nacelle, or possibly by careful design of the wing for a wing-mounted propfan. Most experiments so far, though, have utilized rear-fuselage-mounted pusher configurations because they minimize aerodynamic interference by the slipstream and minimize cabin noise.

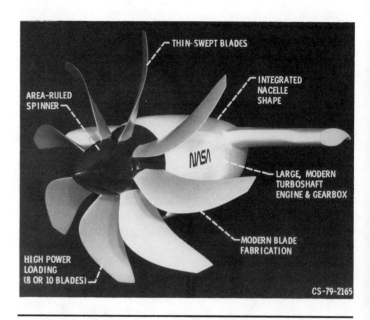

Pic. 12.18 NASA advanced propfan research engine (NASA Lewis)

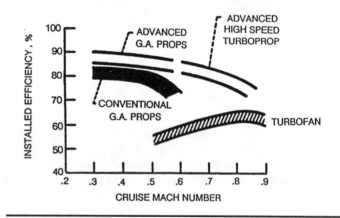

Fig. 12.27 Potential efficiency advantage for aircraft powered by advanced propellers (NASA Lewis)

I. Why Not Ask the Birds?

Sometimes it appears that birds have been using all our brightest "discoveries" for many years. They sweep their nicely tapered wings forward upon landing to prevent tip stall. They open slats and use feather turbulators in the leading edges of their wings for optimized low speed properties. They use wing twist and tails for stability. Even the common pigeon has variable wing sweep for optimized gliding at many speeds. Sweepback and washout are employed for stability in their high speed glides. The true gliding birds, especially those that operate over water such as the ring-billed gull and the albatross, have long and narrow (high aspect ratio) wings. Soaring land birds such as the condor, which need more maneuverability, use five or six wingtip feathers to minimize vortex formation and thereby reduce induced drag. (Some aircraft in recent years have sprouted five or six projections around the wingtip to try to

Pic. 12.19 General Electric's "Unducted Fan" installed on a Boeing 727 (Photo courtesy of NASA Lewis)

produce the same benefits—but winglets may be better for artificial birds.) The new sheared tips have been on birds for many centuries, of course.

By flying in a close "V" formation, wild geese minimize their wingtip vortices and maximize their energy efficiency. A calculation by Lissaman and Shollenberg (Ref. 7) suggests that 25 birds could have a range increase of about 70% compared with a lone bird. A line abreast formation favors the inner birds, but stringing birds back behind a lead bird in a "V" shape will distribute the energy savings rather uniformly and so these optimum positions would be easily sensed by any level-headed goose. Individualists among the flock are happy when they discover that synchronized wing flapping is not necessary. The optimum formation speed is somewhat lower than that for a solo goose.

In fairness it must be noted that using a single unit to produce both lift and propulsion, as birds do, has never worked satisfactorily for us. And they certainly misled us with their small tails and their lack of a vertical stabilizer and rudder; it appears that birds "invented" active controls a few millenia before we got around to trying them. We should be profoundly grateful, on the other hand, that scale effect prohibits birds from growing too large; power-required goes up faster than power-available as size and weight increase and so the largest birds have a stalling speed greater than zero and require a takeoff and landing run, as do we.

Summary

The reduced efficiency of a finite-span wing over an in-finite-span airfoil was shown to be an inevitable result of the mixing at the tips of higher-than-atmospheric-pressure air below the wing with lower pressure air above the wing whenever the wing is developing lift. Alternately, from an energy standpoint it can be seen that an additional drag force (induced or vortex drag) must result from the creation of unproductive wingtip vortices. Or, from an airfoil standpoint, it can be seen that the effective angle of attack is increased by an angle equal to the net downwash angle. Fortunately, most of the dependence of induced drag on wing shape can be calculated based on the aspect ratio of the wing, with the remainder of the dependence incorporated in an efficiency factor e, which is itself a maximum for an elliptical lift distribution. Wingtip vortices can cause heaps of trouble for following aircraft that encounter them, especially during takeoff or landing operations.

Induced drag depends directly on the square of the lift coefficient and so it is the predominant drag at low speeds and high altitudes. At high speeds and low altitudes the profile drag of the airfoil is far greater than the induced drag, so that wings of smaller area are favored.

Because of the large downwash angle, an aircraft with a low aspect ratio wing requires a large angle of attack to obtain its maximum lift coefficient; this effect is counteracted to some extent during landing by the proximity of the ground ("ground effect"). Low wing aircraft with short landing gear exhibit the most noticeable effects. Various wingtip treatments seek to increase the effective aspect ratio without incurring the weight, cost, and structural penalty of simply increasing the span; winglets are a recent innovation that accomplish this very effectively. The "sheared" wingtip, with both taper and rearward sweep, is currently the hottest new idea in wingtips. Vortex turbines on wingtips may be used to recover useful power from wingtip vortices.

An aerodynamically and structurally efficient, tapered wing shape tends to have unsatisfactory stall characteristics because the tips stall first and this results in a loss of lateral control at the stall and incipient autorotation (i.e., a spin). The usual fixes include slots, stall strips, fences, washout, changes in airfoil with span, notches, and the provision of extra-high lift devices near the tip or the aileron. All these stalling problems are aggravated, for swept-wing aircraft, by a pitch-up tendency at the stall.

Indicated or equivalent stalling speeds are a function of aircraft weight, wing area, and the maximum lift coefficient but are **not** affected by altitude or air density changes. However, true stalling speeds are significantly dependent on the air density.

Landing speeds can be reduced by 40% or more while maintaining high cruise speeds by incorporating high lift devices such as flaps, slats, and blown flaps into the wings.

A swept wing encounters wave drag at a higher speed than the same wing without sweep because the component of flight speed along the chord of the wing is thereby reduced. Recent attempts to utilize this principle without some of the typical disadvantages include the oblique-wing and the forward-swept wing; prototypes are now in development.

High speed aircraft with rearward sweep can obtain substantially larger lift coefficients by using a sharp leading edge on either the wing itself or a forebody strake to generate a vortex which produces even lower pressures over the wing than could be produced by simple unseparated flow. This vortex lift is accompanied by large drag increases because it is obtained at very high angles of attack.

Area Rule is a design principle for minimizing transonic and supersonic drag on a wing/fuselage combination.

Propellers are limited by the same aerodynamic problems as are wings, since they are but rotating airfoils, and therefore most of the same bright new ideas are being tried with them; there is convincing evidence that turboprop-powered aircraft with new technology propellers (propfans) can be significantly more fuel efficient than the current jet transports. About the only thing we have tried that the birds haven't (we think) is exceeding the speed of sound.

Symbol Table (in order of introduction)

AR	aspect ratio (of a wing) $= (span)^2/(wing\ area) = b^2/S$
ε	wing downwash angle
e	wing efficiency factor
eAR	effective aspect ratio
D_i	induced drag force
C_{Di}	induced drag coefficient
α_{3D}	effective angle of attack of a wing
α_{2D}	effective angle of attack of an airfoil
$V_{\infty, STALL}$	true airspeed at the stall
$V_{E, STALL}$	equivalent airspeed at the stall
$V_{CAL, STALL}$	calibrated airspeed at the stall
$M_{CR, wing}$	critical Mach number for a wing
$M_{CR, airfoil}$	critical Mach number for an airfoil
Λ	angle of sweep for a swept wing

Review Questions

1. Significantly less power is required to maintain altitude at 1 ft above the runway than at 10 ft because the ___ drag of the aircraft is less at the lower altitude.

2. If you are southbound and about to pass just behind an eastbound aircraft, you can expect shortly to begin to
   a. pitch up,
   b. pitch down,
   c. roll clockwise, or
   d. roll counterclockwise.

3. If you are flying in formation just behind another aircraft but then slide off to your right, you can expect shortly to begin to
   a. pitch up,
   b. pitch down,
   c. roll clockwise, or
   d. roll counterclockwise.

4. From an energy viewpoint, induced drag is the penalty for giving air unwanted ___ energy.

5. Induced drag is zero whenever the ___ developed by the aircraft is zero (i.e., hardly ever!).

6. Wingtip vortices are strongest for a ___ (heavy or light) aircraft with a ___ (low or high) aspect ratio that is moving at its ___ (maximum or minimum) speed.

7. Is induced drag more important in cruise or climb? In low or high altitude flight?

8. If a rectangular (Hershey bar) wing is replaced with a tapered wing of the same aspect ratio and the same area, the span of the new wing will be (greater than, less than, the same as) the span of the original wing and the root chord of the new wing will be (greater than, less than, the same as) the chord of the original wing.

9. Suppose that the wings of a certain aircraft are clipped with the hope of gaining extra top speed through the reduction of ___ drag. The modified aircraft will have an (increased, decreased, unchanged) stall speed and an (increased, decreased, unchanged) stall angle of attack.

10. For what actual flight maneuvers is the induced drag exactly zero?

11. In flight, the aerodynamic angle of attack is (greater than, smaller than) the geometrical angle of attack and the difference is greatest in ___ flight.

12. The aerodynamic efficiency factor for a wing is most closely related to its
   a. span,
   b. maximum chord,
   c. minimum chord,
   d. area, or
   e. shape (rectangular, tapered, or elliptical).

13. The difference between the aerodynamic angle of attack and the geometrical angle is just the ___ angle.

14. The maximum lift coefficient for a given airfoil on a wing of aspect ratio 8 can be expected to be ___ (greater than, less than, about the same as) on a wing of aspect ratio 6.

15. The stalling characteristics of a straight (untwisted) wing tend to be poor because the ___ part of the wing stalls first, especially on a ___ type of planform; the usual solution is to twist the wings so that the tips have a ___ (greater, lesser) angle of attack than the inboard section; this is called ___.

16. Winglets provide many of the benefits of a simple wing extension but have the advantage of producing a smaller ___ ___ at the fuselage.

17. Wingtip design is probably most important for a ___ wing planform.

18. The high aspect ratio of sailplane wings is particularly important and appropriate during ___ and in minimizing the ___ angle.

19. Aerodynamically, the best location for the wing relative to the fuselage is thought to be ___ with ___ in second place; in a low wing design, at least, interference drag probably would be minimized with the use of ___.

20. A one-half scale Cessna 152 would stall at (a higher, a lower, the same) airspeed compared to the full-scale version, assuming the same wing loading (W/S).

21. What is the desired stall progression for an airplane wing and why?

22. The stall pattern of a wing can be altered after it is built by adding ___ ___ to the leading edge; these greatly reduce the airfoil radius at the point of installation.

23. Both forward- and rearward-swept wings exhibit a ___ tendency at the stall.

24. Sweepback has been used to compensate for a too ___ location for the center of gravity as well as enhancing the capability for doing good ___ ___ in aerobatic aircraft.

25. Sketch in cross-section the general shape of the slot that is used on the Cessna Cardinal's stabilator.

26. ___ tend to increase the maximum lift coefficient for a wing by shifting the lift curve upward; ___ tend to increase the maximum lift coefficient by extending the lift curve to higher angles of attack.

27. Why does Boeing use a slot on the outboard part and a leading edge flap on the inboard part of its wing on the 727?

28. A "dead spot" in the control response has been a problem in using ___ for lateral control; the biggest reason for using them is to increase the span and therefore the effectiveness of the ___.

29. If you are in a steady climb to altitude at one speed and then level off at just twice this speed, the induced drag force will ___ (increase or decrease) by a factor of ___.

30. Forward sweep has the advantage over rearward sweep of tending to cause the stall to begin at the wing ___ and this should give it maneuverability advantages; in the past, forward sweep hasn't been used much at all because of ___ a problem, which can now be solved by the use of ___ structures.

31. When an aircraft is flying in level flight there is ___ ahead of the wing, ___ behind the wing, ___ beyond the wingtips, and a net ___. (upwash or downwash)

32. Lowering the flaps always produces a ___ pitching moment on the wing but the overall effect on the aircraft's pitching moment is strongly dependent on the wing and tail locations.

33. A tight "V" formation (for birds or for aircraft) reduces the ___ drag; it is favored over a line-abreast formation because it ___.

34. Thin flat plates aligned in the chord-wise direction are sometimes fastened to the top of a swept wing to impede span-wise flow and thereby improve stall characteristics. These plates are called ___.

35. A flap which utilizes excess (turbine) engine air to increase its effectiveness is known as a ___ flap.

36. The oblique-wing aircraft is an attempt to gain the normal advantages of adjustable sweepback plus a ___ advantage.

37. Which new technologies are now being applied to propeller manufacture?

38. The angle of attack yielding ___ lift is the same for airfoils and wings, for a given airfoil.

39. We can account for nearly all of the effects of using a wing (rather than an airfoil) by specifying the geometrical property called the ___ ___ of the wing.

40. One way of looking at the origin of induced drag is by recognizing that the net ___ produced by the wing must change the effective ___ ___ ___ of the wing.

41. The most efficient wing is one that has an ___ lift distribution over it. For this kind of wing the efficiency factor is ___.

42. If we are flying along using a lift coefficient of 1.0 and then suddenly haul back on the control wheel so as to generate a lift coefficient of 2.0, the amount of induced drag will (increase or decrease) by a factor of ___.

43. If we simply extend the wings on an aircraft to get a longer span, the induced drag force will (increase or decrease) and the profile or parasite drag force will (increase or decrease), for a given cruise speed.

44. A tapered wing is structurally (more, less) efficient, is aerodynamically (more, less) efficient, has a (better, worse) roll response, and has intrinsically (better, poorer) stall characteristics compared with a rectangular wing.

45. All the various wingtip designs try to move the wingtip ___ as close to the ___ as possible.

46. It is important to use efficient, high aspect ratio wings on an aircraft that must make a lot of ___ ___ like sailplanes or pylon racing aircraft, or must fly efficiently at high altitudes.

47. The stalling speed is normally determined by reducing the airspeed smoothly while keeping the ___ ___ constant and equal to ___.

48. The ___ stall speed of any particular aircraft does depend on the ___ of the aircraft when it is stalled but it does not depend on any property of the air (such as air density or temperature).

49. The landing speed of an aircraft can be reduced by decreasing its ___ or by increasing the ___ of its wing.

50. A wing should be designed so that the stall begins at the ___ part of the wing.

51. A ___ in front of an aileron or on a stabilator can keep it from stalling.

52. The two primary design methods for achieving satisfactory stall characteristics for a wing are wing ___ and ___ ___.

53. A wing with ___ is susceptible to sudden pitch-up as it nears its stalling speed because the wing ___ are stalling first.

54. A ___ in the leading edge of a wing or stabilizer will generate a strong vortex that can solve some separation or buffeting problems, as recent jet fighters have shown.

55. ___ ___ are commonly attached to wings to look for separation and specifically to see how separation spreads over a wing as the angle of attack is increased.

56. USB, which stands for ___ ___ ___, uses jet engine exhaust and is one of the most effective of all the high lift devices. The biggest problem is dealing with the possibility of an ___ failure.

57. The normal reaction of all low wing aircraft to the deployment of wing flaps is to ___ ___.

58. A disadvantage of slots and suction for producing greater maximum lift coefficients is that these are only obtained at very high ___ ___ ___.

59. The ___ of flaps is one reason that a high wing aircraft will pitch up when flaps are deployed; another possible reason is related to the ___ angle.

60. Of the plain, the split, the slotted, and Fowler flaps, which gives the greatest increase in the maximum lift coefficient? Which is the most complex?

61. The effect of an aileron on the lift of a wing is not usually symmetrical about the zero deflection point because the ___ on which it is used is usually not symmetrical.

62. Flaps not only increase the maximum lift coefficient but they also increase the lift coefficient for any given ___.

63. One of the problems with using spoilers for roll control (so that the whole trailing edge is free to be used with a ___) is the appearance of a ___ ___ in the response to spoiler actuation.

64. The critical Mach number of a swept wing is greater than that for the basic airfoil section because the wing responds to the ___ of the airspeed that is ___ to the sweep line.

65. The ___ wing obtains variable sweep with a single pivot.

66. An aircraft whose control surfaces are moved electrically under computer control without any direct mechanical or hydraulic link to the pilot is called a ___ aircraft.

67. The X-29A (the experimental forward-swept fighter aircraft) and other recent jet fighters gain efficiency and maneuverability by being designed with intrinsic ___.

68. Stub wings are as good or better than swept wings at ___ speeds.

69. ___ appear to be as effective as wing fences is inhibiting span-wise flow on swept wings; they have the further advantage of ___.

70. The newest wingtip treatment is the ___ tip. It combines a leading edge sweep of ___ degrees with a taper down to about ___ % of the

chord. It is said to aid the ___ in remaining effective through the stall as well as generating a stabilizing pitch-down moment.

71. From Fig. 12.11, what is the difference in section characteristics for a plain versus a split flap, when a lift coefficient of 1.0 is being generated? What is the significance of this for landing and takeoff performance?

72. What type of wing planform, in its unmodified form, allows the ailerons and the very ends of the wings to stall at the highest angle of attack, based on Fig. 12.8?

73. Compare the stalling angles of attack for the 20% aileron section of Fig. 12.10 for (a) the basic section, (b) an aileron deflected 50° upward, and (c) an aileron deflected 50° downward.

74. What flap deflection/angle of attack combination for the airfoil of Fig. 12.15 to 12.17 yields the largest pitching moment?

75. In England, the preferred term for induced drag is ___ drag.

Problems

1. Determine the AR of a delta wing with a 30 ft span and a maximum chord of 20 ft. Answer: 3

2. What is the aspect ratio for a delta wing made with equal length sides? Answer: 2.31

3. The Chance Vought XF5 V-1 of 1948 was an experimental United States Navy aircraft that used a nearly circular ("pancake") planform. What was its aspect ratio? Answer: 1.27

4. Suppose that the aircraft of problem 3 used the 4412 airfoil with the properties depicted in Fig. 11.1. Estimate its stalling angle of attack, assuming a wing efficiency factor of 0.7. Answer: About 48°!

5. Supposing that the aircraft of problem 4 landed with its wing about 3/10 of its span above the ground (on the average), estimate its maximum angle upon landing. (Note: In fact, the aircraft in question had a landing gear that tilted the airplane at an absurd-looking angle of about 30° when it was sitting on the ramp.) Answer: About 42°

6. Estimate the magnitude of the induced drag on a 1500 lb Cessna 152 in level flight at an equivalent flight speed of 55 kt. (Wing area is 160 ft^2; effective aspect ratio is about 5.) Answer: 87 lb

7. The area of an ellipse is $\frac{1}{4}\pi bc$ where b is the long dimension and c is the short one. Obtain an algebraic expression for the AR of a wing with an elliptical planform. Utilize this expression with the listed wing span and aspect ratio for the Spitfire fighter (Table 12.2) to calculate an approximate value for the wing chord at the fuselage.
Answer: 7.6 ft

8. Currently, an ultralight must have a calibrated stall speed of 27 mph or less or else it must be licensed as an aircraft. What maximum lift coefficient must the CGS Hawk (at a gross weight of 550 lb and with a wing area of 140 ft^2) obtain from its flapped UI-1720 airfoil to achieve this? Answer: 2.11

9. What is the true stall speed of the ultralight of problem 8 at a density altitude of 10,000 feet? Answer: 31 mph

10. A Cessna 182Q has a listed flaps-up calibrated stall speed of 56 kt and a flaps-down stall speed of 50 kt for the most rearward location of the cg at a weight of 2950 lbs. With a wing area of 174 ft^2, what maximum lift coefficient is its wing developing? Answer: 1.6, 2.0

11. If the aircraft of problem 10 could achieve the kind of lift coefficients that have been achieved by USB, what would its stalling speed be? Answer: About 22 kt

12. A certain aircraft using the basic LS(1)-0417 airfoil has a no-flap stall speed of 43 kt under the conditions of Fig. 12.14 through 12.17. Estimate the new stall speed with 40° flap deflection, assuming that spoilers are used for lateral control and the whole wing is flapped.

13. Calculate the net downwash angle for the Cessna 152 at maximum speed and also at minimum speed, under SL StAt conditions. (Refer to Section D for speeds. Assume e = 0.8.) Answers: 0.84°, 5.5°

14. The 6% thick wing on the North American F-100 Super Sabre had 45° sweep. If its airfoil had a critical Mach number of 0.74, estimate the critical Mach number of the wing. Answer: 0.80

15. Estimate the drag divergence Mach number of the AD-1 in its swept-wing configuration of Fig. 12.19 if it is using an MS(1)-0313 airfoil (Figs. 10.26 and 10.32). Answer: About 0.85

16. A P51-D Mustang has a wing surface area of 233.19 ft^2 and a wing span of 37.31 ft. What is its MGC? Answer: 6.25 ft

17. Estimate the flaps-up calibrated stall speed for the aircraft of problem 10 when it is at a weight of 2200 lbs. What is the percent change in the stall speed with flaps? Answer: 48 kt, about 10%

18. What is the percent change in the stall angle, the stall speed, and the stall section drag coefficient for an aircraft that deploys the Fowler flap whose characteristics are shown in Fig. 12.11?
Answer: −33%,−28%,+157%

19. How much tail download is needed to bring the Cessna 152 no-flap stall speed into agreement with the book value of 48 kt, assuming a maximum wing lift coefficient of 1.5? Answer: 12%

20. What minimum wing surface area is needed if an ultralight at a gross weight of 460 lb is to satisfy the maximum stall speed requirement of 24 kt, power off? Assume that the UI-1720 airfoil of Sec. E in Chap. 11 is used, and that the wind tunnel value for the maximum lift coefficient is also obtained in flight. Answer: 120 ft^2

21. The Piper PA-38-112 Tomahawk is a 2-seat trainer that uses the LS(1)-0417 airfoil. At a weight of 1670 lb, it has a power-off, no-flap stall speed of 51 kt (calibrated airspeed) if only the outboard stall strips are installed (52 kt if the inboard ones are also installed, as on later models). (a) At about what RN is this aircraft operating when it lands, under SL StAt conditions? The Tomahawk has a rectangular wing with a wing chord of 3 ft 8 inches. Answer: 2.0×10^6 (b) From Fig. 9.7, estimate the maximum lift coefficient for this airfoil at this RN. (c) At its maximum weight, the Tomahawk has a wing loading of 13.39 lb/ft^2. What effective maximum lift coefficient for the wing does this imply? (Use the 51 kt stall speed.) Answer: 1.52 (d) If the maximum in-flight lift coefficient is actually 10% less than the 2-D wind tunnel value and if there is a 10% download on the tail when landing, estimate the expected stall speed. Answer: 55 kt

22. The Voyager around-the-world aircraft had a wing span of 110.8 ft, a wing area of 363 ft^2, a canard span of 33.3 ft, and a canard area of 61 ft^2. (a) What was the aspect ratio of each of these two flying surfaces? (b) The design lift coefficient for the canard was 0.9. If this was in use while cruising at 100 kt, how much lift was developed by the canard with natural transition and also with the BL suddenly tripped? (Use Fig. 11.10.) Answers to (b): 1860 lb, 1220 lb

23. While approaching to land at 71 miles/hr, a certain Quickie has a total drag force acting on it of about 46 lb. Of that, about 30 lb is induced drag. (a) Estimate the total drag when this aircraft flares above the runway at this same speed, assuming the lifting surfaces are about one-quarter span above the runway. Answer: 39 lb (b) The canard is about 1/30 of its span above the ground at touchdown. Estimate the total drag force at 71 mi/hr for this height, using Eq. 12.14 and assuming an efficiency factor of 0.9. Answer: 19 lb

24. Calculate and plot the variation of L/D (which is the same as the C_L/C_D ratio) with flap angle for the slotted flap of Fig. 12.15.

25. Determine the percent increase in lift, drag, and pitching moment when the flap of the airfoil of Fig. 12.15 is lowered from 0° to 30°.

26. Determine the chordal location of the center of pressure for the airfoil of Fig. 12.15 for flap deflections of 0° and 40°. Answers: $0.50\,c$ and $0.52\,c$

27. For a Piper Cadet ($S = 170$ ft^2) travelling at 106 kt under SL StAt conditions ($q_\infty = 38.1$ lb/ft^2), what lift and drag would you expect if its entire wing possesses the characteristics of Fig. 12.15 and the flap deflection was 10°? 30°? Answers: For 10° : 7708 lb; for 30°, 142 lb

28. Estimate the geometric angle of attack for the airfoil of Fig. 12.15, for the 20° flap condition, if it is on the entire wing of an aircraft with an effective aspect ratio of 5. How about for the 40° flap condition? Answer for $\beta = 20°$: 6.6°

29. For the airfoil of the previous question, estimate the three-dimensional drag coefficient. Answer for $\beta = 20°$: 0.246

30. Estimate the stalling angle of attack of the flapped airfoil of Fig. 12.17 if it is on a wing with an effective aspect ratio of 6. Answer: 20.6°

31. Estimate the drag coefficient at the stall of the flapped airfoil of Fig. 12.17 if it is on a wing with an effective aspect ratio of 6.5. What then is the percent of total drag that is induced drag? Answers: 0.767, 78%

References

1. "Advanced Technology Airfoil Research," Volume I, NASA CP-2045, Part 2, March 7–9, 1978. (Available from NTIS as N79-19989 through N79-20007.)

2. Betz, A., "History of boundary layer control research in Germany," *Boundary Layer and Flow Control, Its Principal and Applications*, Vol. 1, edited by G. V. Lachman, Pergamon Press, Oxford, England, 1961.

3. Whitcomb, R. T., "A Design Approach and Selected High Speed Wind Tunnel Results at High Subsonic Speeds for Wing-Tip Mounted Winglets," NASA TN D-8260, July, 1976.

4. Armstrong, Neil and Reynolds, Peter T., "The Learjet Longhorn Series: The First Jets with Winglets," Twenty-Second Symposium Proceedings of the Society of Experimental Test Pilots, September 27–30, 1978. (Available from NTIS as N79-18893.)

5. Nicks, Oran W., "The Science of Low Speed Flight," Motorless Flight Research, NASA CR-2315, 1972. (Available from NTIS as N74-10051.)

6. "Safety Board Studies Transport Incident,", *Aviation Week & Space Technology*, October 5, 1981, p. 90. "Safety Board Cites Pilot in 727 Incident," *Aviation Week & Space Technology*, October 12, 1981, p. 93. (This is the report by the National Transportation Safety Board of the slat-related accident of a Boeing 727.)

7. Lissaman, P. B. S., and Shollenberger, Carl A., "Formation Flight of Birds," *Science*, May 22, 1970, p. 1003.

8. Stinton, Darrol, *The Design of the Aeroplane*, Van Nostrand Reinhold Company, 1983. (This very readable, interesting, up-to-date text by an aeronautical engineer/test pilot presents a fine introduction to the design of high lift devices, tip shapes, and winglets, as well as many other topics in low speed aerodynamics. The frontispiece photo of a gannet gets things started in fine fashion. Highly recommended!)

9. Hurt, H. H., Jr., *Aerodynamics for Naval Aviators*, U. S. Navy, 1960. (A useful, qualitative introduction to aerodynamics, circa 1960.)

10. Wentz, W. H., Jr., Seetharam, H. C. and Fiscko, K. A., "Force and Pressure Tests of the GA(W)-1 Airfoil with a 20% Aileron and Pressure Tests with a 30% Fowler Flap," NASA CR-2833, June, 1977. (Available from NTIS as N77-25083.)

11. Wentz, W. H., Jr. and Ostowari, C., "Summary of High-Lift and Control Surface Research on NASA General Aviation Airfoils," SAE #810629, April 7-10, 1981. (Available from the Society of Automotive Engineers, Inc., 400 Commonwealth Drive, Warrendale, PA 15096.)

12. Van Dam, Cornelius P. G. and Griswold, Michael, "Comparison of Theoretical Predicted Longitudinal Aerodynamic Characteristics with Full-Scale Wind Tunnel Data on the ATLIT Airplane," NASA CR-158753, July, 1979. (Available from NTIS as N79-26018. Presented here are the disappointing results of modifying a Piper PA-34-200 Seneca twin by replacing the untapered 64$_2$-415 wing possessing partial-span 20% plain flaps with a tapered GA(W)-1 wing possessing full-span 30% Fowler flaps.)

13. Cole, James B., "Airplane Wing Leading Edge Variable Camber Flap," Proceedings of the 14th Aerospace Mechanisms Symposium, NASA CP-2127, May, 1980. (Available from NTIS as N80-23514; discusses development of the variable camber leading edge flap for the 747 that would complement its triple slotted trailing edge flap.)

14. Montoya, Lawrence C., "KC-135 Winglet Flight Results," Advanced Aerodynamics and Active Controls, Selected NASA Research, NASA CP-2172, Feb., 1981. (Available from NTIS as N81-19011. The wind tunnel–predicted 7% drag reduction with winglets was confirmed.)

15. Yip, Long P. and Coy, Paul F., "Wind-Tunnel Investigation of a Full-Scale Canard-Configured General Aviation Aircraft," 13th ICAS Congress/AIAA Aircraft Systems and Technology Conference, August 22–27, 1982. (Wind tunnel study of a strake and winglet-equipped VariEze.)

16. Van Dam, Cornelius P., Holmes, Bruce J. and Pitts, Calvin, "Effect of Winglets on Performance and Handling Qualities of General Aviation Aircraft," *Journal of Aircraft*, Vol. 18, No. 7, July, 1981, p. 587. (Points out that winglets do improve climb and cruise performance but that possible adverse effects on handling, stability, and spin recovery must be investigated.)

17. Weisshaar, Terrence A., "Divergence of Forward Swept Composite Wings," *Journal of Aircraft*, Vol. 17, No. 6, June, 1980, p. 442. (Divergence is precluded, according to theory, through proper tailoring of a composite structure.)

18. Katz, Joseph and Levin, Daniel, "Measurements of Ground Effect for Delta Wings," *Journal of Aircraft*, Vol. 21, No. 6, June, 1984, p. 441. (Vortex lift is also increased by ground effect.)

19. Cross, Ernest J., Jr., Bridges, Philip D., Brownlee, Joe A. and Livingston, W. Wayne, "Full Scale Visualization of the Wing Tip Vortices Generated by a Typical Agricultural Aircraft," NASA CR-159382. (Available from NTIS as N81-12019. Pink dust ejected from the wingtips was used to study the trajectories of wingtip vortices, which are of considerable interest in aerial application of agricultural chemicals.)

20. Oehme, H. and Kitzler, U., "On the Geometry of the Avian Wing," NASA TT-F-16901, 1976. (Available from NTIS as N76-19778. Wing profiles for 14 species of birds were studied.)

21. "Low Subsonic Aerodynamic Characteristics of Five Irregular Planform Wings with Systematically Varying Wing Fillet Geometry Tested in the NASA/Ames 12 foot Pressure Tunnel," NASA CR-144600. (Available from NTIS as N76-271714. Double delta or cranked leading edge wing planforms were tested as part of the space shuttle program.)

22. Jones, Robert T., "Wing Plan Form for High-Speed Flight," NACA Report No. 863, 1947. (First U.S. aerodynamicist to realize the high speed advantages of swept wings.) Jones, Robert T. and Nisbet, James W., "Transonic Transport Wings—Oblique or Swept?," *Astronautics and Aeronautics*, January, 1974, p. 851. (R. T. Jones was a very early promoter of the oblique or skewed wing as an optimum

swept planform. The collected papers of this distinguished, mostly self-trained aerodynamicist are available from NTIS as N76-19059 through N76-19122.)

23. Rutkowski, Michael J., "Aeroelastic Stability Analysis of the AD-1 Manned Oblique-Wing Aircraft," NASA TM-78439, October, 1977. (Available from NTIS as N78-13037. The AD-1 is a proof-of-concept, jet-powered, composite-structure oblique wing aircraft.)

24. "Oblique Wing Testbed Flight Evaluation Continues," *Aviation Week & Space Technology*, July 6, 1981, p. 18.

25. Heyson, Harry H., Riebe, Gregory D. and Fulton, Cynthia L., "Theoretical Parametric Study of the Relative Advantages of Winglets and Wing-tip Extensions," NASA TM-X-74003, January, 1977. (Available from NTIS as N77-16989.)

26. Flechner, Stuart G. and Jacobs, Peter F., "Experimental Results of Winglets on First, Second, and Third Generation Jet Transports," NASA TM-72674, May, 1978. (Available from NTIS as N78-23052.)

27. Holmes, Bruce J., van Dam, Cornelius P., Brown, Philip W. and Deal, Perry L., "Flight Evaluation of the Effect of Winglets on Performance and Handling Qualities of a Single-Engine General Aviation Airplane," NASA TM-81892, 1980. (Available from NTIS as N81-12012. Winglets on an A36 Beechcraft Bonanza were found to yield a cruise speed increase of 5.6% at a density altitude of 13,000 feet, a rate-of-climb increase of 6% at 5000 feet, and improved handling qualities.)

28. Van Dam, C. P., "Investigating the Effects of Wing-Tip Modifications on Performance, Stability and Control of General Aviation Aircraft," NASA CR-169738, December, 1982. (Available from NTIS as N83-15282. This analysis shows that winglets may degrade lateral stability.)

29. Sims, Kenneth L., *An Aerodynamic Investigation of a Forward Swept Wing*, thesis, December, 1977. (Available from NTIS as ADA 048898.)

30. Hertz, Terrence J., Shirk, Michael H., Ricketts, Rodney H. and Weisshaar, Terrence A., "On the Track of Practical Forward-Swept Wings," *Astronautics and Aeronautics*, January, 1982, p. 40.

31. Mordoff, Keith F., "Forward-Swept-Wing Aircraft Advances," *Aviation Week & Space Technology*, April 5, 1982, p. 49. (Construction of the X-29A.)

32. Ross, J. C. and Matarazzo, A. D., *Low-Speed Aerodynamic Characteristics of a Generic Forward-Swept-Wing Aircraft*, SAE Paper No. 821467, October 25–28, 1982. (Available from SAE; see Ref. 11.)

33. "Swept Wing Parts Shown in Cutaway," *Aviation Week & Space Technology*, October 18, 1982, p. 46.

34. Uhuad, G. C., Weeks, T. M. and Large, R., "Wind Tunnel Investigation of the Transonic Aerodynamic Characteristics of Forward Swept Wings," *Journal of Aircraft*, Vol. 20, No. 3, March, 1983, p. 195.

35. "Grumman Completes Assembly of X-29A Forward-Swept Wing Aircraft," *Aviation Week & Space Technology*, November 7, 1983, p. 27.

36. "NASA Studies New Civil Aircraft Technology," *Aviation Week & Space Technology*, March 5, 1984, p. 52. (Included in these studies are natural laminar flow wings with supercritical airfoil sections, high aspect wings, forward-sweep, pusher turboprop propulsion, and fly-by-wire controls.)

37. Yeager, Jeana and Rutan, Dick, with Phil Patton, *Voyager*, Alfred A. Knopf, New York, 1987.

38. Wentz, W. H., Jr. and Seetharam, H. C., "Development of a Fowler Flap System for a High Performance General Aviation Airfoil,"

NASA CR-2443, December, 1974. (Tests of two slotted-flap configurations on the LS(1)-0417 airfoil.)

39. Sirells, James C., "Maximum-Lift and Stalling Characteristics of Wings," NACA-University Conference on Aerodynamics, June 21–23, 1948.

40. Whitford, Ray, *Design for Air Combat*, Jane's Publishing, Inc., 1987. (A wealth of information is presented in this up-to-date book. Highly recommended for any one interested in the design of fighter aircraft.)

41. Van Dam, C. P., "Swept Wing-Tip Shapes for Low-Speed Airplanes," SAE Technical Paper 851770, October, 1985. (sheared wingtip)

42. Naik, D. A. and Ostowar, C., "Experimental Investigation of Non-Planar Sheared Outboard Wing Planforms," Paper 88-2549, AIAA 6th Applied Aerodynamics Conference, June 6-8, 1988.

43. Suh, Young B. and Ostowari, Cyrus, "Drag Reduction Factor Due to Ground Effect," *Journal of Aircraft*, Vol. 25, No. 11, November, 1988.

44. Lachman, G. V., editor, *Boundary Layer and Flow Control, Its Principle and Applications*, Pergamon Press, Oxford, England, 1961. (A multi-national review of boundary layer control and practice as of 1960.)

45. Campbell, James F., Osborn, Russell F. and Foughner, Jerome T., Jr., editors, *Vortex Flow Aerodynamics*, Volume I, NASA CP-2416. (Available from NTIS as N86-27190 to N86-27208.)

46. Moore, M. and Frei, D., "X-29 Forward-Swept Wing Aerodynamics Overview," AIAA Applied Aerodynamics Conference, July 13–15, 1983. (Available from AIAA as AIAA 83-1834.)

47. Sefic, Walter J. and Maxwell, Cleo M., "X-29A Technology Demonstrator Flight Test Program Overview," NASA TM-86809, May, 1986. (Available from NTIS as N86-26328.)

48. Trippensee, Gary A. and Lux, David P., *X-29A Forward-Swept-Wing Flight Research Program Status*, NASA TM-100413, Nov. 1987. (Available from NTIS as N88-17644.)

49. "Turbines Recover Power by Dissipating Induced Drag from Wingtip Vortices," *Aviation Week & Space Technology*, September 1, 1986, page 199.

50. Facey, John R., "Return of the Turboprops," *Aerospace America*, October, 1988, pp. 14ff.

51. Vijgen, P. M. H. W., van Dam, C. P. and Holmes, B. J., "Sheared Wing-Tip Aerodynamics: Wind-Tunnel and Computational Investigation," *Journal of Aircraft*, Vol. 26, No. 3, March, 1989.

52. Van Dam, C. P., Vijgen, P. M. H. W. and Holmes, B. J., "Experimental Investigation on the Effect of Crescent Planform on Lift and Drag," *Journal of Aircraft*, Vol. 28, No. 11, November, 1991.

53. Cornish, J. J., III, and Tanner, R. F., *High Lift Techniques for STOL Aircraft*, SAE Paper No. 670245, April 5–7, 1967. (This is a very useful survey of methods that can be employed to increase the maximum lift coefficient of a wing.)

54. Kroo, Ilan, Gallman, John, and Smith, Stephen, "Aerodynamics and Structural Studies of Joined-Wing Aircraft," *J. Aircraft*, Vol. 28, No. 1, January, 1991.

55. Gerhab, George and Eastlake, Charles, "Boundary Layer Control on Airfoils," *The Physics Teacher*, March, 1991. (The authors present some nice smoke tunnel photographs showing the effects of suction and blowing in reducing BL separation. They don't indicate the tunnel Reynolds number, however, and they draw conclusions for full size aircraft that are generally invalid because they have ignored the scale effect.)

Chapter 13

Lift, Drag, and Power for the Complete Aircraft

It was apparent that a propeller was simply an aeroplane (aerofoil) traveling in a spiral course. As we could calculate the effect of an aeroplane traveling in a straight course, why should we not be able to calculate the effect of one traveling in a spiral course? At first glance this does not appear difficult, but on further consideration it is hard to find even a point from which to make a start; for nothing about a propeller, or the medium in which it acts, stands still for a moment. The thrust depends upon the speed and the angle at which the blade strikes the air; the angle at which the blade strikes the air depends upon the speed at which the propeller is turning, the speed the machine is traveling forward, and the speed at which the air is slipping backward; the slip of the air backward depends upon the thrust exerted by the propeller, and the amount of air acted upon.[1]

A calculation indicated that 305 revolutions of the propeller would be required to produce 100 pounds thrust. Later, actual measurement showed that only 302 instead of 305 propeller turns were required. . . . The propellers delivered in useful work 66 per cent of power expended.[2]

Everything except stall speed is a function of span loading, not wing loading. And if you want an elegant airplane that is nimble and willing, then you reduce the span loading. That just makes good flying, efficient airplanes.[3]

A. Review of the Laws of Linear and Angular Motion

The link between the quest for speed and the history of airplane development has been noted more than once in the preceding pages. Yet speed is completely **relative** and **visual**. It is relative, for example, because a 600 mi/hr fighter aircraft has zero speed as viewed from another fighter aircraft flying in close formation. And speed is visual because, with eyes closed, we can detect only a change in motion, a property of the motion we call **acceleration**.

Everywhere we look there are speed limits for both cars and airplanes but, for this wonderfully visceral property we call acceleration, the only limits are those imposed by nature. Therefore powerful cars and airplanes (i.e., those with a high power-to-weight ratio) are prized for the "gut" sense of ac-

celeration they can produce—and sometimes this is a very satisfactory substitute for pure speed. (To some extent, though, our bodies are even more tuned to the rate at which the acceleration itself is changing, something that has been called the **jerk** value for a given motion.)

Acceleration is defined as the rate at which the velocity is **changing**. Because velocity is a **vector** property of motion, acceleration is occurring (**a**) when the **speed** is changing, (**b**) when the **direction** of the motion is changing, or (**c**) when **both** the speed and the direction of the motion are changing. Therefore the acceleration resulting from only a change in direction, as in an aircraft flying a constant-speed, constant-bank, level turn, is just as much accelerated motion as the takeoff or landing roll—and unbalanced external forces are responsible for the acceleration in each case.

We must measure distances and speeds and accelerations if we are to expand our knowledge beyond the merely qualitative, and this means a decision on the units to be used. We note that we have a hierarchy of **distance, speed, acceleration, jerk,**

[1]Wright, Orville, "How We Made the Flight," *Flying* (December 1913).
[2]Kelly, Fred C., *The Wright Brothers* (New York: Ballatine Books, 1956).
[3]Griswold, James E., *AOPA Pilot* (June 1988).

each with units appropriate to the fact that every parameter beyond the first is the rate of change with time of the parameter preceding it. (In calculus terms, speed is the derivative of distance with respect to time, acceleration is the derivative of speed with respect to time, etc.) We have chosen to use the foot (ft) as our primary unit of distance, feet per second (ft/s) as our primary unit of speed, and therefore our primary (and consistent) unit of acceleration should be ft/s^2.

The one acceleration with which we are all somewhat familiar is the rate of change in the speed of a freely-falling object (i.e., one falling with negligible drag force, essentially caused only by the force of gravity), and yet even this acceleration doesn't mean too much to us until the motion it causes is stopped by some other force. The acceleration of gravity is about 32.2 ft/s^2 (or about 22 mi/hr per second) near the earth's surface; in practical terms, this means that if you can reach a count of three after jumping and before hitting the ground, you're already exceeding the double-nickel highway speed limit.

Because the acceleration due to gravity is so common, it has become a secondary unit of acceleration. Therefore we can refer to the hard landing which changed our vertical speed from 80 ft/s to 0 ft/s in 0.4 s as a little 6-g landing rather than telling the unvarnished truth and admitting to an average deceleration of 200 ft/s^2 [because $a = (0 \text{ ft/s} - 80 \text{ ft/s})/(0.4 \text{ s}) = -200 \text{ ft/s}^2$].

It is a real shame that we don't have three-dimension accelerometers (acceleration meters) in our cars and planes so that we could speak as comfortably and confidently and intuitively regarding accelerations as we do of speeds. Aerobatic aircraft, at least, do carry a one-dimension accelerometer that measures instantaneous and maximum accelerations in the direction perpendicular to the aircraft's longitudinal axis. These accelerometers or "g-meters" provide a quantitative measure of the stresses on the wings and the airframe due to variations in lift and ground impacts; they are normally calibrated in g units.

To properly appreciate the success of airplanes, we need to bring our understanding of **angular** motion up to the level of our understanding of linear motion. The hierarchy for angular motion is **angle, angular speed, angular acceleration,** and . . . ? Angles can be measured in terms of complete revolutions of an object, degrees of revolution, or radians of revolution. The unit of revolution is appropriate for a propeller or an aerobatic slow roll and the degree is appropriate for compass headings or pitch angles. Yet the degree is an entirely arbitrary unit, and so ranks right in there with the yard as a unit for the technologically disinclined. The **radian,** though, is a fundamental unit for measuring angles because it is the ratio of the length of a circular arc to the radius of that arc. Therefore one complete revolution is equal to $\frac{2\pi r}{r} = 2\pi$ radians and one radian is equal to $\frac{180}{2\pi}$ radians or about 57.3°.

Angular speed (or angular velocity, because rotation always has a direction associated with it, just as does linear

motion) is usually given either in units of revolutions per minute (RPM), in degrees per second, or in radians per second. The fundamental, consistent unit is again the **radian per second** (rad/s) and all equations assume this unit unless they have been specifically modified for use with inconsistent units. (Engineers who make certain calculations frequently will use the units most commonly available and incorporate the appropriate conversion constant right into the equation. Physicists who are trying to understand — or explain to others — how fundamental laws of nature yield these equations disapprove of the practice.) Finally, **angular acceleration** can also be specified in rev/s^2 or deg/s^2 or **rad/s^2** — but the last one is again the most fundamental unit.

Having discussed the most important terms for describing motion, we turn next to the **causes** for that motion. Newton's first law describes the fact that either no motion or constant-speed motion in a straight line is the natural state for any object subject to no outside forces, **or** subject to forces that balance each other out. (An "outside" or "external" force is one that originates outside the object being described; forces due to objects that touch the plane, such as air molecules and the ground, are included along with the gravitational force which doesn't require contact.) It is change in speed or direction (i.e., the **acceleration**) that requires the presence of at least one outside force, or unbalanced outside forces, acting on the object of interest. Because our sense of motion, other than through sight, is based on detecting accelerations, we naturally associate these accelerations with the forces that produce them — with gravitational acceleration an interesting exception because it isn't a contact force.

The **effectiveness** of an unbalanced force in producing motion depends on the resistance of the object in question to motion, a property we call **inertia** and measure as the **mass** of the object (see the discussion of mass in Chap. 3). If we now specialize to an airplane, we can say that the acceleration of the airplane is directly proportional to the magnitude (i.e., the strength) of the unbalanced external force acting on it, and inversely proportional to its total mass, as stated by Newton's second law of motion:

The vector sum of the external forces acting on an aircraft equals the mass of the airplane multiplied by the linear acceleration of the *cg* of the aircraft.

In symbols:

$$\sum F_{\text{EXT}} = ma_{\text{CG}} \qquad (13.1)$$

THE SECOND LAW OF MOTION

When we specify that the operative forces are the external forces, we mean to exclude all internal forces such as those involved in rotating engine parts or wandering passengers. Although these internal forces **do** alter the location of the airplane's *cg*, they do not alter the path of the airplane's *cg* through the sky. This is similar to what happens when an irregularly shaped object such as a football is tossed through the air; the motion appears irregular but the *cg* of the football follows a nice parabolic path much like a tossed baseball.

Because Eq. 13.1 is a **vector** equation, it represents **three** separate equations in three mutually perpendicular directions, in the most general case. Fortunately, many of the most interesting airplane events (including those to which we will apply it) are either one-dimensional (the takeoff roll) or two-dimensional (a level, banked turn or a loop).

This second law of motion can be applied to a whole airplane or to any part of it; the trick is to accurately identify the **source** of the external forces acting on the chosen object. The mass involved will always be the total mass of the object and the acceleration will always be the acceleration of the center of mass of the object (which is coincident with the center of gravity, for airplanes). If the chosen object is the whole airplane, the external forces are **gravity** (i.e., the weight of the airplane), the **thrust** force due to the propeller or jet engine, the **aerodynamic forces** due to differential air pressures on the aircraft's surface, and (when on the ground) forces on the landing gear **tires**.

When the second law is applied to the pilot, the external forces (in upright, level flight) are generally just the pilot's weight, the upward and sometimes forward force of the seat on the pilot, and the generally upward force of the floor on the pilot's feet. The force provided by the seat allows the pilot to sense what the airplane is doing and respond to it — in the long tradition of "seat-of-the-pants" flying. The acceleration resulting from these forces is that of the pilot's *cg* and is usually (desirably!) the same as the acceleration of the aircraft as a whole.

What now can we say about cause and effect for angular motion? First, we surmise that the cause of angular motion is an unbalanced external moment or torque acting on an airplane, as measured from an axis passing through the airplane's *cg*.

Second, the result of any such unbalanced moment will be angular **acceleration**, just as an unbalanced force produces linear acceleration. But what property of an airplane measures its resistance to angular acceleration? A little thought reveals that this resistance must depend on **both** the amount of mass and where this mass is located relative to the axis of rotation. This property of an object is called its **moment of inertia**, *I*, in obvious analogy to the inertial mass used for linear acceleration. Unlike the mass of an object, though, its moment of inertia is different, in general, for every possible choice of axis of rotation, just as the moment of a force depends on the assumed point of rotation. Even for the restricted case of rotation about axes through the *cg*, every aircraft has at least three important moments of inertia.

The moment of inertia of a simple object is easily calculated, as is explained in most every introductory college physics textbook. First, the moment of inertia of a tiny bit of mass (a "point mass"), *m*, located a distance *r* from the axis of rotation is $I = mr^2$. Second, for a complex object such as an airplane, it is possible to calculate the moment of inertia by dividing up the total mass of the airplane into N little bits of mass m_i (where N is a very large number) and then, using the distance of each

little mass from the axis of rotation, r_i, add up all of these separate contribution in a simple sum:

$$I = \sum_{i=1}^{N} m_i r_i^2 = m_1 r_1^2 + m_2 r_2^2 + \ldots + m_N r_N^2 \quad (13.2)$$

MOMENT OF INERTIA OF A COMPLEX SOLID OBJECT

where the Greek capital sigma, $\sum$, is used in the standard mathematical shorthand to represent a summation.

Equation 13.2 can be solved exactly for objects with considerable symmetry and with uniform mass density. Thus the moment of inertia of a thin uniform rod of mass *M*, radius *R*, and length *L* is (a) $\frac{1}{12} ML^2$ for an axis through its center and perpendicular to its length but (b) only $\frac{1}{2} MR^2$ for an axis along its length and through its center.

The moment of inertia strongly depends on where the mass is located, as evidenced by the fact that the square of the distance from the axis of rotation is involved. Therefore we can predict that the resistance of an airplane to rolling about its longitudinal axis is radically greater for an aircraft with engines mounted on the wings and with tip tanks than for a similar aircraft with fuselage-mounted engines and tanks. But once that wing with engines on it starts to roll, it also resists stopping! Similarly, a wood propeller will rev up noticeably faster than an otherwise similar metal prop.

Because an aircraft is not a point object, these external forces may not be pointing directly toward or directly away from the center of gravity of the aircraft. Therefore there exists the possibility of angular acceleration about the center of gravity as well as linear acceleration of the center of gravity. By applying the second law to these distributed forces, it can be shown that the magnitude of any angular acceleration for an aircraft can be determined from the following principle:

The vector sum of external moments acting on an aircraft about any given axis through the *cg* is equal to the moment of inertia of the aircraft about that axis, multiplied by the angular acceleration about the *cg*. In symbols,

$$\sum M_{\text{EXT,CG}} = I_{\text{CG}} \, \alpha_{\text{CG}} \quad (13.3)$$

LAW OF MOTION FOR ROTATION

The moment of a force (also commonly known as the torque due to the force) is computed (a) by multiplying the force by the effective distance of the force from the chosen axis or, equivalently, (b) by multiplying the actual distance of the force from the axis by the effective component of the force. (See Section C in Chap. 3 for additional discussion of this point.)

Thus in Fig. 13.1 the moment contributed by the lift force about an axis perpendicular to the side of the fuselage and through the center of gravity can be calculated from either of

(a) $M_{\text{LIFT FORCE ABOUT CG}}$ = Effective Distance × Force

$$= -(d \sin \theta) L \quad \text{or}$$

(b) $M_{\text{LIFT FORCE ABOUT CG}}$ = Distance
$$\times \text{Effective Force Comp.}$$

$$= -d(L \sin \theta)$$

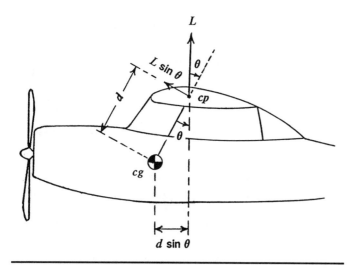

Fig. 13.1 Calculating the moment about the center of gravity of the lift force

Note that "effective distance" here means the perpendicular (or shortest) distance of the axis from the line of action of the force and "effective force component" here means the component of the force that is perpendicular to the line drawn from the chosen axis. The negative sign has been inserted because, as you should recall from Chap. 3, moments that tend to **increase** the angle of attack of the wing are taken as **positive**.

You should convince yourself at this point that these definitions really do measure the tendency of a force to produce rotation about a given axis. In any actual calculation, choose the easier of the two methods. (All of the concepts of force, mass, center of mass, center of gravity, and moments were discussed in some detail in Chap. 3. You may wish to review that material.)

These definitions have stressed the fact that only **external** forces and torques acting **on** the aircraft are being considered. Internal forces, such as the forces between the occupants and the aircraft structure or between the engine and the engine mounts, need be considered only if internal stresses or internal relative motion is being studied. Internal forces, even the most violent, can only change the position of the center of gravity relative to some chosen external point. Thus a passenger who moves forward in the cabin of an aircraft also moves the center of gravity of the aircraft closer to its nose. However, wherever the center of gravity of the aircraft might be located, the motion of that point is determined solely by forces exerted by external objects on the aircraft.

Except for the gravitational force, the external forces on an aircraft can be thought of as contact forces. Thus the force of the ground on the aircraft when it is taxiing is not just a force on the airplane; it is a force directly on the tires of the aircraft, the part that is in contact with the earth. When an aircraft is being braked, the ground pushes **backwards** on the tires because the tires are pushing forward on the ground. Even when the brakes are not in use, there is always some rolling friction from tire distortion which results in a rearward force on the tires when on the ground.

In this discussion of forces we have implicitly used Newton's third law of motion:

For every force exerted by one object on a second object, there is a simultaneous, equal, and opposite second force exerted by the second object on the first object.
THE THIRD LAW OF MOTION (LINEAR MOTION)

Thus, if the earth pushes **up** on an aircraft's tires with a force of 2000 lb, the tires must be simultaneously pushing **down** on the earth with a similar force of 2000 lb. And an aircraft can maintain level flight (or a steady climb or a steady glide) if and only if its wings can somehow manage to push **down** on the air with a force equal and opposite to the weight of the aircraft. A propeller can "propel" an aircraft only to the extent that it is able to exert a backwards force on the air around it. So if we can determine how hard the aircraft pushes on the air, we have also determined how hard the air pushes on the aircraft. This viewpoint is in many ways the clearest explanation for how a wing or a propeller works. It also suggests an explanation for the fact that a decrease in air density causes a decrease in lift and thrust: less force can be exerted on the air when the intercepted air has less mass to be accelerated.

In calculating the lift of an airfoil in earlier work, it turned out to be convenient to use Bernoulli's pressure/speed relationship because the surface pressures are commonly measured. However the lift can also be obtained experimentally by measuring the reaction force required to prevent motion perpendicular to the relative wind. In any case, the three laws of motion are still being used since Bernoulli's equation can be derived from them. And it should be emphasized that all our definitions of angular quantities and laws of angular motion simply follow from an application of the three laws of motion to the extended mass of a rigid body; they arise because we don't want to have to consider the object in terms of all of its constituent atoms when we are trying to describe its motion.

One of the objects that commonly exerts forces on aircraft — the earth — is so massive that the effect of the reaction forces described by the third law is not obvious. When a pilot inadvertently tries to make an airplane-shaped depression in the runway, for instance, the earth also rebounds from the impact; although the resulting acceleration of the earth is too small to measure, it does occur.

The third law of motion for forces has an angular counterpart which follows logically from application of the third law to the constituent particles of an extended (non-point-mass) object such as an airplane:

For every moment exerted by one object on a second object, there is a simultaneous, equal, and opposite second moment exerted by the second object on the first.
THIRD LAW OF MOTION (ANGULAR MOTION)

The force of the propeller on the air generates the thrust force in the airplane but these forces also result in a constant moment on the propeller shaft that tends to slow it down and the engine must provide an equal and opposite torque. And if the rpm is to increase, the engine must provide additional torque for the acceleration.

An aircraft is said to be in **equilibrium** whenever it possesses neither linear nor angular acceleration. (Strictly speaking, all of our measurements of velocity and acceleration should be with reference to an unaccelerated reference point; practically, for our purposes, we can make the surface of the earth our reference without incurring significant error — but the earth is clearly not an adequate reference for something like the space shuttle.)

Therefore the following are all examples of equilibrium conditions for the **center of gravity** of an aircraft: (**a**) a stationary aircraft on the tie-down ramp, (**b**) an aircraft taxiing at constant speed in a straight line, (**c**) a climb to altitude at constant speed, constant heading, and with a constant rate of climb, (**d**) level, upright or inverted, cruising flight with a constant speed and constant heading, (**e**) level flight with a constant speed, constant heading, and a constant rate of roll (this would be a very high quality slow roll), and (**f**) a stabilized glide or dive with constant airspeed and heading.

Evidently we can learn a great deal about aircraft performance while staying within the constraints of equilibrium conditions. However, the initial takeoff and landing phases, as well as the transition from one equilibrium phase to another, all represent accelerated motion. So does the dynamic response of an aircraft to outside disturbances, as is shown in the analysis of aircraft stability. Even a constant-bank, constant-airspeed, constant-altitude turn is accelerated motion because the **direction** of the velocity vector is changing.

We should note again that our two equations of motion are **vector** equations; they are usually solved by considering separately the motion along each of three perpendicular axes. Thus there are at least **six** equations that must be solved simultaneously in the most general description of aircraft in motion. There is good reason for the popularity of computers in the aeronautical world! Indeed, a maneuver such as a spin is so complex that analytical methods are still not able to adequately describe or predict important aspects of the maneuver — such as a sure-fire recovery technique for a particular aircraft. Therefore we will generally restrict our attention to one- or two-dimensional maneuvers. If we wish to describe the individual motions of an object whose mass distribution can change, we must write the second law (Eq. 13.1) in the more general form given below:

$$\sum F_{\text{EXT}} = \textbf{rate of change with time of } \left(m V_{\text{CG}} \right)$$
$$= \frac{\Delta (m V_{\text{CG}})}{\Delta t} \tag{13.4}$$

MORE GENERAL LAW OF MOTION, LINEAR MOTION

If the mass of the described object is constant, then Eq. 13.4 reduces to the form of Eq. 13.1 because a_{CG} is defined as the rate of change with time of the velocity vector of the object. However, we can see now that it is the **product** of the mass and the center of gravity velocity that must stay constant if the vector

sum of external forces is zero. This product, a vector quantity, is defined as the linear momentum of the object:

$$p \equiv m V_{\text{CG}} \tag{13.5}$$

LINEAR MOMENTUM

The principle of conservation of linear momentum, which follows directly from inspection and contemplation of Eqs. 13.4 and 13.5, is the following:

The momentum of an object is constant in any direction for which any external forces have a vector sum of zero.

PRINCIPLE OF CONSERVATION OF MOMENTUM

We can apply this principle immediately to our passenger who walks up the aisle of an aircraft while it is flying in level flight; because he or she must exert a backward force on the aircraft floor in order to start to move forward and because this backward force is equal and opposite to the forward force on the passenger (Newton's third law), and acts over the same period of time, the momentum that the passenger gains is just equal in magnitude to the momentum lost by the aircraft. Thus the total momentum of the aircraft and its cargo is not changed by these internal motions or forces. The really neat part of using momentum concepts is that we can calculate forces, or at least average forces, and sometimes the new velocity after a given interaction, without going into the details of the exact nature of the forces.

EXAMPLE 13.1. Equation 13.4 is in great favor for the analysis of jet engines, where air is taken in, heated by the combustion of fuel, and expelled at a much higher speed. One jet engine, for example, is able to combine 161 lb of air (which by Eq. 3.4 represents a mass of 5.0 slugs) with 9 lb (0.28 sl) of fuel every second, producing exhaust gases that leave the engine at 3300 ft/s relative to the airplane. The resulting thrust at, e.g., a flight speed of 600 mi/hr (880 ft/s) can be calculated from Eq. 13.4 as

$$\text{Thrust} = \frac{\Delta (m_{\text{AIR}} \, v_{\text{AIR}})}{\Delta t} + \frac{\Delta (m_{\text{FUEL}} \, v_{\text{FUEL}})}{\Delta t}$$

$$= \frac{(5 \text{ sl})(3300 \text{ ft/s} - 880 \text{ ft/s})}{1 \text{ sec}} + \frac{(0.28 \text{ sl})(3300 \text{ ft/s})}{1 \text{ sec}}$$

$$= 12{,}100 \text{ lb} + 924 \text{ lb} \approx \textbf{13{,}000 lb}$$

Also, later we will see how the thrust produced by a propeller can be estimated from the change that it produces in the momentum of the air going through it.

The moment of inertia of an object can change as well as its mass, and then we must rewrite Eq. 13.3 in a form analogous to Eq. 13.4 for angular motion:

$$\sum M_{\text{EXT}} = \textbf{rate of change with time of } \left(I \omega \right) \tag{13.6}$$

MORE GENERAL LAW OF MOTION, ROTATIONAL MOTION

where ω, the Greek small letter omega, is the symbol for angular speed. If there are no external moments acting on an object, or if their vector sum is zero, the angular momentum (product of angular inertia and angular speed) will not change:

$$L = I\omega \qquad (13.7)$$

ANGULAR MOMENTUM OF AN OBJECT

In the same way as for linear motion, Eqs. 13.6 and 13.7 imply the principle of angular momentum conservation:

**The angular momentum of an object is constant
in any direction for which
any external moments have a vector sum of zero.**

PRINCIPLE OF CONSERVATION OF ANGULAR MOMENTUM

Angular momentum is a **vector** quantity; its direction is along the axis of rotation and pointing toward the direction where the rotation is clockwise. Thus, in the United States, the angular momentum of the propeller is a vector through the center of the propeller shaft pointing directly forward. (European engines and propellers generally prefer to rotate counterclockwise, as viewed from the cockpit; this has some interesting consequences which will be remarked on shortly.)

Life in general and airplanes in particular would be infinitely duller if they didn't possess angular motion and thereby angular momentum. When cruising along in level flight, the angular momentum of an airplane is that of its propeller/propeller shaft/crankshaft combination (or the turbine wheels for a jet-propelled aircraft) and is constant. Every change in power setting, though, produces an unbalanced moment on the rotating elements, causing the angular speed and the angular momentum to change until the sum of the external moments or torques is again zero. Even when the propeller or turbine wheel has a constant angular momentum, it should be noted again that it is **not** because there are no external torques. Instead, the air pressures on the propeller or turbine wheels generate a resisting torque which is just equal to the torque generated by the engine, at equilibrium.

In changing the angle of attack of airplane, the pilot produces an unbalanced moment about the lateral axis of the airplane by changing the lift generated by the elevator. The plane responds by accelerating angularly until the external moments are again equal; in a stable airplane, the restoring moments increase with angular displacement from the equilibrium angle and the resulting negative angular acceleration stops the angular motion at a new equilibrium angle. In performing a loop, though, the pilot causes the angular speed to remain relatively constant throughout the maneuver; unbalanced moments cause the angular acceleration that initiates the loop and the angular deceleration that concludes it.

Nature's always successful fight to uphold Eq. 13.6 and the principle of conservation of angular momentum results in some seemingly bizarre behavior. This is especially true when a system that has angular momentum is subjected to a moment that does not simply try to change the magnitude of the angular momentum but instead tries to change the direction of the object's angular momentum. The result is that the axis of rotation is changed in the direction of the applied torque but the

magnitude of the angular momentum may not change, a seemingly unexpected behavior called **precession**. For example, if you mount a wheel on an axle, hold it out in front of you by holding on to one end of the axle, and rotate the wheel rapidly clockwise, it will **not** fall down but will rotate its axis to your left—in a counterclockwise direction as viewed from above. If you are nimble and prepared and keep inside of it, you can continue to support it on just one end while it does complete compass turns. You must provide an upward force equal to the weight of the wheel and axle, so there is no linear acceleration. However the weight of the wheel/axle combination is at some distance from where you are supporting it, and this generates an unbalanced moment directed toward your left. This causes the axis of rotation to move that direction but, because the direction of the moment rotates with the change in axis direction, the precessing motion continues until friction or other outside forces stop it.

Riders of big, heavy motorcycles are commonly under the conceit that they steer their machines by turning the handlebars in the direction they want to go or by leaning over toward the side in the direction they want to turn, like a bicycle. Ha! Fat chance that a 500 + lb hulk cares much what the puny rider on its back is doing with his or her weight. In fact, what happens is that the rider banks the machine into a turn by putting pressure on the handlebars as if to turn the machine in the **opposite** direction; the machine responds by precessing into a bank in the opposite direction which continues until the differential pressure is removed.

A propeller is unable to simply push backward on the air; for a U. S. type, the air swirls around the left side of the fuselage and then over the top of the fuselage. (Try taking off on a muddy field to verify this?) The swirling of the air generates a constant rolling moment that can be nullified by some combination of careful rigging of the flying surfaces, offset of the engine thrust vector, and offset of the vertical stabilizer. The greater problem, though, occurs when a powerful engine turns a heavy propeller and the power is suddenly increased while flying at a slow speed. For example, some folks have discovered to their eternal sorrow that the ailerons (and their ability to push on the rudder) on a P-51 Mustang are not powerful enough at low speeds to counter the torque produced by sudden spurring of the 1600 or so horses under the cowl; the result can and has been an involuntary roll to the inverted when a simple go-around from a landing approach was all that was intended. If the ailerons are not used at all, the airplane will (momentarily at least) gain just as much angular momentum as the propeller, but in the opposite direction, so that the total angular momentum is conserved.

In World War I, the most popular engine was the so-called rotary engine in which the cylinders rotated with the propeller. The total angular momentum of the engine was so great that it was practical to roll the plane in only one direction.

When the propeller is **not** pointing into the oncoming air, as in a climb, the blade passing through the right side has a greater angle of attack than that on the left side, causing the airplane to try to turn to the left and requiring application of right rudder for compensation. (After once getting used to a propeller that turns in the European direction, it is evil fun watching other pilots who "automatically" apply right rudder when they think it will be needed.) This isn't a torque problem, of course, but it does add to the torque problem faced by the pilot who is making a takeoff in a tailwheel-type aircraft. At the beginning of the takeoff roll, the propeller is pointing up and pulls more strongly on the right side; this is resisted by the use of right rudder which at low speeds is mostly effective by virtue of a sideways force generated by the tailwheel to which it is connected. When the tailwheel is lifted off the ground, though, the rudder must suddenly make up for what the tailwheel was doing and a little joggle in the heading commonly occurs. On some light, low power aircraft (such as the Piper J-3 Cub), the rudder is powerful enough that the tailwheel can be forced off the ground almost immediately without control problems; on others, this technique leads to an excursion off the runway (commonly known as a ground loop). If the propeller has a significant amount of angular momentum, another potent contributor to the left-turning tendency appears when the tail is quickly raised, for then precession to the left results from the downward change in the direction of propeller rotation. Here we would be better off with an airscrew that turns in the other direction, as in England.

Helicopter pilots soon learn to deal with the problems of angular momentum conservation. When the main rotor is first started, the ground can provide the exactly equal and opposite torque needed to keep the cockpit from rotating in the opposite direction. When the helicopter lifts off, though, the torque provided by the ground better be exactly zero or the cockpit becomes a merry-go-round. Even after it has reached the proper speed, torque must be provided to the main rotor shaft by the engine to counter the air forces resisting the rotor; an opposite torque is generated (traditionally) by use of a tail rotor rotating in a vertical plane or, in some modern helicopters, with aerodynamic forces on one side of the fuselage. In any case, the pilot has to adjust the opposing torque every time the main rotor's angular momentum is changed, which is often. An interesting solution to the problem is to turn the rotor with jet exhausts from the tips; then the exhaust carries away just as much angular momentum as the rotor gains (or from a moment standpoint, the shaft doesn't have to provide any torque) and so no torque compensation is required. This is the reason at least one attempt at a human-powered helicopter has used propellers at the rotor tips.

The concepts of energy and of energy conservation unite the otherwise disparate disciplines of physics, engineering, chemistry, biology, and botany. A well-grounded understanding of the **forms** of energy and the ways in which energy is **transformed** from one form to another and **transferred** from one object to another permits the user to draw important conclusions in many flight situations, conclusions often inaccessible to those who restrict themselves to the usual form of the Newtonian laws and to momentum principles. (When we use the energy tool, though, we do sacrifice the possibility of following motion in time because energy, unlike vector properties, depends only on position or location and not on direction.)

Energy was little more than the "force of life" for romantic poets before scientists became convinced that there was something in nature that stayed constant even through apparently violent interactions, and they gave the name "energy" to this something. The breakthrough came when it was realized that heat was a form of energy, and indeed that large-scale mechanical energy could disappear and become completely transformed into heat energy. The principle of the conservation (or constancy) of energy can be stated as follows:

The total energy of the universe (or of an isolated part) does NOT change with time.

or

Energy can be transformed from one form into a different form or it can be transferred from one object to another.

but

The SUM TOTAL of Energy Does NOT Change.

PRINCIPLE OF THE CONSERVATION OF ENERGY

The forms that energy can assume include (**a**) energy of motion which is called kinetic energy, (**b**) gravitational potential energy, (**c**) elastic potential energy (as is momentarily stored in a spring steel landing gear during a hard touchdown), (**d**) chemical energy, and (**e**) heat energy. (Other forms include electrical energy and mass energy, but we don't need them here.)

We say that **work** is done when, and only when, energy is transferred from one system (an object or a group of objects) to another system. Work is done through the agency of a force exerted by one system on another system, but that force must actually accomplish an energy transfer—it must act during the motion of that system in the plus or minus direction of the force—or no work is done. The amount of work done by a force can be calculated from

Work Done
= (Component of Force **in** the Direction of Motion)
×(Distance Moved)

or, in symbols,

$$W = F_D\, d \qquad (13.8)$$

WORK DONE AND ENERGY TRANSFERRED BY A FORCE

Thus the upward force of the ground on the tires of an aircraft do **not** do work on the tires, either when it is stationary or when it is taxiing on level ground, because there is no motion in the direction of the force. However, the force of friction on a sliding tire when the brakes are locked is a force that acts in a direction parallel and opposite to the direction of motion; this frictional force does do work on the tires, and it does transfer and transform motional energy of the aircraft into heat energy in the tires. Or, if the tires are not locked, most of the energy transfer is from energy of motion into heat energy in the brake system by virtue of the internal work done there by frictional forces. For strictly **translational** motion (no rotation), the energy of motion, kinetic energy, can be calculated from:

$$KE = \frac{1}{2} m V_{CG}^2 \qquad (13.9)$$

KINETIC ENERGY, LINEAR MOTION

Note that kinetic energy is a **scalar** property of an object, depending only on the magnitude of the speed and not on its direction.

Gravitational potential energy is energy that is "stored" when an object moves farther away from a source of gravitational attraction such as the earth. It can be thought of as work done "against" gravity because it requires an external force in a direction opposite from that of the gravitational force. This stored potential energy is normally converted into additional kinetic energy when an aircraft dives to a lower altitude; it is the function of the pilot and of the wings to ensure that the energy of motion near the ground is a result of motion parallel to the ground and not due to motion perpendicular to the ground, but in either case energy will not be lost. The "zoom climb," on the other hand, is a conversion of kinetic energy into additional potential energy.

The gravitational potential energy possessed by an aircraft is appropriately called potential energy because it has the potential for being converted into work done by the aircraft on the air or into work done by the aircraft on the ground. Normally, the lowest possible altitude outward from the earth's center is taken as having **zero** potential energy and then the potential energy is always a positive quantity and can be calculated for any instant of time from

$$PE = mgh \qquad (13.10)$$

GRAVITATIONAL POTENTIAL ENERGY

where h is the **vertical** distance or altitude above that lowest point and g is the acceleration of a freely-falling object (that is, one for which air drag is negligible). It is usually possible to safely neglect the actual variations of g with location on the earth and with altitude and use its average value at sea level. (See the discussion in Sec. B of Chap. 2.)

As you will have guessed, work can be done by external moments as well by external forces. The work done by a moment is just the magnitude of the moment times the angular displacement, θ, accomplished by the moment:

$$W = M\theta \qquad (13.11)$$

WORK DONE AND ENERGY TRANSFERRED BY A MOMENT

An example of rotational work would be the work done by a pilot in moving a control stick through a given angle. (The same value will be calculated if either this equation or the previous one for work done by linear forces is used but it may be easier to use this equation because the displacement may follow a curved path.) Another example is the work required to lift the nosewheel off the ground by pushing on the rear fuselage

(Prob. 3.5). And friction in wheel bearings produces a torque that resists rolling.

The angular kinetic energy is again one-half the inertial resistance multiplied by the square of the speed:

$$KE = \frac{1}{2} I \omega^2 \qquad (13.12)$$

KINETIC ENERGY, ANGULAR MOTION

A helicopter with enough rotational kinetic energy can make a safe landing even in the event of an engine failure because a skillful pilot can transfer the kinetic energy stored in the rotor into work done on the air to stop the descent just before touchdown. (This is known as an **autorotation** to a flare and landing.)

Airplanes provide a rich environment for exploring and using the energy viewpoint. In level cruising flight, the drag force is doing work on the aircraft to remove energy from it but the air is doing an equal amount of work in pushing forward on the propeller blades or turbine wheel, keeping the airplane's kinetic energy constant. In a steady climb the kinetic energy is constant but the potential energy is increasing and this stored energy can be converted into additional kinetic energy later. In a "zoom" climb such as at the beginning of a loop, kinetic energy is intentionally traded for additional potential energy – but all too soon one finds that what one no longer has can't be traded. In a normal landing any vertical speed represents kinetic energy that **is** going to be transformed, either into potential energy in a spring-type landing gear and immediately returned as kinetic energy or into heat energy in the tires and shock absorbers. The horizontal speed on landing represents kinetic energy, too, and it must be completely converted into heat energy before the aircraft is stopped; it can go into random kinetic energy of the air (atmospheric heat) through work done by aerodynamic drag or reverse thrust, or it can go into heat in the brakes and tires. Within the aircraft, chemical energy (i.e., electrical energy in the atoms) is converted into rotational kinetic energy. There are no energy credit cards – all of an airplane's energy must be earned first. Only potential energy can be stored for later use.

The last of the mechanics concepts that we need to review in preparation for the study of aircraft performance and maneuvering is **power**. It is one thing to reach a high speed or a high altitude; it is quite another thing to do it quickly. Quickness requires power; much quickness requires much power. Power, then, is defined as the rate at which work is being done, i.e., the rate at which energy is being transferred. It can be calculated from:

$$\text{Power} = \frac{\text{Work Done}}{\text{Time Required to Do the Work}}$$

or, in symbols,

$$P = \frac{W}{t} \qquad (13.13)$$

DEFINITION OF POWER

If the work is being done at a constant rate, the power is also constant in time; if the rate varies, the above equation will yield the **average** power for the time period involved. Because work is done by a force acting through a distance, the rate at which work is done is

$$\text{Power} = \text{Force}_D \times \frac{\text{Distance Travelled}}{\text{Time Required}}$$

$$= (\text{Force Component}) \times (\text{Speed}) \quad \text{or}$$

$$P = F_D V \quad (13.14)$$

ALTERNATE DEFINITION OF POWER, LINEAR MOTION

where F_D is the force component in the direction of motion. This result, that the power developed is the product of the component of the force that is in the direction of the motion and the speed of that motion, is a **fundamental** relationship for aircraft performance studies. It will shortly allow us to calculate the magnitude of the total drag force on any aircraft for which we know the power used and the true airspeed.

The power provided by a moment is the product of the moment and the angular speed:

$$P = M\omega \quad (13.15)$$

POWER REQUIRED FOR ANGULAR MOTION

This also is a key relationship for airplanes because our engines do produce rotation and we do need to be able to monitor the power being developed. In a simple aircraft with a fixed-pitch propeller, we get used to setting the power simply by looking at the tachometer (which reports the angular speed in rev/min). But, even with a fixed-pitch propeller, the resisting moment provided by the air (and therefore the moment provided by the engine when the speed is stable) depends on the density altitude and so we have to consult a power chart to see how much power we're actually developing at a given angular propeller speed. The charts are useless, though, if the propeller we're using isn't just the same as the one used in developing the charts. Then we require another gauge telling us the torque developed by the engine as well.

It isn't easy to measure engine torque while the engine is in use but fortunately the pressure in the intake manifold is a measure of this torque. When the engine is not running the pressure is the absolute (actual) value of the outside pressure; as the throttle is advanced, the manifold pressure becomes smaller and smaller because of the pumping action of the engine. For an airplane with a controllable or constant-speed propeller, the power must always be set by reference to both of these gauges.

No calculations are meaningful unless proper attention has been paid to units. We propose to stay in the familiar British or engineering system: force in **pounds**, mass in **slugs**, moments of inertia in **slug-feet**^{2}, distances in feet, speeds in **ft/s**, linear acceleration in **ft/s**^{2}, angular speed in rad/s, angular acceleration in **rad/s**^{2}, moments or torques in ft-lb, momentum in **slug-ft/s**, energy and work in ft-lb, and power in **ft-lb/s**. The

equations that have been presented in this section can be used **only** if a consistent set of units, such as these, is used.

You will commonly encounter speed given in knots, angular speed given in RPM (revolutions per minutes, as for propellers), and power given in horsepower. Convert them to the proper units with the conversion factors in Table 13.1.

1 kt = $\dfrac{1 \text{ nautical mile}}{1 \text{ hour}} = \dfrac{1.151 \text{ statute mile}}{1 \text{ hour}} = \textbf{1.689 ft/s}$	
1 RPM = $\dfrac{1 \text{ rev}}{1 \text{ min}} = \left(\dfrac{2\pi \text{ radians}}{\text{minute}}\right)\left(\dfrac{1 \text{ min}}{60 \text{ sec}}\right) = \textbf{0.1047 rad/s}$	
1 horsepower = **1 hp** = $\dfrac{550 \text{ foot-pounds}}{1 \text{ sec}} = \textbf{550 ft-lb/s}$	

Table 13.1 Conversion Factors for Linear/Angular Speeds and Power

Refer to Table 13.2 for a summary of the complementary properties of linear and angular motion.

PROPERTY	LINEAR	ANGULAR
DISPLACEMENT	s in ft	θ in radians
SPEED	v in ft/s	ω in rad/s
ACCELERATION	a in ft/s^2	α in rad/s^2
INERTIA	m in slugs (sl)	I in sl ft^2
2nd LAW	$\sum F_{\text{EXT}} = ma_{\text{CG}}$	$\sum M_{\text{EXT,CG}} = I_{\text{CG}}\,\alpha_{\text{CG}}$
MOMENTUM	$p = mv$	$L = I\omega$
IMPULSE	$\Delta p = F\,\Delta t$	$\Delta L = M\,\Delta t$
WORK	$W = Fd$	$W = M\theta$
KINETIC ENERGY	$KE = \frac{1}{2}mv^2$	$KE = \frac{1}{2}I\omega^2$
POWER	$P = Fv$	$P = M\omega$

Table 13.2 Analogous Properties, Linear and Angular Motion

How shall we contrast the place of the lift force and power in allowing airplanes to do their thing? A first attempt might be to suggest that lift is what holds an aircraft in equilibrium perpendicular to its flight path (i.e., in level flight or in a constant-speed climb or descent) but power is what is required to keep it there! If power is reduced, it is often possible to compensate and maintain lift and equilibrium by increasing the angle of attack, but there must always be some power expenditure from the engine to balance that lost to work done by the drag force, or else it is not an equilibrium situation.

B. Thrust from Propellers and from Jet Engines

As the Wright brothers realized, a propeller may be thought of as a rotating airfoil. If the resultant force vector due to the air pressure distribution on this rotating airfoil contains a forward component, then the propeller is said to be producing **thrust.** (The thrust force can also be considered as the forward reaction force to the rearward force of the propeller on the air, by the third law.) Figure 13.2 depicts a propeller on an aircraft that is in level flight. The section view shows a section of the propeller which has a pitch angle β; it also shows a simplified diagram of the vector velocities involved.

$V_{SC} \equiv$ velocity of the propeller section
relative to the **center** of the prop

$V_{CA} \equiv$ velocity of the center of the propeller
relative to the oncoming air
$=$ the true airspeed of the aircraft

$V_{SA} \equiv$ velocity of the propeller section
relative to the oncoming air

Evidently the angle α in Fig. 13.2 is the angle of attack for this propeller blade section if we define the angle of attack in the same way as we have done for a wing section, since V_{SA} is

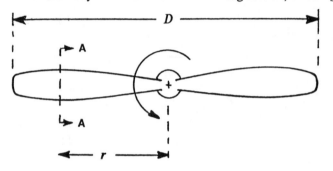

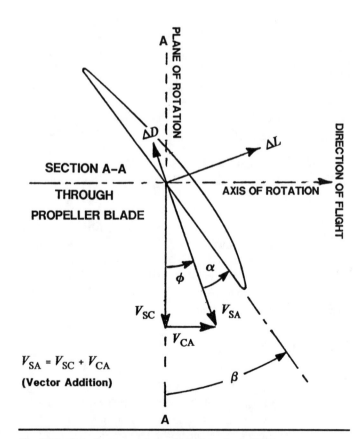

$V_{SA} = V_{SC} + V_{CA}$
(Vector Addition)

Fig. 13.2 Propeller geometry and blade angles in flight

the velocity vector for the oncoming air relative to the blade section. The element of lift, ΔL, produced by the blade section shown has a component in the direction of flight and so it is producing thrust. The elements of lift, ΔL, and drag, ΔD, also have components in the rotational plane of the propeller; these produce a moment that opposes the rotation.

(The vector diagram of Fig. 13.2 is simplified in the same sense that the vector velocities for an airfoil section in two-dimensional flow conditions are not the same as for a wing of finite length in three-dimensional flow conditions. That is, the air that has passed the actual wing or actual propeller must possess an altered direction of motion and this alteration can be thought of as adding an induced increment, α_I, to the two-dimensional angle of attack, α. The result is an effective tilt of ΔL toward the axis of rotation and an effective tilt of ΔD toward the plane of rotation, with the amount of the tilt depending on the amount of lift produced. These added complexities are not significant for our present needs, fortunately.)

If we are to characterize the effectiveness of propellers in general, we need to find a non-dimensional parameter that determines the air flow conditions for any propeller. Normally a propeller blade is twisted so that the pitch angle β is a function of distance from the axis; this is done so that each section may have an optimum angle of attack, since the angle of attack depends on V_{SC} which in turn depends on the distance r from the axis of rotation. The vector diagram of Fig. 13.2 suggests that we might choose to use the angle ϕ as our characteristic parameter; knowledge of ϕ and β will then uniquely specify the angle of attack for any part of a propeller under any forward flight condition.

V_{SC} is the linear speed of any propeller blade section relative to the hub of the propeller. Because angular properties are converted to linear quantities by multiplying by the radius,

$V_{SC} =$ (angular speed in radians/s)
$\times$ (distance from the axis of rotation)

or $\qquad V_{SC} = \omega r = 2\pi n r \qquad\qquad (13.16)$

where n is the rotational speed of the propeller in revolutions per second (because 1 revolution $= 2\pi$ radians). For the propeller **tip**, this equation yields, since $r_{TIP} = \frac{1}{2}D$,

$$V_{SC[TIP]} = \frac{2\pi n D}{2} = \pi n D \qquad (13.17)$$

where D is the propeller diameter. Then, from Fig. 13.2,

$$\tan\phi = \frac{V_{CA}}{\pi n D} = \frac{V_\infty}{\pi n D} \qquad (13.18)$$

Therefore the angle ϕ is determined if we specify the ratio $\left(V_\infty / nD\right)$. This quantity is known as the advance ratio for the propeller/airplane combination.

$$J \equiv \text{advance ratio} = \frac{V_\infty}{n D} \qquad (13.19)$$

ADVANCE RATIO

where V_∞ is the true airspeed of the aircraft. The length and the time units used to calculate J must be consistent; for example, V in ft/s, n in rev/s, and D in feet can be used. If instead the aircraft true airspeed is given in knots, the propeller rotational speed in revolutions/minute (RPM), and the propeller diameter in inches, as is commonly encountered, then it is more convenient to calculate the advance ratio from

$$J = \frac{1215 \times V_\infty (\text{in knots})}{n (\text{in RPM}) D (\text{in inches})} \qquad (13.20)$$

where 1215 is the constant that corrects for the inconsistent set of units being used.

We now define the effectiveness or **efficiency** of a propeller as the ratio of the energy actually transferred to the aircraft to the energy given to the propeller by the powerplant. The rate of energy transfer to the aircraft is, using Eq. 13.14,

Delivered Power = (Propeller Thrust) (True Airspeed)

or, in symbols, $P_{\text{DEL}} = T V_\infty$ (13.21)

Therefore the propeller efficiency, symbolized by the Greek letter eta (η), is

$$\eta \equiv \text{propeller efficiency} = \frac{P_{\text{DEL}}}{P_{\text{IN}}} = \frac{T V_\infty}{P_{\text{IN}}} \qquad (13.22)$$

PROPELLER EFFICIENCY

Note from inspection of this equation that the efficiency is **zero** when the aircraft is not moving relative to the air ($V_\infty = 0$) because no energy is being given to the aircraft. This is true even though the static thrust is usually greater than the in-flight thrust. The difference $[P_{\text{IN}} - T V_\infty]$ is the rate at which the energy delivered to the propeller is being wasted in rotational and translational kinetic energy of the air molecules. Figure 13.3 presents typical variations of propeller efficiency with the advance ratio. Note that a fixed-pitch propeller is normally biased toward low speed (climb) performance, at the expense of cruise efficiency. (A fixed-pitch propeller is a propeller whose pitch cannot be changed even though the pitch normally varies with location along the propeller blade. Some old propellers are ground-adjustable only while others can have their pitch directly changed by the pilot while in flight. However the most common type of variable-pitch propeller is the constant-speed propeller; with this type of propeller, hydraulic pressure is used to change the pitch continuously so as to maintain propeller rotational speed at the value set by the pilot, independent of the throttle setting, to the extent the power setting and the available pitch range permit.)

Note from Fig. 13.2 that the angle of attack, α, for the blade element is given by

$$\alpha = \beta - \phi \qquad (13.23)$$

so that the angle of attack decreases as the aircraft speed increases (because ϕ increases as V_∞ increases, from Eq. 13.18). Thus every propeller has a forward speed at which its blade elements generate zero lift and the maximum cruise speed must surely be less than this! Figure 13.3 shows this

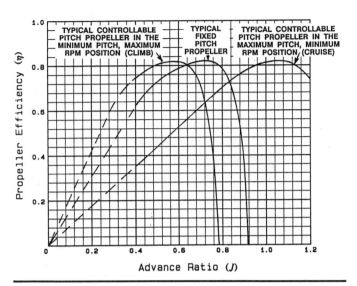

Fig. 13.3 Typical propeller efficiency curves

speed, in terms of J, as the value for which η falls abruptly to zero.

On the other hand, below the efficiency peak the thrust decreases as the blade angle of attack approaches the stall angle. Normally a propeller is used at rotational speeds such that the tip speeds are not supersonic; excessive tip speeds produce both noise and lowered efficiency because of shock wave formation (wave drag). But this tells us that a design tradeoff must be made since engines tend to deliver their greatest horsepower/weight ratio at a high crankshaft speed. High revving motorcycle or automobile engines that are converted for aircraft use usually use relatively inefficient, small diameter propellers that are partially blanked by the fuselage; also, small propellers always tend to produce less static thrust than larger propellers, for a given power input. The old radial engines, developing their maximum power at around 2000 RPM, and the modern turboprop engines, which have to be geared down to drive the propeller, can and do use very large and efficient propellers.

A well-designed propeller can have a peak efficiency of about 0.85 but a value of 0.80 is normally considered to be a conservative first estimate for most propellers in the cruise configuration.

EXAMPLE 13.2. The following are specifications and performance properties of the 1977 turbocharged Cessna 210 Centurion aircraft, as taken from the pilot's operating handbook:

Rated power: 285 brake horsepower at 2600 RPM
Propeller: Diameter of 80 inches
Low pitch setting of 12.4° and a high pitch setting
  of 28.5° at 30 inches from center
Performance at 18,000 feet in the StAt:
  Power: 62% of rated horsepower
  Propeller speed: 2200 RPM
  True Airspeed: 172 kt

(a) Determine the advance ratio for the specified flight condition. From Eq. 13.20,

$$J = (1215 \times 172) / (2200 \times 80) = \mathbf{1.19}$$

(b) What propeller efficiency does the maximum pitch curve of Fig. 13.3 suggest for this advance ratio? From the figure, $\eta \cong \mathbf{0.75}$

(c) Using this estimate of propeller efficiency, estimate the thrust of the propeller for the specified flight condition.

Solution: With the help of Table 13.1,

$$P_{IN} = (0.62)(285 \text{ hp})(550 \text{ ft-lb/s/hp}) = 9.72 \times 10^4 \text{ ft-lb/s}$$

Using Table 13.1 again,

$$V_\infty = 172 \text{ kt} \times (1.689 \text{ ft/s/ kt}) = 290 \text{ ft/s}.$$

From Eq. 13.22, $\quad T = \dfrac{\eta\, P_{IN}}{V_\infty} = \dfrac{(0.75)(9.72 \times 10^4 \text{ ft-lb/s})}{290 \text{ ft/s}}$

$$= \mathbf{251\ lb}$$

(d) Estimate the angle of attack for the blade section (neglecting the induced angle correction) at a point 30 inches from the center of the propeller, assuming the maximum pitch setting of 28.5° is being used.

From Eq. 13.16,

$$V_{SC} = 2\pi n r$$

$$= (2)(\pi)[\ (2200 \text{ rev/min})$$
$$\times (1 \text{ min/60 s})][\ (30 \text{ in}) \times (1 \text{ ft/12 in})]$$

$$= 576 \text{ ft/s} = \mathbf{341\ kt}$$

From Eq. 13.18,

$$\tan(\phi) = \frac{V_\infty}{V_{SC}} = \frac{172 \text{ kt}}{341 \text{ kt}} = 0.504 \qquad \text{so} \qquad \phi = \mathbf{26.8°}$$

Therefore, from Fig. 13.2,

$$\alpha = \beta - \phi = 28.5° - 26.8° = \mathbf{1.7°}$$

(e) Determine the propeller tip speed relative to the on-coming air and the Mach number for the tips.

From Eq. 13.17 and using Table 13.1,

$$V_{SC(tip)} = \pi n D$$

$$= (\pi)[\ (2200 \text{ rev/min}) \times (1 \text{ min/60 s})]$$
$$\times [(80 \text{ in}) \times (1 \text{ ft/12 in})] (1 \text{ kt/1.689 ft/s})$$

$$= \mathbf{455\ kt}$$

From Fig. 13.2,

$$V_{SA} = \sqrt{V_{SC}^2 + V_{CA}^2} = \sqrt{(455 \text{ kt})^2 + (172 \text{ kt})^2} = \mathbf{486\ kt}$$

From Appendix B, the speed of sound at 18,000 feet in the StAt is 619 kt. Therefore

$$M_\infty = (486 \text{ kt})/(619 \text{ kt}) = \mathbf{0.785}$$

There are two problems with trying to use propeller efficiency curves like those in Fig. 13.3. The first is that they are not generally available for the propellers actually in use and the second is that the efficiency theoretically depends on other variables in addition to J.

A general expression for the theoretical propeller efficiency can be derived by idealizing the air flow through the propeller and using momentum principles presented in the previous section. Figure 13.4 depicts this idealized flow. It is assumed that the flow speed well in front of the propeller is just the flight speed V_∞, that the air flow speed through the propeller is V_1, and the flow well behind the propeller is V_S (the **slipstream** velocity). It is also assumed that the pressure well in front of the propeller is the static pressure, p_∞, that the pressure decreases to p_1 immediately in front of the propeller, undergoes an abrupt increase to p_2 immediately behind the propeller, and finally decreases to the static pressure well behind the propeller in the slipstream. All of the air flowing through the propeller disc is assumed to be contained within a streamtube that narrows down as it approaches and passes through the propeller; the rotation of the air in the streamtube is neglected.

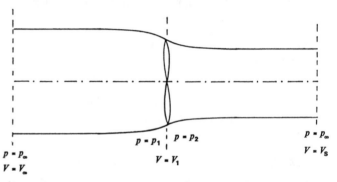

Fig. 13.4 Idealized air flow through a propeller

Then the thrust force generated by the propeller is

$T =$ thrust force

$\quad =$ (change in pressure at the propeller disc)
$$\times (\text{disc area of the propeller})$$

$$= \left(p_2 - p_1\right)\left(\tfrac{1}{4}\pi D^2\right) \tag{13.24}$$

from the definition of pressure, where D is the propeller diameter. But from Eqs. 13.4 and 13.5 we also have

$$T = \frac{\Delta p}{\Delta t} = \frac{\Delta m\,\Delta V}{\Delta t} = \frac{\Delta m}{\Delta t}\left(V_S - V_\infty\right) \tag{13.25}$$

Bernoulli's equation can be used to relate the speeds and pressures in front of the propeller and, separately, the speeds and pressures behind the propeller. Combining these expressions with the streamtube assumption yields a theoretical value for the airspeed through the prop, V_1. (See references 2 through 4 for the details of the derivation.)

The power at the propeller shaft is the thrust force times the actual airspeed through the propeller disc, so the theoretical propeller efficiency is

$$\eta = \frac{P_{DEL}}{P_{IN}} = \frac{T V_\infty}{T V_1} = \frac{V_\infty}{V_1} \tag{13.26}$$

THEORETICAL PROPELLER EFFICIENCY

The expression for η that results from this momentum theory is a little tedious to use on a calculator but easy enough on a computer, and it will be used in succeeding pages to estimate the thrust available from an engine. The steps in the calculation are these:

(1) Calculate $V_P = \left(\dfrac{P}{\rho D_P^2}\right)^{1/3}$ (13.27)

where P is the power the engine delivers to the propeller shaft, ρ is the air density, and D_P is the diameter of the propeller. V_P has the dimensions or units of a speed and can be considered to be a characteristic propeller speed. (Obtain the cube root on a scientific calculator with the y^x function key, where x is $1/3$.)

(2) Calculate the ratio of the flight speed to V_P:

$V_R = \dfrac{V_\infty}{V_P}$ (13.28)

(3) Calculate the quantity X, where

$X = \sqrt{1 + \dfrac{2\pi}{27} V_R^3}$ (13.29)

(4) Calculate the theoretical propeller efficiency η_{THEORY} from

$\eta_{\text{THEORY}} = \left(\dfrac{\pi}{4}\right)^{1/3} V_R \left[\left(1 + X\right)^{1/3} - \left(-1 + X\right)^{1/3}\right]$ (13.30)

THEORETICAL PROPELLER EFFICIENCY

Figure 13.5 presents a graphical short-cut for steps (3) and (4) for those using non-programmable calculators.

The propeller efficiency calculated from this theory is expected to be optimistic because (a) the drag losses on the blades have been ignored, (b) the blades operate in air disturbed by the other blades, (c) the rotational energy imparted

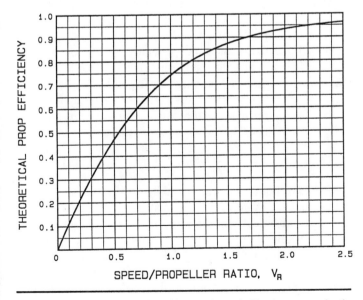

Fig. 13.5 Theoretical propeller efficiency from airplane/prop speed ratio

to the air has been ignored, and (d) the effect of the engine nacelle in reducing the pressure difference over the propeller disc has been ignored. Realistically, we can expect the actual propeller efficiency in **cruise** to be just about **85%** of this theoretical efficiency:

$\eta_{\text{CRUISE}} \approx 0.85\, \eta_{\text{THEORY}}$ (13.31)

If the propeller isn't well matched to the airplane's speed, or if we are diving it to a speed well beyond the cruising speed, or if we are well below the cruising speed (as in climbing flight but especially in the takeoff roll), Fig. 13.3 makes it clear (based on blade element theory) that the propeller efficiency in these off-design conditions will be significantly less than even Eq. 13.31 predicts.

EXAMPLE 13.3. Calculate the theoretical and estimated cruise propeller efficiency for a Piper Cadet at a density altitude of 2000 ft, assuming that it is using 75% of its rated power of 160 hp. The propeller diameter is 73 inches and the resulting true airspeed should be 109 kt.

Solution: (a) First, convert the data to a consistent set of units, using Table 13.1:

$\rho = 0.002241$ sl/ft^3 (from StAt tables)

$P = (0.75)(160\text{ hp}) = 120\text{ hp} = 66,000$ ft-lb/s

$D_P = 73" = 6.08$ ft

$V_\infty = 109$ kt $= 184$ ft/s

(b) Step (1):

$V_P = \left(\dfrac{6.60 \times 10^4}{(0.002241)(6.08)^2}\right)^{1/3} = \left(7.967 \times 10^5\right)^{1/3} = 92.7$ ft/s

(c) Step (2):

$V_R = \dfrac{184\text{ ft/s}}{92.7\text{ ft/s}} = 1.985$

(d) Step (3):

$X = \sqrt{1 + (2\pi/27)(1.985)^3} = \sqrt{1 + 1.820} = 1.679$

(e) Step (4):

$\eta_{\text{THEORY}} = \left(\dfrac{\pi}{4}\right)^{1/3} V_R \left[\left(1 + X\right)^{1/3} - \left(-1 + X\right)^{1/3}\right]$

$= \left(\dfrac{\pi}{4}\right)^{1/3} (1.985)\left[\left(1 + 1.679\right)^{1/3} - \left(-1 + 1.679\right)^{1/3}\right]$

$= (0.923)(1.985)(1.389 - 0.879) = \mathbf{0.93}$

(f) Estimated actual cruise propeller efficiency (Eq. 13.31)

$\eta_{\text{CRUISE}} = 0.85\, \eta_{\text{THEORY}} = \mathbf{0.79}$

The propeller momentum theory also provides us with an estimate for the static thrust which is useful in calculating takeoff performance:

$T_{\text{STATIC, THEORY}} = \left(0.5\, \pi\, \rho\, D_P^2\, P^2\right)^{1/3}$ (13.32)

THEORETICAL PROPELLER STATIC THRUST

but this can be very optimistic, especially for a fixed-pitch propeller in which much of the blade may be stalled. The author has observed one example of where a certain propeller produced a lower static rpm than another but produced **more** rpm and better climb performance in the air, presumably because of stalled blade elements at low speeds.

The predicted dependence of the static thrust on propeller diameter should not be overlooked. It is more efficient to accelerate a large mass of air a small amount than to accelerate a smaller mass of air a greater amount. The big, slow-turning propellers on 1930s aircraft did give impressive takeoff and climb performance for the power available; our newer engines require smaller propellers because they develop their power at a higher rpm (for direct-drive installations) and because they are much closer to the ground to minimize weight and aerodynamic drag.

The same momentum theory predicts the in-flight thrust to be

$$T_{\text{THEORY}} = \left(0.5\right)^{1/3}\left[\left(1+X\right)^{1/3} - \left(-1+X\right)^{1/3}\right]T_{\substack{\text{STATIC}\\\text{THEORY}}} \quad (13.33)$$

THEORETICAL CRUISE THRUST FROM A PROPELLER

EXAMPLE 13.4. Estimate the static and cruise thrust for the Piper Cadet of Example 13.3, assuming that the actual thrust is 85% of that predicted by propeller momentum theory.

(a) $T_{\text{STATIC, THEORY}} = \left[(0.5)\,\pi\,(0.002241)\,(6.08)^2\,(66000)^2\right]^{1/3}$

$= \left(5.67 \times 10^8\right)^{1/3} = \textbf{828 lb}$

so $T_{\text{EST, STATIC}} = (0.85)(828\text{ lb}) = \textbf{704 lb}$

(b) $T_{\text{THEORY}} = \left(0.5\right)^{1/3}\left[\left(1+1.679\right)^{1/3} - \left(-1+1.679\right)^{1/3}\right]$

$\times T_{\text{STATIC}} = (0.794)(1.389 - 0.879)(828\text{ lb}) = \textbf{335 lb}$

so $T_{\text{EST, CRUISE}} = (0.85)(335\text{ lb}) = \textbf{285 lb}.$

In contrast to reciprocating engines, jet engines are rated for the thrust they develop rather than for power, and this thrust has a relatively weak dependence on aircraft speed. The static thrust of a jet engine depends on the effective disc of air that the engine intercepts; thus the same theory that tells us that a **large** propeller has greater **static** thrust than a small propeller also tells us that a turbofan engine (with its large ducted fan outside the core engine) has greater static thrust than the turbojet.

On the other hand, the turbojet remains effective to higher speeds than the turbofan engine and so predominates on very high speed jets. Current turboprops suffer a considerable loss in thrust as the airspeed increases and are not now used at speeds greater than about 400 kt. But with the new propellers developed by NASA and engine manufacturers, turboprops are considerably more fuel efficient at speeds up to about Mach 0.8, so they will begin to appear when their fuel economy more than offsets the extra cost required to develop, certify, and manufacturer the first new generation.

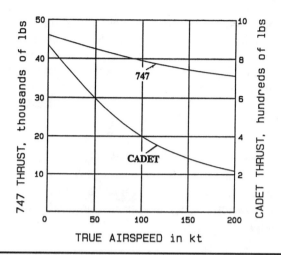

Fig. 13.6 Thrust from a jet engine compared with thrust from a propeller

Figure 13.6 suggests how different is the thrust developed by propellers and jet engines, and why jet engines propel the fastest airplanes. The thrust developed by the four turbofan engines on the Boeing 747 varies only about 25% over the speed range from 0 to 200 kt (based on an equation given in Ref. 3) but the thrust from the 180 hp Lycoming engine on the Piper Cadet decreases by about 75% over the same speed range. The Lycoming thrust was estimated by assuming that the thrust is 85% of that calculated from Eqs. 13.32 and 13.33. Of course, the Cadet isn't supposed to exceed 153 kt at SL, but you get the idea.

C. The Lift and Drag Characteristics of a Complete Aircraft

In theory, the in-flight performance of an aircraft can be completely predicted if the engine thrust and the aircraft's mass distribution and lift and drag characteristics are known. So our next step is to model the interdependence of lift and drag forces for the **complete** aircraft. In Sec. C of Chap. 9 we saw that airfoils (especially turbulent airfoils) show a nearly parabolic (proportional to the square) dependence of drag coefficient on lift coefficient, centered about the design lift coefficient. Then in the last chapter we found that a wing adds an induced drag term which is also proportional to the square of the lift coefficient. Therefore we wrote the drag coefficient for a **wing** (Eq. 12.9) as

$$C_D = c_d + \frac{C_L^2}{\pi\,e\,\text{AR}} \quad (13.34)$$

where c_d is the two-dimensional drag coefficient for the airfoil in use and varies with the angle of attack.

Now we wish to write an expression for the drag coefficient for the whole aircraft and it is certainly not unreasonable to hope that the contributions of the fuselage and tail (and other appendages such as the landing gear) will simply increase the **parasite drag** (the first term) and the lift-dependent or **induced**

drag (the second term) without adding any new dependencies. And so we write

$$C_D = C_{D,min} + \frac{C_L^2}{\pi e \, AR} \qquad (13.35)$$

where we now assume that $C_{D,min}$ is a **constant**, independent of C_L. We can expect the efficiency factor e for the complete aircraft to be a little smaller (worse) than for the wing alone because of wing-fuselage interference and other surface effects. It is also found experimentally that e tends to be smaller for low wing aircraft than for high wing because the critical upper surface of the wing is much more closely coupled to the fuselage boundary layer for the former.

A distinctly troubling feature of Eq. 13.35 (with a constant value for $C_{D,min}$) is that it implies equal drag for equal positive or negative lift. How can this be? After all, one good reason for using a cambered airfoil is to move the minimum in the drag coefficient from zero lift (where we can't use it) to the (positive) cruise lift coefficient. It would seem much more reasonable to expect the drag coefficient for the whole airplane to vary with lift in the following fashion:

$$C_D = C_{D,min} + \frac{(C_L - C_{Li})^2}{\pi e \, AR} \qquad (13.36)$$

where C_{Li} would then be defined as the design lift coefficient for the complete airplane rather than just for the wing. Indeed, this equation **is** sometimes used and is especially appropriate when powerful camber-enhancing devices such as slotted flaps are being used, as in landing and takeoff. With readily available computers to do the grunt work, the additional complexity of Eq. 13.36 over Eq. 13.35 is not convincing as a deterrent. In fact, a computer makes it easy to add additional terms (i.e., terms proportional to the third and higher powers of C_L) if necessary to obtain a good fit to experimental data.

And yet Eq. 13.35, the simple polar approximation, remains popular. It often fits experimental data rather well, especially within the lift coefficients used in normal cruising flight. However it must remain suspect at the minimum C_L values (maximum speeds at low density altitudes, where cambering should be most profitable) and especially at large C_L values. Certainly both Eq. 13.35 and Eq. 13.36 are inaccurate and unusable for angles of attack beyond the stalling angle because we know that the drag coefficient increases rapidly at that point, even as the lift coefficient is decreasing rapidly.

However, in the work ahead we will use Eq. 13.35 for cruise and climb calculations while trying to keep in mind its limitations. In the next chapter we will check out some experimental data (including types of data which you can easily acquire in the aircraft of your choice) to see how well the equation does its job.

In proceeding from an airplane drag coefficient to an airplane drag **force**, we assume the same equation as for the wing by itself:

$$D = C_D q S = \left(C_{D,min} + \frac{C_L^2}{\pi e \, AR} \right)(qS) \qquad (13.37)$$

and so

$$D = C_{D,min} q S + \frac{C_L^2 q S}{\pi e \, AR} \qquad (13.38)$$

ASSUMED DRAG FORCE ON THE COMPLETE AIRCRAFT

where q is the flight dynamic pressure and S is the projected **wing** area of the aircraft, as before. It turns out to be convenient to continue to use the wing as the reference area for both the lift and the drag of the complete aircraft, but we'll get into a lot of trouble if we really believe that the wings are responsible for all the parasite drag (the first term in Eq. 11.38)! The drag produced by the fuselage, by the landing gear, and by the tail surfaces certainly should be referenced to their respective surface areas (and this is done when performance projections are being made for a new design). But it is simpler mathematically to use just the wing area for both the lift and drag forces, thereby allowing the drag coefficient to give the correct total drag force without identifying its source. Also, the parasite drag force includes contributions from interference drag (i.e, flow disturbances caused by fuselage-flying surface intersections) and cooling drag, both of which possess no simple or obvious reference area associated with them.

The parasite drag of the complete airplane is often given in terms of the area of a surface having a drag coefficient of **1.0** and having the same drag. This area is known as the "**equivalent flat plate area**" and is given the symbol f (even though a flat plate oriented perpendicular to the airstream has a drag coefficient closer to **1.28!**). This definition of f allows us to write:

$$f \equiv C_{D,min} S \qquad (13.39)$$

DEFINITION OF EQUIVALENT FLAT PLATE AREA

and then rewrite Eq. 13.38 as

$$D = fq + \frac{C_L^2 q S}{\pi e \, AR} \qquad (13.40)$$

and finally rewrite Eq. 13.35 for the drag coefficient:

$$C_D = \frac{f}{S} + \frac{C_L^2}{\pi e \, AR} \qquad (13.41)$$

The next vital step in predicting airplane performance is the calculating of the **power** required for flight from the assumed form (Eq. 13.40) of the lift/drag relationship.

In level flight at constant speed the external forces acting on an aircraft in directions parallel and perpendicular to the flight direction must add to zero. Therefore the lift force for the complete aircraft is equal and opposite in direction to the weight force and the thrust force is equal and opposite in direction to the drag force for the complete aircraft:

$$L = W = C_L q S \qquad (13.42)$$

and

Thrust Force on an Aircraft $\equiv T$

$$= D = C_D q S \qquad (13.43)$$

Therefore the **power** required for this **level** flight condition, using Eq. 13.14, is

$$\text{Power Required} = DV = C_D q S V$$

$$= fqV + (C_L^2 q S V)/(\pi e \, \text{AR}) \qquad (13.44)$$

If we substitute for the lift coefficient from Eq. (13.42), for the flight dynamic pressure from Eq. 7.9 and for the aspect ratio from Eq. 12.1, we obtain a fundamental equation for the power required for level flight (within the assumption of Eq. 13.40):

$$P_{\text{REQ}} = \tfrac{1}{2}\rho f V^3 + \left(\frac{1}{e}\right)\left(\frac{2}{\pi}\right)\left(\frac{W}{b}\right)^2\left(\frac{1}{\rho V}\right) \qquad (13.45)$$

POWER REQUIRED FOR LEVEL FLIGHT
(Parabolic Drag Polar Approximation)

The first term in this equation is the power required by parasite drag (skin friction, pressure, interference, and cooling) while the second term is the power required by the lift-producing surfaces.

We should pause to ponder the implications of this important power/speed relationship. Note first that the altitude dependence of the required power is explicitly given through the dependence on the density of the air. The great importance of the equivalent flat plate area in determining the top speed and the cruising speed at low altitudes is clear from the fact that the first term depends on the **cube** of the true airspeed.

Another striking feature of Eq. 13.45 is that the power required to overcome induced drag depends on the **span loading** (W/b) and not on the aspect ratio. This makes it appear that we should simply increase the wing span to the largest possible value consistent with strength requirements, without regard to any increase in wing area that results. Not so! The part of the **first** term which is due to the parasite drag of the wing is directly proportional to the wing area S. Therefore the best wing geometry, requiring the least power overall, is that which possesses a minimum value for S and a maximum value for eb^2; i.e., we wish to maximize $\left(\dfrac{e\,b^2}{S}\right)$ — which is just the effective aspect area! (See Appendix D for a computer solution to Eq. 13.45.)

In level **equilibrium** flight, the power used must just equal the power required at that speed. And the power used, for a propeller-driven aircraft, is

$$\begin{pmatrix}\text{Power} \\ \text{Used}\end{pmatrix} = \begin{pmatrix}\text{Fraction of} \\ \text{Rated Power} \\ \text{Used}\end{pmatrix} \times \begin{pmatrix}\text{Rated} \\ \text{Power}\end{pmatrix} \times \begin{pmatrix}\text{Propeller} \\ \text{Efficiency}\end{pmatrix}$$

$$P_{\text{USED}} = \left(\text{Fraction Used}\right) \times P_{\text{RATED}} \times \eta \qquad (13.46)$$

POWER USED, PROPELLER-DRIVEN AIRCRAFT

For a jet-powered aircraft, on the other hand, the power used is just the thrust of the engine times the true airspeed, as was used in deriving Eq. 13.44:

$$P_{\text{USED}} = T_{\text{USED}} \times V_\infty \qquad (13.47)$$

POWER USED, JET AIRCRAFT

For a propeller-driven aircraft, there are two unknowns in Eq. 13.45 (f and e) and one unknown in Eq. 13.46 (the propeller efficiency). The propeller efficiency can be estimated from Eqs. 13.30 and 13.31 if otherwise unknown. Therefore we can calculate e and f for any aircraft for which we have available two power-speed data points (because this yields two simultaneous equations with two unknowns)! Then the way is clear to predict the performance of the aircraft for other conditions for which we do not have data.

Suppose that we measure a true airspeed of V_1 for an air density ρ_1, propeller efficiency η_1, and a power setting P_1, where P_1 is the power delivered to the propeller **shaft** (i.e., the fraction of available power that is used times the rated power). Then Eqs. 13.45 and 13.46 yield

$$P_{\text{USED}} = \eta_1 P_1 = P_{\text{REQ}} = \tfrac{1}{2}\rho_1 f V_1^3 + \left(\frac{1}{e}\right)\left(\frac{2}{\pi}\right)\left(\frac{W}{b}\right)^2\left(\frac{1}{\rho_1 V_1}\right)$$

for this flight condition. If we measure a second power-speed combination and use a subscript 2 to identify these values, we obtain similarly

$$P_{\text{USED}} = \eta_2 P_2 = P_{\text{REQ}} = \tfrac{1}{2}\rho_2 f V_2^3 + \left(\frac{1}{e}\right)\left(\frac{2}{\pi}\right)\left(\frac{W}{b}\right)^2\left(\frac{1}{\rho_2 V_2}\right)$$

These two equations can be solved with algebra by solving one equation for one unknown, substituting in the other equation, and repeating for the other unknown; I'll even do it for you:

$$f = \frac{\rho_1 V_1 \eta_1 P_1 - \rho_2 V_2 \eta_2 P_2}{2\left(q_1^2 - q_2^2\right)} \qquad (13.48)$$

$$e = \left(\frac{2}{\pi}\right)\left(\frac{W}{b}\right)^2\left(\frac{q_1^2 - q_2^2}{\rho_2 V_2 q_1^2 \eta_2 P_2 - \rho_1 V_1 q_2^2 \eta_1 P_1}\right) \qquad (13.49)$$

where q_1 and q_2 are the flight dynamic pressures for the cruise points.

In using these equations, try to find two flight conditions that emphasize first one term in the power equation (a high speed at low altitude, emphasizing the parasite drag term) and then the other (a lower speed at high altitude, emphasizing the induced drag term) in order to minimize errors due to subtracting nearly equal large numbers.

EXAMPLE 13.5. Estimate the efficiency factor and the equivalent flat plate area for a Cessna 182Q using two performance values from the pilot's handbook. The gross weight of the aircraft is 2950 lb, the wing span is 36.0 ft, the rated power is 230 hp, and the propeller diameter is 82 inches.

Cruise speeds (true airspeeds):

At a density altitude of 2000 ft and 75% power:

$V_1 = 136$ kt $= 229.7$ ft/s

At a density altitude of 12000 ft and 44% power:

$$V_2 = 113 \text{ kt} = 190.9 \text{ ft/s}$$

Solution: (a) Estimate the propeller efficiency from the propeller momentum theory (Eqs. 13.30 and 13.31):

$$\eta_{1,\text{theory}} = 0.958 \text{ so we can estimate that } \eta_1 = 0.814$$

$$\eta_{2,\text{theory}} = 0.944 \text{ so we can estimate that } \eta_2 = 0.803.$$

(b) Calculate f from Eq. 13.48, using

$$P_1 = (0.75)(230) \text{ hp} = 172.5 \text{ hp} = 94{,}875 \text{ ft-lb/s}$$

$$P_2 = (0.44)(230) \text{ hp} = 101.2 \text{ hp} = 55{,}660 \text{ ft-lb/s}$$

$$\rho_1 = 0.002241 \text{ sl/ft}^3$$

$$\rho_2 = 0.001646 \text{ sl/ft}^3$$

$$q_1 = \tfrac{1}{2}\rho_1 V_1^2 = 59.1 \text{ lb/ft}^2$$

$$q_2 = \tfrac{1}{2}\rho_2 V_2^2 = 30.0 \text{ lb/ft}^2$$

$$\rho_1 V_1 \eta_1 P_1 = (0.002241)(229.7)(0.814)(94875) = 39754$$

$$\rho_2 V_2 \eta_2 P_2 = (0.001648)(190.9)(0.803)(55660) = 14061$$

$$f = \frac{(39754) - (14061)}{2(59.1^2 - 30.0^2)} = \frac{39754 - 14061}{2(2593)} = \frac{25693}{2(2593)}$$

$$= \mathbf{5.0 \text{ ft}^2}$$

(c) Calculate e from Eq. 13.49:

$$\rho_2 V_2 q_1^2 \eta_2 P_2 = (0.001646)(190.9)(59.1)^2(0.803)(55660)$$

$$= 49{,}053{,}440$$

$$\rho_1 V_1 q_2^2 \eta_1 P_1 = (0.002241)(229.7)(30.0)^2(0.814)(94{,}875)$$

$$= 35{,}778{,}453$$

$$e = \left(\frac{2}{\pi}\right)\left(\frac{2950}{36.0}\right)^2 \left(\frac{59.1^2 - 30.0^2}{49{,}053{,}440 - 35{,}778{,}453}\right)$$

$$= (0.637)(6715)\left(\frac{2593}{1.327 \times 10^7}\right) = \mathbf{0.84}$$

Note that, because of the loss of precision resulting from the subtractions, we have chosen to round off the final answers to only two significant figures.

Although we've succeeded in determining e and f from two power-speed data points, it is much better engineering practice to make **many** measurements and then choose the e and f that **best** fit these data. The process is greatly facilitated by rewriting Eq. 13.45 in a slightly different form. First multiply both sides of the equation by the density ratio (ρ/ρ_0) and then substitute for the true airspeed, V, in terms of the equivalent airspeed, V_E (Eq. 7.8). This yields

$$\boxed{P_{\text{REQ}} \sqrt{\frac{\rho}{\rho_0}}\, V_E = \tfrac{1}{2} f \rho_0 V_E^4 + \left(\frac{1}{e}\right)\left(\frac{2}{\pi}\right)\left(\frac{W}{b}\right)^2\left(\frac{1}{\rho_0}\right)} \quad (13.50)$$

EQUATION FOR EXPERIMENTAL DETERMINATION OF f AND e

The great advantage of this equation is that we can experimentally determine the quantity on the left side of the equation while the right side depends only on known constants plus e, f, and V_E^4. If we plot experimental values of the left side on the vertical axis as a function of V_E^4 (on the horizontal axis), the polar approximation and algebra now tell us that **all** the points should fall on a **straight line** with a slope of $(\tfrac{1}{2} f \rho_0)$ and with an intercept on the vertical axis of $\left(\frac{1}{e}\right)\left(\frac{2}{\pi}\right)\left(\frac{W}{b}\right)^2\left(\frac{1}{\rho_0}\right)$.

(Recall that the equation of a straight line is $y = mx + b$, where b is the intercept on the y-axis and m is the slope of the line, defined as the opposite side over the adjacent side when a right triangle is formed with the straight line as the hypotenuse. Here the x-axis variable is V_E^4.) Therefore if we draw the best possible straight line based on experimental measurements, the slope of that line gives us the best-fit value for f and the y-intercept gives us the best-fit value for e. A computer is very handy for doing both the calculations and graphing both the data points and the best-fit straight line, to show us the goodness of the fit.

In lieu of experimental data, we can always try pilot handbook performance tables again. Figure 13.7 presents a plot of Eq. 13.50 for all of the handbook's standard-temperature, power-speed values for a Cessna 152. The available power was calculated with the help of Eqs. 13.30 and 13.31 for the propeller efficiency. The fit is a very fine one — suggesting that the handbook values were probably generated from the same theory after fitting to experimental data. (Perhaps Eq. 13.31 is somewhat optimistic for this fixed-pitch propeller; if the propeller has only 80% of the theoretical efficiency in the cruise condition, the fit is just as good and the calculated efficiency of the airframe improves to $e = 0.761$ and $f = 5.07 \text{ ft}^2$, for example.)

Now that we've estimated e and f for the 152, we can plot the whole-airplane C_D versus C_L with the help of Eq. 13.41. Figure 13.8 shows this calculated drag polar for the 152. Note the characteristic polar shape, so that the minimum C_D is

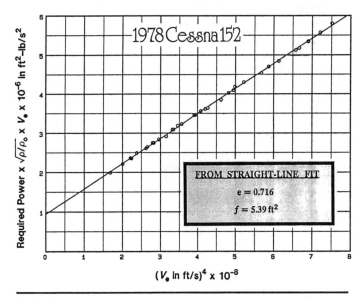

Fig. 13.7 Fit of handbook performance data to a drag polar equation

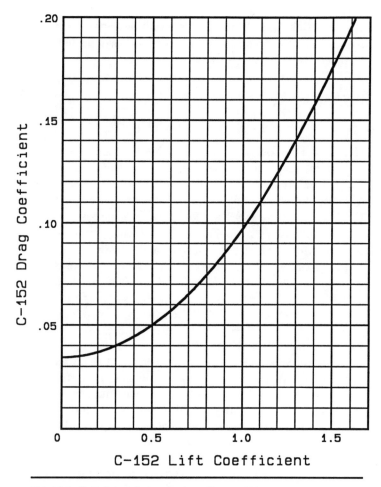

Fig. 13.8 Calculated drag polar for a Cessna 152

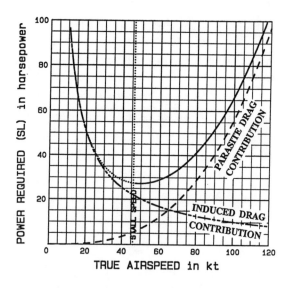

Fig. 13.9 Calculated power required for level flight at SL (C-152, 1670 lb)

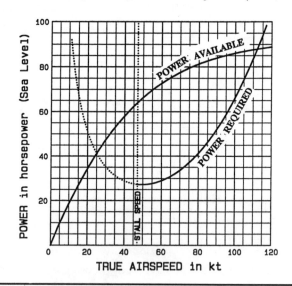

Fig. 13.10 Calculated power required and power available, C-152 under standard sea level conditions and at 1670 lb

predicted to occur (unnaturally) at $C_L = 0$. (It is customary in the engineering business to plot C_L as a function of C_D but we do it the other way because that is the way Eq. 13.41 is written.) The advantage of using this kind of a plot to compare airplanes is that the dependence on size and power and air density is **removed,** just as for airfoil coefficients; the most efficient airplanes are simply those with the smallest $C_{D,min}$ and the largest effective aspect ratio.

The next logical step is to use this e and f to calculate the power required as a function of airspeed for the 152, under SL StAt conditions, using Eq. 13.46. The result is shown in Fig. 13.9 along with, using dashed lines, the individual contributions of parasite drag and induced drag of which it is composed. Just as we've surmised earlier, parasite drag causes most of the need for power at high speeds and induced drag causes most of the need for power at low speeds, at least for level flight at SL air density.

Finally we compare the power available from the engine (thrust force times true airspeed) with the power required (Fig. 13.10). Note that the power available at zero airspeed is **zero** because no energy is **transferred** to the airplane then (as, for example, when you try to taxi after forgetting to remove the tire chocks . . .). It is the dramatic reduction in the available thrust from a propeller as the speed increases (Fig. 13.6) that is

responsible for the bending over of the power-available curve, limiting the top speed.

Note that the power required has a predicted minimum at a speed of only slightly greater than the power-off stall. As suggested earlier, the polar drag approximation breaks down here and is not able to model the **abrupt** increase in drag at lift coefficients greater than the maximum (which correspond to speeds **less** than the stalling speed, in level flight); the power-required curve should increase much more rapidly than shown in the figure, for these low speeds.

A wealth of performance information is available from graphs such as Fig. 13.10. The power-available curve is for 100% of the rated power at SL and so the predicted top speed for the 152 is 112 kt, close to the handbook value of 110 kt. Similarly, if we redraw the power-available curve for 70% power, the intersection of that curve with the power-required curve will give the cruising speed for that power setting. The difference between the power-available and the power-required curve, for any given airspeed, is a measure of the excess power that is available for climb. Note that the maximum excess power in Fig. 13.10 occurs at a speed of about 62 or 63 kt, also close to the handbook value of about 66 kt for the best rate-of-climb speed, with the difference perhaps attributable to deficiencies of the polar approximation at slow speeds.

The fact that Fig. 13.10 predicts a significant rate of climb at the stalling speed is **not** entirely in error; there may well be climb capability at and even **below** the stall speed. The problem is that a pilot can't take advantage of it because one wing always tries to stall a little before the other, making wings-level flight impossible.

It is very instructive to redraw Fig. 13.10 for a much higher density altitude, say 10,000 ft. The altitude dependence of the power required by the airframe is explicitly contained in the density dependence shown in Eq. 13.45, but we need to be able to calculate the altitude dependence of the maximum power available from the powerplant as well. Reference 2 suggests the following relationship for a non-supercharged piston engine:

$$P = \frac{\left(\dfrac{\rho}{\rho_0} - 0.12\right)}{0.88} P_{SL} \qquad (13.51)$$

THEORETICAL ALTITUDE DEPENDENCE OF PISTON ENGINE POWER

Figure 13.11 then presents calculated power curves at a density altitude of 10,000 ft but still at the gross weight of 1670 lb. Note how the available range of operating speeds has been dramatically reduced at this low air density; the stall speed has increased and the top speed has decreased (although the top speed can now be the cruising speed because the power available is less than 75% of the SL rating). This is related to the way

in which the power-available curve has slumped down and the power required has slumped **up** at low speeds (because the induced drag has increased) while slumping **down** at high speeds (because the parasite drag goes down with fewer molecules of air to hit).

It would be very nice to be able to calculate a cruising speed for any desired density altitude, for any weight, and for any power setting. It would also be nice to be able to calculate the effect of increasing the span or some other change. What is needed for this nice capability is the ability to solve Eq. 13.45 for the speed, V. This is not a simple task for a typical calculator but a little computer program that will do the job is available in Appendix D.

Ultralights have given the FAA the "opportunity" to get into the performance-prediction business. (Recall that the advantage of being declared an ultralight is that no FAA testing or certification is required, and the pilot does not have to possess a medical certificate or pilot's license.) Federal Aviation Regulation Part 103 defines an ultralight vehicle, among other things, as one that has a maximum power-off stall speed of 24 kt and a maximum level flight speed of 55 kt calibrated airspeed. In lieu of flight tests to confirm that a particular vehicle conforms to these requirements, the FAA in Advisory Circular 103-7, January 30, 1984, has provided formulas for the prediction of these speeds.

The maximum level flight speed is estimated by (a) determining a drag factor based on the projected areas of the wing and stabilizer and the degree of streamlining of the cockpit and the fuselage, (b) adding 20% to account for induced and interference drag, and (c) consulting a graph of power versus drag factor and maximum speed.

The power-off stall speed (Eq. 12.16) is estimated based on the wing loading (defined as the empty weight plus 170 lb

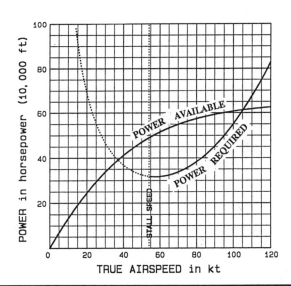

Fig. 13.11 Calculated power required and power available for level flight, C-152 at 10,0000 ft in the StAt and at 1670 lb

Pic. 13.1 An example of an early ultralight design: open-framework cockpit, single-surface airfoil, and spoilers for lateral control

for the pilot plus 30 lb for fuel, divided by the wing area) with extra credit given for double surface wings with 0.07c camber or greater and for high lift devices. Sounds reasonable, doesn't it?

D. When Only Speed Counts: Clip the Wings?

If we simply **have** to make our Cessna 152 go faster on the same power, Eq. 13.45 and Fig. 13.9 suggest that we should concentrate on reducing the flat plate (parasite) drag because that first term in the power-required equation clearly predominates at the highest speeds, being proportional to the **cube** of the speed. In fact, we eagerly note from Eq. 13.39 that f, and therefore the whole parasite drag term, is directly proportional to S, the wing area. Therefore we must simply requisition hacksaws and casually remove one or two inner bays in each of the wings (leaving the ailerons intact). This will certainly increase the induced-drag term in the power required, the second term in Eq. 13.45, because that term is proportional to the square of the wing span, but the parasite drag is so much larger at high speeds that it seems likely that the tradeoff will give us the higher greater top speeds that we seek—and let us win the race. We realize that the reduction in wing area is going to up the stall speed (Eq. 12.16), but race pilots are all aces so that should be no problem at all.

Perhaps—just maybe—it would be a good idea to cut down the wings with the computer before starting our sawing. It's not that we doubt our need to clip the 152's wings, of course, but simply that the computer did get us this far and perhaps it can tell us just how much of the wings to remove for optimum results—for clearly there is a point at which the induced drag from little wings is going to be as important as the parasite drag.

First, we naturally throw out all the excess weight: The pilot is chosen for bravery **and** light weight, the copilot and his seat are tossed, and only the minimum amount of fuel is loaded. We figure we can thereby reduce our Cessna's racing weight from its certified gross weight of 1670 lb to only 1300 lb. When we ask the computer to estimate the increase in top speed from our weight reduction program, it quickly redraws Fig. 13.10 for this lower weight. But to our surprise, the predicted increase in top speed (full throttle at SL) is less than 1 kt! Oh well, what can you expect when only the induced drag depends on weight.

So, we tell the computer to chop down the wings, one foot off each side at a time. The top speed keeps building very nicely until, with 6 feet removed from the 33.3 ft wing, it reaches a maximum of 127 kt (Fig. 13.12). Wow! We've gained 15 kt over those foolish folk still flying stock machinery.

Yet—before we break out the hacksaws and celebrate—could we have forgotten anything? Oops. We have been hacking away on the wings as if the wing alone was responsible for all the parasite drag. Just because the parasite drag coefficient, $C_{D, MIN}$, of Eq. 13.39 uses the wing area, S, as the **reference area** does **not** mean that the wing really is responsible; recall that it is just a mathematical convenience. Eq. 13.39 tells us that the

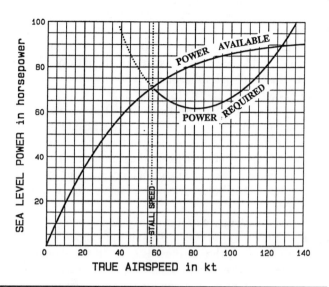

Fig. 13.12 Calculated power required and power available for an all-wing C-152 with extended wings, at 1300 lb

value of f we obtained by fitting 152 cruise performance data corresponds to a $C_{D, MIN}$ of

$$C_{D, MIN} = \frac{f}{S} = \frac{5.39 \text{ ft}^2}{160 \text{ ft}^2} = 0.0337$$

but Fig. 11.2 tells us that the 152's NACA 2412 airfoil should contribute no more than 0.010 of this, even it is rough and flexible. The great majority of the parasite drag (about 70%) is thus due to skin friction or pressure drag on the fuselage and landing gear and tail surface, to cooling drag, and to interference drag where surfaces are joined—and cutting down on the wing span won't reduce this part of the parasite drag at all.

When the parasite drag is separated into a wing part and a fuselage part, and the reduction in S applied only to the wing part, the computer suggests that reducing the wing span provides negligible increase in top speed. So don't be the first on your block with a clipped-wing 152.

If we had started with an aerodynamically clean, streamlined aircraft, one with retractable landing gear, the wing parasite drag would have been a large part of the total parasite drag and **then** the wing clipping would have worked. Even if it should yield a higher top speed, though, Fig. 13.12 reveals penalties at lower speeds. The stall speed has increased from about 42 kt to about 57 kt—but even more importantly, if we lift off just above stall speed there is no excess power available to allow us to climb out of ground effect. If we do let the speed build up even higher, we still get hurt on climb rate and climb angle.

How about sponsoring a race at Reno on a hot, humid day—one with a density altitude of 10,000 ft, for example? The parasite drag goes down very nicely at this high altitude (because it is directly proportional to the air density) but the induced drag goes up (being inversely proportional to the air density). Figure 13.11 presented the predicted performance of the stock 152 at 10,000 ft in the StAt. The power-available curve slumped down because (by Eq. 13.51) the power available is only about

$$\frac{P}{P_{SL}} = \left(\frac{(\rho/\rho_0) - 0.12}{0.88}\right) = \frac{(0.001756 / 0.002377 - 0.12)}{0.88}$$

$$= 0.70 = 70\%$$

of its sea level value. (That's what supercharging fixes.) Yet the parasite drag is sufficiently less at high altitude that our top speed is still predicted to be 104 kt. Our operating range of speeds has shrunk considerably as the true stall speed has increased from about 41 kt to about 48 kt.

At this altitude, **adding** 4 ft to each wing will give us about one more kt of top speed, reduce the stall speed to about 43 kt, and noticeably improve climb capability (Fig. 13.13). Eventually increasing the wing span causes the weight to increase too much, if the wing is to remain strong in bending and torsion, but efficient high altitude airplanes do have long, narrow wings.

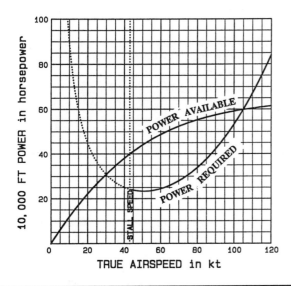

Fig. 13.13 Calculated power required and power available at 10,000 ft in the StAt for a C-152 with extended wings, at 1300 lb

Summary

The Newtonian laws of motion and the concepts of energy and power form the starting point in describing the takeoff, flight, and landing characteristics of aircraft. Angular motion is as important as linear motion; the equations for angular motion, energy, and power are obtained from their linear counterparts by realizing that the moment of inertia, the moment of a force, angular speed, and angular acceleration are analogous to mass, force, linear speed, and linear acceleration for linear motion. The linear motion of a given point on a rotating body is obtained from its angular motion by multiplying by the distance from the axis of the rotation (i.e., the radius).

The parameter that is related to the average angle of attack of a propeller blade, and which thereby relates directly to the efficiency of that propeller, is the advance ratio. Propeller efficiency is defined in terms of the energy actually transferred to the aircraft and so it varies from zero when the aircraft is not moving up to a maximum of about 85%. At high speeds the propeller efficiency is limited by an unfavorable lift/drag ratio for its blades which may be related to Mach effects for its tips. The thrust force and efficiency of a given propeller can be estimated by applying the momentum form of Newton's second law to a simplified picture of the air flow through the propeller disc.

Knowledge of the overall lift versus drag for an aircraft is the key to predicting its performance. A theoretical derivation and experimental confirmations suggest that the drag force can often be considered to be composed of one term (parasite drag) unrelated to lift and a second term (induced drag) that is proportional to the square of the lift force. The magnitude of the first term is directly related to a parameter called the equivalent flat plate area (f); the magnitude of the second term is directly related to a parameter called the airplane's efficiency factor (e). Both of these parameters can be determined for a particular aircraft by mathematically determining which numerical values for the parameters will best fit experimentally-determined power versus speed data. The results of such an analysis, as applied to the Cessna 152, were given.

Calculated power curves were used to estimate the effect of air density and wing span on aircraft performance, especially stall speed and top speed.

Symbol Table (in order of introduction)

I	moment of inertia of an object
α	angular acceleration
p	linear momentum of an object
L	angular momentum of an object
F_D	the force component in the direction of the motion
W	work done by a force
KE	kinetic energy, linear and angular
PE	gravitational potential energy
P	power
J	advance ratio for a propeller
η	propeller efficiency
T	thrust force produced by a propeller or jet engine
e	airplane efficiency factor
C_{Li}	airplane design lift coefficient
f	equivalent flat plate area

Review Questions

1. When Newton's second law is applied to aircraft, the acceleration involved is the acceleration of the ___ of the aircraft.

2. If the vector sum of the external___ acting on an aircraft about any axis through the center of gravity is not zero, then the aircraft is experiencing ___ acceleration.

3. The property of an aircraft that is a measure of its resistance to linear acceleration is its ___ while the property of ___ measures its resistance to angular acceleration.

4. The moment of a force can be computed either as the product of the ___ force component times the actual distance from the chosen axis or as the actual force times the ___ distance from the axis.

5. The more general form of the second law tells us that the ___ of an aircraft will not change in any direction for which the external forces sum to zero.

6. In the most general case, ___ equations are required to describe the translational motion of an aircraft and an additional ___ equations are required to describe the angular motion.

7. A stable climb or descent is considered to be an example of force equilibrium because, although there is motion, there is no ___.

8. Work is done on an aircraft only if there is an external force acting on the aircraft and if there is ___ in the direction of that force. Therefore the force of the air on the propeller blades does work on the aircraft but the ___ component of the aerodynamic force on the aircraft does not do work while in level flight.

9. In general, it can be said that no work is done by a force until ___ is transferred.

10. Conservation of energy tells us that energy can be ___ from one system to another but it cannot be created or destroyed.

11. In a stable glide, only the ___ force is doing work on the aircraft. This work is equal to the ___ energy lost by the aircraft.

12. ___ is the quickness of energy transfer—how much divided by how long.

13. The principle that tells us that the force exerted by a propeller on the air is just equal and opposite to the force exerted by the air on the propeller is stated in ___ Law.

14. The power delivered by a propeller or by a jet engine is the product of the ___ force and the ___ of the aircraft.

15. The power required to maintain level flight is the product of the ___ force and the ___ of the aircraft.

16. The advance ratio is a reasonable parameter to use in discussing propeller efficiency because it is directly related to the ___ of the blades for any particular flight condition.

17. The efficiency of a propeller is the ratio of the ___ delivered to the aircraft to the ___ supplied to the propeller.

18. The average value for the efficiency of a propeller in cruising flight is about ___.

19. The diameter of a propeller has to be chosen on the basis of the rotational speed at which the powerplant delivers its maximum power, as well as for adequate ground clearance. The absolute maximum rotational speed is the speed at which the tips reach ___ speed.

20. A large propeller, other factors being equal, delivers more ___than does a small propeller. The same holds true for turbofan engines versus turbojet engines, for the same reason.

21. Jet engines are rated for the ___ they provide while reciprocating engines are rated for the ___ they develop.(It is better to accelerate a large mass a little bit than a small mass a large amount.)

22. The drag for a complete airplane is expected to have the same form as that for a wing by itself; it consists of a profile or parasite drag

term and an ___ drag term that varies as the square of the ___ force being developed.

23. The parasite drag of a complete aircraft is often specified in terms of the ___ of a hypothetical flat plate possessing a drag coefficient of ___ and oriented perpendicular to the air flow direction.

24. The parasite drag term dominates in the determination of ___aircraft speeds while the second term limits the climb rate at ___ aircraft speeds ("minimum" or "maximum").

25. What happens to the energy that is being lost by an aircraft in a steady glide?

26. What information is needed, in general, to directly calculate the power being developed by a piston engine?

27. Engine manufacturers sometimes rate their piston engines for the power developed at sea level **before** necessary accessories such as an alternator and a vacuum pump are added. What, if any, would be the effect of this practice on the calculated e and f that would be obtained by applying Eqs. 13.48 and 13.49 to a given aircraft?

28. What limits the thrust available from a propeller at low speeds, in addition to that predicted by momentum theory?

29. What **units** should be used for angular displacement and angular velocity if they are to be directly related, through the radius distance, to the linear displacement and linear speed of a single point on the rotating body?

30. What is the definition of "equivalent flat plate area" for an airplane? Is the actual drag force on a flat plate of this area exactly equal, or somewhat smaller or larger than the parasite drag force on an aircraft it describes?

31. Maximizing the wing span reduces the ___ drag part of the drag force on an aircraft; minimizing the wing area reduces the ___ drag part of the drag force.

Problems

Note: Use $g = 32.2$ ft/s^2. Refer to preceding problems for data or answers, as needed.

1. A Cessna 152 at 1600 lb is braked to a halt in 485 ft after touching down at 41 kt (stall speed). (a) How much work was done by the ground on the aircraft, assuming that the work done by the air on the aircraft through air drag is negligible? Use the principle of conservation of energy. (Most of this work went into heating up the brake discs.) (b) What was the average frictional force of the ground on the tires during the landing roll? Answers: 1.19×10^5 ft-lb, 246 lb

2. What was the power being developed by the brakes when the speed was 10 kt? Answer: 7.54 hp (Isn't this impressive for such little things! No wonder they get hot.)

3. What is the advance ratio for the aircraft specified below? Answer: 0.88

Specifications for the 1977 Cessna 182Q
Rated Power: 230 brake horsepower at 2400 RPM
    Propeller: Diameter of 82 inches (maximum)
    Low pitch setting of 15.0° and a high pitch setting of 29.4° at 30 inches from the center
Performance at 8000 ft in the StAt:
    Power setting: 74% of rated brake horsepower
    Propeller speed: 2400 RPM
    True airspeed: 143 kt

4. If the propeller efficiency for the above aircraft is 0.8, what thrust is being developed? Answer: 310 lb

5. Estimate the angle of attack for the blade section for the above aircraft at a point 30 inches from the center of the propeller, neglecting

the induced angle correction and assuming that an intermediate pitch setting of 22.0° is being used. Answer: 1.0°

6. Determine the propeller tip speed relative to the oncoming air and also the Mach number for the tips. Answer: 529 kt, 0.82

7. Using Fig. 13.6, estimate the maximum power developed by the four Pratt and Whitney JT9D-7A jet engines on the 747 at 100 kt. Answer: 1.2×10^4 hp

8. The 1978 Cessna 152 has a wing span of 33'4" and a 160 ft² wing surface area. For a gross weight of 1600 lb, what lift coefficient is in use when it is loafing along on 63% power, at 2300 RPM, and at 95 kt at 4000 ft in the StAt? Answer: 0.37

9. Using the drag polar parameters given in the text for the C-152, estimate the fraction of the total drag that is induced drag for the flight condition of problem 8. Answer: About 20%

10. For the aircraft of problem 8 but at a gross weight of 1670 lb and for cruise at 45% power, 2100 RPM, and 81 kt at 12,000 ft in the StAt, what lift coefficient is in use? Answer: 0.68

11. Estimate the fraction of the total drag that is induced drag for the flight condition of problem 10. Answer: Almost 1/2 (Note: This emphasizes the importance of effective aspect ratio if the induced drag is to be kept small for good high altitude performance. Military surveillance planes such as the U-2, after all, have the wing planforms of gliders.)

12. The Long-EZ, at a weight of 1100 lb, a wing span of 26.1 ft, a wing area of 94.8 ft², and with a 115 hp Lycoming 0-235, promises a speed of 167 kt at SL on 100% power and 155 kt at 15,000 ft on 55% power. (a) For an estimated propeller efficiency of 0.80, estimate the equivalent flat plate area and the efficiency factor for this aircraft. Answer: 1.82 ft², 0.86 (b) The pusher propeller in this canard aircraft probably isn't as efficient as a tractor propeller because it operates in the disturbed wing/fuselage wake. Recalculate e and f for a 70% efficient propeller. (Hint: Reference 4 in Chap. 17 says that $f = 1.6$ ft² for this aircraft.)

13. For the aircraft of problem 12, what wing lift coefficients were being developed at the two altitudes, if canard lift is ignored?

14. Assuming 70% propeller efficiency, what percent of the total drag is induced drag for the aircraft of problems 12 and 13, at the two altitudes? Answers: About 4% and 12%, respectively—but the induced drag due to canard lift is actually significantly larger than this.

15. What propeller diameter should you choose for your PT-17 Stearman, which cruises at about 82 kt at 1750 RPM, if the tip Mach number at 2000 ft is not to exceed 0.75? Answer: 107" (An 8.5 ft propeller is commonly used.)

16. Using Fig. 13.3, estimate the efficiency of the 74" diameter propeller on the Piper Cherokee PA-28-180 at a cruise RPM of 2450 RPM and a cruise speed of 109 kt. Answer: 0.82

17. The Piper PA-24-260 Comanche (single-engine retractable) has a 260 hp engine and claims a 194 mph top speed at SL and a 162 mph cruise with 55% power at 15,500 ft in the StAt at its design weight of 3100 lb. Its wing span is 35.98 ft and its wing area is 178 ft².

   a. Assuming 80% propeller efficiency, determine the power required by this aircraft for these two flight conditions. Answer: 1.144×10^5 ft-lb/s, 6.292×10^4 ft-lb/s
   b. Determine values for e and f from these data. Answer: $e = 0.508$, $f = 3.675$ ft²
   c. Using these calculated drag polar parameters, estimate the percentage of total drag that is lift dependent for the two flight conditions. Answer: 12%, 42%

18. Estimate the propeller tip Mach number for a Centurion when it is climbing out at SL at 2700 RPM and 90 kt with an 80" propeller. (Yes, it is rather loud.) See Example 13.2 for data. Answer: 0.86

19. Using the symbols in Fig. 13.2, obtain expressions for the thrust and torque generated by the aerodynamic forces on a propeller blade element that is located a distance r from the axis of rotation. Note that the thrust is not simply the lift generated by the airfoil section. Partial answer: $T = (\Delta L) \cos \phi - (\Delta D) \sin \phi$.

20. If the altimeter setting is 29.92" Hg (2116.2 lb/ft²) at a pressure altitude of 2000 ft, what should the manifold pressure gauge read (in inches of Hg) **before** the engine is started (i.e., when it is vented to the atmosphere)?

21. Using 85% of the results of propeller momentum theory, estimate the actual static and in-flight thrust for the Cessna 182Q of problem 3 at 8000 ft. Answer: Static thrust of 903 lb, in-flight thrust of 315 lb

22. Estimate the propeller efficiency for the 182Q cruise condition given in problem 3. Answer: 81%

23. Estimate the static thrust, the in-flight thrust, and the propeller efficiency for the two cruise conditions given for the Comanche of problem 17. The propeller diameter is 77 inches. Assume 85% of theoretical values. Partial answers: At SL, estimated static thrust is 1245 lb, in-flight thrust is 412 lb, and propeller efficiency is 0.82.

24. The reference wing area for the Boeing 727-200 airliner is 1560 ft² and its aspect ratio is 7.5. A 727 in climbout configuration (10° of flaps) has been modeled (Ref. 8) with the equation $C_D = 0.073 + 0.0966(C_L - 0.49)^2$. (a) What is the design lift coefficient for this configuration? (b) Estimate the equivalent flat plate area for this configuration. (c) Assuming a weight of 160,000 lb, what is the drag force for SL conditions and 140 kt speed? Answer: 1.87×10^4 lb (d) How much of the drag at this speed is induced drag? Answer: ≈60%

25. Reference 1 describes the experimental determination of the drag polar for a 1970 Cessna Cardinal. The effort was aided by propeller efficiency charts provided by the manufacturer. The Cardinal has a wing span of 36.0 ft, a wing reference area of 175 ft², and a gross weight of 2500 lb. Experimental data fitted to an equation equivalent to Eq. 13.50 resulted in values of $e = 0.564$ and $f = 4.67$ ft² in the clean configuration and values of $e = 0.546$ and $f = 8.09$ ft² with full flaps. Write the drag polar equations for these two configurations.
Partial Answer: $C_D = 0.0267 + 0.0762\ C_L^2$ (clean)

26. Reference 9 describes the experimental determination of the drag polar for a Beechcraft Super King Air both before and after icing encounters. Before ice, the drag polar was $C_D = 0.024 + 0.033\ C_L^2$ with $C_L = 0.257 + 3.67\alpha$; after ice, the drag polar was $C_D = 0.041 + 0.115\ C_L^2$ with $C_L = 0.242 + 4.17\alpha$. The aircraft has a gross weight of 12,500 lb, a wing span of 54.5 ft, and a wing area of about 303 ft². (a) Assuming the aircraft is cruising at 300 kt at a density altitude of 7,000 ft when it encounters this kind of icing, estimate the cruising speed after the ice has accumulated. (Assume the delivered power is unchanged.) Answer: About 240 kt, with the help of trial-and-error substitution or Appendix D (b) What angles of attack are involved?

27. Reference 10 describes the theoretical determination of the drag polar for an open cockpit ultralight. The actual empty weight turned out to be 272 lb (too heavy!), the wing area was 150.9 ft² and the wing span was 36.0 ft. The calculated drag polar for zero control deflection was $C_D = 0.1059 + 0.0421\ C_L^2$. A typical cruise profile for this aircraft would be 39 kt at a density altitude of 1000 ft. (a) Estimate the power required for this cruise condition, assuming a 175 lb pilot and 5 gallons

(30 lb) of fuel. Answer: 11.1 hp (**b**) How much of the total drag is induced drag? Answer: about 14% (**c**) What is the efficiency (L/D) of the aircraft in cruise? Answer: 5.1

28. The 300 hp Glasair III is a 2-seat cross-country aircraft currently available to homebuilders. It promises a top speed of 294 mi/hr at SL and 252 mi/hr at 50% power at 17,500 feet. The wing span is 23.3 ft, the wing area is 81.3 ft^2, and the wing loading is 29.52 lb/ft^2. Assume a propeller diameter of 72 inches. (**a**) Estimate the actual static and in-flight thrust at both SL and 17,500 ft. Partial answer: 1310 lb and 321 lb at SL (**b**) Estimate the propeller efficiencies at cruise. Answer: 0.84 at SL and 0.83 at 17,500 ft (**c**) Estimate the equivalent flat plate area and the efficiency factor for this aircraft, using the drag polar approximation. Partial answer: $e = 0.60$

29. The 160 hp, 2-place Lancair 320 is another fast homebuilt design. At a gross weight of 1685 lb, it promises a top speed of 217 kt at SL and 210 kt at 8000 ft with 75% power. The wing span is 23.5 ft and the wing area is 76 ft^2. Assuming a propeller efficiency of 80%, estimate the drag polar for this aircraft.

Answer: $C_D = 0.0148 + 0.051\ C_L^2$

References

1. Kohlman, David, "Flight Test Data for A Cessna Cardinal," NASA-CR-2337, 1974. (Available from NTIS as N74-14752.)

2. Crawford, Donald R., *A Practical Guide to Airplane Performance and Design*, Crawford Aviation, 1979. (Nomographs for the homebuilder who doesn't have computational skills, a FORTRAN program for those with computers.)

3. McCormack, Barnes W., *Aerodynamics, Aeronautics, and Flight Mechanics*, John Wiley & Sons, 1979. (This recent text includes considerable material on propellers, jet engines, and performance; it is intended for engineering students.)

4. Perkins, Courtland D. and Hage, Robert E., *Airplane Performance, Stability and Control*, John Wiley & Sons, 1949. (A classic text, still available and very useful.)

5. Hale, Francis J., *Introduction to Aircraft Performance, Selection, and Design*, John Wiley & Sons, 1984. (A more recent text, information on turboprop and turbojet propulsion is included.)

6. Covert, Eugene E. (Editor), *Thrust and Drag: Its Prediction and Verification*, American Institute of Aeronautics and Astronautics, Inc., New York, 1985.

7. Crigler, John L. and Jaquis, Robert E., "Propeller-Efficiency Charts for Light Airplanes," NACA TN 1338, July, 1947.

8. Dietenberger, Mark A., Haines, Patrick A., and Luers, James K., "Reconstruction of Pan Am New Orleans Accident," *J. Aircraft*, Vol. 22, No. 8, August, 1985.

9. Cooper, William A., Sand, Wayne R., Politovich, Marcia K., and Veal, Donald L., "Effects of Icing on Performance of a Research Airplane," *J. Aircraft*, Vol. 21, No. 9, September, 1984.

10. Blacklock, Carlos L., Jr. and Roskam, Jan, "Summary of the Weight and Balance, and The Drag Characteristics of a Typical Ultralight Aircraft," SAE Paper 841021, West Coast International Meeting and Exposition, San Diego, California, August 6–9, 1984.

11. Weick, F. W., *Propeller Design*, McGraw-Hill, 1930. (This was a landmark description of the results of NACA propeller investigations in the previous decade or so by the aeronautical engineer who would later become well-known for his Ercoupe and Pawnee ag-plane designs.)

12. Vincenti, Walter G., "Data for Design: The Air-Propeller Tests of W. F. Durance and E. P. Lesley, 1916–1926," Chapter 5 in *What Engineers Know and How They Know It*, McGraw-Hill, 1990. (An interesting historical account of the development of propeller theory and experiment.)

1. WING LOADING — the **weight** that must be supported by the **lift** of each square foot of the airplane wings. For example, a 2800-pound airplane (fully loaded) with 305 square feet of wing area has a wing loading of 9.2 pounds per square foot.

Pic. 13.2 The U. S. Army Air Force illustrated its definitions.

2. POWER LOADING — the weight that must be moved by each horsepower of the engine. The 2800-pound airplane with a 235-horsepower engine has a power loading of 11.9 pounds per horsepower. Horsepower is figured at sea level.

Pic. 13.3 Another U. S. Army Air Force aerodynamics illustration

Chapter 14

Aircraft Performance

Bianchi rang me up and told me he had done several runs across White Waltham Aerodrome but the Demoiselle showed no sign of lifting. . . . A somewhat dispirited conference took place, and out of it came a decision to improve the wing section by modifying the leading edge and to get an engine with more power. Construction of a new set of wings was also put in hand, giving an extra foot of span to each wing. . . . While these modifications were being done a lot of thought was put into the apparent discrepancy between Santos Dumont's successful flying with his machine and the complete inability of the replica to take off with approximately the same weight and horsepower. . . . At last, one obvious solution dawned on us: Santos Dumont was a very small and slight man, weighing only 8 stone (112 lb) . . . . Miss Joan Hughes was invited to do the next series of test flights . . . and managed to do several flights across the aerodrome at heights around 10 ft . . . and took the Demoiselle up to a height of about 250 feet, which did not seem to be the maximum ceiling of the airplane.[1]

Each airplane must have a steady rate of climb at sea level of at least 300 feet per minute and a steady angle of climb of at least 1:12 for landplanes or 1:15 for seaplanes.[2]

Pic. 14.1 A Demoiselle replica

[1]Allen H. Wheeler, *Building Aeroplanes for "Those Magnificent Men,"* John W. Caler, 1965.
[2]Federal Air Regulations, Part 23, Section 65.

Pic. 14.2 Performance question: What is the tale behind this taildragging C-140? [Answer: It left its tailwheel on the fence at the end of a muddy field.]

Pic. 14.3 A 1930s Ryan STA begins its takeoff roll.

A. The Takeoff Roll

Takeoff is the most critical phase of flight for light, single-engine aircraft because any problem that occurs is likely to occur at relatively high speeds. Most of the concern relates to the runway length and then to obstacles that must be cleared after liftoff. In this section we look at the factors affecting the takeoff roll. In the next section the question of climb ability after liftoff is addressed.

Just what is the wing doing as the airspeed builds from zero (in no-wind conditions) to the stall speed and then to the liftoff speed? Generally we try to align the plane's longitudinal axis with the direction of motion to minimize aerodynamic and frictional drag. In a tailwheel-type aircraft, the angle of attack begins at a value close to the stalling angle but the tail is raised to streamline the aircraft as soon as there is enough elevator and rudder authority to do so. In a nosewheel-type aircraft, the angle of attack usually begins at a negative value and the nosewheel is raised slightly off the ground somewhat later in the takeoff roll to minimize both aerodynamic and frictional drag.

A normal liftoff involves very little pitch change as the lift gradually builds, letting the airplane "fly itself" off the ground. A short field liftoff, on the other hand, usually involves a significant pitch-up just as the best angle-of-climb speed is approached. Because the Reynolds number is zero at the beginning of the roll, the aircraft passes through a low RN range in which the wing may stall at a noticeably smaller angle of attack at speeds below the normal airborne stall speed, giving the pilot an incentive to streamline the airplane quickly. Liftoff typically is initiated at about 1.1 or 1.2 times the stall speed. Although many aircraft can be forced off the ground at slower speeds, they often have insufficient power to climb out of ground effect.

The accelerated motion represented by an aircraft in process of becoming airborne presents a nice opportunity to use Newton's second law of motion (Eq. 13.1). Although an exact analysis is complex, approximate methods that are instructive with regard to both the physical factors affecting the takeoff and the practical application of classical mechanics to aircraft performance will also be used.

Figure 14.1 depicts the external forces acting on an aircraft during the takeoff roll. Although a trainer aircraft is shown, the same general force diagram applies to multi-engine or jet aircraft as well; the differences lie in the source of the thrust force and in the greater possibilities for asymmetric thrust for the complex aircraft.

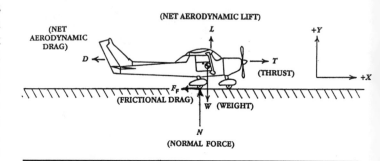

Fig. 14.1 External forces acting on an aircraft during the takeoff roll

In applying Eq. 13.1 to the takeoff roll, we choose two perpendicular directions and require that the stated equality hold separately for forces (or force components) in each of these directions; then the vector nature of Newton's second law reduces to plus or minus signs in front of the magnitude of the forces to indicate their directions relative to the positive directions of the axes. We choose our axes to be perpendicular and parallel to the ground; this choice produces the simplest equations because all of the linear acceleration is then along one axis—parallel to the ground.

First, note that the external forces acting on the airplane in the vertical direction (for which the acceleration is zero) are (a) the upward force of the ground on the tires, (b) the whole-aircraft aerodynamic lift force (which is usually upward), and (c) the downward force of the earth on the aircraft (its weight). In symbols,

$$\sum F_{\text{VERTICAL}} = L + N - W = m\, a_{\text{VERTICAL}} = 0 \qquad (14.1)$$

EQUATION OF MOTION PERPENDICULAR TO THE GROUND

in which, from Eq. (3.3), $W = mg$ and N is the total **upward** force of the ground on both the main gear and (at least at the beginning of the takeoff roll) the nosewheel or tailwheel.

Next, sum the external forces on the aircraft that are parallel to the ground: (a) thrust force, (b) aerodynamic drag force, and (c) rolling friction force:

In symbols,

$$\sum F_{\text{HORIZONTAL}} = T - D - F_F = ma \qquad (14.2)$$

EQUATION OF MOTION PARALLEL TO THE GROUND

Note that the actual force of the ground on the tires is the **vector sum** of F_F and N; we work with the components because it allows us to deal with the vertical and horizontal motion separately.

The force of rolling friction, like that of static or sliding friction, can be modeled fairly well as the product of a coefficient of friction (which depends almost entirely on the nature of the two surfaces in contact, but **not** on the contact area) and the force pushing the two surfaces together (to which we have already given the symbol N). Therefore

$$F_F = \mu N \qquad (14.3)$$

where μ (Greek letter mu) is the coefficient of rolling friction. The appropriate value for μ varies between about 0.02 for properly inflated tires on a smooth, hard surface to about 0.2 on moderately tall grass and up to about 0.3 on very soft ground.

During most of the takeoff roll, there are no significant angular accelerations; in other words, the pilot is holding the aircraft at a nearly constant angle of attack. According to Eq. 13.2, this means that the sum of the moments about the center of gravity must be zero. Inspection of Figure 14.1 reveals the interplay of forces and moment arms that provides this angular equilibrium.

Before the takeoff is initiated, the lift, drag, and thrust forces are negligible. Thus the normal force of the ground on the main gear tires produces a nose-down pitching moment about the *cg* that keeps the nose gear tire on the ground; the greater moment arm of the nose gear tire allows it to balance the aircraft with a much smaller force than that on the main gear tires.

However, on a hard surface runway the pilot (of a non-canard type aircraft) is able to make most of the takeoff roll with negligible weight on the nose gear. The primary control that accomplishes this is the elevator. Often propwash on the elevator makes the elevator effective well before flying speed is reached. Figure 14.1 shows the elevator slightly raised relative to the horizontal stabilizer. This creates a downward force on that surface (not shown but assumed to be included in the lift force shown) which can provide a powerful nose-up pitching moment because of the long moment arm. (An aircraft with a T-tail—i.e., with a horizontal surface on top of the vertical surface—does **not** have its elevator immersed in propwash on takeoff and so the nose cannot be raised until considerably later in the takeoff roll.) Figure 14.1 shows that the thrust, wing lift, and drag force also can contribute pitching moments.

For a tricycle (nosewheel) type aircraft with a conventional tail, as in Fig. 14.1, the weight on the nosewheel is normally reduced early in the takeoff roll in order to minimize both rolling friction and aerodynamic drag by increasing the angle of attack of the wing from negative to neutral or slightly positive. For a "conventional" gear (tailwheel) type aircraft, however, the tailwheel is normally lifted off the ground early in the takeoff roll by applying forward pressure on the control stick or wheel; this also reduces rolling friction and aerodynamic drag as well as preventing liftoff right at the stall speed. These "normal" procedures generally yield the most efficient performance on hard surface runways and are assumed to have been followed in the recording and calculating of takeoff performance contained in owners' manuals. On a very soft runway, however, there may be enough rolling friction that a nosewheel digs into the ground and it is literally impossible to achieve normal liftoff speed; in this case the aircraft can usually be forced off the ground as soon as possible and then allowed to accelerate to a safe climbing speed in ground effect (assuming it isn't a laminar airfoil outside its drag bucket, perhaps?).

The problem we face in trying to use Eqs. 14.1 through 14.3 to describe the takeoff roll is that the forces are **not** constant and therefore the acceleration itself cannot be expected to be constant. In Eq. 14.2, the thrust force provided by either propellers or jet engines will vary during the takeoff roll. The aerodynamic drag is zero at the beginning of the takeoff roll (in no-wind conditions) but increases as the square of the speed during the roll. The rolling friction is a maximum at the beginning of the takeoff roll and zero at the end because $L = W$ at liftoff so Eq. 14.1 tells us that $N = 0$ then and Eq. 14.3 in turn tells us that $F_F = 0$ at that point. Surprisingly, the increasing aerodynamic drag nearly compensates for the decreasing rolling friction, and the propeller thrust force does not vary so very much either.

If the actual variation of these forces with speed is known, Eqs. 14.1 through 14.3 can be numerically solved to obtain accurate takeoff distances and times. The procedure (as in Simpson's rule) is to take a small interval of time (e.g., 0.01 second) and estimate the average acceleration during this time interval by using the actual forces and the computed accelerations at discrete times within this small time interval. The increase in speed during the time interval is then determined from this average acceleration, as is the distance covered. The steps are repeated with additional small increments of the time until the liftoff speed is reached. Implementation of this "bootstrap" procedure is well suited to the abilities of a computer.

If we assume that the drag force in Eq. 14.2 is adequately modeled by the simple parabolic polar approximation of Eq. 13.40, and if we also assume that $e = 1$ because of the proximity of the ground, we can rewrite Eq. 14.2 as

$$T - \left(fq + \frac{C_L^2 q S}{\pi \, \text{AR}}\right) - \mu N = m \, a \qquad (14.4)$$

Then substituting for N from Eq. 14.1 yields

$$T - \left(fq + \frac{C_L^2 q S}{\pi \, \text{AR}}\right) - \mu \left(mg - L\right) = m \, a \qquad (14.5)$$

or, rearranging terms and substituting for L from Eq. 12.3

$$T - f q - \left(\frac{C_L^2 q S}{\pi \, \text{AR}} - \mu C_L q S\right) - \mu m g = m \, a \qquad (14.6)$$

The term in parentheses reveals the force tradeoff when lift is developed during the takeoff roll; the lift produces induced drag but it also reduces the frictional drag. The optimum takeoff involves developing just the right amount of lift to make the term in parentheses a minimum; calculus tells us that this will occur when $C_L = 0.5 \mu \pi \, \text{AR}$. The pilot will have to identify this optimum lift condition by sensing the pitch attitude that yields maximum acceleration; on a soft field, toward the end of the takeoff, the proper choice is especially easy to detect. If we assume this optimum lift coefficient, substitute for the dynamic pressure and the aspect ratio from their definitions, and collect terms, the final form of the takeoff equation of motion is obtained:

$$a_{\text{T-O}} = \frac{T}{m} - \frac{1}{m}\left(f - \tfrac{1}{2}\mu^2 \pi b^2\right)\left(\tfrac{1}{2}\rho V^2\right) - \mu g \qquad (14.7)$$

INSTANTANEOUS ACCELERATION DURING TAKEOFF ROLL

An interesting equation this! It tells us that the acceleration during the takeoff roll will indeed decrease because (a) the thrust decreases with increasing speed and (b) the speed-dependent term changes from zero (when $V = 0$) to a negative value as the takeoff proceeds, but its maximum value is independent of density altitude because it depends on the constant dynamic pressure required for takeoff. We will use the propeller momentum theory of the previous chapter to estimate the propeller thrust during the takeoff but it will be less than the 85% of the theoretical value assumed for cruise because of the inefficiency of the blades at high angles of attack. In fact, $\eta = 0.58\,\eta_{\text{IDEAL}}$ yields the handbook value for the SL takeoff of the Cessna 152 (725 ft) when Eq. 14.7 is integrated on a computer. So we'll use this value to gain insight into the factors affecting the takeoff roll. (For the Cessna 182, which has a constant-speed propeller, $\eta = 0.62\,\eta_{\text{IDEAL}}$ yields the published SL takeoff distance.)

Figure 14.2 presents the takeoff acceleration predicted by this theory. The coefficient of rolling friction for a hard surface (and well-pumped-up tires) is usually taken as $\mu = 0.02$, and for this case we see that the rolling friction reduces the accelera-

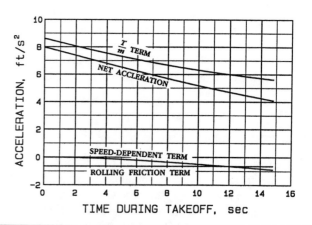

Fig. 14.2 Calculated takeoff acceleration for the C-152 (hard surface, SL, gross weight)

tion by only about 0.75 ft/s². The V^2 term (parasite drag and induced drag) becomes important only in the last half of the takeoff roll or so. The resulting acceleration is predicted to decrease almost linearly from about 8 ft/s² (0.25 g) to about one-half of that value at liftoff.

Figure 14.3 shows the predicted effect of additional friction and higher density altitudes on takeoff performance. Note that the takeoff time and distance increase about two-and-one-half times for the combination of $\mu = 0.1$ (short wet, or long dry, grass) and a 5000 ft density altitude, a notable, even spectacular, effect.

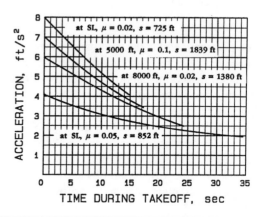

Fig. 14.3 Calculated effect of ground friction and air density on takeoff performance

Figure 14.4 continues the comparisons by plotting the acceleration versus speed, so now all of the plotted quantities end at the recommended liftoff speed of 52 kt. These accelerations decrease even more linearly with speed. Note how a headwind does **not** affect the acceleration very much and so produces a much shorter takeoff because the airspeed **starts** at the headwind value.

The traditional way to plot accelerated motion, as distance versus time, shows the expected upward curving lines (Fig.

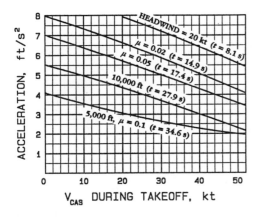

Fig. 14.4 Calculated effect of a headwind, ground friction, and air density on takeoff performance

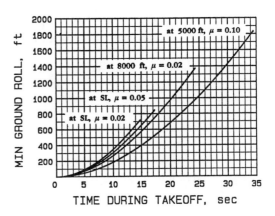

Fig. 14.5 Calculated distance covered as a function of time, takeoff roll

14.5). Note that the **slope** of the lines is the instantaneous speed and so all the lines have the same slope when they end at the liftoff speed.

The effect of wind on the takeoff roll deserves its own graph (Fig. 14.6). Equation 14.7 is adjusted by replacing

$$V^2 \quad \text{with} \quad \left(V - V_{\text{WIND}}\right)^2$$

where the wind speed is a positive number if a headwind and negative if a tailwind. A tailwind helps the plane accelerate, initially, but the plane has to reach a speed just that much greater than the normal liftoff speed before it will leave the ground, too! Note that you have a choice of about 500 ft for your ground roll if you chose to use a 10 kt headwind but almost 1300 ft if you choose the tailwind, even with optimum technique—**spectacular** differences for such **light** winds! You **better** have a good reason before you choose to take off downwind!

That good reason for taking off downwind sometimes can be provided by a runway that is not level; such sloping runways are readily available in hilly or mountainous country. Figure 14.7 estimates the tradeoff conditions, for the 152. What can't be appreciated from this graph is the difference to the occupants of everything **except** the takeoff distance. For example, the graph suggests that the same ground roll will result if, at a density altitude of 10,000 ft and a runway slope of 80 ft per thousand feet, you choose to either takeoff downwind and downhill with a 16 kt wind, or upwind and uphill with the same wind. But the takeoff up the runway is excruciatingly slow, requiring 27 seconds because the acceleration decreases to only about 1 ft/s² at liftoff. On the other hand, if you take off down the sloping runway and accept the tailwind, your acceleration at liftoff is a marvelous 5 ft/s² and takeoff time is only 15 seconds. You find out whether you are going to make it much sooner when you choose the downhill, downwind option!

After having had all this fun with a computer, it is still possible to make useful takeoff predictions with only a calculator, as long as **average** accelerations for the whole takeoff roll are used, rather than trying to calculate them throughout the roll. The three equations describing the interrelationships

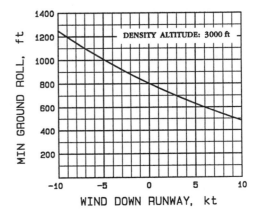

Fig. 14.6 Calculated effect of a headwind or tailwind on the takeoff roll

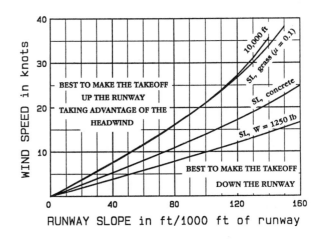

Fig. 14.7 Tradeoff dividing lines between taking off downwind or taking off down a sloping runway

of acceleration, speed, time, and distance for constant or an average acceleration are derived in every introductory college physics text so we'll just list them here.

Kinematic Equations of Motion
(for a Constant or an Average Acceleration)

I. The distance covered by a moving vehicle
  = (initial speed)×(elapsed time)
  + ½ (constant or average acceleration)×(elapsed time)^2

or
$$s = V_0 t + \frac{1}{2} at^2 \tag{14.8}$$

II. Final speed = Initial speed + [constant or average acceleration]×[elapsed time]

or
$$V_F = V_0 + at \tag{14.9}$$

III. (Final speed)^2 = (Initial speed)^2
  + 2×[constant or average acceleration]
  ×[distance covered]

or
$$V_F^2 = V_0^2 + 2as \tag{14.10}$$

These kinematic equations, which follow from the definitions of speed and acceleration, can be applied to the takeoff problem either by estimating the average acceleration from an estimate of the available thrust and the drag or by estimating the corrections to known data for non-standard conditions. We'll give an extended example of the latter here.

EXAMPLE 14.1 The Piper Cadet, at its gross weight of 2325 lb and with zero flap deflection, has a listed minimum takeoff roll of about **980 ft** under SL StAt conditions, using a liftoff speed of 50 kt. Estimate the takeoff roll (a) at a weight of 1800 lb, (b) on a soft field with a rolling coefficient of friction of $\mu = 0.1$, (c) at a density altitude of 4000 ft, (d) with a headwind of 12 kt, and (e) with a tailwind of 12 kt.

Solution: (a) From Eq. 14.10, the **average** acceleration during the takeoff roll under standard conditions is

$$a = \frac{V_F^2 - V_0^2}{2s} = \frac{(50 \times 1.689)^2 - 0^2}{2(980)} = 3.64 \text{ ft/s}^2$$

By Eq. 14.7, the acceleration is inversely proportional to the mass (except for the rolling friction term). This suggests that, in a first approximation, we should adjust the acceleration by multiplying the standard acceleration by the weight ratio:

$$a = (3.64 \text{ ft/s}^2)(2325/1800) = 4.70 \text{ ft/s}^2$$

The liftoff speed can also be reduced because the stall speed depends on the square root of the weight (Eq. 12.15):

$$V_F = 50 \text{ kt} \times \sqrt{1800/2325} = 44.0 \text{ kt}$$

Therefore a first estimate of the ground roll at this reduced weight is

$$s = \frac{V_F^2 - V_0^2}{2a} = \frac{(44 \times 1.689)^2 - 0^2}{2(4.70)} = \mathbf{588 \text{ ft}}$$

which is reasonably close to the handbook value of 520 ft. We could have anticipated the pessimistic result because the reduced mass increases the sum of the first two terms in Eq. 14.7 before the rolling friction term is subtracted.

(b) For fully inflated tires and a hard surface, we have previously assumed a rolling coefficient of friction of $\mu = 0.02$, so that the μg term in Eq. 14.7 and Fig. 14.2 was calculated as $\mu g = (0.02)(32.2) = 0.64 \text{ ft/s}^2$. When μ increases to 0.1 on the soft field, this term increases to $\mu g = (0.1)(32.2) = 3.22 \text{ ft/s}^2$. Therefore we might expect to subtract about half of this, or an additional $(0.5) \times (3.22 - 0.64) = 1.29 \text{ ft/s}^2$ from the standard average acceleration:

$$a = 3.64 \text{ ft/s}^2 - 1.29 \text{ ft/s}^2 = 2.35 \text{ ft/s}^2$$

Then
$$s = \frac{V_F^2 - V_0^2}{2a} = \frac{(50 \times 1.689)^2 - 0^2}{2(2.35)} = \mathbf{1517 \text{ ft}}$$

A slightly different assumption regarding the proper friction coefficient makes a very large difference in the resulting number; for example, the distance more than doubles if the effective friction coefficient is doubled. Anyone who has tried takeoffs from a really soft field might easily believe the larger number. Because the initial thrust and the initial acceleration are greater than their later values, the author can vouch for the fact that it is possible to find a soft field where one can accelerate very close to, but not quite up to, the takeoff speed. Fortunately, airplanes also stop very quickly on this soft a field!

(c) The effect of density altitude enters through the reduction in thrust and the increase in (true) airspeed required for liftoff. First, from Eq. 13.51 we estimate that the power (and the thrust) are reduced by the factor

$$\frac{\left(\frac{\rho}{\rho_0}\right) - 0.12}{0.88} = \frac{\frac{0.002111}{0.002377} - 0.12}{0.88} = 0.873$$

This suggests an acceleration of only

$$a = (3.64 \text{ ft/s}^2)(0.873) = 3.18 \text{ ft/s}^2.$$

Second, by Eq. 12.15, the true airspeed at liftoff should increase to

$$V_F = \left(50 \text{ kt}\right) \times \sqrt{\frac{0.002377}{0.002111}} = 53.1 \text{ kt}$$

Therefore we estimate that

$$s = \frac{V_F^2 - V_0^2}{2a} = \frac{(53.1 \times 1.689)^2 - 0^2}{2(3.18)} = \mathbf{1265 \text{ ft}}$$

The handbook value is about 1600 ft, so this average method is considerably more optimistic. As with the weight adjustment, we could obtain slightly better agreement if we had applied the thrust-reduction adjustment to just the first term in Eq. 14.7.

The Cadet handbook predicts a nearly linear increase in takeoff roll with density altitude while the Cessna 152 handbook predicts a slowly increasing factor. These values are presumably based on averages to a substantial bank of experimental data. For the Cadet, the handbook indicates a 15% to 16% increase over the SL value for every additional thousand feet of density altitude.

(d) The effect of a headwind or tailwind is principally the change in the ground speed at liftoff. For a headwind of 12 kt, a true airspeed of 50 kt corresponds to a ground speed of 38 kt. Therefore we estimate that

$$s = \frac{V_F^2 - V_0^2}{2a} = \frac{(38 \times 1.689)^2 - 0^2}{2(3.64)} = \textbf{566 ft}$$

whereas the pilot handbook indicates about 800 ft.

(e) For a tailwind of 12 kt,

$$s = \frac{(62 \times 1.689)^2 - 0^2}{2(3.64)} = \textbf{1506 ft}$$

while the handbook indicates about 1640 ft.

Note that we must use the equivalent airspeed in determining the aerodynamic lift or drag but the ground speed must be used in the kinematic equations because the accelerations are calculated relative to the ground.

Although adjustments to the average takeoff acceleration at SL haven't resulted in numbers that agree very well with handbook values, it should be noted that there are also great uncertainties in the ability of a specific pilot/airplane combination to match the specifications. Takeoff is an especially good time to be conservative in making the go/no-go decision. Also, in many cases, the total distance required to clear an obstacle is more important than just the takeoff distance.

To add a little spice and scientific excitement to your flying, try obtaining takeoff data in your choice of flying machine. The only equipment you need is a portable tape recorder and a stopwatch. Sound off each time the aircraft comes abreast of a known point such as a runway light. When you play back the tape, you'll have distance/time data. It is not hard to get data but it is hard to get **good** data — because of variable winds and pilot technique and sloping runways. Anyway, Fig. 14.8 shows some experimental data and a fitted curve for a Piper Tomahawk. Note that the calculated acceleration decreased from 8.5 ft/s² to about 4.0 ft/s² — which is in line with the relative decrease that was calculated and displayed in Fig. 14.2 for a similar but different aircraft.

The airspeed readings during the takeoff roll can also be transmitted to a tape recorder to obtain speed/time data of somewhat lesser quality, especially because the airspeed indicator doesn't even register at the beginning of the roll. Experimental data for the same Tomahawk are shown in Fig. 14.9 along with a different fitted curve.

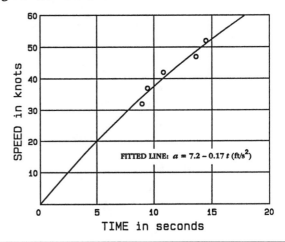

Fig. 14.9 Experimental takeoff data for a Piper Tomahawk

B. Climbing Flight

After takeoff the pilot allows the aircraft to accelerate to a climb speed appropriate to the airport conditions and the flight plan. Most commonly, a relatively high (cruise climb) speed is chosen, one that provides reasonable altitude gain while maintaining good cockpit visibility and good engine cooling; this type of climb also yields the most options if turbulence or a power loss is encountered.

If there are obstructions ahead, though, the pilot must choose an initial climb speed only slightly greater than the stall speed, one that yields the best (maximum) **angle** of climb; the

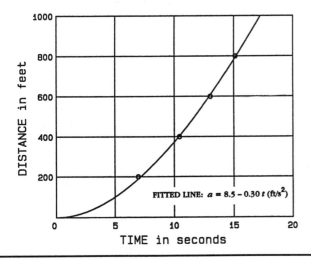

Fig. 14.8 Experimental takeoff data for a Piper Tomahawk

Pic. 14.4 High powered Cubs, with their USA 35B airfoils and a low wing-loading, make good load carriers.

FAA uses the symbol V_X for this speed, and its value is always listed in the flight manuals of modern aircraft.

A third airspeed that is sometimes important to know is that which yields the best (maximum) **rate** of climb; the FAA uses the symbol V_Y for this speed. V_X is the most important of these speeds from a safety standpoint. V_Y is used to most quickly climb over hills surrounding an airport; it also is used at high altitudes where the barely available climb ability quickly goes away at other speeds. Our aim in this section is to gain insights into what factors enter into the rate and angle of climb that is achieved. For a particular aircraft, the recommendations of the flight manual should be followed.

Figure 14.10 depicts the forces acting on an aircraft during a constant-speed (and therefore also a constant-angle) climb. In this figure, the thrust force T is depicted as being directed parallel to the chord line of the wing; this is often not quite the case, but the error resulting from such an assumption is normally small enough to neglect. The net aerodynamic lift L is the combined effect of the wing lift, a small negative lift provided by the tail (note the deflected elevator surface), and perhaps a small positive lift provided by the fuselage.

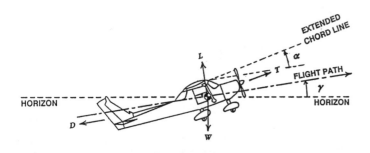

Fig. 14.10 External forces on an aircraft in a steady climb

The angle of attack of the wing, α, is the angle between the relative wind (or the flight path) and the chord line, as usual. The climb angle γ (Greek letter gamma) is the angle between the flight path and the horizon (Chap. 2). The aerodynamic drag D is the combined effect of the induced drag of (mostly) the wing and the parasite drag of the whole aircraft. After making the traditional resolution of the resultant aerodynamic force into the two perpendicular lift and drag components for the **whole** airplane, it is then convenient to apply Newton's second law along these directions. The forces must sum to zero in any direction because the climb is assumed to be at a constant speed and at a constant angle (force **equilibrium** − see Sec. A of Chap. 13).

Parallel to the flight path:
$$\sum F_{PAR} = T \cos \alpha - W \sin \gamma - D = 0 \quad (14.11)$$

Perpendicular to flight path:
$$\sum F_{PERP} = L + T \sin \alpha - \cos \gamma = 0 \quad (14.12)$$

Figure 14.11 instructs us regarding what use to make of these equations. In particular, note that the rate of climb,

V_{VERT} (where the speeds are all **true** airspeeds), is just $V \sin \gamma$ (the vertical component of the speed), where γ (the Greek letter gamma) is the climb angle:
$$\text{Rate of Climb} = V_{VERT} = V \sin \gamma \quad (14.13)$$

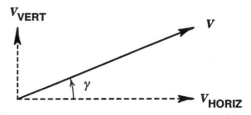

Fig. 14.11 Horizontal and vertical components of the climb velocity

Therefore it appears that we should solve Eq. 14.11 for $\sin(\gamma)$ and then multiply it by the flight speed:
$$V_{VERT} = V \left(\frac{T \cos \alpha - D}{W} \right) = \frac{TV \cos \alpha - DV}{W} \quad (14.14)$$

From Eqs. 13.21 and 13.44 we see that this can also be written as
$$V_{VERT} = \frac{P_{AVAIL} \cos \alpha - P_{REQ}}{W} \quad (14.15)$$

Therefore the **maximum** rate of climb, for a given weight, occurs at the speed for which the numerator has a maximum value and the denominator a minimum value:
$$V_{VERT,MAX} = \left(\frac{P_{AVAIL} \cos \alpha - P_{REQ}}{W} \right)_{max} \quad (14.16)$$
MAXIMUM RATE OF CLIMB

The $\cos \alpha$ term appears because the aircraft nose is pointing higher up in the sky that is the direction of flight. We expect α, the angle of attack, to be quite small at high climb speeds but it will approach the stall angle (perhaps 16° to 20°) as the climb speed is reduced to the best angle-of-climb speed. Nevertheless, it is a common practice to assume that α is sufficiently small that $\cos \alpha \approx 1$. Then
$$V_{VERT} \approx \frac{P_{AVAIL} - P_{REQ}}{W} \equiv \frac{\text{Excess Power}}{W} \quad (14.17)$$

and the really nice feature of this equation is that it tells us to look at the P_{AVAIL} and P_{REQ} curves of Chap. 13 (e.g, Fig. 13.10 at 1670 lb and at SL); the airspeed yielding the best rate of climb should then be just about the speed at which there is a maximum **difference** between the P_{REQ} and P_{AVAIL} curves.

Figure 14.12 presents calculations of Eqs. 14.15 and 14.17 using the fitted f and e obtained for the 1978 Cessna 152 in Chap. 13, the same theory that produced the P_{AVAIL} and P_{REQ} curves of Fig. 13.10. Note that the approximate equation is really quite good; it predicted a maximum rate of climb of 878 ft/min at $V_Y = 65.5$ kt while the more accurate Eq. 14.15 predicted a maximum climb rate of 869 ft/min at $V_Y = 66$ kt.

(For comparison, the flight manual gives 715 ft/min at 67 kt under these conditions. Limitations of the drag polar approximation and the propeller momentum theory probably account for the differences.)

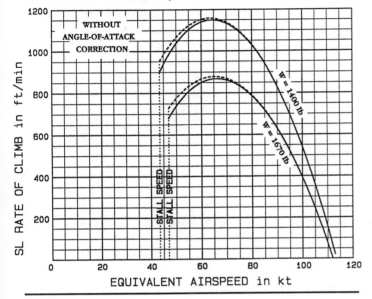

Fig. 14.12 Calculated rate of climb for a C-152 at two different weights

The angle of attack, needed to use the more accurate Eq. 14.15, was estimated by fitting a straight line to the Cessna's 2412 airfoil section characteristics (Fig. 11.1) to obtain $c_\ell = 0.25 + 0.103\,\alpha_{2D}$, where α_{2D} is the section angle of attack in degrees. This section angle of attack was converted to a wing angle of attack with the help of Eq. 12.11:

$$\alpha_{3D} = \alpha_{2D} + \frac{57.3\,C_L}{(\pi)\,(0.8)\,(6.9)} = \alpha_{2D} + 3.3\,C_L$$

so that $\alpha_{2D} = \alpha_{3D} - 3.3\,C_L$

and, after some algebra,

$$C_L = 0.187 + 0.0768\,\alpha_{3D} \qquad (14.18)$$

The C_L required for each airspeed was obtained from Eq. 14.12:

$$C_L = \frac{L}{qS} \approx \frac{W}{qS}$$

and then an estimate for the angle of attack was obtained from Eq. 14.18.

EXAMPLE 14.2. (a) Estimate the increase in the rate of climb at SL and at the gross weight of 1670 lb if a 125 hp engine is substituted for the 110 hp engine more commonly used in the 152. (b) Estimate the increase in the rate of climb at SL with the standard engine if the weight is reduced to 1450 lb.

Solution: (a) Using the factory-supplied maximum rate of climb of 715 ft/min, we estimate that the excess horsepower with the 110 hp engine (from Eq. 14.17) is

$$\text{Excess Power} = V_{\text{VERT}} \times W$$

$$= \left(\frac{715\ \text{ft}}{\text{min}}\right)\left(\frac{1\ \text{min}}{60\ \text{s}}\right)(1670\ \text{lb}) = 1.99 \times 10^4\ \text{ft-lb/s}$$

$$= 36.2\ \text{hp}$$

If we estimate that the additional 15 hp will be given to a propeller that is only 70% efficient in the climb, then

Additional Excess Power
$$= (0.70)(15\ \text{hp}) = 10.5\ \text{hp} = 0.58 \times 10^4\ \text{ft-lb/s}$$

So we estimate that the maximum rate of climb at SL will jump from 715 ft/min to

$$V_{\text{VERT}} = \frac{\text{new excess power}}{W}$$

$$= \frac{(1.99 \times 10^4 + 0.58 \times 10^4)\ \text{ft-lb/s}}{1670\ \text{lb}}$$

$$= 15.4\ \text{ft/s} = \textbf{922 ft/min}$$

This is a very nice (29%) gain in climb rate for such a modest (14%) increase in power. We'll see a little later that the increase in cruise speed for this power increase is much less noticeable.

This calculation also warns us that it doesn't require a very sick engine to bring the rate of climb down to unpleasantly low levels.

(b) If we ignore the decrease in power required at the lower weight, the rate of climb is inversely proportional to the weight, by Eq. 14.17, so we expect that

$$\left(V_{\text{VERT at 1450 lb}}\right) \ge \left(V_{\text{VERT at 1670 lb}}\right)\left(\frac{1670}{1450}\right)$$

$$= (715)(1.152) = \textbf{823 ft/min}.$$

The climb angle also increases by about 15% as well, as it turns out. This is the explanation for the improved climbout performance that a gleeful first-solo pilot will report after kicking out the clangorous 170 lb bulk in the right seat. If we had included the reduction in required power in our calculation, we would have calculated a rate of climb of about 885 ft/min, a relatively small error. At higher density altitudes, though, the power-required and power-available curves close in on each other and the percentage error from ignoring the reduction in power required becomes very significant.

Notice that Fig. 14.12 predicts a very nice 675 ft/min climb rate just as the power-off stall speed of 47 kt is reached. This is undoubtedly optimistic (because the parabolic drag polar approximation doesn't model this region well). But the real problem with trying to climb at the stall speed is **not** that there isn't some excess power available but that the pitch and bank angles are no longer controllable after one or both of the wings stall, so the excess power isn't usable. This characteristic proves particularly treacherous when a pilot is trying frantically to clear an obstacle and keeps raising the nose "above" the obstacle—right up until the plane rolls and dives into it. I've watched it happen.

At altitude, a continued attempt to "fly" at or below the stall speed with full power is likely to lead to an uncontrollable roll off into a power spin, as one wing always stalls before the other.

Figure 14.13 explores the effect of density altitude and weight on the rate of climb, using the theoretical dependence of P_{REQ} and P_{AVAIL}. The top curve shows that the reduction in weight has not only increased the maximum climb rate, as expected, but the speed to obtain this has decreased. It works the other direction, too: If you overload an aircraft, the optimum climb speed will also increase.

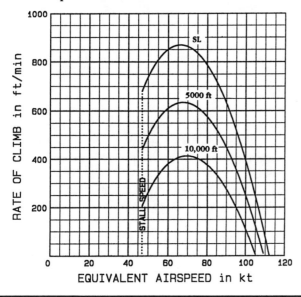

Fig. 14.13 Calculated effect of density altitude on the rate of climb for a C-152

The increase in density altitude has a powerful effect on climb rate, as expected from Eq. 14.17 and power curves such as Figs. 13.10 and 13.11. One useful fact pointed out by Fig. 14.13 is that the maximum rate of climb is a nearly **linear** function of density altitude:

$$\begin{pmatrix}\text{Max Rate of Climb}\\ \text{at altitude } h\end{pmatrix}$$
$$\approx \begin{pmatrix}\text{Max Rate of Climb}\\ \text{at Sea Level}\end{pmatrix} - (\text{constant})\begin{pmatrix}\text{Density Altitude}\\ \text{at altitude } h\end{pmatrix}$$

EXAMPLE 14.4. The 152 pilot manual specifies a maximum rate of climb of 715 ft/min at SL and 465 ft/min at a density altitude of 6000 ft. Estimate the maximum rate of climb at 10,000 ft from these data.

Solution: The rate of climb decreased $(715-465)$ ft/min = 250 ft/min in the 6000 ft, so it changes about $(250 \text{ ft/min})/6 = 41.7$ ft/min for every thousand feet of change in density altitude. Therefore in another 4 thousand feet it should decrease about $\left(4 \times (41.7 \text{ ft/min})\right)$ or 167 ft/min, from 465 ft/min to **298 ft/min**. Alternately, in equation form:

$$V_{VERT, MAX} = 715 - (41.7)\begin{pmatrix}h \text{ in thousands}\\ \text{of feet}\end{pmatrix}$$
$$= 715 - (41.7)(10) = \mathbf{298 \text{ ft/min}}$$

For comparison, the flight manual specifies 300 ft/min as the climb rate at 10,000 ft.

The **service ceiling** of a particular aircraft is defined as the altitude for which the aircraft has a maximum rate of climb of only **100** ft/min; at these high density altitudes the pilot must carefully maintain V_Y if any perceptible gain in altitude is to be observed; also, the engine must be very carefully leaned for maximum power. The service ceiling of the 152 is listed as a density altitude of 14,700 ft but it is unlikely that any reader has ventured that high in this aircraft (oxygen?).

Even though the rate of climb steadily decreases as the density altitude increases (for a normally aspirated or non-supercharged engine), the true airspeed for 75% power continues to increase because the dominating drag is parasite drag which is proportional to the air density. For the recommended 75% maximum cruise power, the highest cruising speed (and the value usually quoted by manufacturers!) occurs at the altitude where **full** throttle and careful leaning produce only 75% of the SL power. Equation 13.51 predicts that this will occur at an air density of 0.001854 sl/ft^3, which corresponds to a density altitude of about 8250 ft.

The other speed, the important best angle-of-climb speed, V_X, is also obtainable from the rate-of-climb curves. A little study of Fig. 14.11 shows why. Because the maximum climb angle occurs when the sine of the climb angle is a maximum, the maximum γ occurs when the ratio $\left(\dfrac{V_{VERT}}{V}\right)$ is a maximum. But then notice that, in Figs. 14.12 and 14.13, we have actually plotted V_{VERT} versus V. Therefore the best angle-of-climb speed is the point on the rate-of-climb curve for which $\left(\dfrac{V_{VERT}}{V}\right)$ is a maximum. To determine this point geometrically, draw a straight line from the origin to the point on the curve which

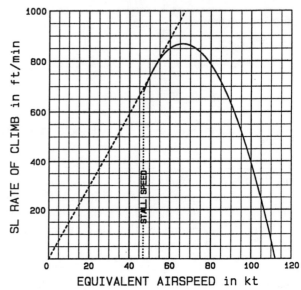

Fig. 14.14 Determination of the best angle-of-climb speed from the rate-of-climb variation

yields the greatest angle for the line (Fig. 14.14). Evidently this speed, V_X, will **always** be **less** than V_Y because V_Y is the speed at the peak of the curve.

EXAMPLE 14.5. Estimate the optimum climb **angle** for a 152 under SL conditions and at 1670 lb for (**a**) no wind, (**b**) a 10 kt headwind, and (**c**) a 10 kt tailwind.

Solution: (a) By inspection of Fig. 14.14,

$$V_{\text{VERT}} \approx 777 \text{ ft/min} = 12.95 \text{ ft/s}$$
$$\text{at } V_X \approx 52.5 \text{ kt} = 88.7 \text{ ft/s}.$$

Therefore, from Eq. 14.13,

$$\sin\gamma = \frac{V_{\text{VERT}}}{V} = \frac{12.95 \text{ ft/s}}{88.7 \text{ ft/s}} = 0.146$$

so that $\gamma \approx 8.4°$

(b) The effect of wind on climb angles was covered in Chap. 2 as a vector problem, but it is important enough to warrant a reprise. The basic idea is that a "pure" wind, one without updrafts or downdrafts, changes only the **horizontal** component of the aircraft's speed relative to the **ground**. (The aircraft itself responds only to gravity and to the speed and direction of the air moving past the plane, irrespective of the plane's velocity relative to the ground; however, the velocity relative to the ground determines important details such as obstacle clearance.)

Using Fig. 14.11, the horizontal component of the aircraft's velocity relative to the ground, in no-wind conditions, is

$$V_{\text{HORIZ}} = V \cos\gamma$$
$$= (88.6 \text{ ft/s})[\cos(8.4°)] = \mathbf{87.6 \text{ ft/s}}$$

With a headwind of 10 kt (16.89 ft/s), this becomes 70.7 ft/s, so

$$\sin\gamma = \frac{12.95}{70.7} = 0.183 \quad \text{and so} \quad \gamma = 10.6°$$

because the rate of climb is not affected by a steady horizontal wind. This is a very pleasant increase in the ability to clear obstacles.

(c) With a tailwind of 10 kt, the horizontal component of the velocity relative to the ground becomes 104.5 ft/s, so

$$\sin\gamma = \frac{12.95}{104.5} = 0.124 \quad \text{and so} \quad \gamma = 7.1°$$

which is not so nice.

For very high performance aircraft, such as modern jet fighters, the climb rate is still of great interest but obstacle clearance is not—because, once they are airborne, many have enough thrust to rotate to a vertical climb angle and continue to accelerate at a zero-lift angle of attack (like a rocket).

C. Cruise for Endurance

There are a few occasions, such as a surveillance mission or when in a holding pattern or when lost over shark-infested water, when a pilot might wish to use the power setting that keeps the airplane in the air for the longest possible time, based on a minimum rate of fuel consumption. If the fuel mixture is properly adjusted, minimum fuel consumption occurs when the power required for level flight is also a minimum. As Fig. 13.9 shows, this minimum power point occurs rather close to the stall speed, when induced drag starts to increase faster than the parasitic drag is decreasing. (The actual minimum power point is probably deeper and may occur at a slightly higher speed because the drag polar approximation doesn't adequately model the partial separation and extra drag which typically occurs somewhat before the stall and complete separation.)

A modest amount of algebra tells us what determines the minimum power/maximum endurance speed for a particular airplane, as well as the factors that determine this minimum power. First, because the net lift force on the aircraft is equal to its weight when in level flight,

$$L = W = C_L q S = C_L \left(\tfrac{1}{2}\rho V^2\right) S$$

and so

$$V = \sqrt{\frac{2W}{\rho C_L S}} \tag{14.19}$$

TRUE FLIGHT SPEED IN LEVEL FLIGHT

In the other dimension, the thrust is essentially equal to the drag, so the power required for level flight is

$$P_{\text{REQ}} = D V$$
$$= C_D q S V = C_D \left(\tfrac{1}{2}\rho V^2\right) S V$$
$$= \left(\tfrac{1}{2}C_D \rho S\right) V^3 \tag{14.20}$$

Substituting Eq. 14.19 into Eq. 14.20 yields

$$P_{\text{REQ}} = \left[\tfrac{1}{2}C_D \rho S\right]\left(\frac{2W}{\rho C_L S}\right)^{3/2}$$
$$= \left(\frac{2W^3}{\rho S}\right)^{1/2}\left(\frac{C_D}{C_L^{3/2}}\right) \tag{14.21}$$

For a given airplane, the minimum required power occurs when everything in the numerator in Eq. 14.21 is as small as we can make it while everything in the denominator is as large as we can make it. While in flight, though, there is little we can do about the variables in the first set of parentheses: the weight, air density, and wing area are adjustable only if we embrace desperate measures such as throwing out baggage. The second set of parentheses contains the complete dependence on airspeed, which is under direct control of the pilot. Therefore

> **Maximum Endurance**
> **(for propeller-driven aircraft)**
> **is obtained at the speed for which**
> $$\left(\frac{C_L^{3/2}}{C_D}\right)$$
> **for the whole aircraft is a maximum.**

Predicting the C_L, the C_D, and the speed at which this maximum endurance will be obtained, using the parabolic drag polar approximation of Eq. 13.41, is a nice calculus exercise.

Pic. 14.5 A Cessna 150, just cruising along

The procedure is to solve for C_L, form $C_L^{3/2}$ from it, divide by C_D, and take the derivative with respect to C_D. To obtain the speed corresponding to this point, differential calculus is similarly applied to Eq. 13.45 for the power. The results are

$$\left(\frac{C_L^{3/2}}{C_D}\right)_{MAX} = \frac{1}{4}\left(\frac{3\pi e b^2}{S}\right)^{3/4}\left(\frac{S}{f}\right)^{1/4} \tag{14.22}$$

and this is obtained at

$$C_L = \frac{b}{S}\sqrt{3\pi e f} \tag{14.23}$$

or at

$$C_D = 4\left(\frac{f}{S}\right) \tag{14.24}$$

and at a true speed of

$$V_{\text{MIN POWER}} = \sqrt{\frac{2W}{b\rho\sqrt{3\pi e f}}} \tag{14.25}$$

MINIMUM POWER, MAXIMUM ENDURANCE
(Propeller Aircraft, Parabolic Drag Polar Approximation)

As anticipated, the calculated speeds are quite slow. For the 152 that was modeled in Chap. 13 with $f = 5.39$ ft^2 and with

$e = 0.716$, $\left(C_L^{3/2}/C_D\right)_{MAX} = 10.44$ at $C_L = 1.26$ or $C_D = 0.135$. At the gross weight of 1670 lb, these coefficient values are obtained at a calculated speed of 49.5 kt — only 2.5 kt above the stall speed for a mid-range cg — so the limitations of the drag polar approximation must be considered. Figure 14.15 shows how the predicted coefficient ratio varies with lift coefficient; evidently the endurance falls off rapidly at speeds greater than the optimum speed. Figure 14.16 presents the variation of the predicted optimum speed with weight, which is an important consideration in any aircraft with very long endurance.

Using the calculated speed in the power-required equation, Eq. 13.45, we find that minimum power to stay in the air at SL and at gross weight is predicted to be only 27.2 hp; with a 70% propeller efficiency, this would require only 39 hp from the available 110 hp, or 35% power. Unfortunately, an unmodified engine cannot run efficiently or without special spark plugs at this low power setting, so this alone would force practical endurance speeds to larger values than those theoretically predicted. The message to pilots is still clear: Reduce to minimum recommended power settings when seeking to endure the most flying on a tank of gas.

Pic. 14.6 This P-51 Mustang seems to be cruising a little low today.

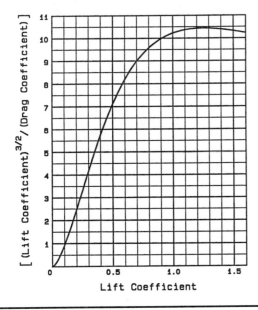

Fig. 14.15 Calculated dependence of the coefficient ratio on the lift coefficient

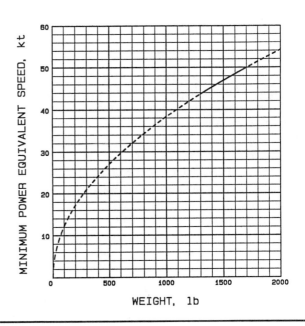

Fig. 14.16 Calculated dependence of minimum-power speed on weight

A jet-powered aircraft avails itself of a power source that produces nearly constant thrust (rather than power) with speed. The consequence is that such an aircraft obtains its maximum endurance at the speed for which $\left(C_L / C_D\right)$ is a maximum (Ref. 8).

D. Cruise for Maximum Speed

Most aircraft are purchased or garage-built in order to go places, **fast**. It is of interest to explore what limits the speed of a particular aircraft and, specifically, the speed benefits of adding additional power.

Under SL conditions a properly designed aircraft will have its maximum speed limited almost entirely by the **parasite drag** part of the power required (Fig. 13.9). Equation 13.45 tells us that the power required to overcome parasite drag varies as the **cube** of the airspeed. Therefore we can expect that an increase in top speed, from V_1 to V_2, produced by an increase in power from P_1 to P_2, will be closely predicted by

$$\frac{P_1}{P_2} = \left(\frac{V_1}{V_2}\right)^3$$

or
$$V_2 = \left(\frac{P_2}{P_1}\right)^{1/3} V_1 \tag{14.26}$$

EXAMPLE 14.6. The Piper Comanche was made in both a 250 hp and a 400 hp version for a while. The machine is aerodynamically pretty clean, with retractable landing gear and a laminar airfoil. Estimate the increase in top speed at SL produced by this **60%** increase in available horsepower.

Solution. From Eq. 14.26, $V_2 = \left(\dfrac{400}{250}\right)^{1/3} V_1 = (1.17) V_2$

or only a disappointing **17%** increase in top speed. This correlates very well with the 16% increase quoted by the manufacturer.

This example points out the importance of not **overpower**ing an airplane. Once adequate climb performance is available, any additional power simply eats away the useful load and the endurance. A much better way to increase the top speed and the cruising speed is to reduce the equivalent flat plate area by minimizing cooling and interference drag. A new, cleaned-up version of this old Comanche may be in the offing.

For decades the graybeards have been advising fledglings to put their aircraft "**on the step**" so as to realize the maximum cruising speeds of which their machines are capable, somewhat as a speedboat or floatplane rises up and rides on just part of its undersurface at high water speeds. Usually the instructions are to set the power at an altitude slightly above the desired cruising altitude and then dive gently down to that altitude. This is supposed to result in an indicated airspeed greater than simply leveling off at the cruising altitude with the same power setting.

For very clean airplanes, it takes a long minute or so after leveling off before the cruising speed settles on its final value, so for these aircraft it should be possible to arrive at the final cruise speed faster (not counting the extra climb time) by diving to it—but the experimental evidence of which the author is aware suggests that the final speed will be the same in either case. Anecdotal evidence is not convincing until it is supported by carefully controlled tests—another good scientific excuse for going flying.

Power-speed curves such as Fig. 13.12 do show that there may be two speeds at which an aircraft can maintain level flight for a given power setting because of the way in which the power-required curve rises at both ends of the speed range. But this represents a significant difference in speeds and in angles of attack and certainly is not what is usually meant by "getting on the step," although the two speeds could be relatively close together (inside the "coffin corner") at very high altitudes (Fig. 7.5).

For jet-powered aircraft, with their nearly linear power-available curves, it is possible for the power-available curve to coincide with the power-required curve through a range of speeds and then it is important for the pilot to keep the plane at the high part of this range.

What is generally being claimed, though, is that there is **hysteresis** in the power-required curve—that the power required (and therefore the drag) is different depending on whether a speed is approached from above or below that speed. Getting "on the step" would then involve following the arrows on Fig. 14.17. This curve is reminiscent of the curves showing the drag when laminar flow transitions to turbulent flow; there is commonly an overlap region and the laminar flow region can be extended well into the turbulent region if the surfaces are very smooth and nothing appears to trigger the transition. (Refer to Fig. 11.20.) We also recall the observed hysteresis in airfoil coefficients at low RN (Figs. 10.36 and 11.13).

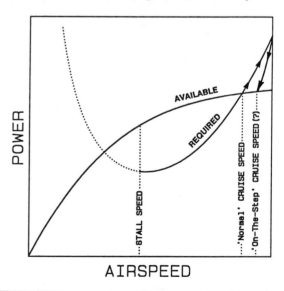

Fig. 14.17 Could there be hysteresis in the drag curve?

And so it can be suspected that one might be able to get an airplane into a lower drag condition by diving it to a smaller angle of attack where more laminar flow is encouraged and then gradually increasing the angle of attack to get enough lift for level flight. But would such a condition, monitored over a period of minutes, be a stable one? Reference 1 suggests strongly that the answer is **NO**.

If the engine is throttled back or if the plane is at high altitude, the cruise C_D could be well above C_{MIN} and then even a gentle dive could produce a significant increase in speed — but no "step" would be involved. Feel free to conduct your own careful experiments and draw your own conclusions.

E. Cruise for Range

From December 14, 1986, to December 23, 1986, pilot/homebuilders Dick Rutan and Jeana Yeager held the rapt attention of the world as they piloted their Burt Rutan–designed Voyager aircraft around the globe nonstop and unrefueled, thereby establishing a record for range that will most probably never be broken again (Pic. 14.7). Voyager was a master's level lesson in both the design and piloting of an aircraft for maximum range. And so we turn now to the factors that determine the range of an aircraft; afterwards it will be possible to return to the Voyager and appreciate the reasons for its special success.

To obtain the maximum distance on a tankful of fuel, it is necessary to maximize the **ratio** of the **speed** to the **fuel consumption**. That is, if we maximize the ratio [miles/hr] ÷ [gal/hr], we also maximize the [miles/gal]. Because fuel consumption is proportional to the power used, the ratio that we want to maximize is

$$\frac{V}{P_{REQ}} = \frac{\left(\dfrac{2W}{\rho\, C_L\, S}\right)^{1/2}}{\left(\dfrac{2W^3}{\rho\, S}\right)^{1/2}\left(\dfrac{C_D}{C_L^{3/2}}\right)} = \frac{1}{W}\left(\frac{C_L}{C_D}\right) \qquad (14.27)$$

where we have substituted for the speed from Eq. 14.19 and for P_{REQ} from Eq. 14.20. From this simple result we can conclude that

> **Maximum range**
> **under no-wind conditions**
> **results when an airplane's speed**
> **is continuously adjusted**
>
> **so that** $\left(\dfrac{C_L}{C_D}\right)$ **for the whole airplane**
>
> **remains constant at its maximum value.**

CRITERION FOR MAXIMUM RANGE (Propeller-driven Aircraft)

This is a most interesting development. Recall that we used this same ratio as a measure of the efficiency of an **airfoil** in previous chapters. Now we must use this information to tell a pilot **how** to fly a given aircraft for maximum range and also to tell the designer how to design an aircraft for good range. And,

just as for the endurance calculations of Sec. C, calculus makes it possible to obtain expressions for the C_D, C_L, and V at which this maximum $\left(C_L / C_D\right)$ will be obtained, within the parabolic drag polar approximation. The results are

$$\left(\frac{C_L}{C_D}\right)_{MAX} = \frac{b}{2}\sqrt{\frac{\pi e}{f}} \qquad (14.28)$$

and this is obtained at

$$C_L = \frac{b}{S}\sqrt{\pi e\, f} \qquad (14.29)$$

or at $\quad C_D = 2\left(\dfrac{f}{S}\right) \qquad (14.30)$

which corresponds to a true speed of

$$V = \sqrt{\frac{2W}{b\,\rho\,\sqrt{\pi e\, f}}} \qquad (14.31)$$

CONDITION FOR MAXIMUM RANGE
(Parabolic Drag Polar Approximation)

Unlike the endurance speeds, the best-range speeds are fairly close to normal operating speeds. The f and e derived for the 152, for example, yield $(C_L / C_D)_{MAX} = 10.8$ at $C_L = 0.73$ or $C_D = 0.067$; at 1670 lb and SL, these correspond to an airspeed of 65 kt. For level flight, though, this still represents less than 40% of the available power.

The calculated variation of the coefficient ratio for the 152 is shown in Fig. 14.18. The variation of the best-range speed with weight is presented in Fig. 14.19. For a truly long range aircraft, the fuel weight is a large part of the useful load and the variation in best-range speed correspondingly greater than that shown.

It is intuitively obvious that a pilot should choose a higher cruise speed by using a **higher** power setting to obtain the maximum range when faced with very **strong** headwinds. The 152 at 3000 ft, for example, can travel at 102 kt (75% power) for

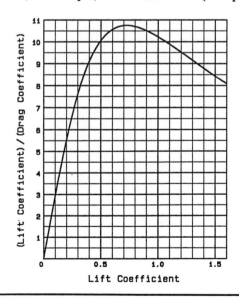

Fig. 14.18 Calculated dependence of the lift/drag ratio on the lift coefficient

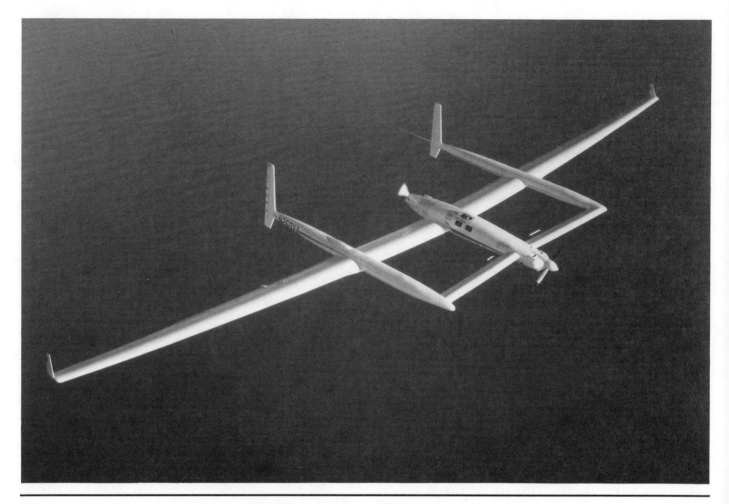

Pic. 14.7 The Voyager aircraft — around the world unrefueled! (Photo courtesy of Jeana Yeager)

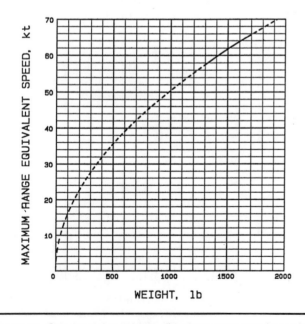

Fig. 14.19 Calculated dependence of the best-range speed on weight

3.38 hours or it can travel at 89 kt (55% power) for 4.47 hours. In the face of an 89 kt headwind, the 75% power setting would yield a 13 kt ground speed while 55% power would merely keep one stationary over the ground! What is **not** obvious (because the best range occurs at such a very low power setting) but what is easily calculated by reference to aircraft performance data is that there is **no** range advantage to cruising above 55% power until the headwind is greater than about **half** of the cruising speed (50 kt for the Cessna 152).

An expression for the total **distance** covered by an aircraft, assumed to be flying at a constant (C_L / C_D) with a constant specific fuel consumption rate Q, had been derived by about 1919 and is usually (perhaps without good reason) associated with the famous French aviator/designer Louis-Charles Breguet. The derivation follows from the use of integral calculus, necessary because the weight of the aircraft changes so much during the flight. The result is

$$\text{Range} \; = \; \frac{\eta}{Q} \, \frac{C_L}{C_D} \ln\left(\frac{W_{\text{T-O}}}{W_{\text{LAND}}}\right) \qquad (14.32)$$

BREGUET RANGE EQUATION
(Propeller-driven Aircraft)

where the units used must be from a consistent set, except the weights can be in any units because it is only the ratio of the takeoff to landing weights that is used. The "ln ( )" means to take the natural logarithm of the quantity in parentheses. (On a calculator, use the $ln\,x$ key.) The propulsion efficiency, η, doesn't have units. The units of Q are (weight of fuel) per (unit of power) per (unit of time), or lb/(ft-lb/s)/s = 1/ft so the distance travelled will be in the unit of ft if these units for Q are used. If Q is specified in lb/hp/hr, as is most common, the range in nautical miles is given by

$$\text{Range} = 325.87\,\frac{\eta}{Q}\,\frac{C_L}{C_D}\,\ln\!\left(\frac{W_{T\text{-}O}}{W_{\text{LAND}}}\right) \qquad (14.33)$$

Range in nm for Fuel Consumption Q in lb/hp/hr
(Propeller-driven Aircraft)

There are several important adjustments to be applied when using this equation. It doesn't consider (a) the climb to the cruising altitude or changes in altitude occasioned by weather or (b) the effect of headwinds or tailwinds. Obviously the maximum range is obtained by using the minimum Q and the maximum (C_L/C_D) — but these may well occur at different speeds!

Everything else being equal, the greater the fuel load the greater the range, of course. What isn't at all obvious is the diminishing returns from adding additional fuel, because of the fact that extra weight reduces the efficiency (see Eq. 14.27) so that the last pound of fuel that is added gives much less distance over the ground than does the first since it is just that much more to carry around before using. Mathematically, this comes out in the calculation because the range depends on the natural logarithm of the takeoff/landing weight ratio. (The natural logarithm of a number is the power or exponent to which the number $e = 2.71828...$ has to be raised to give that number. Because $e^0 = 1$, $e^1 \approx 2.718$, and $e^2 \approx 7.389$, $\ln(1) = 0$, $\ln(2.718)$ ≈ 1, and $\ln(7.389) \approx 2$, for example.) Thus, if $W_{T\text{-}O} = 2\,W_{\text{LAND}}$, we use $\ln(2) \approx 0.693$ and if $W_{T\text{-}O} = 4\,W_{\text{LAND}}$, we must use $\ln(4)$ ≈ 1.386 so increasing our weight ratio by four times has only **doubled** the range! This is why Voyager's doubling of the previous record for unrefueled flight is so very impressive; it certainly earned its present place of honor in the Smithsonian Air and Space Museum, alongside the Wright *Flyer* and the Lindbergh *Spirit of St. Louis*.

EXAMPLE 14.7. Voyager's takeoff weight on its around-the-world record flight was 9694 lb and its landing weight was 2699 lb, which included 106 lb of remaining fuel. It had an estimated $(C_L/C_D)_{\text{MAX}}$ of 36.5 with laminar flow, but only 28 without. The rear engine, used for most of the flight, was a new 130 hp liquid-cooled Continental which gave better fuel consumption than air-cooled engines, about 0.4 lb/hp/hr. Propeller efficiency was about 80%. Using an average (C_L/C_D) of 30, estimate the still-air range for this aircraft, using its actual takeoff and landing weights on the world record flight.

Solution: From Eq. 14.33,

$$\text{Range} = (325.87)(0.8/0.4)(30)\,\ln(9694/2699)$$
$$= \textbf{25,000 nm}$$

The Voyager actually travelled about 23,000 nm over the ground. It was helped by winds, overall, but was hurt by some loss of fuel after takeoff, by climbs over weather, and by the decision to make the final leg at a faster speed than optimum with both engines running, since adequate fuel was available.

As you can readily verify with your calculator, the range of the Voyager could have been increased by 2000 nm by either (a) adding **1044** lb to the takeoff weight or (b) subtracting **262** lb from the landing weight. The importance of weight control in obtaining an efficient airplane cannot be overemphasized! And now you can appreciate why Rutan instructed builders of his designs not to add a single item to the airplane until after it had passed the following test: Toss it up in the air and if it comes down, don't use it.

The design of Voyager required the solution of enormously difficult structural and control problems. Aircraft with a useful load (fuel + cargo) just equal to their empty weight are considered to be very efficient aircraft but Voyager took off with a load over four times its empty weight. The designer, Burt Rutan, originally sketched out a flying wing design, using a small canard for control. This is the most efficient way to carry a huge load because the load can then be distributed evenly throughout the lifting wing, minimizing any bending loads. (This is structural span loading.) However flying wings have always had problems obtaining good stability and control properties — and, at the end of the flight, it was determined that the flying wing would have needed only about 10 hp and would have stalled at only 25 mi/hr, putting it at the mercy of winds. So a "catamaran" structure was chosen; the twin booms placed much of the fuel load at approximately the mid-point of the wings rather than in the fuselage and near the wing roots where the bending moments would have been enormous. The canard not only contributed positive lift and pitch control, it also made a structural box with the booms. Even so, the wingtips transitioned from dragging on the runway to an upward deflection of many feet during the famous record-making takeoff.

The light weight and the strength that made such a structure possible were achieved through the use of modern composites: graphite fiber/epoxy over a honeycomb interior. The 110.8 ft wings, longer than those on a DC-9 airliner, were designed to deflect safely up to ± 30 ft.

Takeoff has claimed many record-seeking aircraft because the power required to take off and climb over obstacles is so much more than that for level flight. An overloaded gas tanker is not a very crash-worthy vehicle. Voyager needed every bit of power from its 250 or so horses as its takeoff roll was 14,200 ft with a liftoff speed of 89 kt. But then the climb rate became a respectable 200 ft/min or so. By the time it had returned to its starting point at Edwards, only about 25 hp was needed to keep it in the air. That is why **two** engines had to be used, so only the more efficient rear one needed to be used after some fuel had burned off.

The flexibility of Voyager's airframe caused aeroelastic stability problems at the highest weights. There was a divergent wing flapping oscillation at 82.5 kt, causing wing deflection to double on every oscillation if not stopped by the pilot. When the Voyager appeared at the 1984 Oshkosh Experimental Aircraft Association Fly-In, it created a sensation because of its unusual and graceful shape—but also for its obvious aeroelasticity. Still, Dick made every landing look easy.

Just as for endurance, a jet-powered aircraft has a different requirement for obtaining maximum range, namely that the speed should be about 32% greater than that for maximum L over D. The Breguet equation also has a different form (Ref. 8).

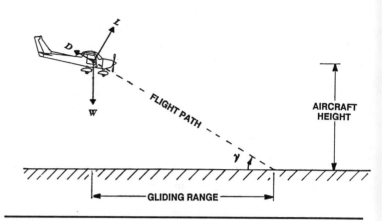

Fig. 14.20 Forces on an aircraft in a power-off glide

F. Gliding Flight

Early Birds such as the Voisin brothers called it a *volplane* and considered it to be a hazardous maneuver indeed. Today, though, we often practice these power-off glides to a landing in our light training aircraft. And soaring aircraft and the space shuttle have no choice but to make their approach and landing without power every time, a situation that still goes by the romantic but misleading name of "dead stick" landing. (The English report that "dead stick" refers to the stationary [wood] propeller on which the pilot may tend to fixate.) Yet there are good reasons why not all aircraft make their landing approach this way, reasons that will become apparent shortly.

The external forces acting on an aircraft in a power-off glide are depicted in Fig. 14.20. Note that the forces are really the same as in climbing flight (Fig. 14.10) except for the absence of a thrust force. Applying Newton's second law, Eq. 13.1, first along the flight path and then perpendicular to it, yields

$$\sum F_{\text{PARALLEL}} = W \sin \gamma - D = 0 \qquad (14.34)$$

$$\sum F_{\text{PERPENDICULAR}} = W \cos \gamma - L = 0 \qquad (14.35)$$

Dividing the first equality by the second produces the basic equation of gliding flight:

$$\tan \gamma = \frac{D}{L} = \frac{C_D}{C_L} \qquad (14.36)$$

POWER-OFF GLIDE ANGLE

Therefore if we know the coefficient ratio for a given speed, we can calculate the corresponding glide angle. Even more fun, if we **measure** glide angles we can obtain the drag polar for any aircraft we fly! You'll see shortly.

Equation 14.36 tells us that, in order to obtain a long power-off gliding range, (C_L / C_D) should be a **maximum** so that $\tan \gamma$ (and therefore γ) will be as **small** as possible:

$$\gamma_{\text{MIN}} = \arctan \left\{ \frac{1}{\left(\dfrac{C_L}{C_D} \right)_{\text{MAX}}} \right\} \qquad (14.37)$$

where $\arctan(x)$ means "the **angle** whose tangent is x." (Obtain this on a calculator by entering the number, pressing the *INV*(erse) key, and then the *TAN* key.)

Isn't it interesting that the criterion for airfoil efficiency becomes the criterion for airplane efficiency, both in terms of powered and unpowered range? And, it should be noted, the considerably greater $(C_L / C_D)_{\text{MAX}}$ for airfoils than for airplanes explains a great deal about the attraction that the flying wing has always held for designers.

The application of Eqs. 14.36 and 14.37 to sailplanes is straightforward but the propeller complicates affairs for powered aircraft. At a certain power setting somewhat **above** the normal idle speed, the thrust of the propeller will be zero; below that rpm the air is doing work on the engine and on the propeller and this causes a significant addition to the drag, reducing the gliding range.

Pilots become **very** interested in their gliding range almost immediately after experiencing a power failure. Most commonly the pilot has merely run out of gas (or emptied one of several tanks) and the propeller is windmilling, turning over the engine at a low rpm. A slightly better situation, from a drag standpoint, is to have the propeller stopped, which happens if an engine seizes or if the aircraft has been sufficiently slowed after the engine is turned off. (The Mooney M20, for example is supposed to have a glide angle of 4.5° with a stopped propeller but a poorer 5.5° with a windmilling propeller.) The next best scenario is a controllable propeller in its minimum rpm, maximum blade angle position (just the opposite of normal power-on approach procedure). Finally, a fully "feathering" propeller in which the blades can be turned directly into the oncoming air produces the very least propeller drag; such propellers are normally used on multi-engine installations so the remaining engine has a fighting chance of keeping the aircraft in the air. Of course, the very best deal of all would be a propeller that you could jettison soon after engine failure; a propeller has no place on a glider, after all.

The gliding ability of both powered aircraft and sailplanes is commonly given in terms of the **glide ratio**, defined as the ratio of horizontal distance travelled per unit of vertical altitude lost. From Eq. 14.36 and Fig. 14.20,

$$\text{Glide Ratio} \equiv \frac{\text{Gliding Range}}{\text{Aircraft Height}}$$

$$= \frac{1}{\tan \gamma} = \cot \gamma = \frac{L}{D}$$

$$= \frac{C_L}{C_D} \qquad (14.38)$$

where "cot" stands for cotangent, defined as the reciprocal of the tangent.

Aircraft manufacturers commonly specify the glide ratio for their aircraft, presumably assuming a windmilling propeller. The 1978 Cessna 152 manual, for example, indicates that the maximum glide occurs at 60 kt calibrated airspeed and yields a gliding range of 18.8 nm per 12,000 ft of altitude above the ground. Therefore the glide ratio for this aircraft, using Eq. 14.38, is

$$\text{glide ratio} = \frac{(18.8 \text{ nm})(6080 \text{ ft/nm})}{12,000 \text{ ft}} = 9.5$$

This should be compared with the calculated $(C_L/C_D)_{\text{MAX}}$ of 10.8 at 65 kt from the previous section; the agreement is reasonably good considering the added drag of the propeller.

On the Long-EZ, a pusher propeller aircraft, the owner's manual suggests a glide ratio of 13.9 with a windmilling or stopped propeller and 17.0 with the engine at idle power, which is very high for a powered aircraft. The glide ratio could be noticeably higher yet if the idle rpm is set too high. This is very fine gliding performance and highly desirable for engine-out situations, but some way must be found to increase the drag when landing, as will be discussed in a later section.

Pilots have been known to look askance at heavy, short-winged aircraft and make snide remarks about the glide angles of bricks. And many pilots figure that they better plan on a much reduced glide ratio when they fly at maximum weight compared to minimum weight. Plus, in thin, high altitude air, the glide ratio surely also deteriorates. Wrong! If you believe the preceding derivation, that the glide ratio depends only on (C_L/C_D), that is, on the basic aerodynamic efficiency of the airplane, then you realize that the optimum glide ratio or glide angle of an airplane is not changed even if you stuff it full of bricks! What is affected by both extra weight and thinner air is the true airspeed at which the optimum glide is obtained.

It works the other direction, too. There's a good chance that you've thought (in the past) that your airplane's gliding ability should improve as your passengers disembark and as fuel is burned off. Not so! The best you can hope for is to just match the glide angle given by the aircraft flight manual for gross weight. and, if you don't reduce your glide speed below that given for gross weight, your glide angle will get poorer and poorer! See Example 14.8.

We can determine at what speed the optimum glide is obtained by solving Eq. 14.35 for V, while using the definition of lift coefficient from Eq. 12.3:

$$V = \sqrt{\frac{2W \cos \gamma}{C_L \rho S}} \qquad (14.39)$$

Then, since $\tan \gamma = \dfrac{C_D}{C_L}$, the cosine of γ must be equal to

$\dfrac{C_L}{\sqrt{C_L^2 + C_D^2}}$ (an identity you can prove for yourself by forming a right triangle with γ as one angle and using the hypotenuse rule to obtain the length of the hypotenuse, and then forming $\cos \gamma$). Substituting this in Eq. 14.39 gives

$$V_{\text{BEST GLIDE}} = \sqrt{\frac{2W}{\rho S \sqrt{C_L^2 + C_D^2}}} \qquad (14.40)$$

EXAMPLE 14.8. Estimate the true airspeed that yields the best glide angle for a Cessna 152 with the propeller windmilling and (a) at an overload weight of 2000 lb at **SL**, (b) at an overload weight of 2000 lb and at 7000 ft density altitude, and (c) at a light weight of 1300 lb at SL.

Solution: (a) Equation 14.40 shows that the best glide speed varies as the square root of the weight. Because 65 kt is the specified speed for 1670 lb, the speed for 2000 lb at SL is

$$V_{\infty, \text{BEST GLIDE}} = \left(65 \text{ kt}\right) \sqrt{\frac{2000}{1670}}$$

$$= \mathbf{71 \ kt} \ (\text{true and equivalent airspeed})$$

(b) Equation 14.40 also shows that the best **true** glide airspeed varies **inversely** as the square root of the air density. Because the 71 kt speed in part (a) is for the SL density of 0.002377 sl/ft^3, we correct to 7000 ft density altitude using the density at that altitude of 0.001927 sl/ft^3:

$$V_{\infty, \text{BEST GLIDE}} = \left(71 \text{ kt}\right) \sqrt{\frac{0.002377}{0.001927}}$$

$$= \mathbf{79 \ kt} \ (\text{true airspeed})$$

Note that the **equivalent** airspeed (and therefore the **indicated** airspeed) for best glide does **not** depend on the air density—but it **does** depend on the weight.

(c) At 1300 lb,

$$V_{\infty, \text{BEST GLIDE}} = \left(65 \text{ kt}\right) \sqrt{\frac{1300}{1670}}$$

$$= \mathbf{57 \ kt} \ (\text{true airspeed})$$

So there is a very significant **14 kt** spread in the best glide speeds for the two weights.

The scientific pursuit of drag polar data is one of the finest excuses ever invented for starting an airplane engine! "All" you have to do is record descent rates in power-off glides at various speeds, for a given aircraft weight. Using in addition the wing surface area, the operative C_L and C_D for each glide condition can be calculated. From this an e and f for the whole airplane can be estimated.

Figure 14.21 shows the relationship between an aircraft's true airspeed and its true vertical speed. The vertical speed can be obtained from the rate-of-climb indicator or (better) from a timed descent through a thousand feet or so of altitude. The vertical speed so obtained should be corrected to a true vertical speed using the ratio of the absolute temperature on the ground to the standard atmosphere absolute temperature for the ground altitude. (This is the air density correction described in Sec. G of Chap. 4 in connection with true altitude.)

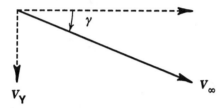

Fig. 14.21   Horizontal and vertical components of the glide velocity

EXAMPLE 14.9. Table 14.1 presents some experimental power-off glide data for a Cessna 152 with flaps up, at an estimated weight of 1600 lb, and at a density altitude of 2100 ft. (The vertical speeds shown have not been corrected for non-standard air density because the data are hardly good enough to warrant it; also, temperature inversion complicated matters.) Obtain values for C_L and C_D for each data point and estimate e and f by fitting C_D versus C_L^2 to a straight line (Eq. 13.41).

Solution: (a) We use the first data point (calibrated airspeed of 43 kt) as an example of the procedure to be followed.

$$V_\infty = \sqrt{\frac{\rho_0}{\rho}}\, V_{\text{CAL}} = \sqrt{\frac{\rho_0}{\rho}} \times 43 \text{ kt}$$

$$= 44.4 \text{ kt} = 75.0 \text{ ft/s}$$

then, using Fig. 14.21,

$$\sin\gamma = \frac{V_Y}{V_\infty} = \frac{\left(\dfrac{650 \text{ ft}}{\text{min}}\right)\left(\dfrac{1 \text{ min}}{60 \text{ s}}\right)}{75.0 \text{ ft/s}} = 0.145 \quad \text{so} \quad \gamma = \mathbf{8.3°}$$

(b) For this part it is convenient to solve Eq. 14.39 for the lift coefficient:

$$C_L = \frac{2\,W\,\cos\gamma}{\rho_0 V_{\text{CAL}}^2\, S} \tag{14.41}$$

So $\quad C_L = \dfrac{(2)(1600)\cos(8.3°)}{(0.002377)(43\times1.689)^2\,(160)} = \mathbf{1.58}$

and $\quad C_D = \tan(8.3°)\times C_L = 0.146\,C_L = \mathbf{0.231}$

(c) After values for C_L and C_D for the rest of the glides are similarly obtained, they can be plotted (C_D versus C_L^2 to see if they fit the simple drag polar of Eq. 13.41 reasonably well. The criterion is that the points should lie on a straight line whose y intercept is $C_{D,\text{MIN}}$ or $\left(\dfrac{f}{S}\right)$ and whose slope is $\left(\dfrac{1}{\pi\,e\,\text{AR}}\right)$.

Figure 14.22 presents the plot of the calculated C_L and C_D of Table 14.1. Note that all the data points fall very close to the straight line except for the one at the very lowest speed where both the airspeed indicator and the parabolic drag polar approximation are suspect, since it is at the stall. But the validity of the parabolic drag polar approximation is certainly strongly supported for both the intermediate and high speeds. (The best-fit straight line was determined by the computer in this case but a good eyeball does nearly as well.)

FLAPS-UP GLIDE DATA		CALCULATED PROPERTIES OF THE GLIDE			
V_{CAL} (kt)	V_Y (ft/min)	γ	C_L	C_D	(C_L/C_D)
43	650	8.3°	1.58	0.231	6.8
46	490	5.9°	1.39	0.142	9.8
49	490	5.5°	1.22	0.118	10.4
53	490	5.1°	1.05	0.093	11.3
57	550	5.3°	0.91	0.084	10.8
60	600	5.5°	0.82	0.079	10.4
65	600	5.1°	0.70	0.062	11.3
69	700	5.6°	0.62	0.060	10.3
78	850	6.0°	0.48	0.051	9.5
88	1100	6.9°	0.38	0.046	8.3
97	1300	7.4°	0.31	0.040	7.7

Table 14.1   Experimental and Calculated Glide Parameters, Cessna 152 (1600 lb)

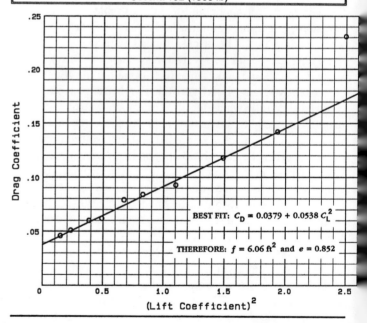

BEST FIT: $C_D = 0.0379 + 0.0538\,C_L^2$

THEREFORE: $f = 6.06 \text{ ft}^2$ and $e = 0.852$

Fig. 14.22   Fit of Table 14.1 data to a parabolic drag polar

Once an f and an e have been determined, it is possible (with some effort) to turn around and calculate sink rate curves for any desired weight. Figure ~~14.23~~ does this for the 152. Both

14.24

speeds are given as **equivalent** speeds in this figure so that the plots are valid for **any** density altitude; note that this means that the optimum glide angle is independent of altitude—but the true airspeed at which it is obtained is not. These calculated curves begin at the stall speed (and undoubtedly overestimate the gliding performance there) and end at the maximum permitted flap speed.

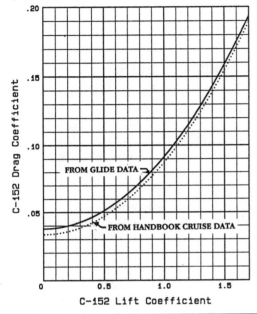

Fig. 14.23 Drag polars from glide and performance data

In zero wind, the glide angle is obtained (using Fig. 14.21 as well) by drawing a straight line from the origin to the curve, for any given airspeed, much as we did earlier for climb angles. The minimum (best) glide angle is where the straight line just touches the curve at only one point. Note how cleverly the sink rate curve shifts around as the weight changes, maintaining the **same** optimum glide angle at any two weights. But the minimum **sink rate** is significantly less at the lighter weight—there has to be some overweight penalty, after all.

In non-zero wind conditions, the glide angle can be determined by drawing the straight line from a point offset from the origin by an amount just equal to the speed of the wind. A twenty knot headwind, for example, means that the straight line should start at 20 kt on the horizontal axis; it is readily apparently then that the glide has steepened dramatically. Similarly, a tailwind means that the straight line should begin on the negative side of the origin. Curves such as these, and their interpretation with winds, are the bread of life for soaring pilots seeking success in cross-country tasks.

Figure 14.23 compares the drag polar obtained from this glide data (for a single aircraft) with the drag polar obtained earlier from the cruise performance specifications in the pilot handbook. As expected, the effect of the windmilling propeller is primarily to increase the equivalent flat plate area without having much effect on the induced drag.

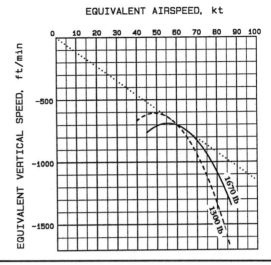

Fig. 14.24 Calculated flaps-up sink rates for glides at two different weights

It is possible to obtain glide data that don't include the drag of a windmilling propeller if just enough power is used to keep the propeller thrust at zero. Experimentally, one might determine this power setting by mounting an airspeed indicator in the propeller wake and adjusting the power until this airspeed was the same as the airplane's airspeed.

A sailplane has no propeller to modify the airplane's lift/drag characteristics and therefore should provide another nice check on the validity of the simple parabolic drag polar approximation for a very aerodynamically efficient aircraft, one possessing a maximum of laminar flow (and probably a "drag bucket"). Reference 5 contains a great deal of glide data on sailplanes; Figure 14.25 shows that a simple parabolic drag polar fits this data very well. Surprisingly, though, the glide data taken after tape was used to artificially trip the boundary layer

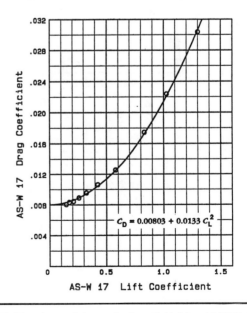

Fig. 14.25 Experimental drag polar for a Schleicher AS-W 20 (clean)

is **not** reasonably fitted by a parabolic drag polar centered at $C_L = 0$; a third term, proportional to C_L must be added (Fig. 14.26). An even better fit can be obtained by adding a fourth term proportional to C_L^3.

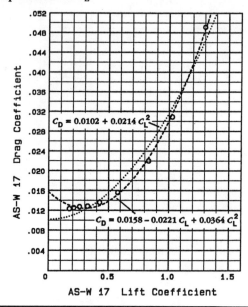

Fig. 14.26 Experimental drag polar for a Schleicher AS-W 20 (simulated bugs)

Figure 14.27 compares the drag polars of some very different aircraft. The sailplane is clearly the efficiency winner with a minuscule equivalent flat plate area as well as the least induced drag. The Cessna Cardinal polar is from Ref. 13.1; the 152 polar is from Chap. 13. The 727 airliner curve comes from Ref. 13.8; obviously a simple parabolic polar cannot be used for this aircraft, at least when high lift devices are employed. The

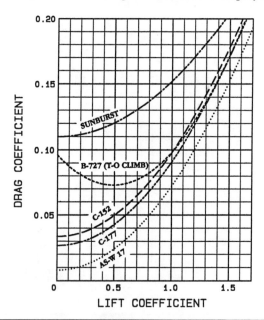

Fig. 14.27 Drag polars: AS-W 17 sailplane, Cessna 177 Cardinal, Cessna 152, Boeing 727, and Sunburst ultralight

ultralight's drag polar is from Ref. 13.10; with its open framework it has an enormous equivalent flat plate area but its induced drag is **very** small.

Feel free to add to the pool of gliding data. The air must be very smooth if reproducible numbers are to be obtained; the data in Table 14.1 were taken in very stable air at night. Even then the accuracy of the data is limited by the quality of the airspeed and rate-of-climb instruments (or the combination of altimeter and stopwatch). Still, usable and interesting results are readily obtainable.

G. Approach to Landing

Suddenly, in the landing pattern, we **want** a high drag airplane (**a**) so we can approach at a steep angle if required to clear an obstacle in front of the runway, (**b**) to slow the airplane when the glide is broken in the "flare" or round-out, and (**c**) to slow the airplane during the landing roll. Slotted flaps at 30° or 40° deflection are great for this. Aerodynamically clean aircraft without high drag flaps can be a real challenge here, requiring **careful** attention to airspeed and good brakes. Trying to fit flaps on canard aircraft presents a special challenge because, normally, the canard doesn't have enough control authority to counter the pitching moment produced by flaps; this problem was cleverly circumvented in the Beech Starship by using a sweepable canard. On the canard-type Long-EZ and the 2-place Quickie (Q2 or Q-200), a "belly board" or **speed brake** is rotated nearly 90° away from the fuselage to approximately double the equivalent flat plate area of the aircraft—and even more drag would often be very welcome.

Some aircraft without flaps can still increase their drag very nicely by flying slightly sideways. This is accomplished by banking the aircraft in one direction while holding just enough opposite rudder to prevent a turn from developing. Because the flight path is between the direction pointed by the lower wing and straight ahead, the maneuver is called a **slip**. The amount of slip is normally limited by the power of the rudder rather than by aileron power. The poorest slipping aircraft are those with strong directional stability (which try very hard to go the direction they're pointed) and with relatively weak rudders. The best slipping aircraft are those with slab-sided fuselages and powerful rudders; these include most tailwheel-type aircraft which need effective rudders for directional control. The Piper Cub and most biplanes can approximately double their glide angle with a good slip, for example. Unlike flaps, slips can be removed almost instantaneously but flaps do have the significant advantage of lowering the stall speed. Flaps and slips don't always get along; some aircraft are restricted against slips with flaps extended because of the altered airflow over the fuselage. Another nice thing about the slip is that, if the airplane is accidentally stalled in the slip, it will tend to roll wings level rather than steepening its bank—a highly desirable response when close to the ground.

A slip that is used strictly to increase the glide angle is called a **forward slip**; the glide path is then offset from the longitudinal axis in the direction of the lowered wing. A slip that is used just to offset a crosswind during the landing approach and flare is called a **side slip**; the glide path is then aligned vertically with the longitudinal axis of the aircraft. A side slip increases the glide angle; if an even steeper angle is desired, a forward slip can be combined with the side slip.

It is very important that somehow the pilot manage to both point the aircraft directly down the runway and force it to go in the direction it is pointing. A side slip can compensate for a crosswind by causing the airplane to slide toward the lowered wing with a horizontal component just equal to the crosswind component of the wing (i.e., the component of the wind's velocity that is 90° to the runway direction). Because passengers may not appreciate the maneuver, it is common to crab the aircraft into the wind to compensate for the crosswind component during the approach and only transition to a side slip during the flare. The only alternative to using a side slip in a crosswind is to keep a crab into the wind until just an instant before touchdown and then "kick" the nose straight with abrupt rudder application just before touchdown. This requires a fine knowledge of when touchdown is going to occur—but in some aircraft, especially low wing aircraft with low slung, wing-mounted engine nacelles or biplanes with short gear, this "kick-out-the-crab" method must be used to avoid dragging a nacelle or wingtip on the ground.

Pic. 14.8 Bob Hoover has lowered the P-51's gear and flaps in preparation for landing.

If you are flying a lefthand landing pattern with a direct tailwind on the base leg, you will tend to arrive on your final approach with too much altitude. As you turn onto final, you may want to initiate a forward slip to lose altitude. This forward slip should be to the left because, if you carry the forward slip down to the flare, it is very awkward to transition from a forward slip with one wing low to a side slip with the other wing low! Additionally, the forward slip is potentially more effective when made into the crosswind because more aileron than usual is needed. In the event of a headwind on a left base leg, on the other hand, you must turn past the runway heading before lowering the right wing in a forward slip.

A simplified force diagram for the forward slip is presented in Fig. 14.28, which is intended to be the view seen by a bug directly on the flight path. Note that this is yet another equilibrium condition for the aircraft because it is assumed that no acceleration in any direction is present! Therefore the forces in any chosen direction and the moments about any point must sum to zero. Specifically, the horizontal component of the lift vector of the banked aircraft is just equal to the drag developed by the sideways motion and the displaced control surfaces.

Stable aircraft try to fly in the direction in which they are pointed. To overcome this, as is necessary in the slip, the pilot must hold aileron into the slip and opposite rudder; thus Fig. 14.28 shows right aileron and left rudder for a side slip to the pilot's right. The raised aileron on the lowered wing and the lowered aileron on the high wing counteract the stabilizing tendency of the wing to generate more lift on the lowered wing

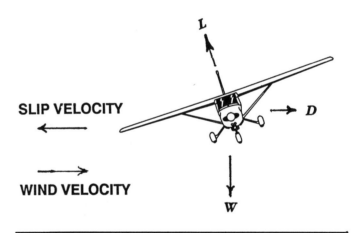

Fig. 14.28 Forces acting on an aircraft in a sideslip in a crosswind landing

and less on the higher one. The bank angle required for a given crosswind component is considerably greater than that expected from simply considering the drag caused by crosswind flow around the fuselage; i.e., the additional drag of the deflected control surfaces and wing require a steeper bank to get the needed sideways velocity component.

A steady crosswind is thus nothing to be feared so long as the ailerons and rudder are sufficiently powerful to produce a slip of equal speed. The real challenge to the pilot is that the wind is seldom steady; there are gusts of wind: up, down, and cross-ways. Then a good landing becomes the difficult matter of effecting a touchdown at the instant when the crosswind compensation is just right.

In trainer aircraft such as the Cadet and the 152, approaches to land are normally made with most or all of the available flap deflection so as to increase the maximum lift and decrease the lift/drag ratio (to obtain a steep glide over obstacles and minimum float in the flare). (Chapter 12 discussed the effectiveness of various types of flaps.) In the Cadet, full 40° flap deflection lowers the calibrated stall speed at 2325 lb from 56 kt to 50 kt, which corresponds to an increase in overall maximum lift coefficient from about 1.3 to about 1.6; in the 152, full 30° deflection of the slotted flaps lowers the calibrated stall speed at 1670 lb from about 47 kt to about 42 kt, corresponding to an increase in overall maximum lift coefficient from about 1.4 to about 1.75.

Figure 14.29 presents a drag polar for the fully-flapped 152, derived from power-off gliding experiments with a somewhat tired 1978 aircraft. (The achievement of good gliding data is hampered by the fact that, in addition to the many other experimental uncertainties and difficulties, the airspeed correction tables may apply to level flight but not to high speed glides.)

Note in Fig. 14.29 that an offset parabolic curve is definitely required to get a good fit to these data. Unfortunately, it is impossible to obtain lift coefficients much close to zero to check this offset because these smaller lift coefficients correspond to glide speeds greater than those permitted by the aircraft's operating limitations. However, whether a simple offset parabola fits the data well or not, it is certainly clear that the drag polar curve must go through significantly greater drag coefficients for corresponding negative lift coefficients because it very hard to imagine an aircraft with full flaps gliding as well inverted as upright!

H. Breaking the Glide

The aim of most pilots is to touch down with zero vertical velocity and a minimum horizontal velocity while at the same time pointing straight down the runway and **tracking** in the same direction as the nose is pointing. The process of bringing the vertical component of the airplane's velocity to zero just before touchdown is the flare or round-out, as mentioned earlier. It requires a fine touch on the elevators and an ability to judge height above the ground, all of which takes a few hours of practice for most student pilots. Yet every landing remains a challenge for most of us because the approach, the runway, competing traffic, and winds are different for every landing. Especially the winds!

Figure 14.30 shows the angles and heights and the force diagrams during the approach-to-land glide and the flare. During the glide the plane is in force equilibrium but as soon as the flare begins the drag is no longer equal to $[\sin(\gamma)]$ and the speed of the aircraft steadily decreases. When the plane has completed its transition to level flight, the drag force is completely unbalanced (note the second force diagram) and directly causing all the deceleration. We can estimate the maximum time available for the flare by assuming that the drag force stays constant at its glide value during the flare; the time will be an estimated maximum because the drag force actually builds up rapidly as the speed is decreased toward the stall.

$$D = W \sin\gamma \quad \text{so} \quad a = \frac{D}{m} = g\sin\gamma$$

then, rearranging Eq. 14.9,

$$t = \frac{V_F - V_0}{a} = \frac{V_{TOUCHDOWN} - V_{GLIDE}}{a}$$

$$= \frac{V_{TOUCHDOWN} - V_{GLIDE}}{g\sin\gamma} \tag{14.42}$$

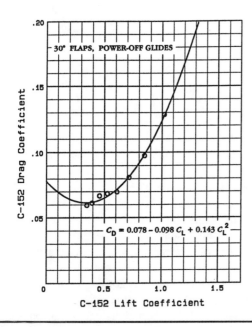

Fig. 14.29 Experimental drag polar for a C-152 with full flaps, based on power-off glides

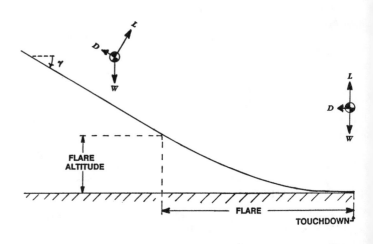

Fig. 14.30 Angles and heights and forces during approach, flare, and landing

EXAMPLE 14.10. Assuming that the touchdown speed, $V_{\text{TOUCHDOWN}}$ (now abbreviated as V_{TD}) is 15% above the stall speed and the speed just before the flare is 30% above the stall, estimate the maximum time that a plane with a true stall speed of 42 kt (70.9 ft/s) and a glide angle of 7.0° can remain in the flare. (These numbers are appropriate for a 152 at gross weight and with full flaps under SL StAt conditions.)

Solution: Substituting $V_{\text{TD}} = 1.15\ V_{\text{STALL}}$ and $V_{\text{GLIDE}} = 1.3\ V_{\text{STALL}}$ in Eq. 14.42,

$$t = \frac{0.15\ V_{\text{STALL}}}{g \sin \gamma} = \frac{0.15 \times 70.9}{(32.2) \sin(7.0°)} = \textbf{2.7 s}$$

It is also useful to consider the flare from an energy viewpoint. Even though there is force equilibrium during the glide to the flare, the aircraft's total energy is decreasing because its potential energy (mgh) is decreasing. The rate at which the PE is decreasing in the glide is the rate at which the height, h, is changing since m and g aren't changing. The rate of change of h is just the vertical component of the aircraft's gliding velocity, so

Rate of Energy Decrease in Glide $= mgV_{\text{GLIDE}} \sin \gamma$

If we assume that this rate is constant, the kinetic energy at touchdown will be the kinetic energy at the ground if no flare occurs minus the rate of energy loss times the time of the flare:

$$\tfrac{1}{2} m V_{\text{TD}}^2 = \tfrac{1}{2} m\ V_{\text{GLIDE}}^2 - \left[m\ g\ V_{\text{GLIDE}} \sin \gamma \right] t$$

Then solving for the time,

$$t = \frac{V_{\text{GLIDE}}^2 - V_{\text{TD}}^2}{2\ g\ V_{\text{GLIDE}}\ \sin \gamma} \tag{14.43}$$

For example, if $V_{\text{GLIDE}} = 1.3\ V_{\text{STALL}}$ and $V_{\text{TD}} = 1.15\ V_{\text{STALL}}$,

$$t = \frac{0.3675\ V_{\text{STALL}}}{2.6\ g \sin \gamma} = \frac{0.14\ V_{\text{STALL}}}{g \sin \gamma}$$

which is essentially the same expression as that derived earlier from the force standpoint.

Large, heavy, fast aircraft are capable of very respectable gliding performance—in the clean (cruise) configuration. However, in the normal landing configuration, their power-off glide angle is so steep and the rate of descent so fearful that it would be very difficult to time the flare (the breaking of the glide angle) successfully. Thus heavy aircraft always make their landing approach with power on.

Biplanes, too, are notorious for having a very poor glide ratio because of the drag associated with flying wires, landing wires, and interplane struts. These aircraft can be volplaned to a landing only if they are on short final with about 800 feet of altitude; in fact such a steep approach is commonly practiced in aerobatic biplanes as a method for giving the pilot the best of the bad possible views of the runway.

These comments should explain the special dangers inherent in a powerplant failure for the very early aircraft, for biplanes, for heavy transport aircraft, and for aerodynamically dirty aircraft in general. The pilot, when faced with a loss of power, must immediately lower the nose way, way down to keep the aircraft from stalling (and the earliest airplanes flew very close to the stall). But then the angle of descent and the perceived rate of descent are so great (and greatly increased by any bank) that the pilot must force himself to maintain it until the instant when the excess kinetic energy must be quickly converted into work done on the aircraft (stopping its rate of descent with a greatly increased lift force). It would seem to be a good thing for a pilot to practice the maneuver under controlled conditions, at least for single-engine aircraft, before having to do it for real. The simulators for large aircraft can simulate this situation quite realistically.

Figure 14.31 compares the general shape of the power available curves for propeller-driven and jet aircraft. Jet engines tend to produce a thrust force that does not depend very much on the aircraft speed; in this approximation, the power available curve will be a straight line through the origin with a slope equal to the thrust ($P = TV = \text{constant} \times \text{speed}$). It can be surmised that jet aircraft have more difficulty in aborting a landing than propeller-driven aircraft because they need to gain speed in order to obtain good climb performance. (A much more important factor is the "spool-up" time, the time required for the turbine to increase its angular speed from a low-power value to a suddenly requested high-power value; this time can be on the order of seconds if the engines are high time and have been idled down.)

The region to the left of the minimum in the power required curve is commonly known as the "reversed command region" or the "back side of the power curve." It is a reversed command region in the sense that, at these speeds, a reduction in speed (additional back pressure on the stick or more up elevator) requires more power and produces less climb or (on a landing approach) a steeper approach angle, unlike what happens at higher speeds. A pilot who gets mired in the back side of the power curve while on final approach will find it difficult to

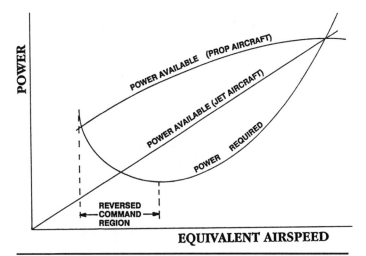

Fig. 14.31 Comparison of power-available curves for jet and propeller-driven aircraft

arrest a steep descent because the aircraft needs to be accelerated through the high drag region to a higher speed first.

Some aircraft (perhaps more so for those with laminar airfoils) seem to delight in lifting off the runway in this reversed command region and then find themselves incapable of accelerating out of it once they are out of ground effect. Such a revolting development is particularly likely when the pilot is psyched by trees at the end of the runway or by the languorous performance of his aircraft on a hot or high altitude airport; a partial solution is to ensure that the proper **indicated** airspeed is reached before allowing the aircraft to leave the ground. Once an aircraft is stuck in ground effect, the reasonable way out is to return to very close to the ground to gain more speed and try again. Short-take-off-and-landing (STOL) aircraft, or aircraft modified for STOL capabilities, tend to operate in this reversed-command region. In smooth air and with good pilot skill, it works well—but watch out for downdrafts and turbulence!

I. Touchdown!

As one cartoon has it, "Flying is the second greatest experience known to man—landing is the first!" And pilots soon learn that they may fly cautiously and with great skill but non-pilots will rather tend to judge them primarily on the quality of their landings.

The touchdown speed is typically slightly greater than the stall speed (we have used 15% greater) because the elevator may not be powerful enough in ground effect to drive the wing to a full stall and because there is more control power to cope with gusty winds at these slightly higher speeds. On a nosewheel-type aircraft, the machine should touch down on the main gear; then it tends to pitch down onto its nose gear (to its non-moving stable configuration) because the *cg* is ahead of the main gear, and this reduces or removes the lift to help ensure that the machine stays on the ground. (On many light aircraft, in stable air, the nosewheel can be spared some strain by holding the nosewheel off the ground as long as possible with continuing nose-up elevator movement.)

A light tailwheel-type aircraft, on the other hand, is commonly "three-pointed" (landed on all three gear tires simultaneously) with the airspeed **very** close to the stall because otherwise the *cg* continues to descend after the wheels touch, causing the angle of attack to increase, which then increases the lift and results in a hop back into the air. If the height of the hop is small, full elevator can be employed at the second touchdown and the pilot can log two landings as the extra energy is dissipated in the landing gear. If the hop height is more than a couple of feet, power should be added to keep a stall from developing at the top of the hop and causing a high rate of descent to a harder touchdown.

An alternate method for landing a tailwheel-type aircraft (and the normal method for a heavy tailwheel aircraft) is to touch down on the main gear with the tail still quite high and then immediately add forward control pressure to keep the weight from increasing the angle of attack. Then the tail is kept

Pic. 14.9 This 1933 Aeronca C-3 ("Flying Bathtub") is approaching the moment of truth—touchdown!

up with constantly increasing forward control movement until the stall speed is reached and then the tail is brought down and full rearward control is used to maximize the effectiveness of the tailwheel during the landing roll. This "wheel" landing is great practice and, when successful, gets the plane on the ground while it still has plenty of control power to cope with gusty winds—but, in a sense, it just postpones the dicey transition to minimum-control tailwheel steering. A very soft touchdown in a wheel landing requires very little forward control movement and the least critical timing. A hard touchdown requires considerably more down elevator at just the moment of touchdown. If the pilot employs the forward control either before or too late after touchdown, the plane will tend to rotate to a higher angle of attack and will "bounce" quite enthusiastically back into the air. If the bounce leads to a stall at the top of the bounce, the result will be a nearly "free fall" to a hard landing; power is always the proper way to recover from a bounce like this. If the flare is made with some power on (which prolongs the flare and makes it easier to obtain a tail-high touchdown) and if the pilot continues to use unsynchronized control movements, the bounces can grow in amplitude to kangaroo-envying proportions—right up until the gear gives up or the propeller strikes the ground. The proper response is to add enough power to stop the vertical motion and then transition to another landing attempt, or add full power and go around for another try.

The tendency to bounce on a hard landing is strongly dependent on the energy dissipation capabilities of the landing gear as well as the location of the third wheel. Spring steel in simple bending returns most of the descent energy as upward kinetic energy but air-oil hydraulic struts and tubes in torsion tend to absorb more of the kinetic energy, converting it into heat energy immediately.

Any aircraft can be bounced if the vertical velocity at touchdown is great enough. If the pilot increases the angle of attack just after touchdown in a nosewheel-type aircraft or just sits there doing nothing after a touchdown on the mains in a tailwheel-type, aerodynamic lift will add to the bounce height provided by the gear. Suppose that we just fly the aircraft into the ground without flaring but pull back immediately after leaving the ground (a particularly bad combination). The maximum kinetic energy from the hard touchdown is

$$KE_{\text{INTO GEAR}} = \tfrac{1}{2}m\left[V_{\text{GLIDE}}\sin\gamma\right]^2$$

and the maximum kinetic energy available from aerodynamic lift is

$$KE_{\text{AERO}} = \tfrac{1}{2}m\left[V_{\text{GLIDE}}\cos\gamma\right]^2 - \tfrac{1}{2}m\,V_{\text{STALL}}^2$$

and if **all** of this goes into potential energy in the bounce,

$$\tfrac{1}{2}m\left(V_{\text{GLIDE}}\sin\gamma\right)^2 + \tfrac{1}{2}m\left[V_{\text{GLIDE}}\cos\gamma\right]^2 - \tfrac{1}{2}m\,V_{\text{STALL}}^2$$
$$= PE = mgh$$

so the maximum height obtainable in the bounce, theoretically, is about

$$h_{\text{MAX IN BOUNCED LANDING}} = \frac{V_{\text{GLIDE}}^2 - V_{\text{STALL}}^2}{2g} \qquad (14.44)$$

If the glide speed is 1.3 times the stall speed,

$$h_{\text{MAX IN BOUNCED LANDING}} = \frac{0.345\,V_{\text{STALL}}^2}{g} \qquad (14.45)$$

EXAMPLE 14.11. Estimate the maximum possible bounce for an aircraft with a stall speed of 57 kt and approaching at a speed 30% greater than its stall speed.

Solution: Using Eq. 14.45,

$$h_{\text{MAX}} = \left(0.345\right)\left[\frac{(57\text{ kt} \times 1.689\text{ ft/s / kt})^2}{32.2\text{ ft/s}^2}\right] = \textbf{99 ft}$$

A particularly inept pilot could do even better by using more speed and increasing the glide angle just before touchdown (i.e., "pounding" the airplane into the ground).

If the plane has negligible lift at the top of this bounce (and power isn't added to arrest the descent), the plane will drop like the proverbial rock, transferring all its potential energy into kinetic energy for another hard landing:

$$KE_{\text{TD \#2}} = \tfrac{1}{2}mV_{\text{TD \#2}}^2 = PE_{\text{TOP OF BOUNCE}} = mgh_{\text{MAX}}$$

so that

$$V_{\text{TD\#2}} = \sqrt{2g\,h_{\text{MAX}}} \qquad (14.46)$$

EXAMPLE 14.12. Calculate the maximum (i.e., no lift) touchdown speed if a plane is dropped in from the top of a bounce to 99 ft, as in Example 14.11.

Solution: Using Eq. 14.46, $V_{\text{TD \#2}} = \sqrt{(2)(32.2)(99)}$

$$= 80\text{ ft/s} = 47\text{ kt} = \textbf{4800 ft/min!}$$

which is guaranteed to fail the gear because it is only required to withstand a vertical speed of 10 ft/s (or 20 ft/s for United States Navy aircraft).

Aircraft certificated under Federal Air Regulations Part 23 are expected to withstand a vertical speed of

$$V_{\text{DESCENT}} \geq \left(\frac{W}{S}\right)^{1/4} \quad \text{except} \quad 7\text{ ft/s} \leq V_{\text{DESCENT}} \leq 10\text{ ft/s}$$

However, Part 23 does allow the designer to assume that wing lift up to two-thirds of the weight of the aircraft continues to exist throughout the landing impact, a considerably kinder assumption than our assumption of zero lift.

Although every pilot has experienced bounced landings, we normally can slink away to our tiedown or hangar with the fervent hope that no one was watching, let alone taking pictures. Not so if you're an airline pilot because these aircraft have flight data recorders that detail just what happens on every landing.

Figure 14.32 shows the altitude and speed variations and the vertical accelerations experienced by a 727 that made an extra hard landing in 1970. (They do show erroneous readings after the second bounce.) The first bounce was mostly an aerodynamic bounce, up to about 50 ft, and did not damage the gear. Instead of adding thrust after the bounce, though, the spoilers were deployed to decrease the lift (not what one wants while still in the air!) and the second touchdown did damage the gear, which gave up completely after the third bounce. Note that this botched landing followed an **un**stabilized approach. "About 3 seconds after touchdown, the captain made a remark commonly used in pilots' parlance to express dissatisfaction with an event or situation."

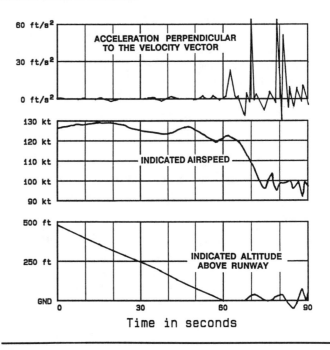

Fig. 14.32 Anatomy of a bounced landing (adapted from NTSB-AAR-72-8)

One of the useful points made by the investigators was that a sudden flare increases the stall speed because the loading on the wing increases. This increases the chances of getting a hard touchdown. The increase in stall speed from a rapid rotation can be calculated from Eq. 12.16; also, the subject will be discussed in some detail in the next chapter in connection with maneuvering loads.

The excitement may not be over just because the plane has touched down with zero vertical velocity. There is still the problem of directional control.

It seems that the Seversky has a bad ground-looping tendency, especially on a runway in a crosswind. I dislike ships that ground-loop. They take away your reserve of safety unnecessarily. Not that ground-looping is likely to result in personal injury, but it can do tremendous damage to a plane—and to your personal dignity.[3]

If a nosewheel-type aircraft touches down on its main gear at a crab angle (Fig. 14.33a), the friction forces of the ground on both the tires produces a turning moment (clockwise in the figure) that tends to straighten out the aircraft; the effect is strongest for the aircraft with the most weight on the nose gear because this increases the moment arm. (It should be noted that you can safely take moments for an accelerating object only if the moments are taken about the *cg*, as done here.) For a tailwheel-type aircraft, on the other hand (Fig. 14.33b), the sliding forces of the ground on the main gear tires create a turning moment than tends to increase the crab angle. If not checked immediately, the plane will skid into a circle of ever decreasing radius, probably failing the gear and often dragging a wingtip. This is the infamous "ground loop." Recovery from an incipient ground loop is best accomplished with power and full rudder so that a correcting moment is aerodynamically produced. Differential braking may be quicker in application and more powerful for the really desperate situations. Normally the tailwheel provides a correcting moment but it may be relatively impotent if the airplane isn't firmly on the ground; also, if the tailwheel has a weak indent and breaks out of the steering mode into the push-backward swivel mode, you have a real ride ahead.

It takes a while for the tricycle-gear-trained pilot to feel comfortable in a tailwheel-type aircraft but the effort is well worth it. Some of the most interesting and fun aircraft sport the little wheel on the tail. Unfortunately, some of them, especially the biplanes and the ones with big radial engines, provide only sideways glances toward the runway as well as being unstable directionally. On these, if you land on a narrow runway and see any part of it, you are already in deep, deep trouble. The advantage of a tailwheel, though, is less weight and less complexity and less aerodynamic drag as well as better handling of round fields. (Some people would love to be able to always fly old biplanes from grass fields because the tires tend to slide on the grass rather than grabbing and turning the plane.) In any case, mastery of the tailwheel should never be claimed, for pride

[3]Charles A. Lindbergh, *The Wartime Journals* (New York: Harcourt Brace Jovanovich, 1970).

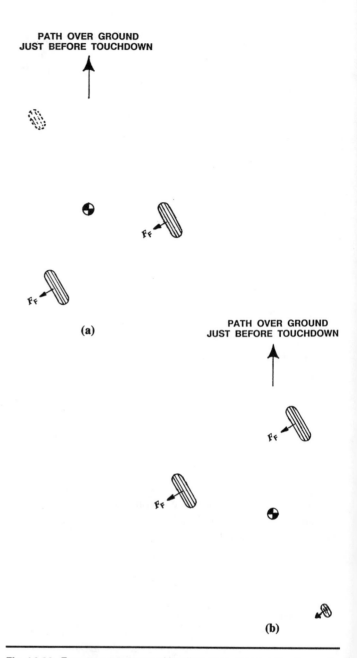

Fig. 14.33 Forces on two types of landing gear when landing in a crab: (a) Nose-gear type aircraft and (b) Tailwheel-type aircraft

and a little inattention commonly precedes the loop. At least ground loops don't usually injure the occupants because the loop occurs at low speed; also, they can be a wonderful trick for dissipating kinetic energy if a forced landing must be made on a too-short field.

It is also possible to ground loop a tricycle-gear aircraft, with a little effort. If a pilot is above the stall speed and tries to help keep the aircraft on the ground with nose-down elevator control movement, the aircraft's weight can be transferred to the nosewheel (as on a **wheelbarrow**) and then it is directionally unstable just like a tailwheel-equipped machine. This can happen on either takeoff or landing. Some aircraft are more susceptible to this because of having extra weight on the nose gear to start out with. In general, though, the nosewheel always tries to straighten things out, within its strength. The relative docility of a nosewheel-type aircraft on the landing roll is a good part of the problem in transitioning to a tailwheel-type; just at the point in the landing where the nosewheel pilot has learned that congratulations on the fine landing can commence, the tailwheel airplane is responding to a little gust of wind and eagerly anticipating a sudden dart for the weeds.

J. The Landing Roll

The total landing distance includes (a) the horizontal distance required to clear any obstacle (commonly taken as 50 ft high), (b) the horizontal distance covered during the flare, and (c) the landing roll. Figure 14.34 depicts the handbook optimum performance of the 152 at a density altitude of 600 ft and with no wind.

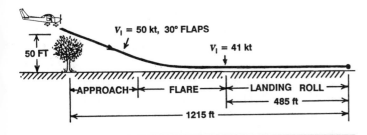

Fig. 14.34 Handbook landing performance of the C-152 at 600 ft density altitude, full flaps

Distance (a) is easily calculated from the glide angle and is the reason for wishing to increase the drag of an airplane when approaching to land. Distance (b) depends strongly on the glide and touchdown speeds as well as pilot technique and airplane drag. Distance (c) is based on a combination of aerodynamic drag and braking power; we now estimate its value.

Aerodynamic drag is very helpful in slowing an aircraft right after touchdown but, because this drag force varies as the square of the airspeed, its effectiveness decreases rapidly as the aircraft slows down. The effectiveness of the brakes, on the other hand, depends very strongly on how much lift the wings are still developing and how much braking can be done without causing wheelbarrowing (for tricycle-gear aircraft) or tail liftoff and a possible propeller strike (for tailwheel-type aircraft). If the brakes are used too vigorously while the wings are still developing lift, the tires will skid rather than roll and this causes a distinct loss in braking force as well as in directional control. Maximum braking is obtained when the tires are just on the point of skidding; some cars and motorcycles and all transport aircraft now utilize computers to keep the braking pressure just at this point — but an unaided pilot can only hope to come close. In any case, full braking effectiveness cannot be realized until some time after touchdown.

The external forces on a tricycle-gear aircraft during its landing roll are shown in Fig. 14.35. The kind of friction we want to use is called static friction because there should be **no** relative motion between the ground and the part of the tire that is instantaneously on the ground (which is the whole idea of a tire!). As in our calculation of rolling friction on takeoff in Sec. A, the frictional force is reasonably well approximated by multiplying a coefficient of friction (symbolized again by the Greek letter mu, μ) times the force pressing the surfaces together, N:

Force of friction between tire and ground $\equiv F_F = \mu N$

FRICTIONAL FORCE DURING BRAKING

For the best of all worlds (a good rubber tire on dry concrete), μ can be as large as 1.0 (static friction) or 0.7 (sliding friction). These values become about 0.7 and 0.5, respectively, if the concrete is wet.

Just as in a car, the vehicle's weight is **transferred** to the forward wheel(s) when the brakes are applied. Some Soviet Union (now CIS) aircraft have brakes on the nosewheel, but such is not true for most aircraft and this significantly limits braking power. To estimate the degree of weight transfer, we take moments about the *cg* in Fig. 14.35:

$$\sum M_{CG} = h \times \mu_2 N_2 + x \times N_2$$
$$+ h \times \mu_1 N_1 - (s - x) \times N_1 = 0$$

where h is the height of the *cg* above the ground, N_1 is the upward force of the ground on the nosewheel, N_2 is the upward force of the ground on the main gear, x is the distance from the main gear to the *cg*, and s is the longitudinal distance between the nosewheel and the main gear, as shown on the figure. Solving this equation for the ratio of upward forces of the ground on the tires,

$$\frac{N_1}{N_2} = \frac{\mu_2 h + x}{s - x - \mu_1 h} \tag{14.47}$$

Normally the nosewheel carries between about 8% and 15% of the total aircraft weight; if it is less than this the nosewheel steering is ineffective and if it is more than this it is difficult to raise the nose on the takeoff roll; if we assume a 10% load on the nosewheel in this case, x will be equal to 0.10 times s. If we also assume that the *cg* is at a height just equal to the separation distance ($h = s$), that the coefficient of friction for

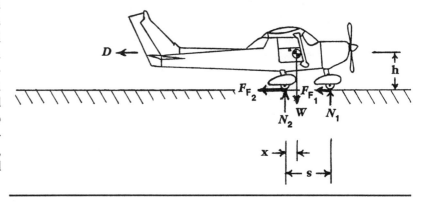

Fig. 14.35 Forces on an aircraft during the landing roll

the main tires (μ_2) is at its maximum possible value of 1.0 while the coefficient of friction for the nosewheel (μ_1) is at its normal rolling friction value of 0.02, Eq. 14.47 becomes

$$\frac{N_1}{N_2} = \frac{1.1\,s}{0.88\,s} = 1.25$$

and because $N_1 + N_2 = W$, where W is the weight of the aircraft, by algebra

$$N_2 = \frac{W}{2.25}$$

so the braking is only about **1/4** as effective as it would be if all the weight were on the wheels that were being braked! Because of this and because it is impossible to always keep the wheels just on the point of sliding, it is common to assume an effective braking coefficient of friction of about 0.2 rather than the theoretical maximum of 1.0 for all the weight on the main gear. This implies a frictional force of $F_F = (0.2\,W)$ and a deceleration due to braking of $a = (F_F / m) = 0.2g$.

Note that it is necessary to sum moments about the cg in this case because the aircraft is (negatively) accelerating. Also, it should be noted that an engineering text would probably place a forward-pointing arrow at the cg and say that it represents the "inertial force" and is equal to ma. There is no real external force there, of course, and the intention is simply to describe the inertial property (Newton's first law), the tendency of the cg to continue at constant speed in a straight line (while the tires, to which it is attached, are being forced to decelerate). If you are **in** the aircraft (i.e., in the accelerated reference frame), it is easy to convince yourself that there **is** a force pushing you forward even though a dispassionate physicist on the ground would prefer to use Newton's second law in the form

$$\sum F_{\text{EXT}} = ma$$

rather than considering the forces to be "balanced" with a fictitious "inertial force" of $-ma$ doing the balancing:

$$\sum F_{\text{EXT}} - ma = 0\,.$$

The 152 has a recommended short-field approach speed over an obstacle with 30° flaps of 54 kt calibrated airspeed, which is just about 1.3 times the stall speed of 42 kt calibrated airspeed. If we assume (conservatively) that touchdown occurs at 1.15 times the stall speed (i.e, at 1.15×42 kt = 48.3 kt = 81.6 ft/s) and use the deceleration of 0.2g estimated above and ignore aerodynamic braking, with the help of Eq. 14.10 we can estimate a ground roll (at SL when the calibrated airspeed equals the true airspeed) of

$$s = \frac{V_F^2 - V_0^2}{2a} = \frac{0^2 - 81.6^2}{(2)(0.2 \times 32.2)} = \mathbf{517\ ft}$$

which is rather close to the handbook value of about 475 ft.

Alternatively, we can involve the computer again. Figure 14.36 presents calculated acceleration terms if the 152 is assumed to be acted on by (**a**) a rolling friction term, $a = -\mu g$, where $\mu = 0.02$, (**b**) a braking term, $a = \dfrac{-fq}{m}$, where $\mu = 0.18$

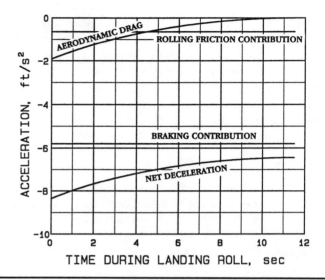

Fig. 14.36 Calculated accelerations in a maximum performance landing roll in a C-152 (600 ft density altitude, 1670 lb, 30° flaps, dry concrete)

(to make a "total" μ of 0.2), and (**c**) an aerodynamic drag force equal to $a = -(fq)/m$, where f is the calculated zero-lift equivalent flat plate area for the 152 when it is using full flaps (from Fig. 14.29). This calculation, because of the addition of aerodynamic drag, predicts a shorter landing roll of 460 ft, even closer to the book value of 485 ft.

But, as always, it is instructive to play the "what if" game. Figure 14.37 presents calculated landing rolls and landing times for various winds, various coefficients of friction and two density altitudes. With a 9 kt headwind, the predicted landing roll is 308 ft versus the handbook value of 436 ft; for a 10 kt tailwind, the predicted landing roll is 600 ft versus the handbook value of 728. These wind estimates were made by simply adding (tailwind) or subtracting (headwind) the wind speeds from the no-wind touchdown speed; evidently Cessna chooses to be more conservative in their figures, perhaps because the wind is seldom really steady. In any case, a downwind landing is clearly

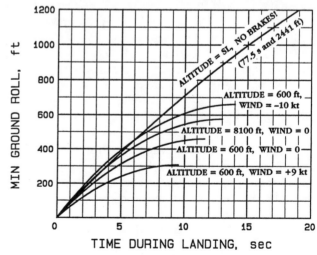

Fig. 14.37 Calculated maximum performance landing rolls in a C-152 (1670 lb, 30° flaps, dry concrete)

an undesirable activity—although an upward-sloping runway could easily make up for it, as we saw in our takeoff calculations.

At a density altitude of 8100 ft, the calculated landing roll is 577 ft—quite close to the handbook value of 605 ft. Finally, if you lose your brakes and have to rely on aerodynamic drag and rolling friction to stop you, find a runway with at least a half mile to roll out on!

Note that, just as with cars coming to a stop, the deceleration on the landing roll is predicted to be independent of the weight of the aircraft, but strongly dependent on the surface properties. Of course, this assumes that your aircraft doesn't sport such goodies as reverse thrust or spoilers or a drag chute!

Summary

It is possible to make estimates of takeoff performance by using an acceleration calculated from estimates of the available thrust, the rolling friction, and the aerodynamic drag. The analysis can be extended to indicate the effects of soft runways, sloping runways, high density altitudes, and surface winds. The rate of climb of an aircraft, if the angle of attack of its wings is not too large, is just equal to the excess of power over that needed for level flight divided by the aircraft weight. Both the best rate-of-climb speed and the best angle-of-climb speed are easily determined from a graph of climb rate versus speed. Climbing performance at high altitudes suffers primarily from the decrease in available power; supercharging can do much to compensate. It can be hazardous for a pilot to let his aircraft get too close to the stall speed (back side of the power curve) while on a landing approach because he may have to accelerate to a higher speed before the aircraft will even begin to climb if the approach has to be aborted or if strong downdrafts are encountered. A similar problem can occur if the pilot forces the aircraft off the ground in ground effect at a very low speed.

The power required for level flight varies as the cube of the true airspeed; thus, if an aircraft is able to cruise near its minimum C_D already, additional power will increase the climb rate significantly but will produce only a small increase in the cruising speed. Maximum endurance occurs at the speed for which $\left(\dfrac{C_L^{3/2}}{C_D}\right)$ is a maximum; however this speed is often so low that it is not practical to use it. Maximum range (under no-wind conditions) occurs at the speed for which $\left(\dfrac{C_L}{C_D}\right)$ is a maximum and this is also the speed which yields the flattest glide angle in a power-off glide (no-wind conditions). The Breguet equation predicts the maximum range of an aircraft. It may be possible to increase the cruising speed by inducing more laminar flow (getting "on the step") but such a state probably is unstable.

Commercial aircraft, from the smallest to the largest, possess a very respectable glide angle—in the sleeky clean cruise configuration—because the angle is the arctangent of the drag-to-lift ratio. The optimum glide angle doesn't depend on either weight or air density, but the equivalent airspeed at which the best glide angle is obtained increases with weight and the true airspeed for the best glide angle increases with density altitude. C_L and C_D (and thereby a drag polar) can be calculated from gliding data: airspeed versus descent rate. For heavy aircraft in the landing configuration or for high drag aircraft such as wire-braced biplanes, the power-off glide angle is very steep and the rate of descent is quite intimidating.

The glide angle can be steepened by increasing the drag, either with flaps or by flying slightly sideways (slipping). A good landing is a matter of bringing the vertical kinetic energy to zero at the moment of touchdown and then tracking straight down the runway; crosswinds and gusty winds makes both aspect difficult. A tailwheel-type aircraft is naturally unstable directionally during ground operations but a tricycle-gear aircraft can also be made unstable by aerodynamically transferring weight to the nosewheel.

The distance required to land over an obstacle depends on the glide angle, the time spent in the flare, and the landing roll—all of which can be estimated with more accuracy than for the symmetrical takeoff operation. It is easier to make good estimates of landing rolls because the frictional forces involved (air drag, rolling friction, and braking friction) are fairly well known. Braking effectiveness is very significantly reduced by weight transfer to the nosewheel on a tricycle-gear aircraft.

Symbol Table (in order of introduction)

F_F	force of friction (rolling, static, or kinetic)
μ	coefficient of friction
γ	angle of climb or (if negative) angle of glide

Review Questions

1. In level cruising flight the lift force is exactly equal and opposite to the ___ force if the ___ force is aligned with the direction of flight.

2. At the beginning of the takeoff roll, the acceleration is limited primarily by the available thrust force and by the ___ force.

3. At the end of the takeoff roll, the acceleration is limited primarily by the available thrust and by the ___ ___ force because ground effect and a small angle of attack reduce the ___ ___ force to a small value.

4. A headwind reduces the takeoff distance primarily because the ___ speed at liftoff is less.

5. Takeoff performance is poorer at high altitudes because the ___ force is reduced and the ___ speed is increased, although the decrease in the ___ force provides a little compensation.

6. In a stable climb, the sum of the external forces on the aircraft in the direction of flight is equal to ___; the sum of the external forces perpendicular to the direction of flight is ___.

7. The simplified equation for rate of climb assumes that the ___ is small.

8. The rate of climb is directly proportional to the ___ power and inversely proportional to the aircraft ___.

9. The ___ climb speed is always less than the ___ climb speed, although the speeds approach each other as the ___ increases.

10. The climb rate at very low speeds is often limited more by problems of ___ than by the available power or the drag increase.

11. If cruising power is limited to 75% of the rated power, the maximum cruising speed (true airspeed) typically occurs at about ___ in the St At.

12. The "back side of the power curve" is the region where any speed reduction requires more ___ to maintain either level flight or the current glide angle.

13. "Hopping up" the engine of an aircraft will greatly increase the ___ but will probably not do very much for its ___ .

14. The maximum value of (C_L/C_D) is the aerodynamically limiting factor in the ___ of the aircraft; it also directly tells us what the optimum ___ angle will be.

15. Getting "on the step"—if it is possible—most likely involves increasing the percentage of ___ flow.

16. A pilot faced with a powerplant failure should remind himself or herself that the aircraft will glide at a steeper angle than before because of the additional drag of the ___.

17. An additional problem for a pilot of a heavy or aerodynamically "dirty" aircraft, when faced with a powerplant failure, is the rapidity with which the ___ decreases, as well as the ___ and the ___ of the descent.

18. Deceleration during the landing roll is dependent on the available ___ force primarily; there is little dependence on the ___ of the aircraft.

19. Reciprocating engines produce maximum ___ that is relatively constant over the usual range of flight speeds; jet engines produce maximum ___ that is relatively constant over the usual range of flight speeds.

20. Lifting the nosewheel off the ground with the elevator well before lifting off from the ground will minimize ___ drag and ___ drag during the takeoff roll.

21. Taking off up a sloping runway into a headwind may take the same distance as taking off down the runway, but you can be sure that the takeoff ___ will be much greater.

22. V_X is the symbol for best ___ of climb speed; V_Y is the symbol for best ___ climb speed; V_X is always (less than, greater than, equal to) V_Y.

23. As an aircraft burns off fuel en route, the C_L required for a given cruise speed (decreases, increases, doesn't change). Therefore the speed at which the maximum endurance or the range is obtained will (decrease, increase, not change).

24. A major problem in actually obtaining the theoretical range or endurance is the problem of getting the engine(s) to run efficiently at a very (high, low) power setting during the flight.

25. For a given power available, cruise speed (increases, decreases) with increasing altitude because the ___ drag increases with altitude (faster, slower) than the ___ drag decreases.

26. The true airspeed at which the optimum glide angle is obtained (decreases, increases, doesn't change) as the density altitude increases; the equivalent airspeed at which the optimum glide angle is obtained (decreases, increases, doesn't change) as the density altitude increases. The optimum glide angle itself (decreases, increases, doesn't change) as the density altitude increases.

27. The indicated airspeed at which the optimum glide angle is obtained (decreases, increases, doesn't change) as an aircraft's weight decreases in flight as fuel is burned off. The optimum glide angle itself (decreases, increases, doesn't change) as the weight decreases.

28. Experimental measurements of ___ and the ___ rate in power-off glides can be used to determine a drag polar for an aircraft.

29. With respect to the two parameters in the simple parabolic drag polar approximation, lowering flaps primarily increases the ___ parameter.

30. Glider pilots add water ballast to their aircraft when embarking on a cross-country task; this allows them to minimize the time travelling between different thermals because the airspeed for optimum glide has (decreased, increased) with the added weight.

31. A ___ slip is used to compensate for a crosswind on landing; a ___ slip is used to steepen the glide angle during the landing approach.

32. If you have a headwind component on a left-hand base leg, your forward slip on the final approach should be to the (right, left).

33. If a tailwheel-type aircraft is landed on the main gear while the plane is well above the stall, the lift developed by the tail must be made (more upward, more downward) to prevent a return to the air.

34. The maximum height to which an aircraft can "bounce" in a botched landing can be estimated from the ___ it has on the touchdown relative to its minimum value.

35. The maximum vertical speed with which an aircraft hits the ground after being stalled above the ground on a poorly-judged landing can be estimated by assuming that all of its ___ energy when it stalls is converted into vertical ___ energy.

36. If a tricycle-gear aircraft during a crosswind approach is landed in a crab on just the main gear, it will tend to
   a. turn in the direction from which the wind is blowing.
   b. turn in the direction toward which the wind is blowing.
   c. go in the direction it is pointed.
   d. straighten out and go the direction it is actually travelling over the ground.

37. If a tailwheel-type aircraft during a crosswind approach is landed in a crab on just the main gear, it will tend to
   a. turn in the direction from which the wind is blowing.
   b. turn in the direction toward which the wind is blowing.
   c. go in the direction it is pointed.
   d. straighten out and go the direction it is actually travelling over the ground.

38. What are the three forces that tend to slow an aircraft on its landing roll?

39. Maximum braking on the landing roll takes advantage of the maximum value of the ___ coefficient of friction between the tires and the ground. It also uses the ___ coefficient of friction between the brake pads and the brake disks.

40. The effectiveness of the brakes on the main gear are greatly reduced by the ___ transfer that accompanies heavy braking.

Problems

1. Estimate the takeoff distance and takeoff time at SL for a Piper Cadet that is 15% over its legal gross weight, using the same assumptions made in the text, in Example 14.1. Include an increase in liftoff speed. Answer: 1300 ft and 29 s

2. Suppose that our Cadet is on a grass runway with moderately tall grass ($\mu \cong 0.2$) at SL. Estimate the takeoff distance and takeoff time under these conditions. Answers: About 4800 ft and 114 s

3. Estimate the takeoff distance and takeoff time for a takeoff with the Cadet for a 6000 ft density altitude. Answer: 1440 ft and 31 s

4. Estimate the takeoff performance of the Cadet at SL and with a 20 kt headwind on takeoff. Answer: 350 ft and 14 s

5. Estimate the takeoff performance of the Cadet at SL and with an 8 kt tailwind on takeoff. Answer: 1320 ft and 27 s

6. Piper specifies a 670 ft/min climb rate at SL for its 2325 lb Cadet, at a calibrated climb speed of 80 kt. How much more power than that required for level flight at 80 kt is available for takeoff?
Answer: 47 hp

7. Suppose that the 160 hp engine in the Cadet is replaced with a 200 hp engine. Assuming a propeller efficiency in climb of 0.75, what would

the new SL climb rate become? (Hint: Just add the additional excess power to the excess power determined in the previous problem.) Answer: 1096 ft/min

8. The maximum speed of the Cadet at SL is listed as 119 kt. What would you expect the new maximum cruising speed to be with the 200 hp engine of problem 7? Answer: 128 kt

9. What is the no-wind climb **angle** for a Cadet which has a listed climb rate of about 472 ft/min at a calibrated airspeed of 80 kt at 4000 ft in the StAt? Answer: 3.1°

10. How much additional distance **after** liftoff is required to clear a 50 ft tree with the climb angle of problem 9? Answer: 910 ft

11. Calculate the corresponding distances for a headwind of 20 kt and a tailwind of 20 kt, for the aircraft of problem 10. Answer: 695 ft and 1125 ft

12. What are the angles of climb for the windy day takeoffs of the previous problem? Answer: 4.1° and 2.5°

13. Using the calculated power curves of Fig. 13.11, what climb rate can the 152 achieve at a density altitude of 10,000 ft, a weight of 1670 lb, and a true airspeed of 95 kt? Answer: About 200 ft/min

14. Estimate the rate of climb for a 152 at 1670 lb and at a true airspeed of 65 kt at a density altitude of 10,000 ft, based on Fig. 13.11. Answer: 400 ft/min

15. Estimate the best angle of climb speed (true airspeed) for the 152 at 1670 lb and at 5,000 ft in the StAt based on Fig. 14.13. Answer: About 61 kt

16. What is the **angle** of climb for this best angle-of-climb speed of problem 15? Answer: 5.4°

17. Using the following performance specifications, estimate the drag force acting on the F-14B at its maximum speed at SL. Answer: 5.62×10^4 lb = 28.1 tons!

Grumman F-14B Tomcat

Powerplant: Two 28,090 lb thrust Pratt and Whitney TF 401-400 afterburning turbofans
Empty weight: 37,500 lb,   Fighter weight: 55,000 lb
Maximum speed at SL: 790 kt
Initial climb at 55,000 lb and at SL: greater than 30,000 ft/min

18. What constant speed in a vertical climb will give the listed climb rate? Answer: 296 kt

19. Using the results of problem 12a of Chap. 13, estimate the true airspeed that will give the greatest range for a Long-EZ at 15,000 ft in the StAt. Answer: 94 kt

20. What is the best glide angle for the Long-EZ, using data from the previous problem?

21. What is the vertical speed under these conditions? Answer: 600 ft/min

22. Reference 10 describes flight tests made on a 12,500 lb, twin-engine, propeller-driven business aircraft. The clean airframe had a drag polar $C_D = 0.024 + 0.033 C_L^2$ while with significant airframe icing the drag polar became $C_D = 0.041 + 0.115 C_L^2$. This aircraft has a wing span of 54.5 ft and a wing reference area of 303 ft². **(a)** Compare the optimum power-off and propeller-feathered glide angles and equivalent glide speeds for these two conditions. Answers: 3.2° and 7.8°; 119 kt and 143 kt. **(b)** Neglecting the weight effects, how much greater is the maximum range in the clean configuration compared to the iced-up state? Answer: About 2.4 times greater. **(c)** For an equivalent speed of 165 kt (which was the maximum attainable at about 18,000 ft for the iced-up airplane), compare the drag force on the aircraft for the two configurations. Answers: 856 lb and 1790 lb. **(d)** What equivalent airspeeds correspond with the maximum range

condition, for these two configurations? Answers: 119 kt and 143 kt **(e)** What equivalent airspeeds correspond with the maximum endurance condition for these two configurations? Answers: 91 kt and 109 kt

23. Reference 9 describes flight tests performed on a 2500 lb, 1970 Cessna Cardinal, which has a 180 hp engine, a wing span of 36.0 ft, and a wing reference area of 175 ft². In the zero-flap configuration, a zero-lift drag coefficient of 0.0267 and an efficiency factor *e* of 0.564 provided the best fit to the experimental data; for 30° flap deflection, these became 0.0462 and 0.545 respectively. **(a)** Compare the flaps-up and flaps-down optimum glide angles (propeller-off) and equivalent glide speeds for this aircraft. Answers: 5.2° and 6.9°; 84 kt and 74 kt **(b)** With flaps up and full throttle at 7500 ft, this aircraft achieved an equivalent airspeed of about 130 mi/hr. What power was the engine/propeller combination delivering at this speed? Answer: 93 hp **(c)** What is the minimum power-required equivalent airspeeds for the two configurations? Answers: 64 kt and 56 kt

24. Reference 10 in Chap. 13 describes the computation of the drag polar for an ultralight with a flying weight of about 500 lb, a wing span of 36.0 ft, and a wing area of 151 ft². Their result, for zero control deflection, was $C_D = 0.1163 + 0.0421 C_L^2$. **(a)** A typical cruise lift coefficient for this aircraft was given as 0.638. What is the corresponding cruise speed and power required under SL StAt conditions? Answers: 39 kt and 12.6 hp **(b)** What is the optimum (propeller-off) glide angle and glide speed for this particular ultralight? Answers: 8.0° and 24 kt **(c)** What is the (propeller-off) glide angle for an approach speed of 40 kt? (Note: This corresponds to a lift coefficient of 0.597.) Answer: 12.4°

25. Suppose that, under SL StAt and no-wind conditions, you are approaching to land on a runway that has a 50 ft government tree at the very end (it happens all the time at small sod strips). Suppose also that you make the approach at 1.3 times the stall speed in the aircraft of Example 14.10. **(a)** How much runway passes below you before you are close enough to the ground to begin your flare? (Assume a minimal 10 ft clearance above the tree and a flare height of 10 ft.) Answer: About 410 ft **(b)** How much additional runway is left behind in the flare, assuming an average speed and using the flare time estimated in Example 14.9? Answer: About 235 ft

26. Make the calculations of problem 25 for a density altitude of 8000 ft. Answer for second part: About 365 ft

27. How much power above that needed to maintain level flight is available to a 152 at its service ceiling of 14,700 ft (at 1670 lb weight)? Answer: 5.1 hp

28. Estimate the power required for level flight for the 152 of problem 27 at its service ceiling by assuming that the available power follows Eq. 13.51 in decreasing from the 110 hp available at SL; assume also a 0.72 prop efficiency at this altitude. Answer: 46 hp

29. One of the special joys of life for a student pilot occurs when he or she successfully contrives to rid the aircraft of the noisesome bulk in the other seat. Suppose on that day that you are in a 152 which is climbing 400 ft/min at a flight weight of 1600 lb, including you and your 170 lb instructor. What kind of climb rate can you expect on your first solo takeoff? Answer: At least 450 ft/min

30. Assuming SL StAt, no-wind conditions, what is the average effective power during the takeoff roll for the 2325 lb Cadet of Example 14.1? (Cf. Eq. 13.13) Answer: 20 hp

31. The Piper Cadet has a wing span of 35.0 ft and a wing area of 170.0 ft². What lift coefficient (theoretically) will minimize the takeoff distance **(a)** on a hard surface runway ($\mu = 0.02$) and **(b)** on a very soft grass strip ($\mu = 0.3$)? Answers: 0.23 and 3.4!

32. What angle of attack for optimum takeoff performance does a μ of 0.02 suggest for the 152 of Eq. 14.18? ($b = 33.3$ ft, $S = 160$ ft^2) Answer: 0.4°

33. The standard 152, at a takeoff weight of 1670 lb and a landing weight of 1540 lb, has a maximum range, using 45% of rated power, of about 430 nm. Estimate the range if the plane were to be overloaded with an additional 20 gallons (120 lb) of fuel. Answer: About 800 nm.

34. Estimate the takeoff time for the Voyager on its record-breaking around-the-world-unrefueled flight. (See Section E for pertinent data and Ref. 4 for a dramatic videotape of the event.) Answer: Over 3 minutes!

35. How much greater was Voyager's optimum glide speed on takeoff than on landing? (Use data in Example 14.7.) Answer: 90%

36. A certain Quickie ($W = 509$ lb, $S = 53.65$ ft^2, $b = 16.67$ ft) demonstrated the power-off gliding performance shown in Table 14.2. The in-flight air density was close to the standard SL value because it was a cold January 2 day when the data were recorded. (a) Calculate C_L, C_D, and γ for the 92 mi/hr data point. Answers: 0.44, 0.037, and 4.8° (b) Calculate and tabulate the remaining C_L and C_D values, plot C_D versus C_L^2, and estimate f and e from your best-fit straight line. Answers: $f \approx 1.2$ ft^2 and $e \approx 0.69$

V_{CAL} (mi/hr)	V_Y (ft/min)
52	580
55	580
60	490
65	595
71	555
76	580
81	690
86	640
92	680

Table 14.2 Experimental Glide Performance of a Quickie

37. Estimate the maximum possible bounce height for an ultralight with a stall speed of 24 kt and an approach speed just 30% greater. Answer: 18 ft

38. What is the maximum vertical speed on impact if the ultralight of problem 37 drops in after the bounce? Answer: 2020 ft/min

39. Part 25 of the Federal Air Regulations requires a landing gear that can handle a 600 ft/min sink speed on touchdown. To what height could an airplane be raised, and then dropped, to simulate this impact velocity? Answer: 1.55 ft

40. The properly supported human body can accommodate accelerations up to 15 g or more for short periods of time. With this kind of deceleration, what would be your landing roll after touchdown at 50 kt? Answer: About 7.4 ft!

41. In Example 14.12, the vertical speed of impact when an aircraft is dropped in from a height of 99 ft was calculated, assuming no lift was being generated by the wings. Using the kinder, gentler assumption of FAR Part 23 ($L = \frac{2}{3}W$), recalculate this vertical speed. (Hint: Because $\sum F_Y$ has changed from ($-W$) to ($L - W$), the acceleration during the fall is significantly reduced; use Newton's 2nd law to find this new acceleration and use it in Eq. 14.10.) Answer: 46.1 ft/s (much better, but still not nice)

References

1. Garrison, Peter, " 'On the Step' Is a Crock," *Flying*, August, 1974. An interesting discussion and attempted verification of "on the step" flying.

2. Norris, Jack, *Voyager—The World Flight*, Jack Norris, Northridge, CA, 1988. (The official log, flight analysis, and narrative explanation of the record around the world flight of the Voyager aircraft.)

3. Yeager, Jeana, Rutan, Dick, and Patton, Phil, *Voyager*, Knopf, New York, 1987. (The origin, construction, and flights of the famous Voyager, with special emphasis on the people who made it work. Included are some nice aerodynamic tidbits.)

4. *The Building of the Voyager*, Paul Harvey Audio/Video Center, EAA Aviation Foundation, Wittman Airfield, Oshkosh, WI 54903-3065. (This is a fascinating video history of the conception, design, and building of the Voyager.)

5. Johnson, Richard H., *The Johnson Flight Tests*, Soaring Society of America, Inc., Santa Monica, CA. (Here is a collection of flight test reports that first appeared in the Society's *Soaring* magazine. Numerous plots of sink rate versus calibrated airspeed for about sixteen different sailplanes, obtained during the period 1974 to 1979, are presented.)

6. McCormick, Barnes W., *Aerodynamics, Aeronautics, and Flight Mechanics*, John Wiley & Sons, 1979. (This is a very useful introductory aeronautical engineering textbook, using the calculus. Although large and high speed aircraft are treated, some data and performance calculations for the Piper Cherokee are also included.)

7. Perkins, Courtland D. and Hage, Robert E., *Airplane Performance, Stability and Control*, John Wiley & Sons, 1949. (A classic text and still available.)

8. Raymer, Daniel P., *Aircraft Design: A Conceptual Approach*, American Institute of Aeronautics and Astronautics, Inc., 1989. (A very recent text that contains useful material on performance and flight mechanics; large and fast aircraft are emphasized.)

9. Kohlman, David L., "Flight Test Data for a Cessna Cardinal," NASA-CR-2337, 1974. (The author details the derivation of a drag polar for a pre-production 1970 Cardinal with the help of factory propeller efficiency information.) Available from NTIS as N74-14752.

10. Cooper, William A., Sand, Wayne R., Plitovich, Marcia K., and Veal, Donald L., "Effects of Icing on Performance of a Research Airplane," *Journal of Aircraft*, Vol. 21, No. 9, September, 1984. (Various experimental drag polars are given.)

11. Pazmany, Ladislao, *Landing Gear Design for Light Aircraft*, Vol. 1, Pazmany Aircraft Corporation, 1986. (Types of landing gears and their components and design requirements are covered.)

12. Currey, Norman S., *Aircraft Landing Gear Design: Principles and Practices*, American Institute of Aeronautics and Astronautics, 1988. (A new textbook, emphasizing landing gear design for large aircraft.)

Chapter 15

Stalls, Dives, and Turns

I eased the throttle back, rolled the ship over in a half roll and stuck her down. I felt the dead, still drop of the first part of the dive. I saw the air-speed needle race around its dial, heard the roaring of the motor mounting and the whistle of the wires rising, and felt the increasing stress and stiffness of the gathering speed. I saw the altimeter winding up—winding down, rather! Down to twelve thousand feet now. Eleven and a half. Eleven. I saw the air-speed needle slowing down its racing on its second lap around the dial. I heard the roaring motor whining now and the whistling wires screaming, and felt the awful rocking of the terrific speed. I glanced at the air-speed needle. It was barely creeping around the dial. It was almost once and a half around and was just passing the three-eighty mark. I glanced at the altimeter. It was really winding up now! I looked at the air-speed needle. It was standing still. It read three ninety-five. You could feel it was terminal velocity. You could feel the lack of acceleration.

. . . I went back down to the hangar and crawled into the ship to do the first two of the next set of five dives. Those were to demonstrate pull-outs instead of speed. Here was where I found out what the accelerometer was for. . . . I took off and went up to fifteen thousand feet and stuck her down to three hundred miles an hour. I horsed back on the stick and watched the accelerometer. Up she went and down into my seat I went. Centrifugal force, like some huge invisible monster, pushed my head down into my shoulders and squashed me into that seat so that my backbone bent and I groaned with the force of it. It drained the blood from my head and started to blind me. I watched the accelerometer through a deepening haze. I dimly saw it reach five and a half. I eased up on the stick, and the last thing I saw was the needle starting back to one. I was blind as a bat. I was dizzy as a coot. I looked out at my wings on both sides. I couldn't see them. I couldn't see anything. I watched where the ground ought to be. Pretty soon it began to show up like something looming out of a morning mist. My sight was returning, due to the eased pressure from letting up on the stick. Soon I could see clearly again. I was level, and probably had been for some time. But my head was hot with a queer sort of burning sensation, and my heart was pounding like a water ram.[1]

A. Stall Speed Reduction with Power

Figure 15.1 shows the forces acting on an aircraft while it is flying in level flight at its maximum lift coefficient (i.e., at the stalling angle of attack) with power applied. This is an equilibrium situation so the forces in any given direction must sum to zero. In the vertical direction,

$$\sum F_{\text{VERTICAL}} = L + T \sin \alpha_{\text{STALL}} - W = 0$$

so
$$L = W - T \sin \alpha_{\text{STALL}} \qquad (15.1)$$

Substituting for the lift coefficient from Eq. 12.3 and for the dynamic pressure from Eq. 5.13,

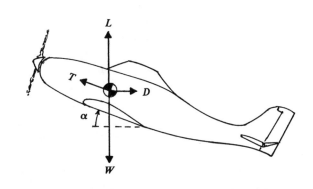

Fig. 15.1 External forces on an aircraft in a level, power-on stall

[1]Collins, Jim, "Return to Earth," *Saturday Evening Post* (February 9, 1935).

$$C_{L,MAX}\left(\frac{1}{2}\rho\,V_{STALL}^2\right)S = W - T\sin\alpha_{STALL}$$

Solving this equation for the stall speed gives us

$$V_{STALL,\,POWER\;ON} = \sqrt{\frac{2\,(W - T\sin\alpha_{STALL})}{C_{L,MAX}\,\rho\,S}} \qquad (15.2)$$

We can combine this equation with Eq. 12.15 to obtain

$$\frac{V_{STALL,\,POWER\;ON}}{V_{STALL,\,POWER\;OFF}} = \sqrt{\frac{(W - T\sin\alpha_{STALL})}{W}}$$

$$= \sqrt{1 - \frac{T\sin\alpha_{STALL}}{W}} \qquad (15.3)$$

Equation 15.1 told us right off that the power-on stall speed was going to be less than the power-off value because the wings don't have to provide as much lift (the aircraft is partially "hanging on its prop"). Equation 15.3 allows us to estimate how much. The effect is really quite small for most civil aircraft but is often exaggerated by airspeed indicator errors.

EXAMPLE 15.1. Estimate the percentage reduction in the stall speed for a C-152 at its gross weight.

Solution: Figure 13.10 shows that the power available at the stall speed (flaps up) of 47 kt at SL is about 64 hp. From Eqs. 13.14 and 13.43, this translates into a thrust of

$$T = \frac{P}{V} = \frac{64\;hp\times 550\;ft\text{-}lb/s/hp}{47\;kt\times 1.689\;ft/s/kt} = 443\;lb$$

The stall angle for this airfoil is about 16°. Substituting these values into Eq. 15.3,

$$\frac{V_{STALL,\,POWER\;ON}}{V_{STALL,\,POWER\;OFF}} = \sqrt{1 - \frac{(443\;lb)\sin(16°)}{1670\;lb}}$$

$$= \sqrt{0.927} = 0.96 \qquad \text{or} \qquad \mathbf{96\%}$$

So the effect does seem to be small for low (power/weight) aircraft. However, we didn't include the effect of increased elevator effectiveness with power on. The 152, and many other aircraft, appear to possess a much larger difference in their power-on/power-off stall speeds to the pilot because the propeller blast on the elevator drives the (whole) wing deeper into the stall and into speeds where the airspeed indicator is increasingly inaccurate.

The download on the tail of a conventional (non-canard) type aircraft can be a significant fraction of the aircraft weight, especially at the minimum speed. This would tend to minimize the difference between the two speeds even more.

B. Weightless in Space

A tossed stone or other dense object (for which aerodynamic drag is negligible compared to its weight) will follow a parabolic path through space. When you jump off a diving board you too (or, rather, your *cg*) follow a parabolic path until splashdown and you have no sensation of weight while in the air. (Just as in space, "zero-g" does **not** mean that the force

of gravity is zero; instead it means that the only unbalanced force is that of gravity. After all, the space shuttle would be a one-time space probe and not a shuttle if it weren't for gravity in orbit.) The same principle can and commonly is used by NASA with an airplane to simulate the apparent weightlessness of space travel. The pilot initiates a high speed climb and then pushes the elevator control forward at such a rate that the lift remains zero. Then the only forces acting on the aircraft are air drag (which is carefully balanced by continuously adjusting the thrust) and the force of gravity by the earth, so the aircraft will make like a stone, and the inhabitants of the aircraft will feel themselves apparently "floating" in space—although they are actually just in free fall, as they can readily tell if they choose to look outside.

EXAMPLE 15.2. Suppose that we manage to get the aerobatic version of the Cessna 152 into a 60° climb at a speed of 110 kt. How much zero-*g* time do we have before the machine is in a 60° dive at 110 kt? (Because we are emulating a stone, there will be complete symmetry in the speeds and the times for the up and down portions.)

Solution: Just like any old stone, our free fall times are determined by the vertical velocity component, which we can calculate as

$$V_Y = V\sin 60° = (110\;kt\times 1.689\;ft/s/kt)\sin 60° = 161\;ft/s$$

Then the time required to reach the highest point in our parabolic trajectory is just the time required for V_Y to become zero due to the force of the earth. Using Eq. 14.9, we have

$$V_Y = 0 = 161\;ft/s - (32.2\;ft/s/s)\,t \qquad \text{so} \qquad t = 5.0\;s$$

so that the total zero-*g* time available before return to our initial altitude and speed is

$$2\times 5.0\;s = \mathbf{10.0\;s.}$$

If the maneuver is done properly there is negligible strain on the aircraft because the only aerodynamic force acting on it is the drag force. There is **never** any question of stalling the aircraft (no matter what the indicated airspeed may be at any one time) because the angle of attack is being held constant at its zero-lift value.

You are strongly urged **not** to try this maneuver, though, unless you are in an aerobatic aircraft and are following Federal Air Regulations governing aerobatic flight. For one thing, a self-teacher is all too likely to push forward too enthusiastically and place negative loads on the aircraft, for which non-aerobatic aircraft are not designed. For another thing, on an aircraft with gravity feed like the 152, the engine will be fuel-starved and quit!

C. Terminal-Velocity Dives

A zero-lift vertical dive through a 10,000 ft loss of altitude was one of the U.S. Navy requirements for its new aircraft up until the 1940s, at which time fighter aircraft had became so sleek and fast that problems with control surface flutter and compressibility (shock waves) made it too risky for all but dive bombers. Nevertheless, such a wild maneuver provides a safe and enjoyable paper exercise. (References 1 and 2 detail some early attempts to calculate this speed.)

Figure 15.2 depicts the forces acting on an aircraft that is performing such a "maneuver." There must be zero net lift for the whole aircraft just to make the dive a truly vertical one (in no-wind conditions). The two separate forces shown on the wing are to remind you that a cambered airfoil produces a pitching moment even when it is not developing lift (which requires a negative angle of attack); this moment must be balanced by a moment produced by a net aerodynamic force on the tail, as shown. The stresses internal to the wing will depend on the lift distribution, as influenced by the planform and any wing twist that is incorporated; in any case, the wing fittings must transmit the resulting net wing moment to the fuselage. The situation is also considerably complicated by possible contributions to drag by the propeller.

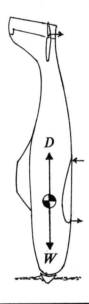

Fig. 15.2 External forces on an aircraft in a terminal velocity dive

Once an aircraft is placed in the proper attitude, the speed will increase ever more gradually (reduced acceleration) until the drag force is just equal to the aircraft's weight (zero acceleration). Then force equilibrium prevails and terminal velocity has been achieved! If the drag force depends on the square of the true airspeed, as is normally assumed, then

$$D = C_D q S = W$$

or

$$C_{D,\,PAR} \left(\frac{1}{2} \rho V^2_{\text{TERMINAL VELOCITY}} \right) S = W$$

and

$$V_{\text{TERMINAL VELOCITY}} = \sqrt{\frac{2W}{C_{D,\,PAR}\, \rho S}} \qquad (15.4)$$

where V is the **true** airspeed and $C_{D,\,PAR}$ is the parasite drag coefficient (as in equivalent flat plate drag) for the whole aircraft (since there is no induced drag). If we know the

equivalent flat plate area of the aircraft, as defined in Eq. 13.39, Eq. 15.4 can be rewritten as

$$V_{\text{TERMINAL VELOCITY}} = \sqrt{\frac{2W}{f \rho}} \qquad (15.5)$$

TERMINAL VELOCITY DIVE SPEED

EXAMPLE 15.3. Estimate the true airspeed terminal dive speed for the Cessna 152, for sea-level air density.

Solution: Using the value of f from Chapter 13,

$$V_{\text{TERMINAL VELOCITY AT SL}} = \sqrt{\frac{(2)(1670\ \text{lb})}{(5.39\ \text{ft}^2)(0.002377\ \text{sl/ft}^3)}}$$

$$= 511\ \text{ft/s} = \textbf{302 kt}$$

The posted never-exceed speed for the 152 is 145 kt, so it wouldn't be healthy to try to verify this calculation, even though propeller drag would doubtless reduce the actual terminal velocity somewhat.

We can easily obtain another formula for terminal velocity that can be applied to any aircraft for which a maximum speed (V_{MAX}) at a given power (P_{MAX}) and a given density altitude (where $\rho = \rho_{MAX}$) is given. We have

$$\text{Power Available} = \eta\, P_{\text{RATED}}$$

$$= \text{Power Required} = D\, V_{MAX}$$

$$= C_D \left(\frac{1}{2} \rho_{MAX} V^2_{MAX} \right) S\, V_{MAX} \qquad (15.6)$$

If we solve Eq. 15.6 for the drag coefficient (assuming it is nearly equal to its minimum value when the aircraft is at its maximum speed) and substitute into Eq. 15.4, we obtain

$$V_{\text{TERMINAL SPEED}} = \sqrt{\left(\frac{\rho_{MAX}}{\rho} \right) \left(\frac{V^3_{MAX}\, W}{\eta\, P_{\text{RATED}}} \right)} \qquad (15.7)$$

EXAMPLE 15.4. At the low end of the scale, the 1910s Curtiss Jenny weighed 2130 pounds and claimed a maximum speed of 65 kt on 90 horsepower. Assuming a propeller efficiency of 0.75, estimate the true airspeed terminal velocity at 2000 feet in the StAt for this aircraft.

Solution: From Eq. 15.7 and Appendix B,

$$V_{\text{TERMINAL SPEED}} =$$

$$= \sqrt{\left(\frac{0.002377}{0.002341} \right) \left[\frac{(65\ \text{kt} \times 1.689\ \text{ft/s/kt})^3 (2130\ \text{lb})}{0.75 \times 90\ \text{hp} \times 550\ \text{ft-lb/s/hp}} \right]}$$

$$= 278\ \text{ft/s} = \textbf{164 kt}$$

We normally think of air drag as an enemy of aircraft efficiency but drag is caused by the same air molecules that allow the wing to develop lift. In free fall, it keeps the terminal velocity of raindrops to only around 27 ft/s when they would otherwise pound you at a speed of 300 mi/hr in dropping through 3000 ft. In sport parachuting, it keeps the terminal velocities of the parachutists between about 110 and 150 mi/hr;

they start with a vertical acceleration of 1 g and "terminate" with 0 g.

When Jim Collins yanked back on the stick to demonstrate a high-g pull-out in the plane he was testing (the chapter's opening quote), he noted that "centrifugal" force squashed him into the seat. This "centrifugal" force is really an imaginary or "inertial" force. There was **no** force pushing him **into** the seat at all! Instead, the seat was pushing **up** on him because his body was trying to continue to go straight down at a constant speed and the plane was being forced to curve around to level flight and it was simply trying to force him to follow along! To some extent the one being subjected to forces during accelerated motion is tempted to use the inertial force idea because he is used to feeling the seat pushing only hard enough to equal the force of gravity; however, for an observer outside the accelerated reference frame it is clear that the pilot is simply being pushed on extra hard because he is accelerating upward. Then the external forces acting **on** the pilot are **not** in balance! (The third law is still valid, of course; the pilot pushes back on the seat just as hard as it pushes up on him or her.) The force that causes an object to change its direction of motion (particularly into a circular arc) is often called a **"centripetal"** (center-seeking) force. A centripetal force is any external force that pushes (or pulls) an object toward the center of a curved arc, preventing the object from following its natural inertial tendency to go straight; a centripetal force is then a real, external force that acts in the opposite direction from the direction normally espoused for the fictitious, imaginary centrifugal force. The centripetal force that causes the pilot to pull out of his dive is the force of the seat on him; the centripetal force that causes the airplane to pull out of its dive is the extra lift developed by the wings.

D. Load Factor

We have seen that the lift force on an aircraft just equals its weight whenever it is in dynamic equilibrium: level flight, a stabilized climb, or a stabilized descent. Now we turn to situations for which the lift force is **not** equal to the weight. These include (**a**) the transitions from one of these equilibrium states to another, (**b**) the steady banked turn, (**c**) the onset of atmospheric turbulence, and (**d**) aerobatic maneuvers.

The **load factor** for an aircraft is defined as the ratio of the lift force to the gravitational force (i.e, the weight); it is commonly represented by the symbol n.

$$n \equiv \frac{L}{W} \qquad (15.8)$$

LOAD FACTOR IN "g's"

Evidently the load factor for equilibrium flight is **1** g, by definition. Any value greater than or less than 1 g represents some positive or negative acceleration, respectively, in the direction perpendicular to the relative wind. A "g-meter" is normally mounted in the instrument panel of an aerobatic aircraft; this instrument has three pointers which present (**a**)

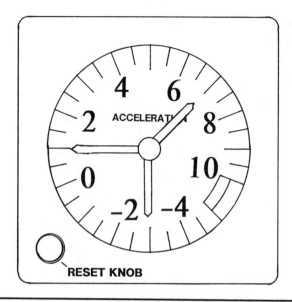

Fig. 15.3 A typical aircraft accelerometer ("g-meter"); this one is indicating a current load factor of $n = 1.0$ ($a = 0$) and maximum load factors (since being reset) of $n = +7.0$ ($a = 6\,g$) and $n = -3.0\,g$

the current load factor, (**b**) the maximum positive load factor since being reset, and (**c**) the maximum negative load factor since being reset (Fig. 15.3).

Since $\Sigma F_{\text{VERT}} = L - W = ma_{\text{VERT}} = (W/g)a_{\text{VERT}}$, where g is the acceleration of a free-falling object, we can also write

$$L = W\left(1 + \frac{a_{\text{VERT}}}{g}\right)$$

or, dividing through by W, $\quad n = 1 + \dfrac{a_{\text{VERT}}}{g} \qquad (15.9)$

and therefore $\quad a_{\text{VERT}} = (n - 1)g \qquad (15.10)$

EXAMPLE 15.5. What acceleration perpendicular to his spine caused Jim Collins to lose his vision for a few seconds?

Solution: Collins' g-meter was reading 5.5 (the last he was able to notice, at least) so
$$a = (n-1)g = (5.5-1)g = 4.5\,g = 145 \text{ ft/s}^2$$

The load factor to which a flying machine should be designed has been a matter of considerable interest for a long time. In 1911, Wilbur Wright wrote Orville, "Some time ago they turned four machines upside down and piled on sand till they broke. The Wright stood 4 times its flying weight, the Farman 2 ¾, the Antoinette 1, and the Bleriot 1 ¼." Today, aircraft are certificated in different categories depending upon their intended use. In general, **transport category** aircraft are those intended for large-scale passenger or freight hauling for hire, **normal category** aircraft are those intended for relatively lighter payloads, **utility category** aircraft are intended for more rigorous training or freight hauling assignments, and **aerobatic ("acrobatic") category** aircraft must be strong and stable enough to safely execute some of the common aerobatic maneuvers. Table 15.1 summarizes the strength requirements for these four categories. The **limit load factors** are those that should never be exceeded in flight; beyond these load factors the structure may be permanently bent out of shape. The **ultimate load factors** are in every case 1.5 times the limit load factors; the structure must withstand these load factors for at least **3** seconds before failure.

CATEGORY	LIMIT LOAD FACTOR	ULTIMATE LOAD FACTOR
Transport	+2.5 and −1.0	+3.75 and −1.5
Normal	$+2.1 + \dfrac{24,000}{W + 10,000}$ and −40% of positive	1.5 × (pos limit factor) and 1.5 × (neg limit factor)
Utility	+4.4 and −1.76	+6.6 and −2.64
Acrobatic	+6.0 and −3.0	+9.0 and −4.5

Table 15.1 The Four U.S. Certification Categories

Because aluminum fails at just about 1.5 times its yield values, metal aircraft can be essentially designed for failure at the ultimate load and will then meet the limit load criteria. However the yield/fail ratio may be very different for composites and other materials. For this reason and because composites normally don't show any permanent deformation before they fail, at the present time they are commonly designed for at least two times the limit load factor.

Transport aircraft have the least severe load factor requirements because they are least affected by air turbulence (as we shall see) and are presumed to be flown by experienced pilots.

The most efficient aircraft, it might be argued, is the one for which every part fails at the minimum allowable value. In practice, most aircraft are stronger than this to allow for production variations, material fatigue, and so forth. But don't count on any significant margin; it has been suggested, for example, that the aerobatic Cessna 150 (the Aerobat) will develop permanent wing wrinkles if the load factor goes just 1 *g* over the limit value. Also, the aerobatic Citabria has shown a tendency to loosen its fabric-to-wing rib attachments if used in aerobatic training. On the other hand, the "aerobatic" B727 of Chap. 12 proved to be significantly stronger than it was required to be, at least with regard to its maneuvering strength.

E. Turns

One of the early achievements of a student pilot is the constant-altitude **coordinated** turn. This is a roll into a banked turn, a steady turn, and a roll out of the bank, in which the pilot feels no tendency to slip in his or her seat toward the inside or the outside of the turn; it is achieved by a dynamically coordinated use of the ailerons and the rudder. Some aircraft require considerable rudder pressure in the direction of the turn (to combat adverse yaw) while rolling into or out of the turn while others require little or no rudder application.

To a student of physics, the steady turn is described as one for which the lift force is exactly perpendicular to the wings and to the pilot's seat, and the horizontal component of the lift force provides just the required amount of centripetal force. (In terms of the vector force diagram, it is identical to the forces in a steady turn on level ground for either a bicycle or a motorcycle, with the force of the ground replacing the lift force. It is also the force diagram for a car going around a banked turn at the precise speed that produces no sensation of sliding down the hill or skidding up the hill.)

The force diagram of Fig. 15.4 shows us that the weight of the aircraft is just balanced by the **vertical** component of the lift force. Then note again that the **horizontal** component of the lift force is **not** balanced; it provides the centripetal force needed to make the aircraft fly in a circular path. Recall (from physics) that the magnitude or amount of acceleration (the rate of change in direction) for something going in a circle at constant speed depends directly on the square of the speed of the turn and inversely on the radius of a turn; thus you require a very large centripetal force if you try to go in a small circle at high speed — and the required force quadruples when the speed doubles and doubles when the radius is halved. In symbols,

$$a_{\text{CENTRIPETAL}} = \frac{V^2}{R} \qquad (15.11)$$

CENTRIPETAL ACCELERATION

So circular motion at constant speed is **accelerated** motion with an acceleration equal to (V^2/R), where V is the true airspeed and R is the constant radius of the turn.

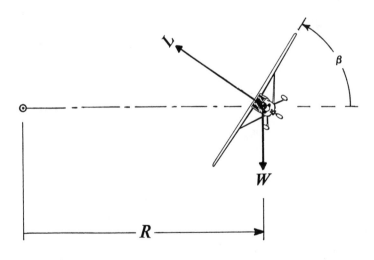

Fig. 15.4 External forces on an aircraft in a steady, level, banked turn

Equating the vertical forces to zero (since the vertical acceleration is assumed to be zero) yields

$$F_{\text{VERT}} = L \cos \beta - W = 0 \quad \text{and therefore}$$

$$L = \frac{W}{\cos \beta} \qquad (15.12)$$

LIFT FORCE IN A STEADY, LEVEL, BANKED TURN

And then we equate the horizontal force to the required centripetal force to obtain

$$\sum F_{\text{HORIZ}} = L \sin \beta = ma_{\text{CENTRIPETAL}} = \frac{mV^2}{R} \quad (15.13)$$

When we substitute Eq. 15.12 in the defining equation for load factor (Eq. 15.8), we see that

$$n = \frac{1}{\cos \beta} \quad (15.14)$$

LOAD FACTOR IN A STEADY, LEVEL TURN

Thus a 60° banked turn, on the legal borderline of aerobatic flight, produces a load factor of $n = 1/(\cos 60°) = $ **2**, and the wings are producing a lift equal to **twice** the weight of the aircraft. Since the pilot and passengers are experiencing the same acceleration as the aircraft, they will feel the same load factors; a 170 pound person suddenly finds himself with an apparent weight of 340 pounds in this bank, for example.

Obviously, a pilot can increase the load factor above the value given in Eq. 15.14 if the plane is forced to climb as well as turn. The combination is an effective method for inadvertently stalling the aircraft, too.

The maximum steepness of a level, banked turn for a given aircraft is determined by the maximum lift coefficient and the available power. In utility or aerobatic aircraft, the maximum lift coefficient normally is reached before the load factor reaches the limit value—if the transition from level flight is smooth and relatively gradual.

EXAMPLE 15.6. Calculate the maximum safe constant-altitude bank for a transport category aircraft.

Solution: From Eq. 15.14 and Table 15.1,

$$n = +2.5 = \frac{1}{\cos \beta} \quad \text{so} \quad \beta = \textbf{66}°.$$

The truly good aerobatic aircraft is quite symmetrical in its flight characteristics, even though its pilot isn't. It is good fun and good practice to make coordinated turns in the inverted flight position, as well as in level, upright flight, and then the load factors are the negative of those given by Eq. 15.14. In level inverted flight, the g-meter should be indicating -1; in a 60° bank the g-meter should be indicating -2, etc. Making coordinated inverted turns by "feel" means using the sense of slip or skid that is transmitted through the shoulder harness that is holding you—quite a bit different from the "seat-of-the-pants" tactile sense you use when making upright coordinated turns!

The banked-turn force diagram that appears in the FAA's *Flight Training Handbook* (Fig. 15.5) is **not** recommended at all. The diagram includes the fictitious "centrifugal" force from the misguided desire to have "balanced" forces when, in fact, they cannot be balanced because the maneuver is an accelerated one. It labels the vector sum of this fictitious force and the weight as the "Load Factor" when in fact the load factor is the

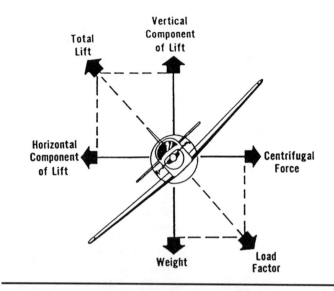

Fig. 15.5 The confusing, erroneous "force" diagram that appears in the Federal Aviation Administration's *Flight Training Handbook*

scalar **ratio** of the lift force to the gravitational force and is **not** a force at all! Please return to Fig. 15.4 and study it a bit before continuing lest the FAA version does permanent harm to your understanding of what is happening in a turn.

If an aircraft is "overbanked" so that the horizontal component of the lift force is greater than $ma_{\text{CENTRIPETAL}}$, the plane will not be travelling in the direction it is pointed, the occupants will feel a tendency to slide "down" the seat, and the turn is said to be in a "slipping" turn. If, on the other hand, the horizontal component of the lift force is less than $ma_{\text{CENTRIPETAL}}$, the plane also won't be travelling in the direction it is pointing and is said to be in a "skidding" turn, much like a car that goes too fast around a banked turn and has to apply a sideways force on the seat of the driver to keep him or her from sliding toward the outside of the turn—i.e., not turning as much as the car.

Most aircraft will happily stay in a medium banked turn (say 30°) **once** they get there but will need rudder in the direction of the turn to get there without slipping because of the "adverse yaw" generated by the differential drag caused by the oppositely deflected ailerons and the angular acceleration of the bank angle. (More on this in Chap. 17.)

An aircraft always stalls at a **higher** indicated (and calibrated and equivalent and true) airspeed when it is stalled in a constant-altitude bank. The reason is that only part of the lift is holding up the airplane; the rest is making it turn. The stalling angle of attack does not change. If we equate the maximum lift available (Eq. 12.15 and Eq. 12.16) to $W/(\cos \beta)$ (Eq. 15.12), we obtain

$$C_{\text{L, MAX}} \left(\frac{1}{2} \rho_0 V_{\text{E, STALL}}^2 \right) S = \frac{W}{\cos \beta}$$

$$\text{or} \quad V_{\text{E, STALL}} = \sqrt{\frac{2W}{\rho_0 \, S \, C_{\text{L, MAX}} \, \cos \beta}} \quad (15.15)$$

and then combining this with Eq. 12.16 yields

$$V_{\text{E, STALL, AT A BANK ANGLE OF }} \beta = \frac{V_{\text{E, STALL, IN LEVEL FLIGHT}}}{\sqrt{\cos \beta}}$$

(15.16)

STALL SPEED IN A STEADY, LEVEL BANKED TURN

From this we see that an aircraft in a 60° bank stalls at $1/\sqrt{0.5}$ or 1.414 times its level flight stall speed! From Eq. 15.14 we can also write this as

$$\mathbf{V}_{\text{E, STALL, AT A BANK ANGLE OF }} \beta = \sqrt{n}\ V_{\text{E, STALL, IN LEVEL FLIGHT}}$$

EFFECT OF LOAD FACTOR ON STALL SPEED   (15.17)

and this is true whether the load factor is increased or decreased by a banked turn, by a pull-up, push-down, or by some combination thereof. (For negative lift, use the negative lift stalling speed and the absolute value of n. For $n = 0$, the equation is not joking; the stall speed is really and truly zero—because a free-falling rock can't stall.) The importance of this relationship for pilots cannot be overemphasized. A pilot who maneuvers abruptly or banks steeply while close to the ground is asking for real trouble in the form of a stall at an altitude insufficient for recovery. (This and bad weather account for most of the fatal accidents with light aircraft.) The fact that properly designed and built canard-type aircraft are characteristically incapable of stalling their main wings, and therefore also cannot spin, should make them considerably safer—but there aren't enough out there to be sure at this time.

A pilot who has extra altitude can safely execute a steeply banked turn onto final approach if the wings are "unloaded" by pushing the nose down throughout the bank so that the airspeed increases while the load factor is maintained at 1 or less.

If an aircraft is stalled with a high load factor while in a skidding turn, it will tend to stall the high wing first and may roll very abruptly to a very steep bank, pitch nose down to the vertical or beyond, and begin a spin entry. Aircraft designed in the 1930s and earlier, or a newer design that is rigged improperly, are most likely to do this. Usually, recovery is easily effected with standard spin recovery technique but a certain amount of altitude will be lost.

Every pilot must be taught that an aircraft stalls at a higher indicated airspeed when the load factor is greater than 1, to help prevent accidental stalls and altitude loss while maneuvering and turning close to the ground. Therefore a standard training maneuver (permissible only for aircraft in the utility category that are not prohibited from doing it) is the "**accelerated**" or "high speed stall." It is usually executed by (**a**) entering a 45° bank, (**b**) gradually increasing the back pressure on the wheel or stick until the airspeed is at or below the maneuvering speed, and (**c**) smoothly but rapidly adding additional elevator pressure until the stall break occurs. The idea is to demonstrate to the pilot that the stall can be made to occur at a higher than normal value and to demonstrate that the stall break may be more violent under those circumstances, without overstressing

the aircraft. We'll have more to say about the maneuvering speed in the next chapter.

With a fixed power setting, an aircraft always slows down in transitioning to a steady banked turn. The culprit is easily identified—it is **induced drag**. Because the wing develops lift greater than the aircraft weight in a banked turn, induced drag and therefore the total drag force increase, slowing the aircraft. The effect is complicated by the additional loading on the propeller caused by the turn, which occurs at the same time and which causes the power delivered to decrease, as evidenced by a reduction in rev/min for an aircraft with a fixed-pitch propeller. If the turn is entered at less than full power, the throttle can be advanced as necessary in the turn to keep the delivered power constant; then the speed loss will be a genuine reflection of the effective span loading of the aircraft.

Equation 13.45, in Chap. 13, specifies the power required for level flight within the parabolic drag polar approximation. It is easily modified to obtain an expression for the power required in a steady, coordinated, banked turn. The lift force, and therefore the lift coefficient, increases as the inverse of the cosine of the bank angle (by Eq. 15.14); because the induced drag term in Eqs. 13.44 and 13.45 varies as the square of the lift coefficient, this term gains a $(1/\cos\beta)^2$ factor in a steady, banked turn:

$$P_{\text{REQ, BANKED TURN}} = \frac{1}{2}\rho\, f\, V^3 + \left(\frac{1}{e}\right)\left(\frac{2}{\pi}\right)\left(\frac{W}{b}\right)^2\left(\frac{1}{\rho V}\right)\left(\frac{1}{\cos\beta}\right)^2$$

(15.18)

Figure 15.6 presents the calculated power required and power available (75% of rated) for various bank angles, for a Cessna 152 at its gross weight and at a density altitude of 4000 ft. It is assumed that the throttle is advanced in the bank as

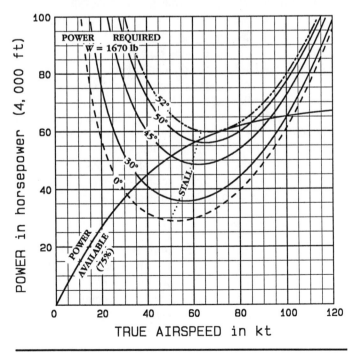

Fig. 15.6 Calculated power curves for a C-152 for various bank angles

necessary to keep the power constant. Note that the true cruise speed is predicted to decrease from 103 kt in level flight to 93 kt in a 45° bank. Also note that the steepest possible level bank for this power setting is predicted to be 52°, at a true airspeed only about 9 kt over the stall speed for that bank angle.

Figure 15.6 is very pretty but it does have one significant defect: it can't be true! Those who have tried know that it is quite possible to maintain a 60° banked turn in this aircraft with even less than 75% of rated power. In fact, a certain C-152 of my acquaintance, one with a well-established "doggy" reputation, still managed to reach the book speed (for an aircraft without wheel fairings) of 85 kt indicated airspeed and 88 kt true airspeed at 56% power at a density altitude of 3950 ft. And then it lost only about 8 kt (indicated) in a steady 45° bank and an additional 17 kt (indicated) in a steady 60° bank. Thus it was indicating 60 kt in the 60° bank even though the calculated calibrated stall speed (by Eq. 15.16) is $(47 \text{ kt})/\sqrt{\cos 60°} = 66$ kt. Evidently additional airspeed errors come into play in steep banks—an hypothesis confirmed by the pilot's handbook. But the aircraft **was** capable of maintaining a 60° bank in level flight, even at a power level below that shown in Fig. 15.6!

Therefore Fig. 15.6 must either show an overestimate of the induced drag increase in a banked turn or else the power available (through the propeller efficiency) is not being modeled properly. Fitting a displaced drag polar with a minimum at a positive lift coefficient to the C-152 cruise speeds did not improve the agreement of theory with reality, as it turned out. Calculating a drag polar from level flight power/speed data suggests that the drag coefficient turns up even faster, at large values of the lift coefficient, than that estimated from cruise performance. It appears that the second term in Eq. 15.18 simply shows a greater dependence on the bank angle than is true for this aircraft, and probably for most aircraft.

Fairly close agreement between predicted and experimental speed loss in this aircraft can be illicitly obtained by either assuming that (a) only one-quarter of the induced drag is affected by the bank, or (b) the dependence on bank angle is proportional to $\cos^{-1}\beta$ rather than $\cos^{-2}\beta$ and only 75% of the induced drag is affected by the bank.

In any case, here is another good scientific excuse to go flying. Good data require patience, repeated measurements, and smooth air, but you can learn much about the span efficiency of the aircraft you're testing. Use the computer program in Appendix D to compare your experimental measurements with the predictions of the parabolic drag polar theory.

F. Rate of Turn

The horizontal component of lift in a turn is the force that is actually turning the aircraft. From it we can obtain the **rate of turn**, that is, the number of degrees turned through in each second. A **standard rate turn** is one that accomplishes a turn through 360° in just two minutes—that is, 3° per second.

The radius of the turn depends on both the steepness of the bank and the speed of the aircraft. We can obtain the exact relationship from the two force equations; if we substitute the

lift force of Eq. 15.12 into Eq. 15.13, we obtain an expression for the radius of our banked turn:

$$R = \frac{V^2}{g \tan \beta} \qquad (15.19)$$

RADIUS OF A BANKED TURN

where V is the true airspeed in the turn.

The **angular velocity** of the turn in radians per second is $\omega = V/R$, so

$$\omega \text{ (in rad/sec)} = \frac{V}{R} = \frac{g \tan \beta}{V}$$

and then, since there are 2π radians in 360°, we obtain finally

$$\textbf{Rate of Turn} \text{ (in deg per sec)} \cong \frac{57.3 \, g \tan \beta}{V} \qquad (15.20)$$

where the speed must be in ft/s and g is in ft/s².

EXAMPLE 15.7. Determine the rate of turn and the radius of the turn for an aircraft flying in a 60° bank at constant altitude and at a true airspeed of 100 kt.

Solution: From Eq. 15.19,

$$R = \frac{V^2}{g \tan \beta} = \frac{(100 \text{ kt} \times 1.689 \text{ ft/s/kt})^2}{(32.2 \text{ ft/s}^2)(\tan 60°)} = \textbf{511 ft}$$

From Eq. 15.20,

$$\text{Rate of turn} = \frac{(57.3)(32.2 \text{ ft/s}^2)(\tan 60°)}{168.9 \text{ ft/s}} = \textbf{18.9°/s}$$

so that 360°/(18.9°/s) or 19.0 s are required to make a complete turn.

We can also solve Eq. 15.19 for $\tan \beta$ and substitute a turn rate of 3°/s to find the angle of bank necessary to obtain a standard rate of turn as a function of true airspeed.

$$\tan \beta = 0.0524 \, \frac{V}{g} \qquad (15.21)$$

BANK ANGLE (in deg) FOR STANDARD RATE TURN

where V must be in ft/s and g in ft/s².

EXAMPLE 15.8. Calculate the bank angle required to obtain a standard rate of turn of 3° per second for (**a**) a true airspeed of 100 kt (single-engine trainer) and (**b**) a true airspeed of 550 kt (jet transport).

Solution: From Eq. 15.21, the 100 kt speed requires a bank angle of

$$\beta = \arctan \left(0.0524 \times \frac{168.9 \text{ ft/s}}{32.2 \text{ ft/s}^2}\right) = \arctan(0.275) = \textbf{15.4°}$$

which is reasonable enough, but for a 550 kt speed, the required bank angle is

$$\beta = \arctan(1.512) = \textbf{56.5°}$$

so that transport aircraft normally use a rate of turn equal to half of smaller aircrafts' "standard" rate!

G. Turns about a Point

One of the best and most challenging of the maneuvers presented to student pilots is the constant-radius, constant-altitude turn about a point on the ground. The successful pilot will focus his or her attention outside the airplane while constantly changing the angle of bank to compensate for the varying effect of the wind. Figure 15.7 displays the vector diagrams for the ground speed at two points. At point 1, directly **downwind**, the ground speed is a maximum and therefore the angle of bank, from Eq. 15.19,

$$\tan \beta_1 = \frac{(V_{A1} + V_W)^2}{gR} \qquad (15.22)$$

is also a **maximum** at this point. We can derive expressions for the angle of bank and the wind correction angle at any point in the turn by examining the velocity vectors at point 2. The requirement for a turn of constant radius means that no component of the resultant ground velocity can be toward or away from the point, so

$$V_{A2} \sin \phi = V_W \sin \theta$$

or $\qquad \sin \phi = [V_W / V_{A2}] \sin \theta$

and, in general,

$$\sin \phi = \frac{V_W}{V_A} \sin \theta \qquad (15.23)$$

where V_A is the true airspeed anywhere in the turn and the requirement for the appropriate amount of centripetal force means that the angle of bank must be (using Fig. 15.3 and Eqs. 15.19, 15.22, and 15.23)

$$\tan \beta_2 = (V_{A2} \cos \phi + V_W \cos \theta)^2 / (gR)$$
$$= V_{A2}^2 \left[\cos \phi + (V_W / V_{A2}) \cos \theta \right]^2 / (gR)$$

and, in general,

$$\tan \beta = \frac{V_A^2}{gR} \left[\cos \phi + \left(\frac{V_W}{V_A} \right) \cos \theta \right]^2 \qquad (15.24)$$

BANK ANGLE DURING A TURN ABOUT A POINT

In the general expressions of Eqs. 15.23 and 15.24, I have substituted a general symbol for the true airspeed relative to the **ground**, V_A, in place of the symbol V_{A2} to remind us that these equations are valid everywhere in the turn.

In practice the aircraft will change its speed during the turn, even at constant altitude, because of changes in induced drag as the bank angle is varied. The actual speed at each point in the turn can be calculated from Eq. 15.18 (with some difficulty) for any given power available. Using this strategy, Fig. 15.8 presents calculated turn variables for the Cessna 152 that was modeled in Chap. 13, at a density altitude of 1700 feet and constant 75% delivered power.

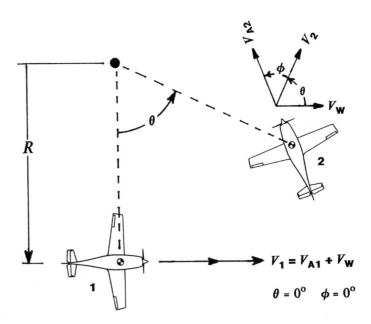

V_W = WIND VELOCITY

V_1 = GROUND SPEED AT POINT 1

V_2 = GROUND SPEED AT POINT 2

V_{A1} = TRUE AIRSPEED AT POINT 1

V_{A2} = TRUE AIRSPEED AT POINT 2

ϕ = WIND CORRECTION ANGLE

Fig. 15.7 Geometry of the turn about a point with wind

Oops! Did you catch what that last calculation says about **downwind turns**? Fig. 15.8 suggests that we've cranked our little airplane over to a 55° bank to keep our perfect circle over the ground in this stiff 20 kt wind, and the extra induced drag has caused our airspeed to decay down to only 76 kt! Have we just stalled and crashed and burned?

The flaps-up, power-off, $n = 1$ stall speed is 47 kt calibrated airspeed (48 kt true airspeed), so the calibrated stall speed in the 55° bank is (from Eq. 15.16)

$$V_{\text{STALL SPEED}} = 47 \, \text{kt} / (\sqrt{\cos 55°}) = \mathbf{62 \, kt}$$

So we're all right! But wait—what if we should encounter a 14 kt wind gust or a severe downdraft just at that point? It could spell trouble, and the possibility is greatly enhanced if we're down at low level attempting to show off for friends on the ground. Surely we'll be frightened by our low speed, though? Oh, no, not at all! We're looking at the ground and it is passing below us at a very comfortable speed of (76 kt + 20 kt) = 96 kt!

The maneuver is supposed to teach us to be wary of turns about a ground object at low altitude while developing the

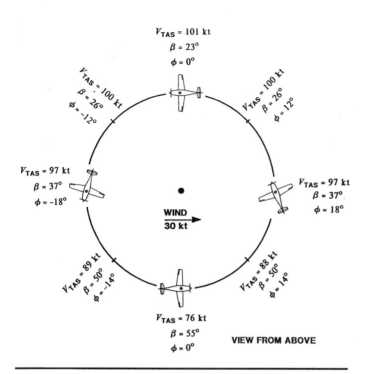

$V_{TAS} = 101$ kt
$\beta = 23°$
$\phi = 0°$

$V_{TAS} = 100$ kt
$\beta = 26°$
$\phi = -12°$

$V_{TAS} = 100$ kt
$\beta = 26°$
$\phi = 12°$

$V_{TAS} = 97$ kt
$\beta = 37°$
$\phi = -18°$

WIND
30 kt

$V_{TAS} = 97$ kt
$\beta = 37°$
$\phi = 18°$

$V_{TAS} = 89$ kt
$\beta = 50°$
$\phi = -14°$

$V_{TAS} = 88$ kt
$\beta = 50°$
$\phi = 14°$

$V_{TAS} = 76$ kt
$\beta = 55°$
$\phi = 0°$

VIEW FROM ABOVE

Fig. 15.8 Calculated speeds, wind correction angles, and bank angles for a turn about a point by a C-152 at 1700 ft and 75% power; the radius of the turn is 571 ft.

ability to make good turns with our attention directed outside the aircraft. Too many bold pilots have stalled and crashed while trying to circle over friends or houses at low altitudes. There is no way the airplane can "know" that we're making a downwind turn since it responds only to the relative wind. The problem is that we are looking outside the airplane and, deceived by the nice ground speed, might allow the aircraft to get too slow and too steeply banked in an attempt to keep the important object below in view. The possibility of upsetting updrafts and downdrafts is extra high at low altitude with some tree or hills around, too.

If a pilot's attention is directed outside the aircraft, it is all too easy to **skid** the turn because a skid increases the turn rate without increasing the bank angle. However, in a skidding turn an airplane will stall with the nose on the horizon (rather than well above it) and many aircraft will roll steeper and into a spin entry at the stall break.

Figure 15.8 was drawn for a steep turn to illustrate the point about downwind turns and stalls in the turn, but in practice a greater radius (calculated to be about 940 ft) would be chosen to ensure a maximum bank of 45° in the turn.

H. On-Pylon 8

One of the most interesting maneuvers (expected for the commercial certificate) is the on-pylon 8. In this the pilot tries to maintain the lateral axis of the aircraft (through its *cg*) pointing alternately at two

particular spots on the ground. He or she soon finds that there is only one altitude, for a given ground speed, at which this can be accomplished. At too high an altitude, the lateral axis will move behind the point, and vice versa. In the absence of wind the flight path over the ground is a partial circle about first one point and then the other, with the bank angle constant except for the transition from one circle to the other.

With a wind, though, the ground speed will depend on the direction of the flight relative to the wind. When heading into the wind the pilot must dive the plane to keep the ground speed about the same while a climbing turn must be made when heading with the wind.

The result is that, from a viewpoint high overhead, a horizontal figure 8 is traced over the ground. The force diagram for the maneuver is shown in Fig. 15.9.

The force diagram is the same as for a level flight turn, so Eq. 15.19 holds here also. But in this case, the geometry is such that the tangent of the bank angle is also equal to the altitude, h, divided by the radius, R, so

$$\frac{h}{R} = \frac{V^2}{Rg} \quad \text{and} \quad h = \frac{V^2}{g} \quad (15.25)$$

PYLON ALTITUDE

and we see that the altitude does indeed depend only on the ground speed, since g is so nearly constant. However, there is an amazingly strong dependence on the speed of the aircraft!

EXAMPLE 15.9. Calculate the pylon altitude for (a) an aircraft flying at a true airspeed of 100 kt (168.9 ft/s) and (b) at a true airspeed of 200 kt.

Solution: (a) $h = (168.9)^2/(32.2) = $ **886 ft** but (b) for an aircraft flying at 200 kt, the proper altitude is way up to **3540 ft** above the ground. This latter altitude is too high to provide any useful practice in wind correction.

Figure 15.10 depicts the geometry and the vector velocities involved in an on-pylon turn, including the effect of wind and, in particular, beginning with a point directly downwind. What differentiates this geometry from that for the turn about a point

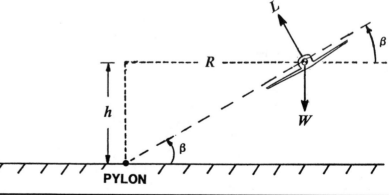

Fig. 15.9 External forces and the geometry for an on-pylon turn

(Fig. 15.7) is that here the aircraft is always pointing perpendicular to a line drawn toward the pylon and both the radius and the height of turn above the ground are constantly changing. Equation 15.25 holds for each point but, because these points are on the ground, the speed involved is the **ground speed** of the aircraft.

$$h_1 = \frac{V_1^2}{g} \qquad (15.26)$$

$$h_2 = \frac{V_2^2}{g} \qquad (15.27)$$

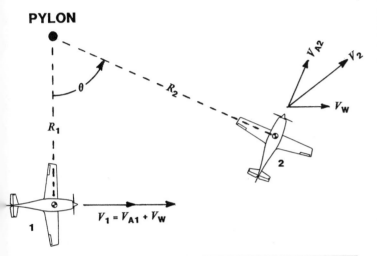

Fig. 15.10 Velocities and radii for an on-pylon turn

From Fig. 15.9, the bank angle is related to the horizontal distance from the pylon and to the pivotal altitude through

$$\tan \beta_1 = \frac{h_1}{R_1} \qquad (15.28)$$

$$\tan \beta_2 = \frac{h_2}{R_2} \qquad (15.29)$$

From Fig. 15.10, the ground speed at point #2 is the vector sum of the wind speed and the true airspeed:

$$V_2 = \sqrt{(V_{A2} \cos \theta + V_W)^2 + (V_{A2} \sin \theta)^2} \qquad (15.30)$$

Combining Eqs. 15.26 and 15.28 yields

$$\tan \beta_1 = \frac{V_1^2}{g R_1} \qquad (15.31)$$

ENTRY POINT FOR THE ON-PYLON TURN

which tells us that (**a**) the distance R_1 from the pylon should be chosen such that the bank angle is reasonable and (**b**) the steepest bank angle occurs at point #1 because the ground speed is a maximum at that point, while contrariwise the smallest bank angle occurs at $\theta = 180°$. Commonly, entry is made at the downwind point, as shown, and the distance is chosen so as to require a maximum bank of 45°.

EXAMPLE 15.10. Determine the proper initial distance from the pylon for an aircraft with a 100 kt cruising speed, a 20 kt wind, and a pilot that desires a maximum bank of 45°.

Solution: Solving Eq. 15.31 for R_1,

$$R_1 = \frac{V_1^2}{g \tan \beta_1} = \frac{[\,(1.689 \text{ ft/s/kt})(100 \text{ kt} + 20 \text{ kt})\,]^2}{32.2 \text{ ft/s}^2 \tan 45°} = \textbf{1276 ft}$$

In practice, the pilot finds that the decreased ground speed at point #2 requires a dive to a lower altitude to maintain the pivotal point. However, this altitude loss involves a loss of potential energy and therefore the kinetic energy and the true airspeed of the aircraft increase. If we consider changes in drag due to changes in speed and bank angle as secondary effects and neglect them so as to make the problem more tractable, we can use the principle of **conservation of energy** to write

$$\tfrac{1}{2} m V_{A1}^2 + m g h_1 = \tfrac{1}{2} m V_{A2}^2 + m g h_2$$

or

$$V_{A2}^2 = V_{A1}^2 + 2 g (h_1 - h_2) \qquad (15.32)$$

Suppose now that we wish to calculate the flight parameters for our 100 kt airplane in the 20 kt wind after **90°** of turn have been accomplished. Equation 15.28 tells us that, for $\beta = 45°$, $h_1 = R_1 = 1276$ ft, and Eq. 15.30 becomes

$$V_2 = \sqrt{V_W^2 + V_{A2}^2}$$

or

$$V_2^2 = V_W^2 + V_{A2}^2 \qquad (15.33)$$

which we should have expected since the vector velocities are at right angles to each other at that point. Combining Eqs. 15.27, 15.32, and 15.33 gives us

$$g h_2 = V_W^2 + V_{A1}^2 + 2 g (h_1 - h_2)$$

or

$$h_2 = \frac{V_W^2 + V_{A1}^2 + 2 g h_1}{3 g} = \textbf{1156 ft} \qquad (\text{at } \theta = 90°)$$

The new true airspeed, from Eq. 15.32, is $V_{A2} = \textbf{113 kt}$
Similarly, after 180° of turn, Eq. 15.30 yields

$$V_2^2 = (V_W - V_{A2})^2 = V_W^2 - 2 V_W V_{A2} + V_{A2}^2 \qquad (15.34)$$

since the aircraft is heading directly into the wind. Then, from Eqs. 15.27, 15.32, and 15.34,

$$g h_2 = V_W^2 - 2 V_W \sqrt{V_{A1}^2 + 2 g (h_1 - h_2)} + V_{A1}^2 + 2 g (h_1 - h_2)$$

which can be solved for this special case (with trial and error or a root-finding computer program) to yield

$$h_2 = \textbf{1007 ft} \qquad (\text{for } \theta = 180°)$$

It would be very nice to be able to calculate the path over the ground and the times involved for this turn about a pylon. Unfortunately, the energy method doesn't directly yield this information. It can be obtained, however, by calculating the airplane's velocity vector for small increments of θ and using the calculated incremental values to obtain new values for the vector distance travelled and thereby the time required and the corresponding bank angles. Figure 15.11 shows the results of this calculation for $\theta_1 = 45°$, $V_{A1} = 100$ kt, and $V_W = 20$ kt.

There is a clear message to the pilot in Fig. 15.11. Just as for the turn about a point, the pylon turn should be entered directly downwind so that the maximum bank in the maneuver is determined immediately.

To transform an on-pylon turn into an on-pylon 8, the pilot should choose two distinctive points on the ground (for example, two road intersections) whose connecting line is perpendicular to the wind. The chosen points should be separated by a distance somewhat greater than twice the pylon distance at the downwind point so that there is a short, straight segment when transitioning between pylons.

Pylon turns have real-world applications as well. They have been used to lower food to the ground and also to direct concentrated gunfire at a given spot on the ground.

I. Efficient Glides and Gliding Turns

An expression for the load factor in a steady glide is easily obtained by using the definition of load factor (Eq. 15.8) and the equation for the equilibrium of forces perpendicular to the glide path (Eq. 14.35). The result is

$$n_{GLIDE} = \cos\gamma \qquad (15.35)$$

LOAD FACTOR IN A STEADY GLIDE

Normally the pilot doesn't notice this because γ is a relatively small angle, making $\cos\gamma$ very close to 1. For a steady 45° dive, though, n = 0.7, which should be noticeable if you can find the time to think about it. For a vertical dive, Eq. 15.35 says the load factor is zero — which is right on because there is no lift force then.

Equation 15.35 is also nearly correct for climbing flight, as inspection of Eq. 14.12 reveals, because T is much smaller than W and $\sin\alpha$ is a small quantity; it is exactly correct for a (zero-lift) vertical climb, though.

If we substitute $L = C_L\left(\frac{1}{2}\rho_0 V_{E,GLIDE}^2\right)S$ for the lift force in Eq. 14.35, the following expression for the equivalent glide speed is obtained:

$$V_{E,GLIDE} = \sqrt{\frac{2W\cos\gamma}{C_L\,\rho_0\,S}} \qquad (15.36)$$

Because the flight variables in this expression (C_L and C_L/C_D through γ) do **not** depend on the weight, the actual weight dependence of the optimum glide speed is just $\sqrt{W}$:

$$V_{BEST\,GLIDE\,AT\,WEIGHT\,W_1} = \sqrt{\frac{W_1}{W_2}}\,V_{BEST\,GLIDE\,AT\,WEIGHT\,W_2} \qquad (15.37)$$

WEIGHT DEPENDENCE OF THE BEST GLIDE SPEED

This is a particularly useful relation because manufacturers often fail to provide optimum glide speeds for other than the maximum (gross) weight. Recall also the calculation of sink rate versus airspeed for the 152 at two different weights in Chap. 14 (Fig. 14.23); that calculation shows that the minimum sink rate increased with increased weight while the optimum glide **angle**

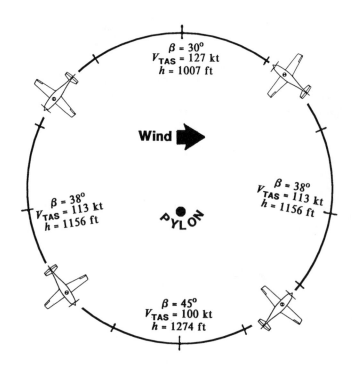

(a) Calculated Path over Ground

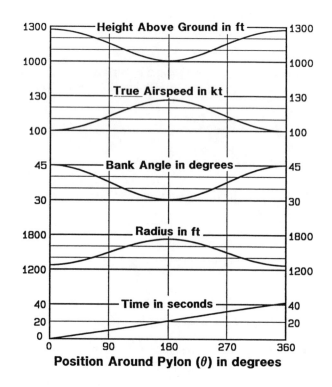

(b) Calculated Pylon Parameters

Fig. 15.11 Calculated on-pylon turn for a 100 kt airplane and a 20 kt wind

was unchanged – although the speed at which the best glide was obtained did increase with weight.

Competition soaring aircraft utilize this weight dependence of the best glide speed in a spectacular fashion: They may add nearly as much water ballast as their empty weight to increase their optimum glide speed between regions of natural lift – but they dump all that is left just before landing to reduce the approach and landing speeds.

EXAMPLE 15.11. A Cessna 152 has a listed best glide speed of 60 kt equivalent airspeed at its maximum weight of 1670 lb. At 1500 lb weight, what is the best glide speed?

Solution: $V_{\text{BEST GLIDE AT 1500 LB}} = \sqrt{\dfrac{1500}{1670}} \; 60 \text{ kt} = 57 \text{ kt}$

so the dependence on weight is not extremely strong for aircraft that have a narrow weight variation. In all cases, though, the optimum glide speed, at reduced weights, is always **less** than its value at maximum weight.

The best glide speed, as an **equivalent** airspeed (and therefore also as an indicated airspeed), does **not** depend on air density so it does not vary with altitude. The **true** airspeed for the best glide angle will **increase** with decreasing air density (increase in altitude) in the usual fashion:

$V_{\text{BEST TRUE GLIDE SPEED AT ALTITUDE}}$

$$= \sqrt{\frac{\rho_0}{\rho}} \; V_{\text{EQUIVALENT FOR BEST GLIDE}} \qquad (15.38)$$

DEPENDENCE OF TRUE GLIDING SPEED ON AIR DENSITY

so the glide angle doesn't change with altitude but you arrive at the ground sooner in low density air.

EXAMPLE 15.12. Calculate the best glide true airspeed for a 152 at its gross weight and at a density altitude of 5000 ft.

Solution: $V_{\text{TRUE AT 5000' DENSITY ALTITUDE}} = \sqrt{\dfrac{0.002377}{0.002048}} \; 60 \text{ kt}$

$$= \mathbf{65 \text{ kt}}$$

The stalling speed increases rapidly as the angle of bank in a level turn increases (Eq. 15.16), so it is important to keep the indicated speed in a banked turn well **above** the zero-bank stall speed. Refer now to the force diagram in Fig. 15.12 for the external forces in the up/down direction for an aircraft in a gliding **turn** with a bank angle β. Note that the upward force component that is perpendicular to the glide path is $L \cos\beta$, from Fig. 15.4. Applying Newton's second law to this direction gives us

$$\Sigma F_{\perp \text{ TO GLIDE PATH}} = L \cos\beta - W \cos\gamma = 0 \qquad (15.39)$$

There is also no acceleration parallel to the glide path because this direction is also perpendicular to a line to the center of the circle; therefore

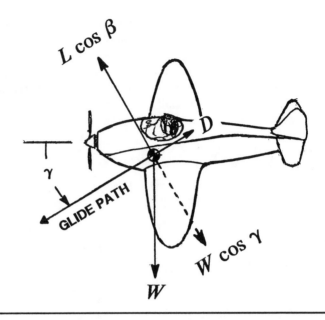

Fig. 15.12 The vertical forces acting on an aircraft in a banked, gliding turn (horizontal view)

$$\Sigma F_{\text{PARALLEL TO GLIDE PATH}} = W \sin\gamma - D = 0 \qquad (15.40)$$

Equation (15.13) is still valid for the forces in the direction from the cg of the airplane to the center of the circle; so the horizontal component of the lift is still providing all the required centripetal force for the continuous change in direction.

Solving for W in Eq. 15.40 and substituting in Eq. 15.39 yields

$$\tan\gamma = \frac{D}{L \cos\beta} = \frac{1}{\left(\dfrac{L}{D}\right) \cos\beta} \qquad (15.41)$$

GLIDE ANGLE IN A POWER-OFF, BANKED TURN

and we see that the best glide angle is still obtained by utilizing the angle of attack that yields the maximum value of L/D.

The load factor in a gliding turn is obtained by solving Eq. 15.39 for L/D:

$$n_{\text{GLIDING TURN}} = \frac{\cos\gamma}{\cos\beta} \qquad (15.42)$$

LOAD FACTOR IN A GLIDING TURN

EXAMPLE 15.13. Estimate the optimum (power-off) glide angle and the load factor for a Cessna 152 that is making a gliding turn with a bank angle of 60°.

Solution: From Eq. 15.41,

$$\gamma = \arctan\left[\frac{1}{(9.5)(\cos 60°)}\right] = \arctan(0.21) = \mathbf{12°}$$

From Eq. 15.42,

$$n = \frac{\cos 12°}{\cos 60°} = \mathbf{1.96}$$

Even though the load factor is indeed reduced by the glide angle, it's not enough to notice in flight. The **doubling** of the glide angle sure is, though, especially close to the ground! This is what causes pilots to almost involuntarily pull back on the stick or wheel (and often stall the airplane) when making steep gliding turns close to the ground after an engine failure.

If we write the lift force in Eq. 15.39 in terms of the lift coefficient,

$$L \cos \beta = C_L \left(\frac{1}{2} \rho_0 V_{\text{GLIDE}}^2 \ S \right) \cos \beta = W \cos \gamma$$

and if we solve this equation for the glide speed,

$$V_{\text{GLIDE}} = \sqrt{\frac{2 W \cos \gamma}{C_L \ \rho_0 S \cos \beta}} \tag{15.43}$$

we see that our glide speed – for a given L/D – is proportional to $(\cos \beta)^{-1/2}$, where β is the angle of bank. Therefore

$$V_{\text{E, BEST GLIDE AT BANK ANGLE } \beta} = \frac{V_{\text{E, BEST GLIDE AT ZERO BANK}}}{\sqrt{\cos \beta}} \tag{15.44}$$

DEPENDENCE OF BEST GLIDE SPEED ON BANK ANGLE

EXAMPLE 15.14. Determine the speed to use in a gliding turn with a bank of (a) 30° and (b) 45° if it is desired to cover the maximum distance over the ground in an aircraft with a best glide speed of 60 kt.

Solution: $V_{30° \text{ BANK}} = (60 \text{ kt})/(\cos 30°) = \textbf{64.5 kt}$
and $\qquad V_{45° \text{ BANK}} = 60 \text{ kt}/(\cos 45°) = \textbf{71.4 kt}$
which is a very significant effect.

The altitude that is lost in a gliding turn is equal to the vertical component of true airspeed multiplied by the time for the turn:

$$h = V_Y \ t_{\text{TURN}} \tag{15.45}$$

ALTITUDE LOST IN A BANKED, GLIDING TURN

But the vertical component of the airspeed is given by

$$V_Y = V \sin \gamma \tag{15.46}$$

where γ is again the glide angle. The time required to make a turn through a compass angle θ (assuming a steady, constant-airspeed, constant-bank turn) is

$$t_{\text{TURN}} = \frac{\theta \text{ in degrees}}{\text{rate of turn in deg/sec}} = \frac{V \theta}{57.3 \, g \tan \beta} \tag{15.47}$$

where the expression for rate of turn was obtained from Eq. 15.20 and where β is the angle of bank.

Substituting Eqs. 15.46 and 15.47 into Eq. 15.45 yields

$$h = \frac{V^2 \sin \gamma}{57.3 \, g \ \tan \beta} \times (\text{angle turned, } \theta, \text{ in degrees}) \tag{15.48}$$

ALTITUDE LOST IN A BANKED, GLIDING TURN

From this expression we can see that the altitude lost is directly proportional to the square of the true airspeed and inversely proportional to the tangent of the bank angle; this tells us to use a very slow speed and a very steep bank angle to minimize the altitude that is lost. However, we recall that a large bank angle goes right along with an increased stall speed and we certainly don't want to stall in the turn, especially if we are trying to glide to a suitable landing area after an engine failure at low altitude. The next thought might be to use the glide speed that would keep L/D at its maximum (and therefore γ a minimum), as calculated earlier. This would give the shallowest glide but **not** the minimum altitude loss because a high speed gives a much greater radius for the turn and just that much more distance over the ground that must be covered to make the turn. If instead a speed of 1.15 times the actual stall speed for each angle of bank is chosen, and if we specialize to the f and e derived for the Cessna 152, it is possible to calculate gliding turn performance based on the airspeed and bank angle used (Table 15.2 and Fig. 15.13). The calculations in the table are for 180° of turn, a 60 kt best glide speed, a stall speed of 48 kt, and a density altitude of 2000 ft at a gross weight of 1670 lb.

Bank Angle (deg)	V_{TAS} (kt)	Altitude Lost (ft)	Radius of Turn (ft)	Glide Angle (deg)
5	55	1073	3113	6.3
10	56	545	1563	6.4
15	56	372	1049	6.5
20	57	290	794	6.7
25	58	243	643	7.0
30	59	215	544	7.3
35	61	198	474	7.7
40	63	188	424	8.2
45	66	185	386	8.9
50	69	187	357	9.7
55	73	195	335	10.9
60	78	210	319	12.4
65	85	234	307	14.5
70	94	273	301	17.7

Table 15.2 Calculated Optimum Gliding-turn Performance, C-152

Note that the least altitude is lost, according to this calculation, at a 45° bank angle, while about **five** time as much is lost at 5° bank angle and about **15%** more at 60° bank angle. (The numbers in the last three columns can be converted to other density altitudes by multiplying them by the ratio of the density at 2000 ft to the actual density. Also, the times and altitudes can be converted to turns other than 180° by multiplying by the ratio [actual number of degrees turned/180°]. For example, twice the time and twice the altitude indicated would be lost in a 360° turn.)

The primary weakness in this calculation is probably that it assumes optimum piloting and ignores the dynamics of the maneuver – the rolling in and rolling out of the bank.

It is very difficult to obtain good experimental data to compare with these theoretical values because of altimeter lag, lateral instability and pitch/bank coupling at large bank angles,

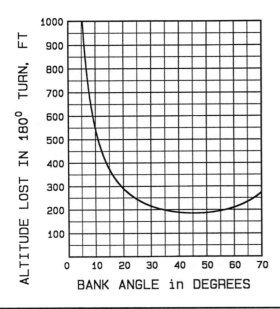

Fig. 15.13 Calculated optimum gliding-turn performance, C-152

and so forth. However, I have recorded an altitude loss of about 200 ft in a 45° bank at 66 kt and a loss of 320 ft in a 60° bank at 78 kt, so the theory can't be too far off. You are cordially invited to make your own measurements—at altitude.

J. Quiet after Takeoff

Light single-engine aircraft glide quite nicely with power off and land quite slowly with or without power, so that a power failure should not be a major problem if a suitable landing area is within gliding range. Engine failure on takeoff is another story, though, because the choice of terrain is often very uninviting. The natural instinct is to try to turn back and land on the friendly airport with its smooth runways that is just behind. This **may** be just fine if considerable altitude has been gained, the wind is light, and the airport is large and multiple runways are available so that the turn is through 90° or less. Historically, though, the attempt to make a 180° turn and land back on the runway after an engine failure has been deadly, beginning with World War I fighter aces or even earlier. The primary problem is that no one practices power-off steep turns close to the ground. When the ground appears to be rushing up at an astonishing rate, it is almost impossible to look at the instruments and the instinctive reaction is to try to slow down the awful rate of descent with "up" elevator—resulting in a high speed stall and, often, a spin entry. Fortunately, the previous section has prepared us to make some best-case calculations regarding the outcome without hazarding ourselves or our machines.

Suppose that we make our takeoff on runway 19 (heading of about 190°, to the south) at Kent State's Paton Airport (see Fig. 15.14 for a diagram of the airport) on a day when the density altitude is 2000 ft and there is a 20 kt wind right down the runway (from 190°). Suppose further that we use our Cessna 152's best

rate of climb until we reach 300 ft above ground level, at which point we are saddened by the unkind, unilateral decision of our engine to take a lunch break. If we immediately establish an optimum gliding turn of 180° with a 30° bank angle, where will we end up at touchdown?

Based on calculations of the previous chapter or the owner's manual, we can estimate a ground roll under these density and wind conditions of 340 ft. The owner's manual tells us that the best rate of climb will occur at a true airspeed of about 67 kt and will yield a 630 ft/min climb rate. Thus it will take an additional time of

$$t = \frac{h}{V_Y} = \frac{300 \text{ ft}}{630 \text{ ft/min}} = 0.476 \text{ min} = \textbf{29 s}$$

to reach an altitude of 300 ft above the ground. The angle of climb, γ, in **zero** wind (Fig. 14.11) can be calculated from

$$\sin \gamma = \frac{V_Y}{V} = \frac{630 \text{ ft/min} \times 1 \text{ min/60 s}}{67 \text{ kt} \times 1.689 \text{ ft/s/kt}} = 0.0928$$

so $\gamma = \textbf{5.33°}$

but with a 20 kt headwind, the horizontal component, relative to the ground, will change to

$$V_X = (67 \text{ kt} \times \cos \gamma) - 20 \text{ kt} = \textbf{46.7 kt}$$

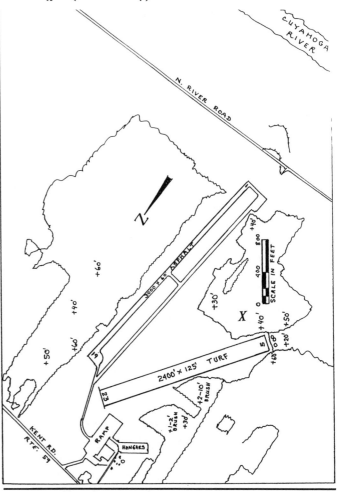

Fig. 15.14 Kent State University's Paton Airport (adapted from State of Ohio Airport Directory, as of 1980)

and the angle of climb relative to the ground will increase to

$$\gamma = \arctan\left(\frac{V_Y}{V_X}\right) = \arctan\left(\frac{630 \text{ ft/min} \times 1 \text{ min/60 s}}{46.7 \text{ kt} \times 1.689 \text{ ft/s/kt}}\right)$$
$$= \arctan(0.133) = \textbf{7.59}°$$

and our aircraft, while climbing to 300 ft above the ground, travels an additional horizontal distance of x, where

$$\tan(7.59°) = \frac{300 \text{ ft}}{x} \qquad \text{so} \qquad x = \textbf{2250 ft}$$

and therefore we are (340 ft + 2250 ft) = **2590 ft** from the beginning of runway 19 (i.e., still over the runway) when our engine fizzes.

The 152's best glide angle, symbolized by γ again, is 6.0° and (at 1670 lb) occurs at an indicated airspeed of 60 kt, which is a true airspeed of about

$$V_\infty = \sqrt{\frac{0.002377}{0.002241}} \; 60 \text{ kt} \cong \textbf{62 kt}$$

Suppose that we manage to achieve the 30° banked performance calculated in the previous section (Table 15.2): time of 16.9 s, altitude lost of 215 ft, and a diameter of turn of 1088 ft. So that puts us 1088 ft to the west of runway 19, assuming we made a right-hand turn after the engine failure, and we are now only **85 ft** above the ground. At our 62 kt, 6° glide, we will have a vertical descent rate of

$$V_Y = (62 \text{ kt}) \sin (6°) = 6.48 \text{ kt} = \textbf{10.9 ft/s}$$

so the landing **must** happen in another

$$\text{time} = \frac{\text{distance}}{\text{speed}} = \left(\frac{31 \text{ ft}}{10.9 \text{ ft/s}}\right) = \textbf{2.8 s}$$

and in this time we will travel a distance of

$$x = (V \cos \gamma) \times \text{time}$$
$$= \left[\left(62 \text{ kt} \times 1.689 \text{ ft/s/kt}\right) \times \cos 6.0°\right] \left(2.8 \text{ s}\right) = \textbf{292 ft}$$

but this is all with respect to the **air**. During the whole (16.9 + 2.8) = 19.7 seconds since engine failure, the wind has been pushing us north relative to the ground; it has in fact pushed us

$$d = vt = (20 \text{ kt} \times 1.689 \text{ ft/s/kt})(19.7 \text{ s}) = \textbf{665 ft}$$

and therefore we end up

$$(292 \text{ ft} + 665 \text{ ft}) = \textbf{957 ft}$$

north of where we had the engine failure and 1078 ft to the west. The airport diagram suggests that we end up roosting with the birds in some bushes or trees near the approach end of runway 5. (X marks the spot!)

Summary

Power reduces an aircraft's stall speed because the thrust augments wing lift in the job of matching the gravitational force. In low powered aircraft, however, the decrease in indicated speed may be dramatic because of increased elevator effectiveness and increased airspeed indicator errors.

An aircraft can simulate the force world of an orbiting satellite (so-called zero-g flight) by climbing steeply and then using control inputs and thrust adjustments to allow the aircraft to follow the parabolic flight path of a dragless stone, for up to a few minutes at least.

To execute a terminal velocity dive, the aircraft must be maintained in a zero-lift, nose-down attitude until the longitudinal acceleration becomes zero. An estimate of the terminal velocity for a given aircraft requires knowledge of the aircraft's equivalent flat plate area or its maximum sea level speed, and is easily accomplished if propeller drag is ignored.

An aircraft's load factor is defined as the ratio of the lift force to the weight; any deviation from one denotes accelerated flight. The various certification categories are characterized by different allowable maximum load factors.

Any turn represents accelerated flight; the load factor in a level, banked turn is proportional to the bank angle. Any load factor greater than 1, either from maneuvering or from turning, results in an increase in the stall speed (but not in the stalling angle of attack if dynamic effects can be ignored). The extra lift required in a level, banked turn increases the induced drag and therefore always slows down the aircraft; the actual speed loss is a measure of the efficiency of the wing planform or the wing-tail/wing-canard configuration.

The rate of turn varies as the tangent of the bank angle and inversely as the true airspeed. The varying bank required when a constant-radius turn about a point is made with a wind present is easily calculated; the maximum bank corresponds to the point of maximum ground speed (i.e., when directly downwind). Downwind turns are potentially dangerous only because the heightened ground speed suggests to the pilot that the airspeed is much greater than it really is; the aircraft doesn't fly any differently.

An on-pylon turn, in which the lateral axis of the aircraft is held fixed on a point on the ground, requires a specific altitude (the *pylon altitude*) that varies strongly with the ground speed. The maximum required bank angle in an on-pylon turn again occurs when directly downwind—so that is again the preferred entry point.

A steady, straight-ahead glide reduces the load factors below 1—but a steady banked turn increases it by the same factor as for level flight. The optimum glide speed varies as the square root of the weight, so reduced speeds are indicated at reduced weights. But the optimum glide speed also varies inversely as the square root of the cosine of the bank angle, so increasingly higher speeds are required as the bank angle in a gliding turn increases.

Minimizing altitude loss in a 180° gliding turn requires a compromise between the need for a high turn rate and the necessity of avoiding a high speed stall. The equations for optimum glides permit the estimation of the best-case result of trying to make a turn after an engine failure on takeoff.

Symbol Table (in order of introduction)

n load factor (L/W)
β bank angle
R radius of turn
ϕ wind correction angle
h altitude above ground during a turn   or altitude lost in a turn

Review Questions

1. The basic reason that the level-flight, power-on stall speed is less than the power-off value is that the ___ is **less** than the ___ for the former.

2. The power-on **indicated** stall speed is generally less than the power-off value; which one or more of the following are responsible?

     a. a change in the stalling angle of attack
     b. airspeed indicator errors
     c. increased elevator effectiveness
     d. a lift force that is less than the weight at the stall
     e. intimidation of the resident gremlin

3. Is it possible to (power-on) stall an aircraft that has a thrust as great as, or much greater than, its weight?

4. How does the drag of the aircraft affect the shape of the trajectory during a reduced-power, zero-g, space simulation maneuver?

5. How could you use an angle of attack meter to execute a zero-g space simulation maneuver?

6. What effect does doubling the entry speed have on the duration of the zero-g space simulation maneuver?

7. What, if any, is the effect of altitude on the duration of the zero-g space simulation maneuver?

8. What is the effect of using power during the zero-g space simulation maneuver?

9. Is the tail surface on a conventional aircraft developing positive, negative, or zero lift during a terminal velocity dive?

10. There is no ___ drag during either the space simulation maneuver or the death-defying terminal velocity maneuver.

11. How large is the drag force on a Cessna 152 in a terminal velocity dive at 10,000 feet in the StAt?

12. What is the load factor during a "greased-on" (smoooooth) landing?

13. Does a g-meter in an aircraft register a large positive or a large negative value after a hard landing?

14. When an aircraft is stalled a few feet above a runway and drops in, how do the bending moments on the wing-fuselage attachments compare with those during a smooth landing?

15. What is the load factor for an aircraft doing the space simulation maneuver?

16. What is the load factor in level, upright flight?

17. A coordinated turn is one in which the horizontal component of the lift force is exactly equal to the mass of the plane times its ___.

18. In a pylon turn, the ___ axis of the aircraft points at the pylon.

19. What is a good entry point for an on-pylon turn? Why?

20. Compare a terminal velocity dive and straight and level flight with regard to the magnitude and direction of the forces acting on the pilot.

21. What is the "0-g" stall speed of an aircraft with a 50 kt unaccelerated stall speed? (Hint: Eq. 15.17 may be of some assistance.)

22. Most biplanes have bracing wires that stretch from an outer panel of the lower wing to the root part of the upper wing and from an outer panel of the upper wing to the lower fuselage. These wires can only support tension loads, of course. Which set should be called "flying" wires and which "landing" wires?

23. What is the acceleration, if any, of an aircraft that is at the very highest point in its trajectory while executing a zero-g maneuver?

24. What is the name given to any real external force that causes an object to traverse a circular path, contrary to its natural inclination?

25. Is it possible for the load factor to remain at one while rolling from level flight to a 60° bank?

26. For a given angle of bank, is the load factor greater or smaller or the same for a heavy aircraft versus a light aircraft?

27. What can cause an aircraft to suddenly roll to a steep bank when it is stalled?

28. An "accelerated" stall is one that occurs when the load factor is ___ (1, greater than 1, less than 1) and the indicated airspeed is ___ (greater than, less than, the same as) the normal indicated stall speed.

29. Is there any maneuver a pilot might make that would allow him to stall an aircraft at a speed **below** its normal stall speed?

30. ___ drag is responsible for the speed decrease when a steady turn follows level flight.

31. The aircraft that experiences the greatest fractional speed loss when transitioning from level flight to a steady bank is one that has the smallest value for its ___ loading.

32. The radius of a turn depends primarily on the angle of bank and the ___ of the aircraft.

33. The great danger in low altitude downwind turns is that the actual airspeed is ___ (less than, greater than) the ___ (indicated, apparent) speed.

34. The pylon altitude is determined by the ___ (indicated, true air, true ground) speed.

35. In a steady, wings-level glide, the load factor is ___ (greater than, equal to, less than) one.

36. The airspeed which yields the optimum (smallest) glide angle ___ (increases, decreases, doesn't change) as the aircraft's ___ is increased.

37. The optimum glide angle ___ (increases, decreases, doesn't change) as the density altitude increases.

38. The true airspeed at which the optimum glide angle is obtained ___ (increases, decreases, doesn't change) as the density altitude increases.

39. The indicated airspeed at which the optimum glide angle is obtained ___ (increases, decreases, doesn't change) as the angle of bank is increased from 0° to 30°.

40. If an engine failure occurs at the end of the runway, a headwind (increases, decreases, doesn't affect) the time required to make a 180° turn.

41. You've decided to give your wings a well-earned rest and try a little bungee jumping. Your first (practice) jump is a simple fall off a 150 ft bridge, using a 50 ft bungee cord that is tied to the bridge at one end and to a harness attached to you at the other end. Assuming that the force of the bungee cord on you takes the place of the lift on the wings of an aircraft, what is your "load factor" when you have fallen 10 ft?

     a. 0
     b. +1
     c. −1
     d. less than −1
     e. greater than +1

42. For this bungee jump (previous question), what is your load factor when your downward motion has just stopped, only inches above the water, and you are about to rebound upward?

a. 0

b. +1

c. −1

d. less than −1

e. greater than +1

Problems

1. Estimate the power-on stall speed for the Cessna 152 when it has the 150 hp engine (giving it about 30% more thrust). Use the owner's manual value of 48 kt for the power-off stall speed.
Answer: About 45.5 kt

2. Assuming Fig. 7.3 is valid for this aircraft, estimate the indicated airspeed for a power-on stall at gross weight.
Answer: About 35 kt or less

3. Estimate the stall speed if this Cessna 152 is given a jet engine with a thrust rating of 1400 lb in place of its reciprocating piston engine.
Answer: 42 kt

4. What is the equivalent flat plate area of a 2 ft by 2 ft plate? (Refer to the definition of f in Chap. 13.) Answer: 5.12 ft^2

5. If the flat plate of problem 4 were deployed as a drag flap on the terminally-diving Cessna of section C, estimate the new terminal dive speed. Hint: Using ratios is the easy way. Answer: 216 kt

6. Estimate the true airspeed terminal dive speed of the Cessna of Section C if it tries the maneuver at 10,000 ft in the StAt.
Answer: 351 kt

7. Estimate the true airspeed terminal dive speed of a twin-engine Cessna 401A at 16,000 ft density altitude and at a gross weight of 6200 lb. (Maximum power available at 16,000 ft from the supercharged engines is 600 hp, yielding a maximum speed at 16,000 ft of 227 kt.) Assume a propeller efficiency in cruise of 80%.
Answer: 681 kt

8. What is the Mach number of the terminal dive speed of problem 7? What type of drag have we incorrectly ignored in that problem?

9. What is the magnitude of the vertical acceleration for a landing that registers a 3.0 on the g-meter? Answer: 64.4 ft/s^2

10. For a given airspeed, how does the radius of a turn with a 30° bank compare with that using a 45° bank? Answer: 1.73 times greater

11. At what bank angle will an ultralight flying at 50 kt obtain a standard rate turn, assuming a coordinated turn? Answer: 7.8°

12. What is the pylon altitude for a Cub, which cruises at about 70 mph? Answer: 327 ft! Worse yet, what is the range of altitudes when pylons are practiced in a 15 kt wind?

13. Suppose a pilot makes a takeoff at KSU airport under the same conditions as for the calculated 300 ft climb and gliding 180° turn to a landing **except** that there is **no** wind and the turn is to the **left**. Assume a ground roll of 840 ft (from the pilot's manual). Estimate where this the landing will take place, giving the distance in feet east of the runway as well as the distance south from the takeoff point. Mark the spot on Fig. 15.14 with your own large "X."

14. Same as previous problem except assume a 30 kt direct headwind. Assume a takeoff roll of 162 ft.

15. Same as previous problem except assume a 10 kt direct tailwind and a turn to the right. Assume a takeoff roll of 1183 ft.

16. For turns about a point at a cruise speed (true airspeed) of 100 mi/hr, what horizontal distance from the point on the ground should be chosen if a maximum bank of 45° is desired (a) in no-wind conditions and (b) with a 30 mi/hr wind?
Answers: About 668 ft and about 1129 ft, respectively

17. What would be the minimum bank for part (b) in problem 16?

18. Find the no-wind pylon altitude for the aircraft of problem 16.

19. What is the range of pylon altitudes for the aircraft of problem 16 in a 30 mi/hr wind? Answer: 327 ft to 1129 ft

20. For the conditions of problem 19, what horizontal distance from the pylon will yield a maximum bank of 40°? Answer: 1345 ft

21. What was the actual design limit load factor for the Wright airplane, according to the test quoted by Wilbur? Answer: 2.7

22. For a stall angle of 20°, what effect does a thrust-to-weight ratio of 0.5 have on the power-on stall speed?
Answer: It reduces it to 91% of its power-off value.

23. If the landing gear on a certain aircraft just triples the equivalent flat plate area, how much is the terminal velocity reduced by the extended gear? Answer: To about 58% of its gear-up value

24. If the average acceleration for the aircraft of Example 15.3 is 0.5 g during the time required to reach 90% of the terminal dive speed, (a) about how much time is required to reach this speed and (b) about how much altitude is lost? Answers: 29 s, 6600 ft

25. For a glide angle of 8°, what angle of bank will allow the load factor to remain at a steady value of 1?

26. For a 1600 lb aircraft that is executing a level, 40° banked turn at a speed of 120 kt, (a) what is the centripetal force acting on it, (b) what is its acceleration, (c) what is the vector sum of the external forces acting on it, and (d) what is the radius of the turn?
Answers: 1343 lb, 27.0 ft/s^2, 1343 lb, 1520 ft

27. If 20% more power must be added to maintain level flight at constant speed in a 50° banked turn in a given aircraft, what fraction of the total power required in level flight was due to induced drag?
Answer: About 16%

28. If 40% more power must be added to maintain level flight at constant speed in a 60° banked turn in a given aircraft, what fraction of the total power required in level flight was due to induced drag?

29. If an aircraft with a no-bank glide angle of 15° maintains a true airspeed of 80 kt in a 35° banked turn through 180°, about how much altitude would be lost? Answer: 658 ft

30. Estimate the terminal dive speed at a density altitude of 6,000 ft for the 160 hp Piper Cadet, weighing it at 2325 lb. Assume a propeller efficiency of 80% when it is achieving its listed 119 kt sea level speed with full power. Answer: 335 kt

31. Surely you would like to know your personal terminal dive capabilities? For a first estimate, use the flat plate drag coefficient of 1.28 and an estimate of your projected surface area for both the spread-eagle position and the head-first hurry-up position. Alternately, Hoerner (Ref. 5.2) provides wind tunnel values of $f = 9$ ft^2 and 1.2 ft^2. Partial answer: For a 170 lb person, Hoerner's parameters yield terminal speeds of 75 kt and 204 kt, respectively.

References

1. Miller, Howell W., "Terminal Velocity," *Aero Digest*, February, 1934.

2. Bennett, T.C., "Propeller Effect in a Dive," *Aero Digest*, May, 1934.

3. Smith, Peter C., *The History of Dive Bombing*, Nautical & Aviation Publishing Company of America, Maryland, 1981.

4. Bahr, Dennis E. and Schulz, Robert D., "Acceleration Forces Aboard NASA KC-135 Aircraft During Microgravity Maneuvers," *Journal of Aircraft*, Vol. 26, No. 7, July, 1989.

Chapter 16

Winds, Loops, Rolls, and Spins

He spared no time that day for talk with other gulls, but flew on past sunset. He discovered the loop, the slow roll, the point roll, the inverted spin, the gull bunt, the pinwheel. . . . When Jonathan Seagull joined the Flock on the beach, it was full night. He was dizzy and terribly tired. Yet in delight he flew a loop to landing, with a snap roll just before touchdown. When they hear of it, he thought, of the Breakthrough, they'll be wild with joy. How much more there is now to living! . . . We can lift ourselves out of ignorance, we can find ourselves as creatures of excellence and intelligence and skill. We can be free! We can learn to fly![1]

The entry into the nonoscillatory spin was started at 24,000 ft. and the steady state spin was identified by an angle of attack of more than 30 deg., a pegged turn needle to the left and an airspeed of 118 kt. The anti-spin controls were fed in at 18,000 ft. after six spins.

Navy procedure dictates than if an F-4 or F-14 pilot is still in a spin at 10,000 ft., or enters a spin below 10,000 ft., the pilot and rear seat naval flight officer are to eject.

The T-2 was at 24,000 ft. and 300 kt. when I pulled the trainer into a vertical climb and pulled the power back to idle. At 200 kt., right aileron was used to perform a 180-deg. vertical aileron roll. The next 180 deg. of roll was combined with full right rudder, which was followed by a quick application of full right forward stick and full left rudder.

I can best describe the T-2 maneuver as a result of those control inputs, as a tumbling football travelling end over end. The aircraft had negative pitch, left yaw and a right roll and approximately zero air speed indication. This time, however, I was securely strapped in the seat. . . . Once the controls were centered, the aircraft stopped tumbling and departed into an inverted spin.[2]

A. Wind Gusts and Microbursts

One of the most important considerations in aircraft design is the provision of sufficient strength to withstand the stresses induced by turbulent air. An aircraft cruising at a true airspeed V and suddenly flying into air with a vertical gust speed of U (either up or down) will experience a sudden change in its angle of attack. Note the vector diagrams of Fig. 16.1.

By similar triangles, the change in angle of attack due to the gust, $\Delta\alpha$, is given by $\Delta\alpha \cong \dfrac{U}{V} = \dfrac{U_E}{V_E}$, in which we have used an equivalent wind gust speed, $U_E \equiv \sqrt{\dfrac{\rho}{\rho_0}}\, U$ in analogy with the definition for aircraft speed. A positive gust (from below, as shown) will then increase the angle of attack, thus increasing both the lift force and the load factor. The lift force changes by

$$\Delta L = (\Delta C_L)\, q\, S = (a\, \Delta\alpha)\, q\, S$$

where a is the slope of the lift curve in radians^{-1} (see Section C in Chap. 9). Substituting for $\Delta\alpha$ from the equation above and for q from Eq. 7.9 yields

[1]Reprinted with permission of Macmillan Publishing Company from *Jonathan Livingston Seagull*, by Richard Bach. Copyright © 1970 by Richard D. Bach and Leslie Parrish-Bach.

[2]David M. North, "Navy Spin Training Heightens Confidence," *Aviation Week & Space Technology* (August 25, 1980).

(a) CRUISING FLIGHT: Smooth Air, Equivalent Airspeed V_E, angle of attack α

(b) CRUISING FLIGHT: Sharp–Edged Updraft of Speed U_E causing new angle of attack $\alpha^{\bullet}$

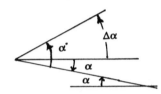

(c) VECTOR DIAGRAM FOR THE VELOCITIES

Fig. 16.1 The increase in wing angle of attack caused by an updraft

$$\Delta L = \tfrac{1}{2} a \rho_0 U_E V_E S$$

and so the load factor just when the wing reaches the wind gust is

$$n = \frac{L}{W} = \frac{W + \Delta L}{W} = 1 + \frac{a \rho_0 U_E V_E S}{2W} \qquad \text{or}$$

$$\boxed{n = 1 + \frac{a \rho_0 U_E V_E}{2 \left(\dfrac{W}{S} \right)} \qquad (16.1)}$$

LOAD FACTOR IN AN ENCOUNTER WITH A SHARP-EDGED WIND GUST

The second term in Eq. 16.1 is the **incremental** load factor increase or decrease relative to the normal value of 1.0 for upright, level flight. This is the change in load factor due to a sudden wind gust. Note that it depends **(a)** almost directly on the equivalent airspeed, **(b)** directly on the density-corrected wing gust speed, and **(c)** inversely on the wing loading (W/S). Factor **(a)** tells us to slow down the aircraft when entering potentially turbulent air, factor **(b)** tells us that the change in load factor is directly proportional to the wind gust speed (it could be worse), and factor **(c)** tells us (as anticipated earlier) that large and heavy aircraft (with high wing loading) are less affected by turbulence than light aircraft.

For those who do aerobatics in turbulent air, it is important to realize that the first term in Eq. 16.1 becomes –1 when flying in level, inverted flight.

EXAMPLE 16.1. Suppose that an aircraft at a weight of 2500 lb and travelling at a speed of 150 kt encounters an upward wind gust of 20 ft/s, causing an instantaneous load factor of 2.5. **(a)** What would the load factor have been if the aircraft had been at a weight of 2000 lb, if the gust had been 30 ft/s, and if the speed had been 175 kt? **(b)** What would the load factor have been if the gust had been a downdraft rather than an updraft? **(c)** What if the 20 ft/s gust had been encountered while flying in level, inverted flight? **(d)** What if the 20 ft/s gust had been a downdraft, encountered while flying in level, inverted flight?

Solution: **(a)** From Eq. 16.1 we see that the incremental (additional) load factor due to the gust was 1.5. The changes in this second term are proportional to the aircraft speed and the wind gust speed but inversely proportional to the aircraft weight. Therefore we can multiply the original values for this term by the **ratios** of the parameters that have changed, without having to worry about the parameters that haven't changed. Thus

$$n = 1 + (1.5) \left(\frac{175 \text{ kt}}{150 \text{ kt}} \right) \left(\frac{30 \text{ ft/s}}{20 \text{ ft/s}} \right) \left(\frac{1}{\dfrac{2000 \text{ lb}}{2500 \text{ lb}}} \right)$$

$$= 1 + (1.5)(1.167)(1.5) \left(\frac{1}{0.8} \right) = 1 + 3.3 = \mathbf{4.3}$$

We see that all the changes increased the load factor.

(b) The second term would then have been –1.5, so
$$n = 1 + (-1.5) = \mathbf{-0.5}$$

(c) The first term would have been –1 and the second term would have been –1.5 (i.e., up relative to the earth but negative relative to the aircraft), so the load factor would have been **–2.5**.

(d) The first term would have been –1 and the second term would have been +1.5 (down relative to the ground but up relative to the normal top of the aircraft), so the load factor would have been **+0.5**.

Using Eq. 16.1, Fig. 16.2 graphically depicts the dependence of the load factor on the airspeed and the gust speed. Note that the load factor begins at 1 (assuming level, upright flight) and then increases (or decreases, for downdrafts) directly proportional to both the equivalent airspeed at the time and the strength of the gust. The example lines start from 0 airspeed but in practice the actual equivalent airspeed could be anywhere between the stall speed and the never-exceed speed.

Updrafts and downdrafts aren't really sharp-edged, and Federal Air Regulation Part 23 takes this into account when specifying certification requirements. Aircraft must be capable of withstanding positive and negative gusts up to 50 ft/s at the maximum cruising speed (V_C, the end of the green arc and the beginning of the yellow arc on the airspeed indicator) and positive and negative gusts of up to 25 ft/s at the design dive speed (V_D, defined in Eq. 16.6 ahead). However, the gust speed is assumed to increase gradually and smoothly from zero to its maximum and then taper off at about the same rate, in a

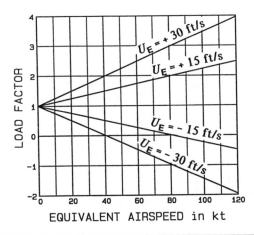

Fig. 16.2 Load factors generated by wind gusts of various magnitudes; this calculation is for a C-152 with a wing loading of 10.4 lb/ft^2.

distance that depends only on the mean geometric chord ($\bar{c}$) of the wing.

Figure 16.3 depicts the shape of the FAA-approved gust for a 50 ft/s updraft and a distance scale appropriate for the wing on the 4-place, 160 mi/hr, Cessna 182. The aircraft figure on this velocity-distance graph is shown at an instant when the plane has penetrated just one plane length into the gust; the updraft velocity at the propeller is about 22 ft/s while the updraft is only now about to affect the tail. The maximum updraft velocity, it can been seen, will be acting on the wings after the plane has travelled another one and a half plane lengths or so. At a speed of 150 mi/hr (253 ft/s) and an aircraft length of 28.1 ft, a C-182 will travel one plane length in [(28.1 ft)/(253 ft/s)] or 0.11 second; thus it requires about one-quarter second for this aircraft to feel the full impact of the updraft—still kind of a sharp-edged gust, if you ask the passengers.

An L-1011 airliner was approaching Runway 17L (170°, left) at the Dallas–Fort Worth airport on the 101°F summer afternoon of August 2, 1985. There were some scattered, small thunderstorms in the vicinity, but nothing like the large thunderstorms that were well to the northeast and associated with a warm front there. However, the aircraft encountered such severe winds on its final approach that it impacted short of the runway. The aircraft had traversed a region of air containing what is now known as a **microburst**.

Figure 16.4 (Ref. 1) indicates the primary features of a microburst, which has a diameter ranging from about 3000 ft to 10,000 ft. When an aircraft first reaches the edge of the microburst, it is met by the outflow and the airspeed therefore suddenly increases, the aircraft responds by pitching **up**, and altitude is **gained**. At the center of the microburst the headwind has vanished and the aircraft is in a strong downdraft. From then on the aircraft experiences a rapidly increasing **tailwind** which severely decreases performance. The fact that the microburst is first encountered as a headwind causes the pilot to remove power in an effort to stay on the glide slope, and puts the aircraft in a particularly poor position to cope with the

Pic. 16.1 An airshow Pitts biplane begins its act with a roll after takeoff.

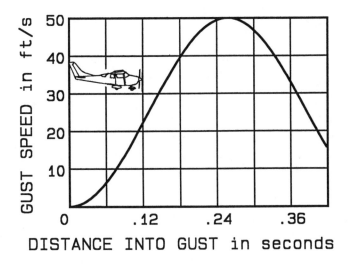

Fig. 16.3 Shape of an FAR Part 23 updraft; the distance scale is appropriate for a 4- to 6-place aircraft such as the C-172 and C-182.

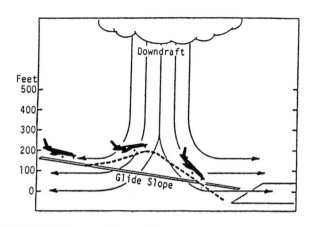

Fig. 16.4 Microburst winds and the resultant glide path on final approach (From Ref. 1. ©AIAA. Reprinted with permission of the American Institute of Aeronautics and Astronautics)

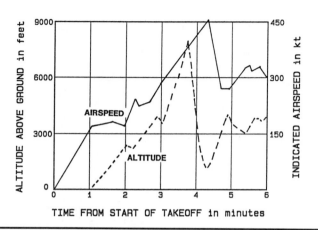

Fig. 16.5 Flight recorder altitude-speed records for an aircraft passing through wind shear after takeoff (Ref. 49)

downdraft and tailwind when they are encountered. In this accident, the engines were reduced to idle even though this meant a significant delay before full power could be delivered because of spool-up time. (Piston engines do not have this problem.)

This particular microburst, apparently only a medium-performance example, resulted in a maximum headwind of 27 kt, a maximum downdraft of 24 kt (**2400 ft/min!**), and a maximum tailwind of 44 kt. Load factors ranged from –2 to +2. The airspeed fluctuated from –44 kt to +20 kt away from the target speed.

The overall wind pattern pictured in Fig. 16.4 is obscured in practice by the small-scale turbulence encountered, yielding continuous fluctuations in load factor and strong rolling moments. Figure 16.5 depicts the variation of indicated airspeed and altitude for a different air transport aircraft that encountered unexpected wind shear some three minutes after the takeoff was initiated, as measured by the flight data recorder (Ref. 49). Note how the altitude increases more rapidly for about 40 second — and then the bottom falls out! Nearly 7000 ft of altitude and about 150 kt in indicated airspeed are lost before recovery is accomplished.

Efforts are continuing on better methods for microburst detection as well as optimum flight crew response. Simulator experiments have indicated that optimum crew response to a microburst includes the use of the maximum available lift capability — even flight at speeds sufficiently close to the stall that the stick-shaker will activate. However, this advice may have to be modified because NASA experiments have confirmed a significant loss in maximum lift coefficient when heavy rain is present.

B. Accelerated Stalls and Snap Rolls

Another way in which the pilot can put extra stress on an aircraft is by suddenly changing the angle of attack by abruptly pulling back or pushing forward on the control stick or control wheel. At high speeds the dynamic pressure is large and very large load factors can be generated.

Consider an aircraft in level flight at a true airspeed V. If the pilot increases the angle of attack to the stalling attack very abruptly, the lift coefficient will quickly reach its maximum value but the airspeed will not have had time to decay appreciably. The maximum wing lift generated by this maneuver is

$$L = C_{L,\,MAX} \left(\frac{1}{2} \rho V^2 \right) S$$

and this lift may be many times greater than the weight of the aircraft. If we solve Eq. 12.15 for the weight and substitute in the above equation, we can see that the load factor in this abrupt stall will be

$$n = \frac{L}{W} = \left(\frac{V}{V_{STALL}} \right)^2 \qquad (16.2)$$

LOAD FACTOR IN AN ACCELERATED STALL

Since ratios of speeds are involved, the expression above is valid for either true or equivalent airspeeds.

EXAMPLE 16.2. A Cessna 152 with a 1-g stall speed of 48 kt is abruptly stalled (by a wind gust or by the pilot) while it is cruising at 100 kt. What is the resulting instantaneous load factor?

Solution: From Eq. 16.2, $n = (100/48)^2 = $ **4.34** which is much too close to its limit load for comfort.

Equation 16.2 usually predicts the cockpit-measured values within 5 or 10 percent. Some of the early NACA reports involve experimental measurements of load factors and comparisons with this formula (Ref. 4). The author has found a close correspondence between Eq. 16.2 and values recorded on g-meters in aerobatic aircraft, also. (At low speeds, though, some aircraft do not have enough elevator power to produce an abrupt stall.)

A **snap roll** (**flick roll** if you're in England) is an aerobatic maneuver which is suitable only for aerobatic, light aircraft. It is entered at about cruise speed or somewhat lower with power on; the aircraft is abruptly stalled while simultaneously being yawed with full rudder in the desired direction of roll. The result is an accelerated stall but with one wing obtaining a greater amount of lift than the other. Thus if the aircraft is yawed to the right upon entry, it will snap roll to the right. (With poorly snapping aircraft such as the 150 Aerobat, using aileron with the roll after entry can sometimes improve the rate of roll.) Good aerobatic aircraft with short wings and powerful elevators can be made to do a vicious snap that will continue for 3 or more complete rolls before the nose drops appreciably below the horizon; a specially trained stomach is also required. Good aerobatic aircraft can also do snap rolls using accelerated **inverted** stalls—but these are even more unpleasant for the pilot.

Recovery from a snap roll is accomplished with full opposite rudder followed by neutral elevator. If recovery is not initiated soon enough, the aircraft will do some wonderful and even glorious gyrations before ending up in a power spin (not good!).

Snap maneuvers do not make the pilot as uncomfortable as maneuvers in which g-loads are sustained over longer periods of times—but the aircraft structure responds more to the instantaneous loads. Also, snap maneuvers put a severe torsional stress on the tail assembly, the fuselage, the engine mounts, and so forth. It has been said that one can recognize a Cub that has been snapped by the misalignment between the wings and the horizontal fuselage.

Figure 16.6 presents the aircraft motions and load factors that were measured in a snap roll in a military biplane fighter back in the 1930s (Ref. 50).

C. Spins—Good and Bad

A spin is entered with the same control movements as for a snap roll but these are initiated just slightly before a power-off stall occurs. The result is that the nose abruptly drops to nearly a vertical attitude—or momentarily beyond—and autorotation (rotation with no aileron input) begins. After one-half turn, or perhaps as many as six turns, the spin stabilizes. In light aircraft the pitch attitude relative to the horizon is normally constant during the spin (see Fig. 16.7) but heavy propeller and jet aircraft usually have an oscillatory mode. For example, in the F-51D Mustang, a World War II fighter aircraft, the flight handbook advises that

> When controls are applied to start a spin, the airplane snaps one-half turn in the direction of spin, with the nose dropping to near vertical. At the end of one turn, the nose rises to or above the horizon and the spin slows down, occasionally coming almost to a complete stop. The airplane then snaps one-half turn with the nose dropping to 50°–60° below the horizon and continues as during the first turn. ...Approximately 1000 feet of altitude are lost per turn.

Figure 16.8 provides a reason for the autorotation. The outside wing has a lesser angle of attack with accompanying

Pic. 16.2 Harold Krier turns on the smoke to begin his aerobatic routine in the modified Chipmunk.

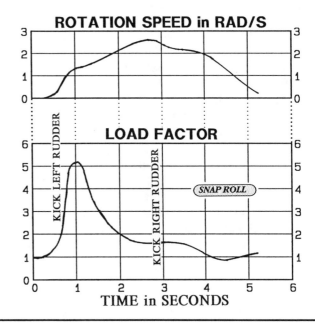

Fig. 16.6 Displacements and load factors in a snap roll (Ref. 50)

greater lift and less drag than the inner one, so it "chases" the inner wing. The large drag of this high angle of attack allows the airspeed in a true spin to be stable at a speed not much greater than the unaccelerated stall speed. (Aircraft that are elevator-limited, such as the Cessna 172 with only the front seat occupied, cannot be held deep enough in the stall to maintain both wings past their stalling angle and so will typically transition into a diving spiral and the speed will rapidly begin to increase.)

The primary recovery control for most single-engine aircraft is the **rudder**. Full opposite rudder is used, followed by

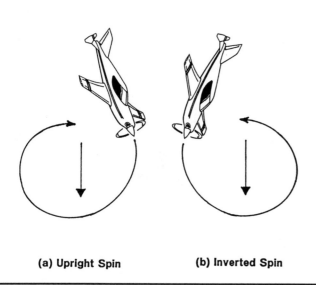

(a) Upright Spin   **(b) Inverted Spin**

Fig. 16.7 Desirable aircraft attitudes in spins

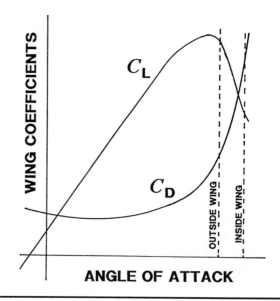

Fig. 16.8 Desirable wing angles of attack for a good snap roll or spin

neutral elevator (or full down elevator in some heavy aircraft); the ailerons usually are not used. The speed will build up when the stall is broken, since the nose is still well down, so the pull-out should be gentle to avoid imposing large load factors at this point. Otherwise the load factors in a spin are generally quite low and the primary consideration is recovering before the ground arrives. A light aircraft may be recoverable in 500 or 1000 feet but military aircraft can require 10,000 feet or more. (The F-111B is said to require 24,000 feet.) Even with plenty of altitude, many aircraft turn out to have an unrecoverable spin mode if they haven't been approved for spinning. Usually this is a **flat spin**.

A normal, steep spin involves an angle of attack in the range of about 25° to 40° while a flat spin is commonly defined as one for which the angle of attack is greater than 60°. In a flat spin, as the name implies, the longitudinal axis of the aircraft is often nearly parallel with the horizon with almost no bank angle. Whereas a normal, steep spin includes both considerable yawing and rolling, a flat spin is almost entirely a yawing maneuver. The yaw rates are astonishingly high — over 200° per second. An aircraft which otherwise spins quite normally can often be made to spin flat by loading it such as to produce a far aft *cg*.

Until recently there hasn't been much interest in flat spins — they were thought of as something to be strictly avoided because they were thought to be unrecoverable. The control surfaces needed for recovery are blanked by the fuselage and the engine commonly quits from fuel starvation. Many aircraft have indeed been lost through stall testing and resultant flat spins and it is still one of the major danger areas in test flying. The descent rate in a flat spin in a light aircraft is typically very low, though, and the resulting crash sometimes hasn't even injured the pilot. (One stunt pilot apparently successfully executed a flat spin for an old aviation movie with the help of an aft *cg*.)

The radius of a normal spin can be estimated by recognizing that the lift force is providing the centripetal force; the descent rate can be estimated by recognizing that the drag force equals the weight.

The last few years have seen a resurgence of interest in aerobatics (formerly called acrobatics) and some of the better (competition) aerobatic aircraft of today are capable of being forced into a flat spin (erect or inverted) through the use of opposite aileron and power. (These aerobatic aircraft differ from ordinary lightplanes in having very powerful ailerons and fuel systems that will continue to supply fuel during strong negative or positive accelerations.) Recovery is effected by removing the power and using aileron into the spin to help increase the rolling moment; just before recovery, though, the spin rate may increase and this can fool the pilot into thinking he has done the wrong thing. See Ref. 5.

Even in these aerobatic aircraft, an accidental flat spin for an untrained pilot is often fatal. A few years ago, a pilot in northeast Ohio tried to force an early recovery from a normal, upright spin for his aerobatic biplane by adding power and lots of down elevator, but this caused the aircraft to snap into an inverted, flat spin from which the pilot was unable to recover.

The **inverted** spin, whether steep or flat, is entered and recovered from in an entirely analogous, if mirror-image, manner to that of an upright spin. However, pilots are not symmetrical top to bottom and an inverted spin is one of the most disorienting of all maneuvers when it is first encountered. It can be difficult to maintain the spin because it is hard to push the stick forward while one is being held in the cockpit only by the shoulder harness; the recovery from inverted spins tends to be quicker than for upright spins in most planes, at least, probably because there is more rudder area above the horizontal stabilizer than below.

To practice an inverted spin, the aircraft is usually slow-rolled to level, inverted flight, power is reduced to idle (with or without an inverted system), the speed is gradually reduced by applying forward pressure on the stick or wheel to obtain a nearly $(-)1$-g stall and, just before the stalls occurs, full rudder in the desired direction of spin is applied and then full down elevator is held. To recover, opposite rudder is held just until the rotation stops (or else an inverted spin in the other direction will ensue!) and then forward pressure is reduced to return flight to the wings-level, inverted mode.

No matter what maneuver an aerobat is attempting, if the elevator is in an extreme position and if the rudder is deflected, there is always a good chance that the maneuver will end up in, or may transition abruptly to, either an upright or an inverted spin. Therefore aerobatic pilots need to be fully aware of the spin characteristics of their aircraft before exploring the aerobatic envelope.

Recently, Gene Beggs, a member of the U. S. aerobatic team and an aerobatic instructor, has been demonstrating that the common competition aerobatic aircraft will recover effectively from all types of spins if (a) the power is set to idle, (b) the hands are removed from the stick, and (c) full rudder opposite to the spin direction is held. The altitude loss is often little or no more than with conventional recovery techniques. Surprisingly, the Cessna Aerobat and the North American AT-6 are two of the few aircraft possessing a spin mode that don't respond to this treatment.

Aircraft that are not certificated for spins may very well possess unrecoverable spin modes. Such has turned out to be the case for the Yankee trainer, for example (Refs. 6 and 7). All that is required by certification regulations is that an aircraft be recoverable from a one-turn spin entry, but not necessarily from a fully developed spin.

Certificated single-engine aircraft must recover from a one-turn spin within one turn after recovery is initiated; in the aerobatic category a six-turn spin must be made (FAR 23.221). But do not make the mistake of assuming that all aircraft of a given model and type will spin identically. Older aircraft, in particular, will often end up being rigged differently than factory specifications recommend and spin characteristics are sometimes strongly affected.

The stall/incipient spin at low altitude historically and currently is responsible for about one-third of all the fatal lightplane accidents (Ref. 8). It can happen to the most experienced pilot if an attempt is made to return to the airport after an engine failure on takeoff; the tendency to tighten up a turn and pull back on the control wheel under such circumstances is nearly irresistible. It can happen to a pilot who decides to make a low altitude pass or circle about a friend on the ground, or circle to maintain spacing from other traffic in the traffic pattern, or circle to land under minimum instrument flight conditions. It can happen to pilots who feel they need to steepen their bank as they turn onto final approach, especially with a crosswind blowing them past the runway; because they are already low, they want to turn quickly and so they add rudder and opposite aileron and then some up elevator to keep the nose up and — over it goes.

Pic. 16.3 The modified Chipmunk begins an outside loop.

D. Stall-Spin Accident Solutions

One possible solution to this safety problem would be to make an airplane that is unstallable. The Ercoupe, designed by Fred Weick (Ref. 42), is just such an airplane; with limited elevator travel and interconnected ailerons and rudder, the aircraft can't be stalled or spun. (However, it has not had as good an accident record as one would expect because it still can be mushed into the ground during a landing attempt. It also doesn't have a good *cg* range, it can't compensate for crosswinds on landing in the normal fashion, it can't maintain a straight track down the runway on takeoff with a strong crosswind, and many pilots don't like being limited by its interconnected, two-control system.

A **second** possible solution is to utilize the canard configuration, which is characteristically incapable of stalling its main wing or spinning, if properly designed and loaded. Canard guru Burt Rutan (VariViggen, VariEze, Long-EZ, Quickie, Defiant, Beechcraft Starship, Voyager, Pond racer ...) explains in Ref. 16 that these considerations started him on his very successful tail-first ways. There is little doubt that some pilots have been spared serious injuries in these aircraft because they retain full lateral control down to their lowest speed and so they can land hard on their gear but they will not uncontrollably roll over and crash wing low.

A **third** possible solution is to make aircraft that are spin resistant. NASA has discovered that adding a drooped leading edge cuff or glove to just the outer part of a wing can change a flat spinner into one that is quite spin-resistant. The drooped part has a stall angle beyond that which the elevator can provide, so the outer part of the wing never stalls and rolling moments at minimum speeds are small; the cruise drag penalty is relatively small. These drooped gloves must form a discontinuous addition to the old wing because the vortex generated at the intersection is vital to reducing or preventing span-wise flow, much like the snags or strakes or vortilons found on swept-wing aircraft. The very promising Questair Venture, a 300 mi/hr, 280 hp, 2-seater now available to homebuilders, uses some of this anti-spin work, along with perpendicular "saw cuts" into the wing at the discontinuity to generate even stronger vortices. Reference 10 contains many of the recent NASA test reports and fixes for general aviation aircraft.

The old NACA developed spin recovery criteria to guide aircraft designers; these criteria are contained in Technical Notes 421, 711, 1045, and 1329. (See also Ref. 11.) The importance of the mass distribution of the airplane and tail design were particularly addressed. A "good" tail had a long moment arm and plenty of both fuselage area and rudder area under the horizontal stabilizer. This was quantified by calculating what was called the Tail Damping Power Factor (TDPF). (This criterion suggests that a T-tail with the elevator above the rudder would be great for recovery from an upright spin and useless in an inverted spin!)

NASA has completed a new decade of spin research and it now appears that some additional criteria are important, perhaps chief among them being the fuselage shape because of the important part the fuselage plays in determining the velocity of the air reaching the elevators and rudder at high lift coefficients. For example, it appears that a rounded upper fuselage and a flat lower fuselage near the tail make spins more likely and promotes the development of flat spins. See Ref. 11 and references contained therein.

NASA uses a special, ballistically-deployable spin parachute on their spin research aircraft. On the Beechcraft Sundowner they attached little rocket engines at the wingtips to give additional control power. They also modify the doors to make egress easier, although even professional test pilots have had difficulty in exiting a spinning aircraft. Without these survival aids, it takes a real hero, with a death wish, to spin test an aircraft that hasn't been cleared for spinning.

There has been a lot of progress recently in predicting spin behavior before the first flight test is made. The aircraft static and dynamic characteristics are first studied in wind tunnels; one of the most useful of the wind tunnel devices is the rotating balance which can study angles of attack up to a full 90°, in actual rotational motion. Another valuable tunnel at NASA's Langley Field and one of only two in the free world is the **spin tunnel** in which a radio-controlled model with the same mass distribution as the full-scale article is stabilized in all the possible spin modes in an upward flow of air. Screening keeps the air velocity greater near the edge of the tunnel, so models don't tend to crash into the walls. (Now you know what happens to old model airplane enthusiasts.) The spin tunnel reportedly has also been used to practice skydiving maneuvers. A final tool is

the computer, which is becoming more and more reliable and useful (Ref. 12 and Chap. 18).

Twin-engine aircraft with engines on the wings present a special spin problem. With one engine out, as in training or after an engine failure, they typically spin rather easily toward the dead engine as the stall is reached—or they may even snap inverted and enter a flat spin. But there is no requirement that they be tested for satisfactory spin characteristics; the large moment of inertia of the outboard engines results in a different kind of spin which typically relies more on the elevators than on the rudder for recovery. However, the Beechcraft Model 76 Duchess is one twin-engine aircraft that has undergone extensive spin tests (Ref. 13).

As suggested by some of the quotes at the beginning of the chapter, military aircraft have a continuing need for ever greater maneuverability and this means that the possibility for departure from controlled flight is always present in new aircraft. The new fighter aircraft, the F-14, F-15, F-16, and later models, are using vortex lift to supplement normal leading edge airfoil lift so that usable lift extends to $\alpha = 45°$ or more. For example, the F-14A has its maximum lift coefficient at about $\alpha = 35°$ and the coefficient smoothly decreases from there to over 90° where it must be zero. The aircraft has been flight tested over the full –90° to + 90° angle of attack range with full stores (Refs. 14 and 15)!

In flight at these very large angles of attack, the stability about the other axes often degenerates significantly. A yaw departure at high alpha is known as nose slice; a lateral departure is known as wing rock.

Transport category aircraft and some military aircraft commonly employ mechanical or electronic means to try to prevent inadvertent stalls and spins. For example, stick shakers and stick pushers are often automatically turned on as the angle of attack approaches the stall angle. (Recall the accident described in Chapter 7, Section B.) With a fly-by-wire aircraft, the computer may be programmed to prevent a departure—although on the F/A-18 it was found necessary to provide an override switch after a pilot discovered a spin mode with an unexpectedly low yaw rate—and the computer stopped him from recovering!

One of the most spectacular aerobatic maneuvers, now commonly demonstrated at airshows, is the **lomcovak**—a weird head-over-tail tumbling maneuver which is usually initiated by doing an inverted snap roll while in a climb. The T-2 in the chapter preface appears to have done just such a maneuver. Conservation of angular momentum and precession are behind the physics of the maneuver. If the controls are held in the entry position, the tumbling motion departs into an inverted spin. The lomcovak is considered to be hard on aircraft structure, particularly the crankshaft of the engine and the engine mounts. See Ref. 22 if you still want to do one.

Figure 16.9 presents the aircraft motions and load factors that were measured in the 1930s for a spinning military biplane fighter aircraft (Ref. 50).

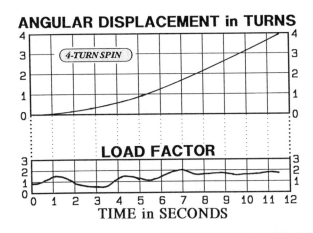

Fig. 16.9 Measured displacements and load factors in a spin (Ref. 50)

Pic. 16.4 The Boeing Stearman, a World War II trainer, is an excellent plane in which to learn aerobatics—especially if you remember to keep your harness tight. . . .

E. Deep and Dynamic Stalls

Ordinarily the loss of downwash on the horizontal stabilizer at the point of wing stall will decrease the negative lift generated by the tail, generating a strong negative pitching moment which tends to return the aircraft to a lower angle of attack. However, it is possible on some aircraft for the machine to become stable at a very high angle of attack and, because of a blanked elevator, be unable to regain normal flight without additional assistance. This is called a **deep stall**.

Perhaps the first aircraft that suffered a distinct tendency toward deep stalling was the Mach 2+ Lockheed F-104 Starfighter which has a T-tail (that is, a horizontal stabilizer mounted at the top of the vertical stabilizer). Lift from the fuselage makes it possible to increase the angle of attack to the point where the horizontal stabilizer is blanked; up to 35,000 ft of altitude is required for recovery. A number of T-tailed airliners and business jets with rear-mounted engines have been lost in deep stalls in the last 15 years or so. The DC-9 Super 80 uses strakes on its engine nacelles for this reason.

Canard-type aircraft will enter a deep stall if the elevator on the canard is powerful enough to drive the rear wing past its stalling angle; this is always possible if the *cg* is well aft of where it should be. (Before the wing is stalled, though, as the *cg* is moved rearward, there usually will be some wing rock—some uncontrolled rolling at the canard stall—because wing stall usually begins at the wingtips.)

It appears that some otherwise docile canard aircraft may enter a deep stall mode under conditions of violent maneuvering. These have involved angles of attack close to 90° and a descent to the ground in a flat attitude with no rotation. See Ref. 52.

EXAMPLE 16.3. Estimate the descent rate of a Long-EZ (Fig. 1.9) that might find itself in a deep stall.

Solution: As a first approximation, we treat the aircraft as a flat plate ($C_D \approx 1.3$) that is descending vertically with the lift vector essentially coincident with the drag vector, and just equalling the weight vector, as for the terminal velocity dives of Chap. 15, Sec. C.

The wing area of the Long-EZ is 82.0 ft², the canard area is 12.8 ft², and the fuselage has a projected area of perhaps 27 ft², so the total drag reference area is about 122 ft². We assume that the weight of the aircraft is 1100 lb (one person, partial fuel). Then

$$D = C_D q S = C_D \left(\frac{1}{2}\rho_0 V_E^2\right) S = (1.3)\left(\frac{1}{2}\right)(0.002377)V_E^2(122)$$
$$= W = 1100 \text{ lb}$$

This can be solved for V_E to yield

$V_E \approx 76 \text{ ft/s} \approx \textbf{4600 ft/min}$

For comparison, in the only known example of a Long-EZ becoming trapped in a deep stall, in 1991, the estimated descent rate (based on a videotape) was only 2900 ft/min, so this aircraft appears to be generating considerably more drag than a simple flat plate of the same projected area. From a momentum standpoint, it can be calculated that the 2900 ft/min descent rate is equivalent to continuously forcing a column of air above the aircraft to descend at 47 ft/s. To some extent the greater drag of a "falling leaf" aircraft over a flat plate might be expected because of the interactions of the horizontal components of the flow over the various surfaces, since the outer contours are so different than a simple flat plate. However, another canard aircraft (a 4-place Velocity) was involved in a deep stall accident into water in which the descent rate was estimated to be less than 1500 ft/min and the pilot was uninjured in the "landing." It has been suggested that these aircraft must be generating an extremely large vortex above them, in order to explain such low descent rates. With very little forward motion and no rotation, the descent rate is the key to survivability.

In the early 1980s, NASA conducted controlled deep stall experiments with a modified sailplane. The horizontal stabilizer was modified to allow it to be positioned up to 72° nose down. The procedure used was to stall the airplane normally and then pitch the whole stabilizer down, which retains its effectiveness as the angle of attack approaches 90° and the aircraft descends almost like a parachute with an L/D of 0.2 and a sink rate of 3000 to 4000 ft/min. Some of the possible applications that have been mentioned include (a) recovery of remotely-piloted vehicles (RPVs), (b) controlled steep descents for fighter-type aircraft, and (c) lightplane recovery from spins. The technique has actually been used since 1943 on model airplanes where it is known as dethermalizing. Loss of funding has since caused the program to be terminated (Ref. 36).

A final type of stall is the **dynamic stall**, included in what is known as "unsteady aerodynamics." This refers to the fact that an airfoil will show greater maximum lift and hysteresis in its lift and drag and moment curves if the angle of attack is changing at a rapid rate. Dynamic stall has always been of importance in understanding the behavior of the main rotor on a helicopter (Ref. 37). It is probably responsible for some of the oscillatory motions of an aircraft being held in a stall.

F.   Maneuvering Speed

The **maneuvering speed** of an aircraft, usually represented by the symbol V_A, is defined as the maximum speed at which full or abrupt control movements can be made without damaging the aircraft. The speed is based on Eq. 16.2, the recognition that aerodynamic loads reach their maximum when the angle of attack is abruptly brought to the stalling angle (without letting the speed decrease to the normal stall speed). Therefore the design maneuvering speed for an aircraft can be calculated from Eq. 16.2 by substituting the design load factor (Table 15.1) for n and the stall speed. This will be done in the last section on the flight envelope.

Is then an airplane completely safe from aerodynamic overload damage so long as the speed is kept below the maneuvering speed? Can the meanest thunderstorm be safely penetrated (at least at altitude) if the pilot merely remembers to slow down first? Well, certainly the pilot **should** slow below maneuvering speed – but not too close to the stall speed – if turbulent weather is anticipated. But the problem is that severe wind gusts can cause the true altitude, the indicated attitude, the bank angle, and the indicated airspeed to fluctuate wildly. The first punch should be survivable all right, but it may just be setting you up for a knockout punch a little down the road.

Many pilots believe that their airplane can safely sustain larger load factors when they fly at weights **less than** the listed gross weight, so that it may even be safe to do aerobatics at these reduced weights. This is often **not** the case! If the reduced weight is because the weight of the fuselage has been reduced, the bending moments at the wing attach points will indeed be less for abrupt stalls at a given speed – so the wings are more likely to stay on when doing abrupt maneuvering at reduced weights. But the reduced fuselage weight and an unchanged maximum wing lift means that the fuselage will be accelerated **more** for any abrupt stall at any given speed, and the engine moments (or the battery mount or the seat attach fittings or ?) may give way at the original acceleration limit, since those weights **haven't** been reduced. This is why most aircraft list a **lower** maneuvering speed (for bumpy air) when they are at reduced weights. A numerical example should help to explain what is happening.

EXAMPLE 16.4. A Cessna 182 at its gross weight of 2950 lb is cruising at a calibrated airspeed of 130 kt when it hits an updraft that suddenly stalls the aircraft. The stall speed for the aircraft in cruise configuration is 56 kt. If the pilot is a 170 pounder and if the engine mounts hold up 400 lb in level flight (a) what are the load factors for the airplane, the engine mounts, and the pilot at gross weight and (b) what would they have been if the plane had been at its minimum weight (with pilot and fuel) of 2000 lb?

Solution: (a). The aircraft load factor (by Eq. 16.2) is

$$n = \left(\frac{130\ \text{kt}}{50\ \text{kt}}\right)^2 = \textbf{5.39}.$$

By Eq. 15.10, the increased lift force results in a normal acceleration of

$$a = (5.39 - 1)g = \textbf{4.39}\,g$$

and this is shared by everything in the airplane.

(b). By Eq. 15.8, the total lift force generated by the wings during the gust penetration was

$$L = n\,W = (5.39)(2950\ \text{lb}) = 15{,}898\ \text{lb}!$$

This same force will be generated at the lighter weight as well but now the stall speed will be reduced (by Eq. 12.16):

$$V_{\text{STALL, AT 2000 lb}} = \sqrt{\frac{2000}{2950}}\ V_{\text{STALL, AT 2950 LB}}$$

$$= (0.823)(56\ \text{kt}) = 46.1\ \text{kt}.$$

Therefore the aircraft load factor at the light weight will be $n = (130/46.1)^2 = \textbf{7.95}$! The normal acceleration is $a = 6.95$ g. Everything in the airplane shares this acceleration and so this 7.95 is the load factor for the pilot and the engine mounts as well. How about the stress on the fittings that attach the wing to the fuselage, transmitting as much force there as necessary to give the fuselage the same acceleration as the wings? The force on the fittings is lift force times the fraction of the total weight that is in the fuselage: $L \times (W_{\text{FUSELAGE}} / W_{\text{TOTAL}})$. If you try some numbers in this you will see that the forces transmitted to the fuselage do go **down** if the reduction in weight is entirely due to loss of the weight in the fuselage, but it will go **up** if all of the weight loss is due to a reduction in the wing weight (due to fuel burned, e.g.).

Equation 16.2 tells us that the following relationship exists between any two equivalent airspeed at two different weights if they are to result in the same maximum load factor:

$$\frac{V_{1,\text{MANEUVERING}}}{V_{1,\text{STALL}}} = \frac{V_{2,\text{MANEUVERING}}}{V_{2,\text{STALL}}}$$

and Eq. 12.13 tells us that equivalent stall speeds are directly proportional to the square root of the weight (and nothing else, if it is the same airplane in the same configuration), so that

$$V_{2,\text{MANEUVERING}} = \sqrt{\frac{W_2}{W_1}}\; V_{1,\text{MANEUVERING}} \qquad (16.3)$$

WEIGHT DEPENDENCE OF THE MANEUVERING SPEED

EXAMPLE 16.5. The listed maneuvering speed for the Cessna 172 at 2300 lb is 96 kt. What could we expect the maneuvering speed to be at a flying weight of 1600 lb?

Solution: From Eq. 16.3,

$$V_{2,\text{MANEUVERING}} = \sqrt{\frac{1600\text{ lb}}{2300\text{ lb}}}\;\left(96\text{ kt}\right) = \mathbf{80\text{ kt}}$$

and this is exactly the value given in the owner's manual.

G. Loops

The loop is probably the most universally enjoyed of all the aerobatic maneuvers. It also contains some delightful physics. When we do them for fun we often aim for a minimum load factor and don't worry about drawing a perfect circle in the sky. In aerobatic competition, though, the judging **is** on roundness of the shape.

As we will see shortly, a properly **executed** upright loop requires only 2.5 to 3.5 positive *g*'s at the entry and about the same or somewhat less on the recovery, but ham-fisted pilots commonly go considerably beyond these values. It has been suggested that an expert can do aerobatics in an airplane not in the acrobatic category and get away with it for a long time; an amateur is taking a large risk; and a beginner is almost certain to buy the farm. The results of overstress are often seen on the upper leading edge of the wing or on the anti-drag structural members in the wing because of the enormous forward (suction) forces there (Fig. 9.1f).

In the Cessna 150 Aerobat the normal loop entry speed is 130 mph (although a very sloppy loop can be made down to about 100 mph entry speed). A 2.5- to 3.0-*g* pull-up is initiated upon reaching the given entry speed in a shallow power-on dive. Back pressure and positive *g*'s are normally maintained throughout the maneuver, with a slight reduction in back pressure at the top. The throttle has to be pulled back considerably during the entry (as on any fixed-pitch aircraft) to avoid overspeeding the engine; full power is added during the second quarter of the loop and all of the power is removed at the beginning of the third quarter, just after coming over the top. Recovery from the back side dive requires about the same 2.5 to 3.0 load factor.

Pic. 16.5 The Christen Eagle, a homebuilt design, is an excellent aerobatic aircraft; as a trainer, though, it makes the maneuvers too easy.

A too-abrupt entry leads to high load factors and can cause a failure to get around the loop because of a high speed stall. On the other hand, a too-gradual pull-up may lead to a tail slide, which few aircraft can safely handle. Even though an experienced aerobatic pilot can consistently do loops with load factors within either the utility or acrobatic category load limits, it is illegal and immoral to do them in an aircraft not rated for them. And any attempt to teach oneself aerobatic maneuvers, even in an acrobatic category aircraft, is really the pits. Even if the wings and tail stay on, the red-line for the prop and the airspeed are almost certainly going to be exceeded and hidden structural damage is very likely. Unfortunately, it is currently very difficult to find affordable aerobatic aircraft in most parts of the U.S. (The International Aerobatic Club, a division of the Experimental Aircraft Association, can provide a listing of currently available aerobatic instructors and airplanes.)

Figure 16.10 shows the geometry of a perfect loop. *T* is the thrust vector, *D* is the total aircraft drag, *L* is the total aircraft lift, and *V* is the entry speed at the bottom of the loop. We can obtain some numerical predictions if we make the assumption that the **thrust** force **equals** the **drag force** throughout the loop. This is true for level flight, of course, but it isn't likely to be continuously correct for a loop in which the pilot is adding power during the ascent and reducing power during the descent. Nevertheless, the average longitudinal force could be close to zero, so the results of such a simplified analysis are interesting and instructive.

The lift force does not contribute to the energy of the aircraft since it always acts perpendicular to the flight path. Thus our assumptions imply constant mechanical energy for the loop; much like a pendulum, the pilot trades kinetic energy for

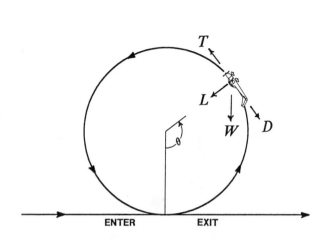

Fig. 16.10   The geometry of a perfectly circular loop of radius R

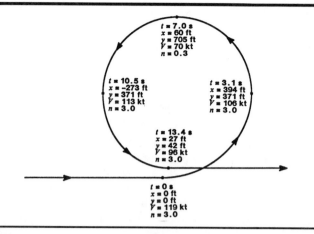

Fig. 16.11   Calculated loop for a C-152 Aerobat at 4000 ft density altitude

potential energy, and then gets it all back again on the way down.

The result of this analysis is a prediction (for the 150 Aerobat) of a loop radius of about 230 ft for the recommended entry speed of 130 mph, a load factor of 6 at the entry and recovery, a minimum speed of about 65 mph at the top, and a total time required of about 14 seconds. Experimental results, obtained from a sequence of about 50 loops of dubious circularity, were for a loop radius that varied from 225 feet to 300 feet, a maximum load factor of 2.5 to 3.0, and a minimum indicated airspeed too low to read at the top of the loop. The correlation between the prediction of the simple theory and this experiment isn't too shabby, considering.

But it is possible to make much more realistic calculations. Suppose we look now at the 152 Aerobat, the more recent version of Cessna's aerobatic trainer. The recommended entry speed for a loop is 115 kt indicated airspeed which works out to a true airspeed of 119 kt at a density altitude of 4000 ft (a reasonable altitude for loops for Ohioans). If we now initiate a 3.0-g pull-up, the instantaneous loop radius is easy to calculate because the excess 2.0-g acceleration is just the centripetal (inward) acceleration, due to the continuous change in direction in going around the loop:

$$\left(\begin{array}{c}\text{centripetal}\\\text{acceleration}\end{array}\right) = \frac{V^2}{R} = \frac{(119\ \text{kt} \times 1.689\ \text{ft/s/kt})^2}{R}$$
$$= 2.0\,g = (2.0)(32.2\ \text{ft/s}^2)$$

which can be solved for the radius R, yielding: R = **627 ft**

If we could maintain this 3.0-g load factor and this speed up to the one-quarter loop point (90° of loop), the centripetal acceleration would then be 3.0 g and the loop radius would be

$$R = (119\ \text{kt} \times 1.689\ \text{ft/s/kt})^2 / [(3.0)(32.2\ \text{ft/s}^2)] = \textbf{418 ft}$$

In actual practice the loop radius usually does begin with a relatively large value and then tightens considerably as the speed decreases.

Figure 16.11 is a computer-calculated loop for a Cessna 152 Aerobat under these same conditions but assuming a constant

3.0-g pull-up with 75% power to the 90° point, then a constant radius loop with 100% power until the load factor builds to 3.0 g on the down side (which occurs at the 260° point), and then a constant 3.0-g recovery with the throttle at idle. In practice one would soften out the load factor toward the end of the loop and then it would be a very respectable loop. (The calculation used an estimated drag polar for the 152 Aerobat along with numerical integration over 0.02 second intervals.) Kershner (Ref. A12, aerobatic books for pilots) suggests that his students perform Aerobat loops starting with 3.5 g's at the bottom, diminishing to 0.7 g at the top, with a loop diameter of 560 ft in about 14 s, on the average — amazingly close to the independently-calculated parameters given in Fig. 16.11.

If you prefer doing things with brute power and splendiferous speed, you will appreciate the parameters of a typical formation loop by the United States Air Force Thunderbirds (Fig. 16.12).

It is useful to point out that the FAA requirements for aerobatic category aircraft are completely inadequate for many of the maneuvers used in the intermediate and advanced levels of aerobatic competition. For example, a normal outside loop (executed optimally) can be expected to require about –2.5 to

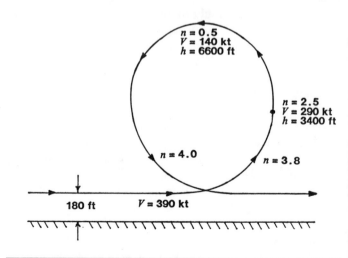

Fig. 16.12   Airshow formation loop by jet fighters

–3.0 *g*'s, but that is the very edge of structural damage if the aircraft is only designed to FAR aerobatic category requirements. A true aerobatic aircraft must possess **actual** limit load capability of plus or minus 7 or 8 *g*'s; world competition aircraft must go up to plus or minus 10 or 12 *g*'s. Even then, the maintenance on aerobatic machines must be performed more carefully and much more often. (The International Aerobatic Club (IAC), a division of the EAA, is a very good source for maintenance experience on various aerobatic aircraft.)

An outside loop is accomplished in much the same manner as for the upright loop except forward pressure on the control stick or wheel is used and a fuel/oil system that will operate inverted is required. A good aerobatic aircraft such as the Pitts or the Christen Eagle will do outside loops with similar speeds and similar (though negative) load factors. Outside loops are much less pleasant for most pilots, though, because the airplane is being forced to cut around the inside of a circle while the pilot's body, on the outside of the circle, is trying to go straight — so a good, five-point shoulder harness is a necessity. A botched outside loop has a very good chance of ending up as an outside spin.

In a strong airplane that doesn't have an inverted system, such as the Stearman, only the first half of an outside loop (called an "English bunt") can be accomplished. The procedure is to slow the airplane and then push the nose down and over — especially great fun in an open cockpit airplane like the Stearman! (You'll never forget the first time you do this maneuver.) If the pushover rate is too great, the plane will momentarily experience a negative-angle stall (even though it may be pointed straight down!) and will just take that much longer, and gain that much more speed, before level, inverted flight is achieved. A half-roll to upright completes the maneuver.

No plane that merely satisfies the requirements of the aerobatic category (Table 15.1) should be used for outside spins or outside snaps or outside loops; rather, the plane should be good for at least a –6-*g* limit load factor to allow a factor of safety — even though all these maneuvers are possible without exceeding –3 *g*'s.

Figure 16.13 presents the displacements and load factors that were measured in an upright loop by that same old 1930s military biplane fighter.

H. Rolls

We have discussed snap rolls and spins and loops already but in practice a pilot has no business doing either loops or snap rolls solo until the spin characteristics of the airplane are thoroughly understood and he can execute a good slow roll. The reason for the spin and slow roll requirement is that a botched aerobatic maneuver results in either a spin or in a high speed, inverted dive, from which a good slow roll is the only safe recovery. (A good slow roll is one for which the aircraft rotates about its longitudinal axis without any motion about the other two axes.) A botched slow roll from level flight, the normal

Pic. 16.6 A Christen Eagle cruises by in its comfortable inverted mode.

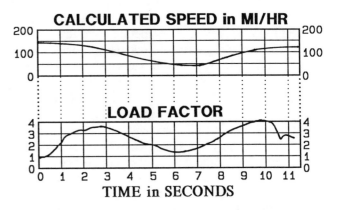

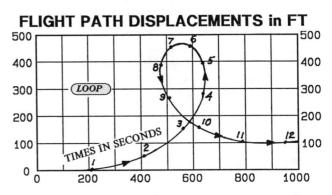

Fig. 16.13 Displacements and load factors in an upright loop (Ref. 50)

result of the first few tries, always results in a nose low, inverted dive.

The only safe way to recover from a high speed, inverted dive, then, is to do a good slow roll to upright (to minimize additional speed gain) and then gently pull out. The novice always tries to recover from inverted flight by pulling back on the control stick or wheel (i.e., executing a "Split-S"), causing a tremendous loss in altitude and an excursion beyond the never-exceed speed where he suddenly becomes a test pilot. A pilot isn't ready to do solo aerobatics, and most certainly not take a passenger in aerobatics, until he or she has attained a sense of what the airplane is doing in every attitude possible and how the controls can be utilized instinctively to make it do what is desired at any time. The controls must be understood in terms of what they do to the flight direction, completely independently of the relative location of the ground.

There are many conditions that must be satisfied before attempting aerobatic maneuvers if safety is to be maintained. First of all, the pilot must have attained aerobatic competency with the help of a qualified instructor. Second, the aircraft must have proven suitability for all the maneuvers that are to be accomplished and there should be an operative *g*-meter in the panel. Third, plenty of altitude below the aircraft must be present. Finally, the aerobatic area must be away from airways and free of other aircraft.

The two primary types of controlled rolls (as contrasted with the snap roll) are the **slow roll** and the **barrel roll**. The slow roll is the roll mostly used in competition and the one most useful in recovery from unusual attitudes and the one maneuver that takes the longest to learn to do even satisfactorily. From an outsider's view, it is a very simple maneuver that involves a purely rolling motion about the longitudinal axis of the airplane without any change in the heading or in the altitude of the aircraft. It is a very beautiful maneuver to watch. Inside the cockpit it is a very, very busy time. The roll must be initiated in one direction with the ailerons and then opposite rudder must be used to prevent a turn from developing. But the rudder deflection must steadily decrease and then go to the opposite direction in the last part of the roll and all the while the elevator must be changing from a nose-up deflection to a nose-down deflection (while inverted) and then back to a nose-up deflection. The load factor is varying during the roll from about +1 to –1 and back to +1, which means that the pilot is really hanging from the shoulder and seat belt harness. The first time pilots are shown the maneuver, their feet will leave the rudder pedals and dangle uselessly, because previous flying hasn't taught them to tense their legs and push on the pedals to keep their feet there. The first few times a pilot tries the maneuver, too many things are happening at once and the roll tends to be stopped in the inverted position with inadequate forward stick, leading to the inverted dive. Then there is a tendency to compensate abruptly for an incorrect location of the controls, placing extra stress on the machine. Oh, but when the maneuver is really learned it is a beautiful thing! It can be combined with portions of a loop to trace out a figure 8 on its side in the sky (the "Cuban 8"). It can be used to view the countryside while going straight up or straight down.

The difficulty of the slow roll maneuver varies inversely with how good an aerobatic airplane you have! A poor aerobatic airplane with a cambered airfoil on the wing, no inverted system for the fuel and oil, and a poor roll rate is a real challenge and tests whether the pilot really knows how to do the maneuver. The cambered airfoil means that there will be a large difference in the nose attitude and in the elevator position when inverted compared to being upright at the beginning and end of the maneuver. The poor roll rate gives the plane plenty of chance to dip its nose or skew off the heading. On the other hand, a good aerobatic aircraft with a high roll rate can be rolled so fast that negligible elevator changes and little rudder are needed. The Citabria and Stearman are examples of good aerobatic trainers for the slow roll. The Pitts airplanes are great aerobatic machines but poor slow roll trainers.

The other primary type of roll is the barrel roll. The idea here is to start a coordinated roll into a banked turn (with the nose rather high) and just keep the coordinated roll going until the aircraft has gone inverted and back to upright flight. The maneuver is nowhere as pretty from the ground because the nose begins well above the horizon, then is on the horizon but well over to one side, then well below the horizon, then well over to the other side, and finally back to its original starting position well above the horizon. The nose traces out a circle relative to the horizon rather than being held on a point as in a slow roll, giving the roll its name. The maneuver really can be a completely coordinated one, one in which the load factors remain positive throughout; airshow master Bob Hoover has filmed a barrel roll in which he continuously pours water into a glass that is resting on top of the instrument panel.

The diameter of the circle traced out by the nose is determined by the rate of roll. If the roll is very rapid, the roll can be started at a relatively low angle above the horizon and the nose never goes so low that the speed really starts to build up; this is sometimes called an **aileron** roll. A slower roll rate means that the maneuver must go far above the horizon to keep from being transformed into a steep dive at the bottom of the maneuver. It is also a rather difficult maneuver to do properly and safely and well but it is much more comfortable for most people because of the positive load factors; you might not believe you were momentarily inverted if your eyes weren't conveying that bit of news.

Because the load factors are all positive and no greater than 2 or so in a high quality barrel roll, this roll can be safely accomplished in almost any airplane if the pilot is experienced and skilled. The prototype Boeing 707 airliner was rolled by test pilot Tex Johnson in this fashion during one of its early flight demonstrations (Ref. 51). Bob Hoover commonly demonstrates barrel rolls in twin-engine, business-type aircraft.

Fig. 16.14 presents the displacements and load factors measured during the execution of a half-loop, half-roll maneuver in the same 1930s military biplane (Ref. 50).

I.   The Flight (Maneuvering) Envelope

The maneuvering flight envelope is a graphical presentation of the load factors within which an aircraft must operate if it is to avoid structural damage. It is defined by the combined requirements of adequate strength for anticipated wind gusts, as discussed in Section A, and adequate strength for an-

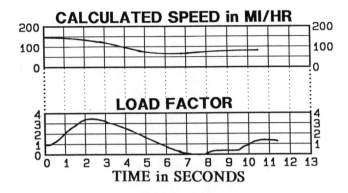

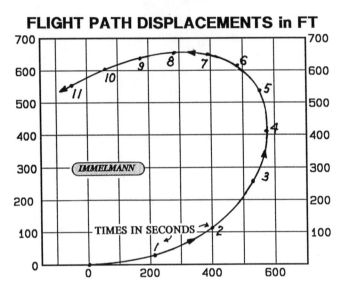

Fig. 16.14 Displacements and load factors in a half-loop and half-roll (Immelmann) (Ref. 50)

Pic. 16.7 This PJ-260 makes inverted flight look easy.

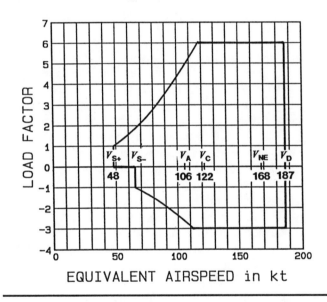

Fig. 16.15 Typical flight envelope (V-g diagram) for an aerobatic category aircraft (C-152 Aerobat)

ticipated maneuvering loads, as discussed in the previous chapter.

The envelope for an aerobatic airplane such as the Cessna Aerobat might look like that shown in Fig. 16.15. The envelope begins at the positive and negative stall speeds and extends to higher speeds with **parabolic** sections. The parabolic shape results because the maximum possible load factor due to maneuvering depends on the **square** of the speed, from Eq. 16.2. For positive load factors, the positive angle of attack stall speed (V_{S+}) is used and for maximum negative load factor the negative angle of attack stall speed (V_{S-}) is used. The sketch is for a positively-cambered airfoil for which V_{S-} is greater than V_{S+}.

V_C is the **design cruising speed,** marked on an aircraft's airspeed indicator as the end of the green arc and the beginning of the yellow arc, and by regulation,

$$V_C \geq 38 \sqrt{\frac{W}{S}} \qquad \text{(Utility Category)}$$

$$V_C \geq 42 \sqrt{\frac{W}{S}} \qquad \text{(Acrobatic Category)}$$

CERTIFICATION REQUIREMENT, DESIGN CRUISE SPEED

for V_C in mi/hr, W in pounds, and S in square feet. For the Aerobat, we can calculate that $V_C \geq 42 \sqrt{1600/157} = 134$ mi/hr, and in fact the manufacturer lists V_C as 140 mi/hr.

As discussed in Sec. F, V_A is the **design maneuvering speed,** the maximum speed in level, upright flight for which abrupt maneuvering or severe wind gusts cannot cause load factors greater than the limit values. From Eq. 16.2 and Table 15.1, we see that V_A can be approximately calculated from

$$\left(\frac{V_A}{V_{S+}}\right)^2 = 4.4 \qquad \text{(Utility Category)} \tag{16.4}$$

$$\left(\frac{V_A}{V_{S+}}\right)^2 = 6.0 \qquad \text{(Acrobatic Category)} \tag{16.5}$$

CALCULATION OF THE MANEUVERING SPEED FOR AN AIRCRAFT

Pic. 16.8 Here's the view from the pilot's cockpit of a Stearman while in an inverted glide.

Pic. 16.9 Bob Hoover pauses in the middle of a roll.

All that a wind gust can do is increase the lift up to its maximum value, just as if the pilot had suddenly pushed the elevator full forward or full back. So the important conclusion is that an aircraft flying at or below maneuvering speed cannot be damaged by a perpendicular wind gust of any magnitude; the wing will stall before the loading exceeds the permissible maximum. Transports have sometimes gotten in trouble by flying too slowly when in turbulent air, so that strong gusts have stalled their wings and caused great loss in altitude. This is not such an easy problem as it may seem to light aircraft pilots because the range of operating speeds narrows continously as the altitude increases; the maximum allowable operating Mach number decreases to lower and lower indicated airspeeds and therefore approaches the stall speed as the altitude increases, boxing the aircraft into what has been called the **coffin corner** (Fig. 7.5). When aircraft have broken up in turbulent air, it is probable that a succession of gusts caused the aircraft to assume an unusual attitude, disoriented the pilot, and thereby caused an increase in the airspeed to a speed above the maneuvering speed.

In the flight envelope of Fig. 16.15 or Fig. 16.16, V_{NE} is the posted never-exceed (red-line) speed for the aircraft. However, for certification every aircraft must be dived to a speed equal to

$$V_D \equiv \frac{V_{NE}}{0.9} \qquad (16.6)$$

DEFINITION OF THE DESIGN DIVE SPEED

Thus, for example, at least one test pilot must have dived an Aerobat to a speed of $V_D = (193 \text{ mi/hr})/0.9 = \textbf{215 mi/hr}$, since its V_{NE} is 193 mph.

In this example the maneuvering requirements completely determine the boundaries of the flight envelope since the gust

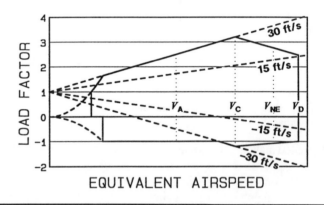

Fig. 16.16 Typical flight envelope for a large transport aircraft (in the cruise configuration, at speeds well below the critical Mach number)

loading requirements are less severe. For a normal category aircraft, the flight envelope can be strongly determined by the wind gust requirements, as is suggested by the flight envelope shown in Fig. 16.16.

The 50% safety factor between limiting (or design) load factors and ultimate load factors allows us to enlarge the diagram of the flight envelope to include all possible flight areas. The region outside the flight envelope in Fig. 16.17 shows where break-up (structural failure) will occur. However, in the region marked **flutter**, destructive control surface flutter may occur first, since the FAA requires flutter testing only up to V_D. The tendency for flutter, a type of aeroelastic resonance, is minimized by static and possibly dynamic balancing of the control surfaces and by making sure the control surface linkage is stiff and without free play. (Static balancing means that, on the ground, net moment about the hinge line is zero. This is normally obtained by placing lead weights a short distance in front of the surface, usually hidden inside the wing but sometimes extending forward on the bottom of the wing. See the next chapter for some additional discussion.)

A balanced design is what counts. Some aircraft are said to have "12-g wings," for example, but that is precious little comfort if you try to pull 6 g's and discover that you have a 5.5-g tail or engine mounts or seat attachments (Ref. 18).

In the process of aircraft design, the flight envelope must be translated into appropriately-sized structural members; there is no point in a structural member being stronger than the weakest link and every extra ounce of weight just subtracts from payload or performance. The load factor, $n = L/W$, as we have defined it, is not quite the load factor used in structural analysis because the lift force acts in a wide range of directions relative to the structure. Instead, two new load factors,

$$n_Z = \frac{F_Z}{W} \quad \text{and} \quad n_X = \frac{F_X}{W}$$

are defined, where F_Z is the component of the resultant aerodynamic force that is perpendicular to the longitudinal axis of the aircraft and F_X is the parallel component. A little vector diagram will show that, if the chord line is parallel to the aircraft's longitudinal axis, F_Z and F_X are readily defined in terms of the lift and drag forces and the angle of attack:

$$F_Z = L \cos \alpha + D \sin \alpha \quad \text{and} \quad F_X = D \cos \alpha - L \sin \alpha$$

while, if the wing is mounted with a non-zero, positive angle of incidence, this incidence angle must be subtracted from α in the equations above.

n_Z is then the larger of the two newly-defined load factors and is what is measured and displayed by the typical g-meter in the instrument panel; however the values of n_Z and our old n (L/W) are very close together because wing incidence angles are small and lift is normally much greater than the drag.

But the new n_X load factor presents some surprises. A careful study of the directions of the resultant aerodynamic force on a wing, as in Fig. 9.1, reveals that n_X is slightly positive (rearward) under low-angle-of-attack conditions (Fig. 9.1c) but strongly negative (forward!) under high-angle-of-attack conditions (Fig. 9.1f). This means that the wing will have to be built to resist a "sweepback" force under high speed cruise and dive conditions but it also must be built to resist a "sweepforward" force that appears under conditions of abrupt stalls or low speed maneuvering. Thus fabric-covered wings usually use two sets of wire or tube bracing, one set of "drag" wires for the first condition and one set of "anti-drag" wires for the second. Metal wings can obtain the needed strength from the spars and the sheet metal covering the wings. (Some early aircraft were designed with wire bracing between the wings and a forward part of the fuselage; pilots were very surprised to find that these wires went slack under maneuvering loads.)

Wings generally have a forward span-wise spar at about the thickest point in the wing to resist most of the flight bending

Pic. 16.10 Remember to wave to the crowd as you fly by.

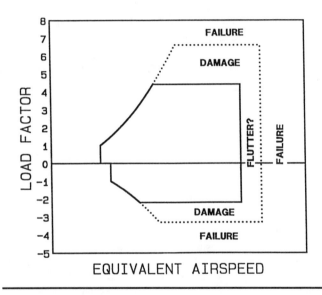

Fig. 16.17 What happens if flight strays outside the flight envelope? (A utility category aircraft is illustrated.)

loads and a rear spar to support the flaps and ailerons. A spar consists of a relatively thin structural member between the upper and lower wing surfaces that resists the vertical (shear) force, called the **shear web**, and reinforced plate material close to and parallel to each wing surface to handle the compression and tension loads associated with the bending moments, the **spar caps**.

Usually the spars must be sized for the maximum compressive loads, rather than for tensile loads, because most structural materials are weaker in compression than in tension. The upper left-hand corner of the flight envelope is where the limit load factor is reached at a high angle of attack, so this places the upper front spar cap in its maximum compressive stress condition because both the bending load and the negative n_X tend to compress it and their effect is additive. Similarly, the upper rear spar cap suffers the most compressive stress during flight in the upper right corner of the flight envelope, the lower rear spar cap during flight in the lower right corner of the envelope, and the lower front spar cap in the lower left corner. See Ref. 53 for details and additional discussion.

Some aerobatic maneuvers do a good job of exploring the flight envelope, in terms of stresses generated, all by themselves. Figure 16.18 illustrates this for an aerobatic category aircraft that is performing the Immelmann of Fig. 16.14. Note that the curve has nothing to do with the flight path but does nicely illustrate the wide range of load factors encountered.

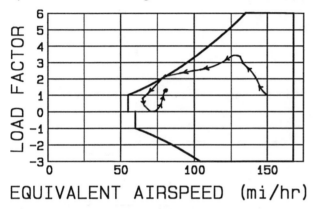

Fig. 16.18 The load factors experienced during an aerobatic maneuver can be plotted on the flight envelope; this is a half-loop and half-roll.

J. Physiological Effect of High Load Factors

Considerable study has been devoted to the effect of high load factors on the human machine as well. In Jimmy Doolittle's pioneering study (Ref. 3), he noted that

The airplane was banked to approximately 70°. It required some time for the pilot to establish steady conditions, but a constant acceleration of 4.7 g was finally obtained. After the steady condition was reached, the pilot gradually began to lose his sight, and for a short time everything went black except for an occasional "shooting star" similar to those seen when one is struck on the jaw. ... Sight returned almost immediately when the acceleration was decreased to normal by restoring the airplane to a condition of steady level flight. ... The effect of this maneuver on the pilot is not particularly uncomfortable. The sensation is that of having a tight band around the forehead and a feeling that the eyeballs are about a half an inch too low in their sockets. ... The pilot experiences no difficulty under instantaneous accelerations as high as 7.8 g.

The reaction of the human body to high load factors varies considerably from one person to another and is strongly affected by the pilot's health and the amount of recent rest. Aerobatic pilots find that their load factor tolerances, both for retaining consciousness and for avoiding airsickness, increases very significantly as the aerobatic season wears on. They, along with pilots of military fighter aircraft, quickly learn to use the M-1/L-1 straining maneuver in which the breath is held and the abdominal and skeletal muscles are momentarily tensed to raise

the blood pressure; this provides an extra 1 or 2 g's of load factor tolerance.

Sustained positive load factors over 4 g's or so provide clues to the pilot that the loading must be reduced or unconsciousness will occur. Because arteries in the eyes require higher blood pressure than in the brain, the first symptom of a sustained load factor is "grayout," in which vision narrows and the view darkens. If the high load factor continues, this is followed by "blackout," in which vision is lost entirely but consciousness is **not**. Finally, consciousness is lost and the pilot involuntarily relaxes pressure on the controls, bringing the load factor back to about 1. However it may take a matter of **tens** of seconds before the pilot is able to regain consciousness and react normally. Afterward the pilot is generally quite unaware of having experienced this sustained period of unconsciousness.

Sustained negative load factor cause the blood pressure in the head to increase, so consciousness is never lost, but most people find negative-g loads to be considerably more uncomfortable than positive ones. Military aircraft have never seen a need for substantial negative-g capability but it is a high scoring factor in aerobatic competition. A plane must be modified for the strength requirements as well to maintain fuel and oil to the engine. Aerobatic aircraft without an inverted system can do rolls and even inverted spins but, when return is made to upright flight, the engine should be started back up at low rpm to minimize engine wear. In the stock Stearman, the large propeller always continues to windmill throughout, which is very nice because the machine isn't equipped with an in-flight starter. However, the author was more than a little dismayed, once, when the similarly no-starter Quickie stopped rotating its little propeller while in the process of executing a substandard barrel roll.

Figure 16.18 presents possible limits of consciousness for various accelerations, explicitly linking the limit to the duration of the acceleration.

In modern aerobatic competition and even in airshow flying, distinct and rapid changes in direction tend to score highest and appear most spectacular to spectators, as well as generating maximum g loadings. A 1988 article in *Sport Aerobatics* suggested that a competition loop in even the Sportsman class should be initiated with a 5-g pull-up. At the world competition level, Soviet Union pilots routinely execute +12/–10-g maneuvers; their Sukhoi SU-26M features a roll rate of 343°/s and includes a seat that reclines at 45° to reduce the component of acceleration along the spine. The Glasair, a fast, homebuilt aircraft made of composite materials, is routinely demonstrated at airshows by a former aerobatic champion; load factors in his demonstration reportedly range from –3 to +8.

The military has always been interested in highly maneuverable aircraft for the advantage they provide in aerial combat. g-suits, which inflate automatically and press against the body to reduce blood flow from the head, are routinely used to help keep the human pilot as strong as the machine. The military uses rotating chairs (a "centrifuge") to simulate and train pilots in high-*g* effects and how to minimize them. The F-15 and F-16 are examples of fighter aircraft that were designed after it became apparent that stand-off missile firing wasn't adequate and enhanced maneuverability for close-in dog-fighting was necessary. But these highly maneuverable aircraft also possess a high rate of *g* increase and this has caused a significant number of pilots and aircraft to be lost. With such high *g* onset rates, the pilot goes almost instantly from total consciousness to total unconsciousness without the usual cues such as grayout or blackout. Unconsciousness follows about 4 seconds after the oxygen supply to the brain is interrupted—but a minimum of 20 seconds and a maximum of minutes is required before full functioning capability is restored. It has been suggested that a computer in the aircraft should monitor the pilot's state of consciousness and take over control when necessary.

The individual at most risk of *g*-induced loss of consciousness is one who is fatigued (and therefore has used up his supply of adrenaline), one who has had a long layoff from high-*g* flying, and one under emotional or other induced stress. The type of aerobatic maneuver that has caused trouble for many sport aerobatic pilots is one in which large, sustained negative loads are followed by large, sustained positive loads, as occurs in a vertical-8 composed of an inverted loop followed by an upright loop. Probably related to this, it has seemed to the author that it is easier on the body to exit from an inverted spin via a half-roll to upright rather than via a half-loop (Split-S) to upright. In any case, all of this just adds to the many reasons for practicing aerobatics at high altitudes.

Even more maneuverability will be available in newer aerobatic and fighter aircraft. The Falkland Islands war, in which the vertical/short-takeoff-and-landing Harrier was used, shows the advantages of **"vectored thrust"** in which the direction of the thrust generated by the jet engine can be quickly changed by pilot in flight. This permits abrupt changes in speed and direction which are very difficult for another aircraft to follow. Vectored thrust is also being used to provide lateral stability and control at very high angles of attack. All new air superiority fighter aircraft will be incorporating some of these ideas and the physiological constraints of the human body will have to be accommodated somehow.

NASA has been exploring the high angle-of-attack regime with a highly instrumented F/A-18. The program is an integrated one, including computational fluid dynamics (theoretical predictions of air flow) as well as carefully measured flows (Ref. 43). Picture 16.11 shows the vortex flow over the wing's leading edge extension (LEX) at an angle of attack of 30°. Smoke is exhausted from the fuselage in front of the wing extension so the vortices are very visible. Yarn tufts are used to show the flow on the wing. With the help of vectored thrust and the fly-by-wire control system, controlled flight has been extended up to an angle of attack of 70°!

Many pilots have found that sport aerobatics is one of the real pleasures of life. At its best, it is a joyous expression of the three-dimensional freedom of flight and it effectively removes the hidden fear that many pilots harbor for the unexplored

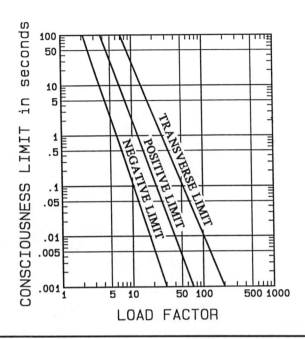

Fig. 16.19 Limits of consciousness as related to the duration, magnitude, and direction of the load factor experienced (for an "average" person)

areas of the flight envelope. You are encouraged to use the references on physiological effects and on aerobatic training to good advantage, if you are interested. There are places where aerobatic aircraft are available for training and rental; *Sport Aerobatics*, the official magazine of the International Aerobatic Club (a division of the Experimental Aircraft Association), is a good place to start your search.

Summary

Wind gusts generate load factor excursions (about the equilibrium value of 1) primarily by changing the instantaneous angle of attack of the wing. The load factor change is directly proportional to the magnitude of the equivalent wind speed and the equivalent airspeed of the aircraft, but inversely proportional to the wing loading. For most aircraft, the greatest hazard from wind gusts is the loss of performance when strong vertical winds or even a "microburst" is encountered on takeoff or landing.

An aircraft can be stalled at any speed by an abrupt increase or decrease in the angle of attack, caused either by a wind gust or by the pilot's use of the elevator control; the resulting load factor is equal to the **square** of the ratio of the entry speed to the stall speed. A snap roll is an aerobatic maneuver involving such an accelerated stall, either positive or negative; the rolling moment is generated by using the rudder to yaw the aircraft as it is stalling, producing more lift from one wing than the other.

A spin also involves autorotation caused by yawing, as well as rolling moments. With power off, the spin is a safe maneuver in many light, single-engine aircraft; maximum load factors are low during the spin but can be relatively large during the

Pic. 16.11 Air flow pattern around an F/A-18 at a high angle of attack is being studied. (Photo courtesy of NASA Ames-Dryden Flight Research Facility)

recovery to level flight if it is made too abruptly. If the power is left on or if the *cg* is too far aft, a flat spin can develop in which the angle of attack is greater than 60° and the motion is almost entirely yawing rather than rolling; such a spin is usually unrecoverable in non-aerobatic aircraft. Multi-engine aircraft and heavy single-engine aircraft generally can't be spun safely because of the extreme altitude losses involved and the lack of design for spin recovery.

Low altitude stalls or incipient spins are still accounting for a large fraction of light aircraft fatalities. Modern aircraft are more spin-resistant than 1920s designs but NASA research has led to wing treatments (cuffs) that offer much greater spin resistance for 1980s designs. The canard configuration is another possible solution to the stall/spin problem.

A deep stall occurs when an aircraft exceeds its normal stalling angle of attack and becomes stable there because of a blanking of the rear control surface; T-tailed aircraft are most likely to suffer from this problem. Dynamic stall refers to the altered stall characteristics when the angle of attack of a wing section is rapidly changed, as is true of rotor blades in a helicopter.

The maneuvering speed of an aircraft is the maximum speed at which abrupt control inputs or strong gusts will not overload the aircraft; its value can be estimated from the stall speed and the certification load factor because the load factor that is generated has a (speed)^2 dependence on the equivalent airspeed. The maneuvering speed is also proportional to the square root of the weight because the acceleration of the fuselage goes up as its weight goes down, by Newton's second law; thus an aircraft in heavy turbulence must be slowed down even more if it is at a weight less than its certificated gross weight.

A loop is a deceptively simple maneuver in an aerobatic aircraft; maximum load factors are typically +3.5 or –3.5 or greater for a positive or negative loop, respectively. The aircraft trades kinetic energy for potential energy in the first part of the loop and then gets it back again on the down side.

Slow rolls and barrel rolls are the two primary types of controlled rolling maneuvers (as contrasted with the snap roll, which is a high speed spin). The slow roll, a pure roll about the longitudinal axis, involves load factors of perhaps ±1.5, at the minimum. The barrel roll is a coordinated (non-slipping and non-skidding) roll which can be executed in almost any aircraft; load factors are all positive and may not exceed 2 or 3 — if the maneuver is accomplished properly.

The flight (maneuvering) envelope is a graphical summary of the load factors and speeds within which a given aircraft can safely operate. It is bounded speed-wise by the upright and inverted stall speeds on the low end and by the never-exceed speed on the high end. It is also bounded load-wise by the maximum available or maximum certificated load factors. Various points on the flight envelope can be calculated. Large aircraft are typically more constrained by wind gust requirements than are small aircraft because of their high speeds and their relatively lower certification load factors.

Pic. 16.12 This is one way the U. S. Army Air Force warned its pilots about the dangers of high load factors.

Large positive load factors, depending on their magnitude and direction, will cause loss of vision and then loss of consciousness for the pilot; large negative load factors are very uncomfortable and cause high blood pressure in the eyes and brain. Ameliorating the adverse effects of high load factors on the pilot is of great importance for both military fighter pilots and competition aerobatic pilots.

Symbol Table (in order of introduction)

U_E Equivalent wind gust speed
R Radius of a loop
V_A Design maneuvering speed
V_C Design cruising speed
V_{S+} Positive load factor stall speed
V_{S-} Negative load factor stall speed
V_D Design dive speed

Review Questions

1. Would you rather have an aircraft rated for a **limit** load factor of 6.0 or one rated for an **ultimate** load factor of 6.0? (Most ultralight manufacturers don't specify which one they are talking about.)

2. A loop entered too slowly will result in a stall or even a ___ ___.

3. At low load factors, the maximum safe speed is usually limited by ___.

4. The maneuvering speed of an aircraft normally decreases with decreasing flight weight because
    a. the wings are then capable of providing greater accelerations
    b. the terminal velocity has increased
    c. the bending moments at the wing roots in level flight are increased
    d. the aerodynamic center moves forward
    e. limit load factors have decreased

5. Failure of primary structure in an aircraft can be expected if the load factors are ___ times the load factor at which permanent deformation occurs.

6. The primary reason that wind gusts increase the load factor on an aircraft is that they momentarily
    a. increase or decrease the angle of attack
    b. increase or decrease the speed of the aircraft
    c. reduce the stability of the aircraft
    d. change the effective aspect ratio
    e. raise the stalling speed

7. During a snap roll or spin,
    a. the inside wing has less lift and less drag
    b. the inside wing has more lift and less drag
    c. the inside wing has more lift and more drag
    d. the inside wing has less lift and more drag
    e. an airsickness bag should be readied for action

8. Jogging may **reduce** your tolerance to high load factors. Do you know why?

9. An uncommanded yaw at high angles of attack is known as "___ ___" and an uncommanded roll oscillation at high angles of attack is known as "___ ___," especially for high performance aircraft.

10. In aerobatics, a maneuver which imposes large ___ load factors followed by large ___ load factors has been found to be particularly hazardous for aerobats.

11. Airfoil stall under rapidly changing angle of attack conditions is known as ___ stall and is a particularly important subject for helicopter ___ ___.

12. A stable, often unrecoverable stall at angles of attack over 30° is called a ___ stall; transport aircraft with engines mounted along the rear fuselage and with a ___ tail have been particularly susceptible.

13. A device that vibrates the control column to warn the pilot of an impending stall is known as a "___ ___" while a device that increases the stick force requirements at high alpha is known as a "___ ___."

14. Experimental spin research usually involves a ___ balance or a radio-controlled model in a ___ tunnel.

15. Recent spin research has demonstrated the importance of the shape of the rear part of the ___ in spin recovery.

16. NASA has found that a discontinuous glove with a ___leading edge, added to the outer half of a wing, will greatly increase spin resistance, for some light aircraft at least.

17. A spin with an angle of attack greater than about 60° is known as a ___ spin.

18. A recent procedure for emergency recovery from "all" types of spins in aerobatic aircraft is to remove any ___, remove the hands from the ___, and use full ___ in the direction ___.

19. Flat spins are sometimes survivable because the ___ rates are so low. However, the ___ rates are much greater than for normal spins and are very hard on internal body organs.

20. The **first** indication to a pilot that he or she is penetrating a microburst is that the nose pitches ___ (up or down) and the altitude ___ (increases or decreases).

21. The normal entry into a spin results in a load factor of about _ .

22. Give three possible solutions to the stall/spin problem for light aircraft.

23. A ___ roll is a coordinated rolling turn that continues through 360° of roll; load factors should all be ___ with a maximum of about ___.

24. A ___ roll is a purely rolling maneuver about an aircraft's ___ axis; when executed from level flight, load factors should range from about –___ to about +___.

25. The (maneuvering) flight envelope is a graph of allowable ___ ___ (on the y-axis) versus ___ ___, for a given aircraft.

26. If the onset of large, positive load factors isn't too rapid, there are distinct stages in their effect on the pilot. Name these stages.

27. The **designer** of an aircraft can increase the load factor tolerance for the pilot by designing the cockpit so that _____.

28. How can the **pilot** increase his or her load factor tolerance while performing an aerobatic maneuver?

29. Will there be a difference in the load factor felt by the occupants in an aerobatic airplane if the passenger is sitting at the aircraft *cg* and the pilot is behind the passenger? If so, which will feel the greater load factor?

Problems

1. A 12-*g* deceleration is no problem for a well-belted pilot. How much distance would be covered in a stop from 100 kt with this deceleration? Answer: 37 ft!

2. Estimate the height gain at the top of a loop for a Piper Cub that enters the maneuver at 100 kt and slows to 38 kt at the top. Answer: 379 ft

3. What is the load factor in a vertical climb? in a vertical dive?

4. What downward force does a 200 pound pilot exert on his seat during a 3.5-*g* snap roll?

5. What is the load factor for a snap roll while doing a vertical climb, assuming that the aircraft is at an airspeed of 150 kt when the maneuver is initiated and that the level flight stall speed is 50 kt?

6. A show-off pilot makes a low pass that is crowned with a 3-*g* pull-up. If the aircraft is a Cessna 152 with an equivalent stall speed of about 48 kt, at what speed must he be travelling to get an immediate stall during the pullup? Answer: 83 kt

7. What is the maximum load factor for a Cessna 152 that is pushing aside the air at an equivalent airspeed of 130 kt when it encounters a sudden wind gust or engages in abrupt maneuvering? Use the stall speed given in Problem 5. Answer: 7.3

8. A typical ultralight might have a wing loading of 4.0 lb/ft², a maximum cruising speed of 55 mi/hr, and a stall speed of 26 mi/hr. Assuming a lift slope of 0.08 per deg, what vertical gust speed under sea level conditions would generate a load factor of 3.0? What is the maximum load factor that a wind gust could generate?
Answers: 18.2 ft/s and 4.5

9. An aircraft in level flight at 100 kt encounters a wind gust that produces a load factor reading of 2.5. What would the *g*-meter have read if the aircraft had been travelling at 140 kt? Answer: 3.1

10. Estimate the height gain for a loop which begins at 150 kt and reduces to 80 kt at the top. Answer: 713 ft

11. Estimate the height gain for a tail slide maneuver that begins at 150 kt. Answer: 997 ft

12. A certain aircraft experiences a load factor of 2 when encountering a certain gust of air. If an ultralight with one-fourth the wing loading and sauntering along at one-half the speed had encountered the same gust, what load factor would it have experienced? Hint: Use ratios.

13. Sketch the maneuvering flight envelope for a Beechcraft Musketeer (utility category) with a stall speed of 72 mi/hr and a never-exceed speed of 171 mi/hr. Indicate as many numerical points on the envelope as you can, including an estimated value for the maneuvering speed.

14. The wings on a Cessna 150 weigh about 213 lb empty; full tanks add 135 lb. The engine and attached accessories weigh about 250 lb. Calculate the force on the engine mounts in an abrupt pull-up (**a**) at the gross weight of 1600 lb with full fuel tanks and (**b**) at a total weight of 1200 lb with nearly empty fuel tanks. Assume that the first pull-up gave a 6-*g* acceleration in the cockpit with full control deflection and that the pull-up was made at the same speed for the second one. Answers: 1500 lb, 2000 lb

15. Calculate the *g* loading for the engine mounts for the two cases of the preceding problem.

16. Use Fig. 11.1 (smooth condition) to estimate the lift coefficient for a 30° angle of attack for the 2412 airfoil if it had **no** stall break. Answer: 3.2

17. Use the result of problem 16 along with the calculated drag polar for the 152 to estimate its descent rate in a spin. Note: $AR = 6.93$, $S = 160$ ft², and $W = 1660$ lb. Answer: 6700 ft/min

18. Using an average lift coefficient of 1.5, estimate the radius of a spin in the Cessna 152. Answer: 180 ft

19. Draw the flight envelope for the Christen Eagle II which has $V_A = 135$ kt, $V_C = 150$ kt, $n_{MAX} = +7$ and -5, and $V_{NE} = 182$ kt. Determine and indicate the stall speed from the given data, assuming that it is the same for upright and inverted flight (symmetric airfoil).

20. Draw the flight envelope for the North American T-6 trainer which has $n_{MAX} = +5.67$ and -2.33, $V_{STALL+} = 78$ mi/hr, and $V_D = 240$ mi/hr.

21. The maneuvering speed of the Cessna 401A at a gross weight of 6300 lb is listed as 180 mi/hr. Estimate a safe maneuvering speed at a weight of 5200 lb. Answer: 164 mi/hr

22. The 7ECA, 7GCAA, and 7GCBC Citabrias have limit load factors of only +5.0 and –2.0 but they are approved for some aerobatic maneuvers because of the old original date of certification. V_A = 120 mi/hr, V_{NE} = 162 mi/hr. Draw the flight envelope.

23. For the aircraft of problem 22, estimate the load factor when doing a (sluggish) snap roll at the recommended entry speed of 85 mi/hr. Answer: 2.5

24. For the aircraft of problem 22, estimate the loop radius when entering a loop at the recommended 140 mi/hr with a 3.0 load factor. Answer: 655 ft

25. The SV4c Stampe (Pic. 2.5) is an open cockpit Belgian biplane designed in the 1930s as a training aircraft. It has a listed upright stall speed of 38 kt, an inverted stall speed of 62 kt, a never-exceed speed of 148 kt, and permissible load factors of –4 to +6. Draw the flight envelope. The maximum recommended entry speed for a flick roll is 65 kt; how does this compare with the estimated maneuvering speed?

References

1. Frost, Walter, Chang, Ho-Pen, McCarthy, John, and Elmore, Kimberly L., "Aircraft Performance in a JAWS Microburst," *Journal of Aircraft*, Vol. 22, No. 7, July, 1985.

2. Fujita, T. Theodore, "DFW Microburst," University of Chicago, 1986.

3. Doolittle, J.H., "Accelerations in Flight," NACA Report No. 203, 1925. (This is an early report on experimental load factors during maneuvering flight by a man who became a famous test pilot. His airplane, the Fokker FW-7 pursuit airplane, and the man himself appear in *Test Pilots*, by Richard P. Hallion, Doubleday and Company, Inc., 1981.)

4. Dearborn, C.H. and Kirschbaum, H.W., "Maneuverability Investigation of the F6C-3 Airplane with Special Flight Instruments," NACA Report No. 369, 1930. (Measured load factors in spins, loops and other maneuvers.)

5. Bairstow, Leonard, *Applied Aerodynamics*, Longmans, Green and Company, 1939. (Chapter V: "Aerial Maneuvers and the Equations of Motion," Speeds and forces on aircraft during loops, spins, and rolls.)

6. Speal, Daniel, "Another Spin Accident," *Sport Aerobatics*, July, 1972.

7. Scholl, Art, "Art Scholl Speaks on Flat Spins," *Sport Aerobatics*, July, 1972. (Early reports on flat spins in aerobatic aircraft. Art, the "Flying Professor," lost his life in an unexplained spin accident.)

8. Sanford, Earl L, "Are. . . Flat Spins Just Another Aerobatic Maneuver?" *Sport Aerobatics*, June, 1976. (Experiences with flat spins in a one-holer Pitts by an airshow pilot are related.)

9. *924 Flight Kit Manual*, Christen Industries, Inc., Hollister, CA. (This flight manual, for the Christen Eagle II, is probably the first to carefully describe how to enter and recover from intentional flat spins.)

10. "Hazards Associated with Spins in Airplanes Prohibited from Intentional Spinning," U.S. Department of Transportation, Federal Aviation Administration, Advisory Circular No. 61-67A, October 8, 1982.

11. DiCarlo, Daniel, Stough, H.P., III, and Patton, James M., Jr., "Effect of Wing Leading-Edge Design on the Spin Characteristics of a General Aviation Airplane," *Journal of Aircraft*, Vol. 18, No. 9, 1981, p. 786.

12. Anderson, Seth B., "Historical Overview of Stall/Spin Characteristics of General Aviation Aircraft," *Journal of Aircraft*, Vol. 16, No. 7, July, 1979. (Valuable survey, starting with the Wright brothers.)

13. Gobeltz, J., "Evolution of Spin Characteristics with Aircraft Structure", August, 1976. (French experiments, 1940 to modern; available from NTIS as NASA-TT-F-17123.)

14. Steinberger, J., "A Study of Lightplane Stall Avoidance and Suppression," February, 1977. (Available from NTIS as AD-A039223.)

15. Shreger, J., "Analysis of Selected General Aviation Stall/Spin Accidents," April, 1977. (Available from NTIS as AD-A040824.)

16. Rutan, Burt, "Design Considerations for Stall/Spin Safety," *Sport Aviation*, September 21, 1974. (Discussion of the considerations that led Rutan to his first canard aircraft, the VariViggen.)

17. Chambers, Joseph R. and Grafton, Sue B., "Aerodynamic Characteristics of Airplanes at High Angle of Attack," NASA TM-7409, 1977. (Available from NTIS as N78-13011.)

18. "Stall/Spin Flight Research Expanded," *Aviation Week & Space Technology*, September 11, 1978.

19. Staff, NASA Langley Research Center, "Exploratory Study of Wing Leading Edge Modifications on the Stall/Spin Behavior of a Light General Aviation Airplane," NASA TP-1589, December, 1979. (Study of a modified Yankee; available from NTIS as N80-13026.)

20. Bradshaw, Charles F., "A Spin-Recovery Parachute System for Light General-Aviation Aircraft," NASA TM-80237, 1980. (Reports the development of a spin-recovery parachute that was used in anger many times; document is available from NTIS as N80-20227.)

21. Stewart, Eric C., Suit, William T., Moul, Thomas M. and Brown, Philip W., "Spin Tests of a Single-Engine, High-Wing Light Airplane," NASA TP-1927, 1982. (Cessna 172 spin tests; document is available from NTIS as N82-16068.)

22. North, David M., "Wing Alteration Boosts Spin Resistance," *Aviation Week & Space Technology*, July 23, 1984. (Report on NASA's T-tailed, modified Piper Arrow.)

23. Kennedy, Donald A., "Spins, Will Your Aircraft Recover?" *Sport Aerobatics*, 1974

24. Taylor, Lawrence W., Jr. and Klein, Vladislov, "Analytical Techniques for the Analysis of Stall/Spin Flight Test Data," SAE Paper No. 810599, April 7–10, 1981.

25. Bihrles, William, Jr. and Barnhart, Billy, "Spin Prediction Techniques," *Journal of Aircraft*, Vol. 20, No. 2, February, 1983, p. 97.

26. Gregg, V.D., "Spin Research on a Twin-Engine Aircraft," AIAA Paper 81-1667, August 11–13, 1981.

27. Martin, Clark W., "F-14 Shows Dogfight Capabilities," *Aviation Week & Space Technology*, June 4, 1973.

28. Bihrle, William, Jr. and Meyer, Rudolph, C., "F-14A High-Angle-of-Attack Characteristics," *Journal of Aircraft*, Vol. 13, No. 8, August, 1976. "F-14 Demonstrates Agile Aerial Combat," *Aviation Week & Space Technology*, November 29, 1976.

29. Ropelewski, Robert P., "F-16 Displays Combat Capabilities," *Aviation Week & Space Technology*, May 28, 1979.

30. "FB-111 Bombers Playing Crucial Role, Strategic Air Command to Form New Training Unit," *Aviation Week & Space Technology*, June 16, 1980. (Discusses U2 and SR-71 high altitude surveillance aircraft.)

31. North, David M., "Navy F/A-18 Demonstrates Dual Mission Performance," *Aviation Week & Space Technology*, August 11, 1980. (These are all reports on spin and performance capabilities of modern military aircraft.)

32. "Fluid Dynamics of Aircraft Stalling," AGARD Conference Proceedings No. 102, April, 1972. (Available from NTIS.)

33. "Aircraft Stalling and Buffeting," AGARD LS-74, February, 1975. "Stall/Spin Problems of Military Aircraft," AGARD CP-199, June 1976. "High Angle of Attack Aerodynamics," AGARD LS-121, March, 1982. (These publications, by the NATO-related Advisory Group for Aerospace Research and Development, deal almost exclusively with military-related stall/spin problems; they are available from NTIS.)

34. Blanchard, W.S., Jr., "A Flight Investigation of the Ultra-Deep-Stall Descent and Spin Recovery Characteristics of a 1/6-Scale Radio-Controlled Model of the Piper PA38 Tomahawk," NASA CR-156871, March, 1981. (Available from NTIS as N81-21084.)

35. Scott, William B., "NASA Researches Aircraft Control During Deep Stall," *Aviation Week & Space Technology*, October 31, 1983.

36. Sim, Alex G., "Flight Characteristics of a Manned, Low-Speed, Controlled Deep Stall Vehicle," NASA TM-86041, August, 1984. (Available from NTIS.) (These references discuss the very recent work with controlled deep stall penetration for full-size aircraft.)

37. McCroskey, W.J., "The Phenomenon of Dynamic Stall," NASA TM-81264, March, 1981. (Available from NTIS as AD-A098191.)

38. Amtech Services, "What Did You Say? . . . 12 G's!" *Sport Aviation*, May, 1971. (Discussion of the flight envelope and the importance of balanced design strength in an aircraft.)

39. Nayler, J.L., "The Effect of Accelerations on Human Beings," *Journal of the Royal Aeronautical Society*, Vol. 36, No. 255, March, 1932.

40. Mohler, Stanley R., *Aerobatic Training: "G" Effects on the Pilot During Aerobatics*, Office of Aviation Medicine, FAA, Washington, D.C., 1972.

41. McNaughton, Grant B., "G-Induced Loss of Consciousness," *Sport Aerobatics*, August, 1983.

42. Weick, Fred E. and Hansen, James R., *From the Ground Up—The Autobiography of an Aeronautical Engineer*, Smithsonian Institution Press, Washington and London, 1988. (This is the fascinating story of a pioneer aeronautical engineer who was instrumental in the development of NACA cowlings, propeller theory, and the tricycle landing gear, as well as the design of various agricultural aircraft, the unspinnable Ercoupe, and the Piper Cherokee series.)

43. Scott, William B., "NASA Adds to Understanding of High Angle of Attack Regime," *Aviation Week & Space Technology*, May 22, 1989.

44. Lorber, Peter F. and Carta, Franklin O., "Airfoil Dynamic Stall at Constant Pitch Rate and High Reynolds Number," *Journal of Aircraft*, Vol. 25, No. 6, June, 1988, p. 548.

45. Hoblit, Frederic M., *Gust Loads on Aircraft: Concepts and Applications*, American Institute of Aeronautics and Astronautics, Inc., Washington, D.C., 1988. (This new text describes the current engineering practice in estimating gust loads on aircraft, placing special emphasis on continuous-turbulence gust loads rather than the simplified "sharp-edged" gusts discussed in this chapter.

46. U.S. Navy, *The Effects of Flight*, McGraw-Hill Book Co., Inc., 1943. (Advice on the effects of accelerations, as applied to naval aviators.)

47. Harvey, Eoin, "Physiological Effects Of Positive G Forces," *Sport Aerobatics*, December 1988, p. 8. (Read this before flying a high-*g* aerobatic maneuver.)

48. Rihn, Richard, "The Effects of High G Loading," *Sport Aviation*, January, 1989, p. 35. (Information needed by every pilot intending to do aerobatics. Rihn concludes with the advice "If a pilot decides to teach himself aerobatics with the aid of a textbook or video tape, he has a fool for an instructor.")

49. Frost, Walter and Camp, Dennis W., "Wind Shear Modeling for Aircraft Hazard Definition," Federal Aviation Administration, March 1977 Interim Report. (Available from the National Technical Information Service, Springfield, VA 22161.)

50. Dearborn, C. H. and Kirschbaum, H. W., "Maneuverability Investigation of an F6C-4 Fighting Airplane," NACA Report No. 386, 1931. (Graphs of position, speeds, and load factors during loops, half loop/rolls, spins, and snap rolls are presented.)

51. Johnson, A. M. "Tex" with Barton, Charles, *Tex Johnston: Jet-Age Test Pilot*, Smithsonian Institution Press, Washington, D.C., 1991.

52. Garrison, Peter, "Deep, Deep Stall", *Flying*, January, 1992. (A description of the deep stalls experienced by two canard-type aircraft and the fact that the low descent rate seem to require an hypothesis of a giant vortex being formed above the aircraft.)

53. Peery, David J. and Azar, J. J., *Aircraft Structures*, 2nd Edition, McGraw-Hill, 1982. (This is the updated version of a standard 1950s college textbook, not entirely an improvement. There is a good discussion of the correlation of the flight envelope with stresses on the wing structure.)

54. Strojnik, Alex, "Structural Testing of Homebuilts," *Sport Aviation*, March, 1992.

Books on Aerobatics for Pilots

A1. Lowery, John, *Anatomy of a Spin*, Airguide Publication, Inc., 1981. (Tail designs, inverted and flat spins, fuselage loaded versus wing loaded aircraft.)

A2. Mason, Sammy, *Stalls, Spins, and Safety*, McGraw-Hill, 1982. (From a test pilot viewpoint; includes a good list of references.)

A3. Williams, Neil, *Aerobatics*, Airlife Publications, 1975. (British test pilot and aerobatic contestant describes the lomcovak and other advanced aerobatic maneuvers such as tail slides; Aresti diagrams.)

A4. Medore, Arthur S., *Primary Aerobatic Flight Training with Military Techniques*, Ardot Enterprises, 1970. (Manual for Primary Aerobatic Training.)

A5. Cole, Duane, *Roll Around a Point*, Ken Cook, 1965. (Another primary training manual, this one from an old-time airshow pilot and master instructor.)

A6. Underwood, John W., *Acrobats In the Sky*, Heritage Press, 1972. (Interesting history of aerobatics.)

A7. Gentry, Everett, *All About Stalls & Spins*, Tab Books, Inc., 1983. (A useful book for non-aerobatic pilots.)

A8. Müller, Eric and Carson, Annette. *Flight Unlimited*, published by the authors, 1983. (A very useful discussion of competition aerobatics by a top Swiss aerobat, and the one who apparently originated the spin recovery technique popularized in the U.S. by Gene Beggs.)

A9. O'Dell, Bob, *Aerobatics Today*, St. Martin's Press, 1984. (Practical advice and instruction for pilots interested in competition aerobatics.)

A10. Thomas, Bill, *Fly for Fun*, published by the author, 1985. (A very nice introduction to aerobatic maneuvers by a master aerobatic instructor.)

A11. Carson, Annette, *Flight Fantastic*, Haynes, 1986. (A masterful 300+ pages of illustrated history of aerobatic flying, starting from the very beginning.)

A12. Kershner, William K., *The Basic Aerobatic Flight Manual*, Iowa State University Press, 1987. (A practical and very well written introduction to aerobatic flight in the Cessna Aerobat, by the premier author of pilot manuals and a very experienced instructor. Excellent illustrations.)

Chapter 17

Stability, Trim, and Control

We . . . resolved to try a fundamentally different principle. We would arrange the machine so that it would not tend to right itself.[1]

"Good" design implies flying characteristics that more or less automatically ensure that an aircraft flies straight and level hands off. In that case the (1916 Morane) Parasol was a thoroughly bad design, for she had only one position to which she automatically reverted and that was a vertical nosedive . . . You grabbed the stick rather hurriedly and pulled it back, whereupon it flicked into the opposite position and the old Parasol stood blithely on its tail! . . . By contrast the rudder was too small and sluggish in action, while the ailerons hardly worked at all. . . . It was a squadron rule never to turn the machine under 500 ft.[2]

In landing, the ship exhibited positive stability during the approach: increasing nose heaviness with decreasing speed. But following the flare-out, increasing elevator stick force reversal resulted in an eventual full forward control position to maintain the holding-off attitude.

In steep turns nose heaviness was exhibited initially, but as the Gs were increased, a point was reached at which the ship would want to root into the turn necessitating forward control pressure to prevent it from doing so.[3]

A. Longitudinal (Pitch) Stability and Trim

An aircraft is in equilibrium when the vector sum of the external forces and the vector sum of the moments caused by these forces are both zero. To give the pilot the ability to produce this equilibrium for all of the desired angles of attack is the problem of **trim**. To give the pilot an aircraft that tends to return to its original pitch angle, heading, and bank angle after a momentary disturbance (by turbulent air, perhaps) is the problem of **stability**.

In a conventional tail-last configuration, the elevator surface (Chap. 1) is the trimming surface that the pilot uses to vote for a desired pitch angle; in a tail-first (canard) configuration, the elevator serves the same function but is located on the trailing edge of the canard.

Suppose that you have built an untwisted wing of rectangular planform using the LS(1)-0417 airfoil section. Being the adventurous sort, you decide that you might as well do some

testing on the wing before you build the rest of the aircraft—so you attach handholds to the bottom of the wing and sally out to the nearest steep hill for some good old-fashioned hang gliding.

You had to do some heavy duty thinking before you knew where on the chord to attach the handholds. Looking at Figure 9.1d, you noticed that an angle of attack of 8.02° would give you a high ratio of lift to drag (to maximize your glide angle) and yet give you a comfortable separation of 11° from the stall angle of 19°. For trim, you know that you must hang below the *cp*, so you attach your handholds just one-third of the way back from the leading edge.

So now it is jumping off time and through some little stroke of luck you manage to begin your flight at the desired angle of attack. However, a little wind gust momentarily increases the angle of attack of your wing to 16.04° (Figure 9.1e). The result is that the *cp* moves forward and this causes an unbalanced, nose-up moment which tends to increase the angle of attack even more—so your wing quickly stalls above you. (A wind gust that reduced the angle of attack would have been no more pleasant; Figure 9.1c shows that this would have moved the *cp* backward so that the pitch-down of the nose would have continued until the negative stall occurred.) Your wing has turned out to be unstable in pitch and you would have been wise to have waited until the rest of the plane was ready!

---

[1]Wilbur Wright, quoted in John D. Anderson, Jr., *Introduction to Flight* (New York: McGraw-Hill, 1978).
[2]Cecil Lewis, *Farewell to Wings* (London: Temple Press Books, 1964).
[3]Moye Stephens, quoted in E. T. Wooldridge, *Winged Wonders: The Story of the Flying Wing* (Washington: Smithsonian Institution Press, 1983).

To make these ideas more concrete, let us design and fly a paper airplane using familiar airfoils and some leftover weightless beams from introductory physics. Figure 17.1 shows the airplane with its LS(1)-0417 [GA(W)-1] wing followed by a NACA 0012 stabilator tail. We have mounted the stabilator high above the wing to keep it clear of the downwash; therefore the dynamic pressure at the tail should be nearly the same as that influencing the wing. We choose an effective aspect ratio of 6.5 for the wing and the 0012 symmetrical airfoil with an effective aspect ratio of 5 for the stabilator. Figure 17.2 shows the assumed airfoil properties of the wing and stabilator; these curves have been derived from the two-dimensional characteristics shown in Fig. 9.7 and Ref. 4 in Chap. 10 and transformed through Eq. 12.11 to three dimensions.

Inspection of the lift curves of Fig. 17.2 shows that they are well represented in the linear regions by the following equations:

$$C_{L,W} = 0.30 + 0.10\,\alpha_W \tag{17.1}$$

$$C_{L,T} = 0.080\,\alpha_T \tag{17.2}$$

where the subscripts w and t refer to the wing and the tail respectively. Note that the 0012 airfoil produces no lift at $\alpha_T = 0$, as must be true for any symmetric section. From Fig. 9.5 we can

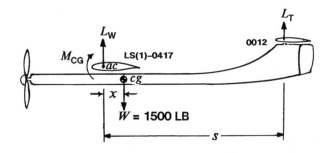

Fig. 17.1 Paper airplane in cruising flight at about 120 kt

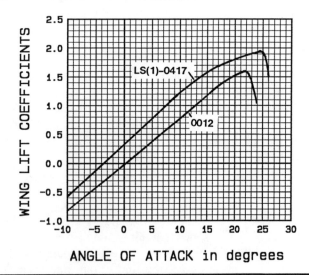

Fig. 17.2 Wing and stabilator lift curves for the paper airplane

see that the moment coefficient for the LS(1)-0417 about its one-quarter chord point is nearly constant at

$$C_{M,AC} = -0.10 \tag{17.3}$$

and we will consider it to be constant in the range of interest; the 0012 has a zero moment for zero lift because of its symmetry and therefore its moment coefficient about its one-quarter chord point is zero. The zero-lift moment for the wing under the assumed conditions is calculated as −3000 ft-lb (Table 17.1), where $\bar{c}$ is the average chord of the wing. (Note that we have shown M_{AC} with an arrow in the clockwise direction and L_T with an arrow in the upward direction; however, this signifies only the **sign conventions** — i.e., an upward force is defined to be a positive force and a clockwise moment is defined to be a positive moment. Since M_{AC} is negative, it is actually a counterclockwise moment and the actual direction of L_T is still to be specified or determined.)

WING PROPERTIES	TAIL PROPERTIES	120 kt CRUISE CONDITION
$S_W = 150\ ft^2$	$S_T = 30\ ft^2$	$q = 50\ lb/ft^2$
$\bar{c} = 4.0\ ft$	$s = 12\ ft$	$q\,S_W = 7500\ lb$
		$q\,S_T = 1500\ lb$
$M_{AC} = c_M q\ S_W \bar{c}$		
$= (-0.1)(50\ lb/ft^2)(150\ ft^2)(4.0\ ft) = -3000\ ft\text{-}lb$		

Table 17.1 Relevant Properties of the Paper Airplane

For **force** equilibrium in the vertical plane we must have

$$\sum F_Y = L_W + L_T + W = 0 \tag{17.4}$$

VERTICAL FORCE EQUILIBRIUM FOR THE PAPER AIRPLANE

For longitudinal torque equilibrium we must have

$$\sum M_{CG} = M_{AC} + L_W x - L_T (s - x) = 0 \quad \text{(at trim)} \tag{17.5}$$

Using Eqs. 8.6, 17.1, and 17.2 and the specified conditions, we can write this as

$$\sum M_{CG} = -3000\ ft\text{-}lb + C_{L,W}\,(7500\ lb)\,x$$
$$- C_{L,T}\,[18{,}000\ ft\text{-}lb - (1500\ lb)\,x] = 0 \tag{17.6}$$

MOMENT EQUILIBRIUM FOR THE PAPER AIRPLANE

where $C_{L,W}$ is the lift coefficient for the wing and $C_{L,T}$ is the lift coefficient for the horizontal tail.

Case Study No. 1: Zero Tail Lift in Cruise

If $L_T = 0$ then $C_{L,T} = 0$ and, also (by Eq. 17.2) $\alpha_T = 0$. Equation 17.4 then tells us that $L_W = W$ (i.e., the wing is doing all the lifting) and therefore

$$C_{L,W} = \frac{W}{q\,S_W} = \frac{1500\ lb}{7500\ lb} = 0.20$$

and, from Eq. 17.1,

$$\alpha_W = \frac{C_{L,W} - 0.30}{0.1} = -1.0°$$

ANGLE OF ATTACK OF WING IN CRUISE

Finally, Eq. 17.6 tells us that this zero tail lift equilibrium state requires that the center of gravity be located at

$$x = \frac{3000 \text{ ft-lb}}{C_{L,W}(7500 \text{ lb})} = \frac{3000 \text{ ft-lb}}{(0.20)(7500 \text{ lb})} = 2.0 \text{ ft}$$

REQUIRED LOCATION OF CENTER OF GRAVITY

which is just 3.0 ft back from the leading edge of the wing (since we are measuring from the *ac* at 1.0 ft back of the leading edge), or at $0.75\,\overline{c}$ in terms of the average wing chord. And so now our paper airplane is cruising along quite comfortably with the *cg* three quarters of the way back from the leading edge of the wing.

But now we must ask whether the equilibrium is a **stable** one. To determine this, we need to write the moment equation for an arbitrary angle of attack, assuming that the airplane has been adjusted or trimmed for level flight. This means that the angle of attack of the wing and the tail are related through

$$\alpha_T = \alpha_W + 1$$

because this is the only way we can have both the assumed zero lift on the tail ($\alpha_T = 0$) and the calculated $\alpha_W = -1.0°$.

With this trim condition, the moment about the *cg* (starting from Eq. 17.6 and using Eqs. 17.1 and 17.2) becomes, in general,

$$\begin{aligned}
\Sigma M_{CG} &= -3000 \text{ ft-lb} + C_{L,W}(7500 \text{ lb})(2.0 \text{ ft}) \\
&\quad - C_{L,T}[18{,}000 \text{ ft-lb} - (1500 \text{ lb})(2.0 \text{ ft})] \\
&= -3000 \text{ ft-lb} + (0.30 + 0.10\,\alpha_W)(7500 \text{ lb})(2.0 \text{ ft}) \\
&\quad - (0.080\,\alpha_T)[18{,}000 \text{ ft-lb} - (3000 \text{ ft-lb})] \\
&= 1500 \text{ ft-lb} + (1500 \text{ ft-lb})\,\alpha_W - (1200 \text{ ft-lb})\,\alpha_T \\
&= (1500 \text{ ft-lb})(1 + \alpha_W - 0.8\,\alpha_T)
\end{aligned}$$

We see that the sum of the moments about the *cg* is 0 for $\alpha_W = -1.0$ and $\alpha_T = 0°$, as before, but we also see that any disturbance that momentarily increases the angles of attack of the wing and the stabilator will also increase the moment to a positive value. Similarly, a disturbance that momentarily decreases the angles of attack below the trim values will change the moment to a negative value. Therefore the equilibrium is **unstable** because the generated moment tends to increase the acceleration in the same direction as the disturbance. Figure 17.3 shows a plot of this moment equation.

Case Study No. 2: Tail Lift = –150 lb in Cruise

If instead we arrange the *cg* so as to produce a download of 150 lb on the stabilator, the wing must generate a lift of 1650 lb (from Eq. 17.4) and so

$$C_{L,W} = \frac{1650 \text{ lb}}{7500 \text{ lb}} = 0.22$$

and, from Eq. 17.1,

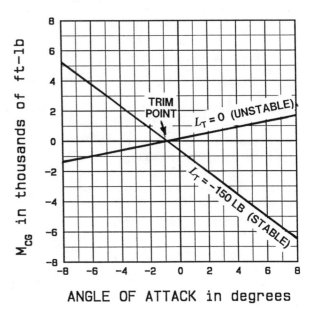

Fig. 17.3 Variation of pitching moment about the *cg* as a function of angle of attack for Case #1 and Case #2 paper airplanes

$$\alpha_W = \frac{0.22 - 0.30}{0.1} = -0.8°$$

while $\quad C_{L,T} = \dfrac{-150 \text{ lb}}{q\,S_T} = \dfrac{-150 \text{ lb}}{1500 \text{ lb}} = -0.10$

and, by Eq. 17.2,

$$\alpha_T = \frac{C_{L,T}}{0.08} = -1.25°$$

For this trim condition we see that

$$\alpha_T = \alpha_W - 0.45°$$

and the *cg*, obtained by solving Eq. 17.6 for *x*, is located at

$$\begin{aligned}
x &= \frac{(3000 \text{ ft-lb}) + C_{L,T}(18{,}000 \text{ ft-lb})}{C_{L,W}(7500 \text{ lb}) + C_{L,T}(1500 \text{ lb})} \\
&= \frac{3000 + (-0.10)(18{,}000)}{(0.22)(7500) + (-0.10)(1500)} = 0.8 \text{ ft}
\end{aligned}$$

which is 1.8 ft back from the leading edge of the wing, or slightly in front of mid-chord, at $0.45\,\overline{c}$, in terms of the average wing chord. The new moment equation becomes

$$\sum M_{CG} = (-595 \text{ ft-lb})(1 + \alpha_W)$$

and now we can see that our paper airplane has been transformed from an unstable machine to a strongly stable one, for now any change from the trimmed condition of $\alpha_W = -0.8°$ will produce a strong restoring moment. This new moment condition is also plotted on Fig. 17.3. Clearly, and this is a general rule, a stable airplane will have a negative slope for its variation of whole-aircraft pitching moment with angle of attack. Also, the steeper the negative slope, the more longitudinally stable is the machine.

The SLOPE of M_{CG} versus α must be NEGATIVE about the trim point where $M_{CG} = 0$.

CONDITION FOR STATIC LONGITUDINAL STABILITY

A slightly unstable aircraft can be flown successfully but it requires a good, attentive pilot. The Wright brothers intentionally chose an unstable configuration for all of their early designs in the belief that this was necessary to give them the control they needed to counter wind gusts such as the one that had killed Lilienthal. Motion pictures of their flights do show distinct undulations in pitch (Ref. 7).

You're probably ready to conclude by now that the only way to obtain an aircraft that possesses positive pitch stability is to arrange the *cg* so as to put a download on the tail in flight. Not quite! The real criterion for longitudinal stability, for our paper airplane (based on using the calculus with $\sum M_{CG}$ from Eq. 17.5), is that

$$a_T \frac{S_T}{S_W} \left(s - x \right) > a_W x$$

SIMPLIFIED CRITERIA FOR LONGITUDINAL STABILITY

where a_T is the lift slope of the horizontal tail and a_W is the lift slope of the wing. For our Case #1, the left hand term is 0.16 and the right 0.20; for our Case #2 the left hand term is 0.18 and the right 0.08, confirming our previous analysis. However, Case #1 (with zero tail lift) can be made stable by increasing the (S_T / S_W) ratio from 0.2 to greater than 0.25. This ratio is commonly about 15% on factory aircraft and was 20% on our paper airplane, but free-flight model airplanes sometimes design for a ratio of 35–40% and thereby obtain stability with a lifting tail. A tandem-winged aircraft like the Quickie has (S_T / S_W) equal to about 50%, both surfaces lift, and the stability is positive so long as the designer's *cg* range is maintained.

The location for the *cg* which yields neutral stability is naturally enough known as the **neutral point** (*np*). For our paper airplane, a little calculus allows one to determine that

$$x_{NP} = \frac{a_T S_T s}{a_W S_W + a_T S_T}$$
$$= \frac{(0.08)(30 \text{ ft}^2)(12 \text{ ft})}{(0.1)(150 \text{ ft}^2) + (0.08)(30 \text{ ft}^2)} = \frac{28.8}{15 + 2.4}$$
$$= \textbf{1.66 ft}$$

where a_T is the lift slope for the horizontal tail surface and a_W is the lift slope for the wing. Thus, in terms of the mean chord, our paper airplane's *np* is located 2.66 ft from the leading edge, i.e., at $\mathbf{0.66\, \bar{c}}$.

A little thoughtful pondering should convince you that the neutral point can be considered as the **aerodynamic center** for the whole airplane, for if the *cg* is located there, any changes in lift will act through that point and no moments (restoring or otherwise) will result.

Any *cg* location ahead of the *np* is stable and the degree of stability is directly related to the ratio of the distance ahead of the *np* to the mean aerodynamic chord (*MAC*) of the wing (which is the same as the mean geometric chord, $\bar{c}$, for simple rectangular wings). This ratio is called the **static margin**.

EXAMPLE 17.1 Calculate the static margin for the Case #1 paper airplane.

Solution: The *cg* was found to be at $x = 2.0$ ft while $x_{NP} = 1.66$ ft. Therefore

$$\text{static margin} = \left(\frac{x_{NP} - x}{\bar{c}} \right) \bar{c} = \left(\frac{1.66 - 2.0}{4.0} \right) \bar{c} = \mathbf{-0.85\, \bar{c}}$$

(or –85% MAC)   where the negative sign reveals the negative stability.

A large static margin improves stability but hurts maneuverability and is inefficient because the large download on the tail must be countered by extra lift on the main wing, with an accompanying increase in drag. A very small static margin is more efficient but makes the pilot work harder — and may give unsatisfactory stall or spin characteristics. The newest jet fighters are trying to get the best of both worlds by using a negative static margin for maximum efficiency and maneuverability while using an interposed computer to give the appearance of stability. We have indeed made a full circle back to the earliest aircraft! Perhaps some day soon this technology will work its way down to aircraft for which the pilot can afford to pay for the fuel.

The extra drag due to the download on the tail of an aircraft is known as **trim drag**. (A canard aircraft has zero trim drag in the sense that both surfaces contribute positive lift — but the canard has to be designed to operate at a greater wing loading (lb/ft²) than the wing (to help ensure that it stalls first, looking at Eq. 12.16) and this typically causes it to have a rather poor [*L/D*] in cruise.) Evidently a conventional aircraft has minimum trim drag when its *cg* is at the very rear of the *cg* range.

The forward limit for the *cg* is usually determined by the need to be able to raise the nosewheel before the stall speed is reached on takeoff in a tricycle-gear airplane, or the need to be able to land in a three-point, fully-stalled attitude in a tailwheel-type aircraft. The author recalls the example of one experimental aircraft that suffered through a heart transplant operation in which it gained a Corvair engine; upon recovery it was asked to make a trip around the pattern without the benefit of a *cg* check and the result was full back stick for the duration of the flight. At least the machine was stable!

"Tuck-under" is a compressibility problem that apparently first showed up on the twin-tail P-38 (a heavy, twin-engine fighter of World War II). In a high speed dive, the *ac* moved rearward as shock waves formed; this increased the stability to the point where the tail, operating in separated air, didn't have enough authority to keep the nose from pitching down even more. Several aircraft were lost due to this tuck-under phenomenon. Drag flaps that also maintained reasonable flow over the tail solved the problem for the P-38; later jet fighters introduced stabilators to increase the power of the pitch control surface.

The instability of a swept-wing aircraft at high angles of attack (causing it to pitch up at the stall onset) results in a moment coefficient curve like that shown in Figure 17.4.

What modifications to our paper airplane must be made to convert it to real life? Two of the major influences which we neglected are the effect of the fuselage and the thrust from the power source. The news is mostly bad, for both are usually destabilizing so that an even more forward *cg* location will be needed for stability, in general.

It is not difficult to see why the fuselage is a destabilizing influence; it is similar to what happens when you point your hand into the oncoming air from a car or an open-cockpit aircraft. Another viewpoint is that the aerodynamic center for a symmetrical fuselage is at about one-fourth of the way back from the nose, and this point is ahead of the *cg*. (Because most of the area of the fuselage is in the front half, the actual situation is somewhat worse than this.) Thus the long forward fuselages of some modern airliners must be countered with larger vertical tails.

A propeller or jet engine inlet is also destabilizing if it is in front of the *cg*. To see why, consider what happens when the aircraft momentarily pitches up in air turbulence. The air will then make a downward turn to get into the propeller or the jet inlet, and by Newton's third law there is a reaction force up on the aircraft—i.e., a destabilizing pitch-up moment is generated. A pusher aircraft such as the Long-EZ or Beechcraft Starship obtains a stability increase from its pusher engine(s), on the other hand. And when the 1948 Northrup YB-35 flying wing bomber was converted from pusher propeller power to jet engine power, becoming the YRB-49A, large vertical surfaces had to be added.

When the pilot pulls back on the control stick or wheel, the download on the tail is increased. This moves the whole-aircraft moment coefficient curve of Figure 17.3 more to the right, so that the aircraft becomes trimmed at a higher angle of attack. If the pilot persists, the main wing (in a conventional aircraft) will eventually stall and the downwash angle will suddenly become much less as lift is lost. For a low-tailed aircraft, the tail will suddenly have air coming at it with a more positive angle of attack and so both the tail and the wing will combine to pitch the nose back down where recovery is possible. With a T-tailed aircraft (having the horizontal stabilizer on the top of the vertical surface), however, the tail will be out of the wake and not affected by the downwash as the angle of angle of attack is being increased. But at the stall the T-tail can suddenly be immersed in separated, low energy air and if the aircraft also has swept wings and the usual pitch-up tendency at the stall, the pitch attitude will increase beyond the stall and become stable at an even higher angle of attack. This is the deep stall discussed

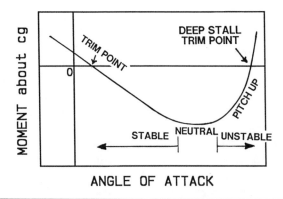

Fig. 17.4 General shape of the pitching moment curve for a simple swept-back wing

in Chap. 16. The rule is that, for any 2-surface configuration, the **rear** lifting surface must **never** be allowed to stall!

Choosing a lower aspect ratio for the horizontal tail surface than for the wing helps ensure that the wing stall first, as inspection of Eq. 12.11 and Fig. 12.6 remind us. Recall too the story of the reverse slot that was added to the Cardinal's stabilator to prevent it from stalling, as mentioned in Sec. F. of Chap. 12.

The advantages of a T-tail include the absence of power effects and greater effectiveness because of the availability of nearly the same dynamic pressure as seen by the wing, which translates into a smaller (and less draggy) horizontal surface. The joker is that the tail cone and vertical fin must be beefed up to handle the increased bending moment, and the result may be a net increase in weight. (See Ref. 2 for an estimate of the pre- and post-T-tail Cherokee Arrow.) Also, the takeoff roll may be lengthened if the machine really needs the extra power-on air blast to lift a heavy nosewheel.

For most aircraft, the horizontal tail surface will be immersed in low energy air in the wake of the wing and fuselage so that the effective dynamic pressure at the tail is much less than that at the wing; in fact, the tail may need to be as much as twice as large as otherwise estimated. It seems to be a rule of thumb that newborn aircraft have a smaller tail than their adult incarnation.

The presence of downwash from the wing also means that a stabilizer apparently pointing in the direction of flight actually has a negative angle of attack. The actual angle of attack for the horizontal tail can be roughly estimated from

$$\varepsilon_{\text{TAIL}} \approx -\left(\frac{4}{AR + 2}\right)\alpha_{\text{WING}} \qquad (17.7)$$

APPROXIMATE DOWNWASH ANGLE AT THE TAIL

See Refs. 2 and 9 for more detailed information.

We have really spoken only of static stability. When a disturbance produces a change in the angle of attack, the statically stable aircraft will generate a restoring moment; however, if left to itself the aircraft will overshoot its trim point, and then tend to return but again overshoot, and so forth. If these pitch oscillations increase in amplitude, the aircraft is said to be statically stable but **dynamically** unstable. In any case, an aircraft must be statically stable if it is to be dynamically stable, and increased static stability tends to promote dynamic stability. Figure 17.5 (from Ref. 8) shows the dynamic stability possessed by the VE-7 advanced-training aircraft (in 1923) and the less satisfactory dynamic stability possessed by an improved version of the JN4h "Jenny." That study concluded that static stability was so much more important than dynamic stability that the latter normally didn't have to be investigated except for a radical new design. Presumably the pilot has no business sleeping through more than one oscillation anyway.

It would be nice to have some measure of the effectiveness of the tail in countering the destabilizing influence of the fuselage and the wing. Specifically, we would like to isolate geometric effects from aerodynamic ones. In making our calculations with the paper airplane, we saw that the moment contributed by the tail was determined geometrically by the product of its surface area (S_T) and its lever arm (s). On the other hand, the moment generated by the wing depends on its surface area (S_W) and its mean chord ($\bar{c}$). Therefore we can reasonably expect that the required size of a horizontal tail should be related to the **ratio** of these factors:

$$\boxed{\textbf{TAIL VOLUME COEFFICIENT} \equiv \frac{S_T s}{S_W \bar{c}}} \quad (17.8)$$

The tail volume coefficient normally has a value of about 0.2 to 1.2, with sport aircraft at the low end and transport aircraft at the high end. A large tail volume coefficient permits the widest range of acceptable *cg* locations and permits the use of high lift devices, at the sacrifice of some efficiency.

Reference 3 suggests that most modern airplanes have a *cg* range of about 0.16 $\bar{c}$ to 0.28 $\bar{c}$. For instance, the Pitts S-2A aerobatic aircraft specifies an allowable range of 0.163 $\bar{c}$ to 0.287 $\bar{c}$ for aerobatics and 0.163 $\bar{c}$ to 0.296 $\bar{c}$ when operated in the normal category.

A separate elevator hinged to a fixed stabilizer is a simpler design in many ways than a stabilator but it is not as effective and will generally require a larger total surface area. The tradeoff with a stabilator is extra complexity. The stabilator is normally mounted near its aerodynamic center to minimize the changes in stick forces with deflection but then a balancing weight must extend forward into the fuselage to provide the mass balance required to forestall flutter. Also a tab linked to the fuselage is used for trimming and to adjust the stick forces.

There are actually two types of longitudinal stability that are important. One is **stick-free** and the other is **stick-fixed**. The first refers to the stability when the pilot is letting the plane fly itself and the second is the stability when the pilot is holding on to the stick but doing nothing with it.

We have seen that stability exacts a drag penalty because the extra lift required of the wing naturally increases its drag also. (The problem is really severe for supersonic airplane because the rearward shift of the aerodynamic center in going

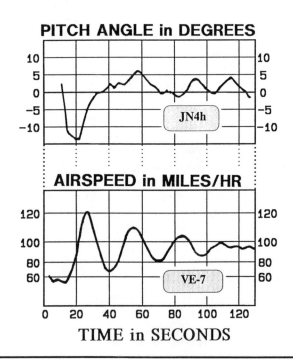

Fig. 17.5 Experimental stability measurements for two old aircraft

supersonic increases the stability greatly—so the *cg* is also usually shifted rearward in flight to compensate.) To minimize this trim drag penalty, both the fly-by-wire aircraft and high performance sailplanes flirt with or even embrace instability to maximize performance or maneuverability. Most aircraft, though, simply accept the trim drag as a necessary part of stability. Even for the marginally stable aircraft, though, there is a considerable amount of fuselage that is doing nothing more than connecting the load-carrying part of the fuselage to the tail, and this leads to the next topic.

B. Canards and Flying Wings

If we place the smaller horizontal surface, with the pitch-control surfaces, at the front of the airplane and the larger surface at the rear, we have created a canard-type aircraft.

Case Study No. 3: A smaller, lifting front wing

Suppose we modify the properties of our paper airplane of Fig. 17.1 by specifying that $S_W = 30$ ft^2, $\bar{c} = 3$ ft, and $S_T = 150$ ft^2. Then the zero-lift moment of the canard becomes

$$M_{AC} = C_M q S \bar{c} \quad (17.9)$$

$$= (-0.1)(50 \text{ lb/ft}^2)(30 \text{ ft}^2)(3 \text{ ft}) = \textbf{-450 ft-lb}$$

Next let us calculate the *cg* position and the moment equation if the lifting front wing (canard) supports 450 lb (canard loading = **15** lb/ft^2) and the rear wing supports the remaining 1050 lb (wing loading = **7** lb/ft^2). Also, ($q S_{CANARD}$) = 1500 lb and ($q S_{REAR WING}$) = 7500 lb. Then we have

$$C_{L, CANARD} = \frac{L}{q S_{CANARD}} = \frac{450 \text{ lb}}{1500 \text{ lb}} = \textbf{0.30} \quad (17.10)$$

$$C_{L, \text{CANARD}} = \frac{L}{q \, S_{\text{CANARD}}} = \frac{450 \text{ lb}}{1500 \text{ lb}} = \mathbf{0.30} \qquad (17.10)$$

using Eq. 17.1, $\quad \alpha_{\text{CANARD}} = \dfrac{0.30 - 0.30}{0.1} = \mathbf{0.0°}$

and, using Eq. 17.5,

$$\sum M_{\text{CG}} = -450 \text{ ft-lb} + (450 \text{ lb})x - (1050 \text{ lb})(12 \text{ ft} - x) = 0$$

or $\qquad\qquad x = \mathbf{8.7 \text{ ft}} \qquad\qquad (17.11)$

which means that the cg is about two-thirds of the way back from the canard to the rear wing. Also,

$$C_{L, \text{REAR WING}} = \frac{1050 \text{ lb}}{7500 \text{ lb}} = \mathbf{0.14}$$

and, from Eq. 17.2, $\alpha_{\text{REAR WING}} = \dfrac{0.14}{0.08} = \mathbf{1.75°}$

so that $\alpha_{\text{REAR WING}} = \alpha_{\text{CANARD}} + 1.75°$

Then, from Eq. 17.5,

$$\sum M_{\text{CG}} = -450 \text{ ft-lb} + C_{L, \text{CANARD}} (1500 \text{ lb}) \, x$$
$$- C_{L, \text{REAR WING}} [90{,}000 \text{ ft-lb} - (7500 \text{ lb}) \, x]$$
$$= -(675 \text{ ft-lb}) \, \alpha_{\text{CANARD}} \qquad (17.12)$$

and we have a stable aircraft once again!

On first sight, it appears that all aircraft should lead with their tails, for we have achieved stability with both horizontal surfaces producing positive lift. However, there are many problems with the canard configuration that have restricted its use. For one thing, the large rear wing is operating in the downwash of the front wing and therefore its efficiency tends to be significantly reduced. This penalty can perhaps be largely removed by very carefully tailoring the part of the rear wing that is beyond the front wing so as to utilize the upwind wingtip vortices to gain extra lift, as Burt Rutan does with his VariEze and Long-EZ.

It is also difficult to obtain good takeoff and landing performance with canard aircraft because the large rear wing cannot be allowed to get too close to its maximum lift coefficient lest it stall and turn the machine into a tailsitter. Performance in turns tends to be degraded because the induced drag builds up more quickly with the heavily loaded canard. Finally, the front elevator normally doesn't have enough power to handle high lift devices, although the new Beechcraft Starship (in the design of which Rutan played a major role) uses a sweepable canard to partially get around this limitation. In any case, the VariEze/Long-EZ aircraft are extremely efficient in cruise with their low weight (thanks in part to having a fuselage which is entirely used) and, with its high aspect ratios, very good high altitude performance. Above all, a properly designed and constructed canard aircraft, loaded properly, can stall only its front lifting surface and therefore can perform steep turns and climbs with the elevator at maximum deflection without fear of losing a lot of lift or of spinning—so it has the potential to be a much safer aircraft. Certainly you can expect to see many more

canard aircraft in the sky in the future (although plans for Rutan aircraft are no longer available).

The Quickie, more of a tandem-wing aircraft than a canard, has the disadvantage of having great sensitivity to the relative angles of incidence of its two wings—and they are bonded in place before flight tests can be made. If the front wing has too little incidence relative to the rear, the machine will lack steering effectiveness upon landing because the rear wing is still lifting and keeping weight off the tailwheel; also, the aircraft will be very elevator limited when rain or bugs produce boundary layer separation on the front airfoil (see Sec. 10E). But if the front wing has too much incidence, performance will suffer from additional drag. A partial solution has been to adjust both ailerons on the rear wing so as to change the pitching moment there. Thus if both ailerons are moved up only 1/8" or so, the elevators (on the canard) will ride noticeably higher in cruise. There is a strong pitch-up trim change, and the longitudinal stability is reduced. Plans for the Quickie are also no longer available.

In 1981 an experimental, 4-place, canard-type, tractor aircraft that showed considerable promise crashed during a test flight. It appears that vigorous (destabilizing) power-on maneuvering near the stall had encouraged the angle of attack to go well beyond the normal stall angle of the canard but that the particular canard airfoil being used started to regain lift there (which is true for many airfoils) and was thereby able to force the aircraft into a stable deep stall with both wings thoroughly stalled. Figure 17.6 presents the results of a full-scale wind tunnel investigation of the pitching moment characteristics of this aircraft along with those of a VariEze (Ref. 10). Note how the power-off pitching moment curve for the tractor canard has a positive (unstable) slope for an angle of attack exceeding about 17° and then it nearly reaches a second trimmed point at an angle of attack of about 33°. Figure 17.7 shows the strongly destabilizing influence of power for the tractor canard and the positive, stabilizing effect for the pusher-powered VariEze. Two more recent (survivable) deep stall accidents have involved canard aircraft getting into stable deep stalls with angles of attack close to an astonishing 80°. (See also Chap. 16.)

Canard-type aircraft are back and surely here to stay. Almost every advanced fighter design incorporates them to aid

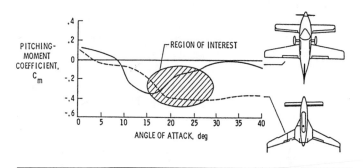

Fig. 17.6 Comparison of pitching moment data for two canard configurations for power off and with neutral controls (Ref. 10)

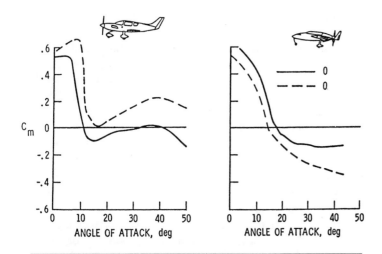

Fig. 17.7 Effect of tractor and pusher power on the pitching moment coefficient for two canard configurations (Ref. 10)

and a tail! Moving part of the wing forward as a fixed lifting canard, while leaving the pitch control surface in the rear, permits the main wing to be mounted farther back on the fuselage, allowing passengers to sit mainly in front of the wing for a better view. It also may have some induced drag advantages over the conventional wing/tail arrangement.

The author's first honorable aerodynamics experiment occurred with the design and construction of a little model flying wing and the early discovery that it would nose up or down very abruptly almost as soon as it was released for gliding flight. Finally perceiving that it seemed to prefer inverted flight, it was determined that low performance (but stable) glides could be accomplished by releasing the model inverted. This shouldn't be too much of a surprise to you at this point but you are encouraged to design and fly a paper airplane for verification, if you wish.

There are much more efficient ways to get a stable flying wing than negative camber airfoils, though. All that is necessary is to obtain a decreasing front-to-rear lift distribution. An airfoil that is turned up at the back (reflexed) or a wing that is swept and perhaps contains wash-in or a wing that has a fancy variation of airfoil with span will do it. However, all these tricks reduce the efficiency of the lifting surface and the benefits of having a single, all-lifting surface are largely lost. The principle is most attractive for very large aircraft where the wing must be very large anyway. Most actual flying wing aircraft have suffered from marginal stability (especially laterally), marginal control, poor takeoff performance, and a very limited *cg* range. Reference 2 is a highly recommended book that provides a panoramic view of the long history of flying wings, paying special attention to the famous Northrup Flying Wings. The second quotation at the beginning of this chapter came from a test pilot experiencing pitch-up while flying one of the early Flying Wings.

In 1989 Northrup rolled out its new flying wing, the B-2 "Stealth" bomber. By avoiding vertical surfaces, as well as any large continuous surface, and using absorptive materials, its radar image is very small—much better even than the original

flying wing bombers. Stability and control are obtained through redundant computers in a fly-by-wire system. Elevons (combined elevators/ailerons) are used inboard on the trailing edge of the wing and two-surface drag rudders/spoilers are used outboard, much like the earlier Northrup Flying Wings.

C. After the Stall Is Over

> After the stall is over,
> After the wheel's full aft,
> What is the nose to do:
> Will it go up? Will it go down?
> That is the question now.
>
> After the stall is over,
> After the pressures are down,
> What is the wing to do:
> Is it to lift? Is it to quit?
> Where should it go from here?
>
> After the stall is over,
> After the sky's the view,
> What is the tail to do:
> Is it to push? Is it to pull?
> Are its troubles now o'er?
>
> After the stall is over,
> After the wings best push,
> What is the air to do:
> Will it go over? Will it go under?
> Does it even care?
>
> After the stall is over,
> After the forces have moved,
> What is the 'd 'amicist to use:
> Will it be *cp*? Will it be *ac*?
> Where will the moment go?
>
> After the stall is over,
> After the buffet is through,
> What is the pilot to do:
> Is she to push? Is he to pull?
> What does it matter now?

Most student pilots approach the stall maneuver with considerable trepidation. Their previous experience with "stalls" probably includes embarrassing instances of momentary engine stoppages while driving a car. Now they are told that the aircraft engine won't stop running but the airplane wing will "stop flying" when a certain critical angle of attack is attained. They are told to perform the maneuver by reducing the power to idle and then trying to maintain altitude by gradually bringing the wheel or stick back toward them, causing the airspeed to begin a steady march toward the beginning of the green arc.

Usually the pilot feels some aerodynamic buffeting of the tail (through the controls and in the fuselage) as the wing root stalls first, but continued back pressure causes enough of the wing to stall to force the nose of the aircraft to abruptly pitch down through the horizon. This is the scary part!

For one thing, the sudden pitching motion, replacing a windshield full of blue sky with one full of ground, reminds us that steep descents to the earth can be hazardous to our health.

For another, the sudden reduction in wing lift causes a downward acceleration as well as a reduction in lateral stability that tends to cause a wingtip to drop. But worst of all, perhaps, the controls have "reversed." The rearward motion of the control column, which in the past always raised the nose, suddenly has caused the nose to drop away!

The student pilot needs to understand that the stall is an important maneuver because every normal landing is at least an approach to a stall, and because the pitching motion and the loss of altitude and the possible roll-off at the stall **are** potentially hazardous if they occur at low altitude.

The recommended solution is to recognize the first indication of a stall and be prepared to quickly recover to nose-level, wings-level flight with minimum loss of altitude. But it is important to know that the stall can occur at any airspeed and at any pitch attitude, requiring only an excursion beyond the critical angle of attack, as we have discussed earlier. Also, an aggravated stall is the precursor of a spin or snap roll entry, an especially likely event in aircraft designed in the 1940s and earlier, and so the debate rages on as to whether all pilots (and not only flight instructors) should be required to demonstrate the ability to enter and properly recover from normal (upright) spins.

From an aerodynamics and physics standpoint, the pitch-down at the stall is of great interest. Evidently, at the stall, the aerodynamic forces change such that the net pitching moment about the cg suddenly decreases from zero to distinctly negative (nose-down). (The only alternatives to this pitch-down at the stall are to provide insufficient elevator power, which significantly limits the cg range and aircraft efficiency, or else to have the nose pitch up at the stall — into a deep stall — leading to the distinct possibility of a stable, unrecoverable departure from controlled flight.)

Let us then perform the stall maneuver with the stable paper airplane of Section A and see how it responds. Figure 17.8 displays the previously calculated locations for the cg and the np. The aerodynamic center, as is usual for low speed airfoils, is located very close to the one-quarter chord point.

We look now at the changing forces on the wing and the tail as our paper airplane transitions from its high speed cruise configuration to an angle of attack beyond the stalling angle, maintaining stable, level flight at each of four angles. First we use the center of pressure approach, employing Fig. 9.1 and the approximation of a constant moment coefficient (about the quarter-chord point) to locate the cp. From the definitions (Chap. 8), the following relationship can be derived:

$$x_{CP} = 0.25\,\overline{c} - \left(\frac{c_m}{c_\ell}\right)\overline{c} \qquad (17.13)$$

where x_{CP} is the distance of the cp from the leading edge. Thus, for the paper airplane in the cruise configuration,

$$x_{CP} = 0.25\,\overline{c} - \left(\frac{-0.10}{0.22}\right)\overline{c} = .7045\,\overline{c}$$

$$= (0.7045)(4\text{ ft}) = 2.818\text{ ft} \quad (1.018\text{ ft from the }cg)$$

and so forth for the other locations.

Figure 17.9 presents this safe, sane stall maneuver. In 17.9a, the cp is behind the cg, so the horizontal tail must generate a negative lift force to make the moments about the cg equal to zero. Figure 17.9b is for a lift coefficient of 1.6, which (by Eq. 17.3) corresponds to a cp location 0.55 ft in front of the cg. With this cp location, the tail must generate a positive lift force to provide angular equilibrium. The stability is still positive, though; it can be determined that the tail moment increases more than the nose moment if a wind gust suddenly increases the angles of attack, and the paper airplane will tend to return to its trimmed angle of attack.

In Fig. 17.9c, the maximum lift coefficient (stall angle of attack) is reached at a value assumed to be 1.9. Thus our paper airplane's stall speed is

$$V_{\text{E, STALL}} = \sqrt{\frac{2\,L}{C_{\text{L, MAX}}\,\rho_0\,S}} = \sqrt{\frac{(2)(1425)}{(1.9)(0.002377)(150)}}$$

$$= 64.9\text{ ft/s} = 38\text{ kt}$$

Figure 17.9c shows why the nose pitches down so abruptly at the stall (to the extent that the whole wing stalls at the same time). As the critical angle is exceeded, the wing lift will drop and the center of pressure will move back, both of which reduce the nose-up pitching moment which it was providing, and so the tail lift, which is undiminished, pitches the nose down.

However, if somehow the lateral instability at the stall can be overcome (as by fly-by-wire) and if the very large drag force can be overcome with a powerful engine, it would be possible to maintain level flight at an angle of attack 2° greater than the critical angle (Fig. 17.9d). Evidently this tail has plenty of available power to lower the nose anytime it wishes.

Even though Fig. 17.9 shows the angles of attack of the tail in slow flight and in the stall to be small, the angles relative to the axis of the fuselage are constantly increasing in a negative sense — i.e., the pilot is indeed pulling the stick or wheel back. In slow flight the tail angle relative to the fuselage is about −10° and at the stall it is about −18°, for example.

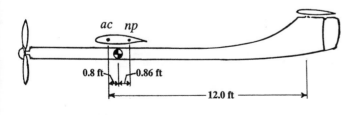

Fig. 17.8 Locations of the center of gravity and the aircraft aerodynamic center, stable paper airplane

The approach to wing and tail forces using the aerodynamic center concept yields equivalent results, as it must, but is so much simpler that you are encouraged to prefer it (Fig. 17.10). Here the lift force act at a fixed point, the aerodynamic center, while the constant moment coefficient ($C_M = -0.1$) about that point provides the compensation for the actual lift force not acting there. However, at angles of attack beyond the stall the moment coefficient increases rapidly in a negative sense, explaining differently but equivalently why the nose pitches down then (Fig. 9.3). Even though the moment coefficient about the *ac* is constant throughout the normal operating range, the moment about that point keeps decreasing as the speed decreases because the dynamic pressure is decreasing ($M_{AC} = C_M q S_W \overline{c}$).

Most aircraft probably spend most of their flying lives with the cg close to the *ac*. The *ac* approach of Fig. 17.10 makes it very clear that this means that the tail load will be negative (a download) **throughout** the normal operating angles of attack because then the tail must balance only the nose-down moment created by the moment coefficient. This tells us plainly enough

that the more forward the cg, the more the download on the tail will become and the more trim drag we can expect.

Symmetrical airfoils or those with negligible moment coefficients behave much like the highly cambered, aft-loaded airfoil around which the paper airplane was designed, except that the *cg* must be in front of the *ac* to obtain positive stability. Even though a tail download in cruise is still therefore a requirement for pitch stability, it turns out that not as much download is required for a given amount of stability, in general.

We have made the important simplification of ignoring the fact that the tail receives deadened air that has a downward component. The effect of this downwash from the wing can be expected to reduce the elevator deflection required but the lower dynamic pressure will tend to increase it. Also, if the tail receives speeded-up air when a power-on stall is executed, the tail will have enhanced effectiveness.

Just what does happen to the lift and drag coefficients if you somehow manage (perhaps while trying to cut falling tp?) to get your airplane into an angle of attack much greater than the stalling angle? Figure 17.11 provides an answer, based on experimental measurements in a wind tunnel of the lift and drag

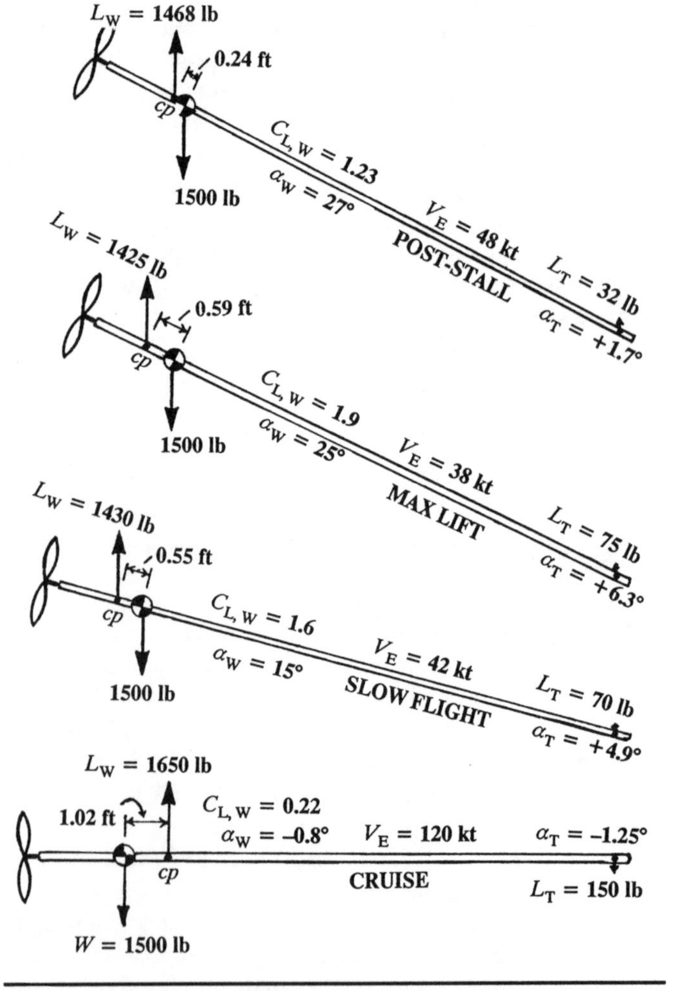

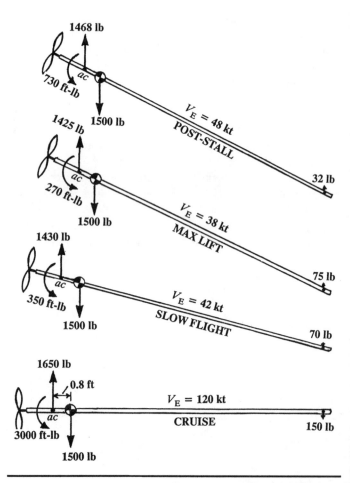

Fig. 17.9 Forces on the wing and horizontal tail of the paper airplane at four different speeds and angles of attack (center of pressure viewpoint)

Fig. 17.10 Forces on the wing and horizontal tail of the paper airplane at four different speeds and angles of attack (aerodynamic center viewpoint)

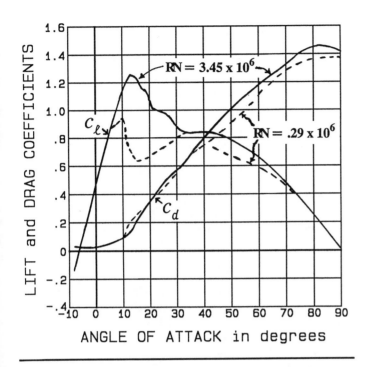

Fig. 17.11 Lift and drag coefficients all the way to 90° (Ref. 17)

on a fuselage-wing combination (no tail) over the whole angle of attack range up to 90°! (See Ref. 17.)

For this aircraft model, the results are strongly dependent on Reynolds number. At low Reynolds number (here equivalent to a 1 ft chord wing or canard at a speed of 26 kt), the BL separates prematurely at an angle of attack of about 10° but it then starts lifting again and reaches a second maximum at an angle of attack of 45°, as anticipated in the previous section. This is sometimes called flat plate lift because it is what a flat plate, with totally separated flow, would do. As expected for "flat plate" lift, there is mirror symmetry in the lift curve about the 45° point. Note that, independent of Reynolds number, the lift falls to zero for a 90° angle of attack and the drag continues to increase, as expected from their definitions as the components of the resultant aerodynamic force.

A related topic is the sensitivity of the stall speed to *cg* location. Some aircraft flight manuals give this information. It should be clear from the discussion surrounding Figs. 17.9 and 17.10 that any forward motion of the *cg* causes the wing load at the stall to increase, but the actual increase depends on the details of the airfoil used and the relative locations of the wing's and the tail's aerodynamic centers, as well as the complication that the manual's "stall speed" may be only a minimum speed because of a reduction in elevator power at the forward *cg* locations. However, it is possible to derive an equation that gives an estimate of the magnitude of the effect, based on how far forward (Δx) the *cg* is moved (Eq. 17.14).

$$V_{E,STALL,2} = \sqrt{1 + \left(\frac{\Delta x}{s}\right)}\ V_{E,STALL,1} \qquad (17.14)$$

APPROXIMATE EFFECT OF CG POSITION ON STALL SPEED

EXAMPLE 15.2 The T-210 pilot's manual specifies a 65 kt (calibrated speed) stall speed at 3800 lb and the most rearward *cg* location. Estimating the distance *s* as 14 feet, estimate the stall speed with the *cg* one foot farther forward.

Solution: $V_{E,STALL,2} = \sqrt{1 + \left(\frac{+1\ ft}{14\ ft}\right)}\ 65\ kt\ =\ 67\ kt$

which is to be compared with a handbook value of **69 kt**. The difference is probably due to lesser elevator effectiveness at the forward *cg* location.

D. Directional (Yaw) Stability

If we push on the rudder so that the aircraft yaws to the right or left, will it tend to return to its former direction of flight when the rudder is returned to the neutral position? If a wind gust yaws the nose over to one side or the other, will the plane tend to straighten itself out or will it become a whirligig? This is the question of directional or **yaw** stability—sometimes also called weather vane stability, for obvious reasons. Yaw stability is provided by the fixed vertical stabilizer on conventional configurations and by the winglets on some canard aircraft such as the Long-EZ. A tail volume coefficient, defined in a fashion similar to that for the horizontal stabilizer, gives guidance regarding the necessary size.

The effect of the fuselage and a forward-mounted propeller or engine inlet is destabilizing again, for the same reason as for pitch stability. This is one of the less obvious problems that suddenly presents itself when a lightweight turboprop engine is used as a replacement for a heavier piston engine. The transplant increases the fuselage side area which is ahead of the *cg*, and is therefore destabilizing.

The vertical stabilizer will lose effectiveness at supersonic speeds and therefore a loss in yaw stability accompanies high speeds. For some aircraft, this is what determines the never-exceed speed.

It is generally more efficient to use a dorsal fin (i.e., a vertical strake) connecting the middle of the upper fuselage to the middle of the vertical stabilizer than to increase the height of the tail. A secondary result of such a low aspect ratio control surface, which was used to especially good advantage in the World War I Fokker D VII, is to allow very large yaw angles (up to 30° for the Fokker) before the vertical tail surface stalls.

On low speed aircraft, a swept vertical stabilizer is less efficient than a straight one; it is swept only for cosmetic reasons.

E. Lateral (Roll) Stability

If an aircraft rolls into a bank due to a wind gust, will it tend to roll back to level flight or will it just keep on steepening the bank? This is the question of lateral or roll stability. Lateral stability is normally obtained by **dihedral**, i.e., having the wingtips higher than the wing root. A low wing aircraft will need at least a few more degrees of dihedral than a high wing aircraft because the high wing machine gains some lateral stability from having its *cg* below the *cp*; this is called the "pendulum effect" because it is similar to the way a pendulum bob tends to return to its lowest spot when disturbed.

Too much dihedral is inefficient because too much of the "lift" force of the wings is then self-canceling in the horizontal direction. Also, too much dihedral combined with insufficient weathercocking stability leads to **Dutch roll**—a side-to-side yaw combined with a rolling motion. A Piper Tomahawk flown by the author would do a very gentle little Dutch roll when trimmed and cruising along in very stable air, tracing out little circles with the wingtips and with the nose. Reference 2 suggests that you may observe Dutch-roll induced rotary motion in your drink if you sit near the tail in a jet airliner.

Before the Wright brothers showed the importance of good lateral control, many of the earliest flying machines provided little or no lateral control, preferring an inherently stable machine obtained through a very large dihedral angle. When the wing is rolling in response to atmospheric turbulence, the down-going wing has a greater angle of attack and generates a rolling moment that tends to return the aircraft to level flight. When the wing is flying wing-low, the lift on the lower wing generates a larger rolling moment than that on the higher wing, again tending to return the plane to wings-level flight (Fig. 17.12).

Theoretically, the dihedral of the wing should follow an elliptical curve but a double joint is practically as good and much easier to build. But even this is too hard for most aircraft builders, so usually the only bend for dihedral effect is at the fuselage.

Rear sweepback contributes positively to lateral stability; forward sweep contributes negatively. Figure 17.13 shows how sweepback makes its contribution to roll stability; the wing that leads the other has a greater drag and a longer moment arm about a vertical axis through the *cg*, tending to straighten out the machine.

High performance jet aircraft sometimes have too much roll stability from sweepback and have to use **anhedral** in their wing or horizontal tail to reduce it to a reasonable value.

If you take a disk such as a paper plate or LP or CD and launch it into the air, its flight is not likely to be graceful or stable because the *ac* is ahead of the *cg* (which is in the center). You can do much better if you launch the disk with lots of spin; then

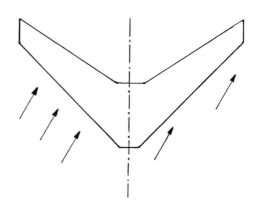

Fig. 17.13 Sweepback contributes positively to lateral stability.

conservation of angular momentum tends to keep the pitch angle of attack of the disk constant. However, just as with a gyroscope or the propeller on a plane, any unbalanced torque perpendicular to the spin axis causes the spin axis to **precess**—i.e., rotate in a plane perpendicular to both the spin axis and the plane containing the force. This means that the spinning disk is unstable in bank; it tends to roll either one way or the other, stopping only when 90° of bank, zero lift, and zero torque are achieved—and its flight path becomes that of a stone.

Can the aerodynamics of the spinning disk be improved? Sure. Somehow the *ac* must be moved rearward so that it can coincide with the *cg*. The Frisbee® possesses a familiar, successful solution to the problem with its curled outer edge. The drawback of the curled edge is the resulting high drag which limits the range. The Aerobee® is a lower drag solution to the spinning disk stability problem. The general shape of a cross-section through its center is shown in Fig. 17.14. The airfoil shape is such that the "spoiler" part is the leading edge of the part of the disk that is currently the nose and therefore little lift is developed there, but at the trailing edge there is a relatively large region of lifting surface before the air flow is spoiled; therefore the overall *ac* is moved rearward so that it essentially coincides with the *cg* over the most important range of speeds(Ref. 16). The Aerobee can be thrown about twice as far as the Frisbee; it has been thrown over a thousand feet in fact!

OUTER EDGE   INNER EDGE

Fig. 17.14 General shape of the "airfoil" of the Aerobee®

F. Aspects of Control

Stability, control, and maneuverability for a given airplane are closely related. A very stable airplane resists any change in pitch, trying very hard to fly wings-level at a constant speed—the trim point. Such an aircraft may be a good machine for

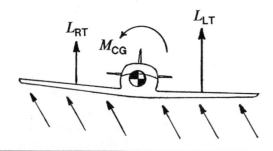

Fig. 17.12 How dihedral provides lateral stability (side slip produces more lift on the down wing)

instrument flying—but gives no pleasure to the pilot in any kind of maneuvering flight. The extra stability also comes at the expense of overall efficiency—a lowered L/D.

The longitudinal stability of an aircraft is directly reflected in the force required to produce a given acceleration in the direction of the lift vector; this is known as the **stick force per g**. Equation 15.10 gives this acceleration in terms of the load factor as $a_{VERT} = (n-1)g$. Therefore you can experimentally estimate the stick force per g for a given aircraft by measuring the pull force required on the stick or wheel to hold the plane in a level, 60° banked turn after it has been trimmed for level flight, since that produces a load factor of $n = 2$. Stick force per g is a measure of stick-free stability (or maneuverability); stick **movement** per g is a measure of stick-fixed stability. A transport or business aircraft is likely to require a stick force of 12 lb or more per g, a training aircraft may be about 8 lb/g, and an aerobatic aircraft as little as 2 lb/g. These values are also lowered for a side-stick control system since the wrist can't generate the forces that the whole arm can; side-stick control system work very well in light homebuilt aircraft where they compensate for the typically low control forces and small deflections while saving space. Pilots tend to quickly adjust to and then very much like a good side-stick control system.

A pilot can always decrease the longitudinal stability of a given aircraft (and increase its maneuverability) by allowing the cg to be at its rearmost point. In an accident in 1977 (Ref. 15), it was later discovered that the twin-engine aircraft involved had been loaded so far to the rear that its cg was 2 to 3 inches **aft** of the approved rearmost location. The plane went out of control shortly after entering instrument flight conditions. In the course of their investigation of this accident, the National Transportation Safety Board arranged to test the effect of cg location on this aircraft with the help of a special variable-stability research aircraft, a B-26. The results, in terms of stick force per g, are presented graphically in Fig. 17.15.

At the aft certified cg limit, 138", this aircraft was found to be behind its stick-fixed neutral point (but ahead of the stick-free neutral point). Pilots tended to overcontrol the aircraft and the workload was high but the aircraft was not considered unsafe. At a simulated 142", though, the plane was considered to be unsafe by all the pilots. When maneuvering at speeds other than the trim speed, control forces were opposite from the normal direction—i.e., the pilot had to actively push or pull to prevent a divergence in pitch, just as for the flying wing that was the subject of the third quotation which introduced the chapter. Because of this instability, load factors in level flight at this cg location varied from about 0.25 g to 1.75 g and the airspeeds varied from −20 kt to +40 kt about the trim speed.

The location of the neutral point for a given airplane can be estimated by measuring the force required to hold the plane in a banked turn for different cg locations within the approved envelope and extrapolating graphically to the cg location giving zero stick force in that bank. Reference 11 suggests another experimental method: Measure the stick force per kt of airspeed away from the trim speed for several airspeeds and cg

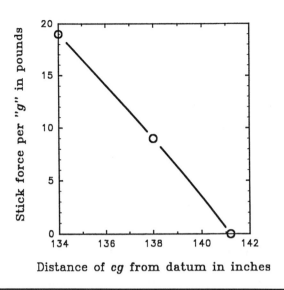

Fig. 17.15 Experimental stick force per g for a twin-engine aircraft

locations and again graphically extrapolate to the cg location where the stick force per kt is predicted to be zero.

An aircraft is said to have "well-harmonized" controls if the forces required for ailerons, elevator, and rudder are well balanced relative to each other. This would be a plane that "feels right" because it requires the forces humans are comfortably and easily and naturally able to provide. It is usually thought that the ailerons should be the lightest control, the elevators should follow with about twice as much required force, and the rudder should have another twice as much or so. Most certified aircraft have heavier and less effective ailerons than most pilots prefer for sport flight, to put it kindly. The first time a pilot flies a homebuilt or an aerobatic aircraft usually provides a revelation of what maneuverability really means.

While an aircraft is being rolled into a banked turn, the upgoing wing has its aileron deflected down, increasing both the lift and the drag of that wing. The drag causes the wing to yaw opposite to the direction of turn; in a worst case (e.g., the J-3 Cub and earlier designs), the airplane will fly indefinitely in a slip if the rudder isn't used to counter this "**adverse aileron yaw**" and make the nose point into the turn. Because most aircraft possess at least some adverse yaw, trainer aircraft probably should too, so that a student pilot gets used to using as much rudder to initiate a turn as necessary.

Adverse aileron yaw is minimized by using differential aileron movement (about 50% greater deflection up than down) and/or by Frise ailerons. If the upgoing aileron on the downgoing wing deflects more than the downgoing aileron, it will increase the drag on this wing, which is what we want.

When spoilers are used instead of ailerons, adverse yaw is generally not a problem because turning is made toward a wing that has both less lift and more drag. One problem with spoilers as an aileron substitute is that considerable overall lift may be lost from the wing if the aircraft is maneuvering in turbulent air near the surface; this has contributed to some airline accidents.

The Frise aileron (named after British aircraft designer Leslie George Frise) minimizes adverse yaw by using an unsymmetrical nose profile on the aileron and having the hinge line behind the leading edge (Fig. 17.16 and Refs. 13 and 14). The upward-deflected aileron has more drag because the nose is allowed to deflect down into the airstream.

A control surface that has some area ahead of the hinge line (as does the Frise aileron) is said to have **aerodynamic balancing**. The purpose is to lighten control forces — but it can be difficult to avoid nonlinearity in the amount of balance per degree of deflection, especially at large deflection angles. Rudders and elevators often employ a "horn balance" (Fig. 17.17) in which the end of the control surface juts well forward of the hinge line, but any surface area ahead of the hinge line contributes to aerodynamic balance. If the hinge line coincides with the center of pressure of the surface, though, overbalance is present and the surface will tend to stay at whatever deflection it finds itself.

If the center of mass for a control surface is behind the hinge line, the inertia of the surface will tend to make it continue to deflect past the initial deflection angle, until the elasticity of the surface to which it is mounted stops it. There is always a speed at which these aeroelastic forces and restoring moments will be in resonance (similar driving and driven frequencies) and the amplitudes of the oscillations will diverge and the control surface will very quickly depart the aircraft, probably taking with it the wing or whatever it was attached to. This is the much-feared **flutter** phenomenon. The wing by itself can do this, too, and this is one of the reasons for hanging podded jet engines well out in front of a swept wing.

For control surfaces, flutter is counteracted by placing the center of mass on the hinge line (or even slightly in front of it),

Fig. 17.16 The Frise aileron

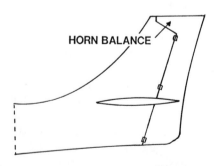

Fig. 17.17 A rudder with a "horn balance" to lighten control forces

Fig. 17.18. This is called **mass balancing** and is mandatory for any aircraft that has a never-exceed speed of perhaps 150 kt or greater. Elevators and ailerons both must be mass balanced, and perhaps also the rudder (depending on the stiffness of the fuselage). The extra mass may be provided by hanging a weight on a rod that extends forward of the hinge line, either into the supporting structure or outside (as often seen on some older aircraft). If the hinge line is behind the leading edge, it may be possible to accomplish mass balancing by just adding lead to the leading edge of the control surface.

Friction in the control system must be minimized to the greatest extent possible so that the "break-out" force isn't greater than the force needed to sustain the deflection, leading to extremely poor control feel and masking a low stick force per *g* stability condition. The other two contributors to flutter problems are slack or "play" in the control linkage or in a trim tab system.

The Lockheed P-80A, America's first jet fighter, had an aileron "buzz" and oscillation problem. At Mach numbers close to 0.8, the trailing edge would oscillate ±1/2" at a high frequency while at higher speeds the whole aileron would oscillate even more. Compressibility shock waves were the cause. Changing the static and dynamic balance, increasing the wing torsional

(a) Unbalanced Control Surface

LEAD WEIGHT ON LEADING EDGE

EXTERNAL BALANCE WEIGHT

INTERNAL BALANCE WEIGHT

(b) Balanced Control Surfaces

Fig. 17.18 Mass unbalanced versus balanced control surfaces (the internal balance is commonly used with stabilators)

stiffness, adding a leading edge counterweight, and changing the contour of the aileron and wing had little or no effect. The solution turned out to be to increase the tension in the aileron cables to a sea level value of **350 lb**. (Hint for very subsonic model airplane builders: you too can get flutter if your control system wires aren't taut.)

For any aircraft there is a speed at which deflecting an aileron will deflect the wing in the other direction enough that **control reversal** results, at which point the response of the airplane for a given control deflection is opposite to that which normally happens. For example, increasing the aileron deflection to roll out of a bank would just increase the bank angle some more; the only way out of this mess would be to reduce aileron deflection and use the rudder. This was a common problem for the earliest flying machines such as the Bleriot and Antoinette and made them unsafe to fly in anything except very stable air. The structural requirement is for **torsional** rigidity and not simply a load factor (strength in bending) requirement.

Some of the early, untwisted wings stalled all at once with no warning. If a little aileron was being used at the same time, the adverse aileron drag caused the machine to break immediately into a spin. Reference 13 details the unsuccessful attempt many years ago to find an aileron design that would provide positive control all the way into the wing stall. Geometric or aerodynamic twists now is used to ensure aileron effectiveness at least up to the stall onset.

Sealing the air gaps between the aileron and the wing (or the elevator and the stabilizer) produces substantially less drag and more control power because any leakage of air tends to equalize the pressures. The new sheared wingtip (Fig. 12.7) may prove useful in providing efficient aileron power up to the stall.

Sometimes, as on the P-51D Mustang, more attached flow over the ailerons can be obtained by reducing the thickness of the wing just before the ailerons so that the BL tends to speed up and reattach over more of the aileron surface.

Ailerons and elevators follow a V^2 law whereby doubling the equivalent airspeed **quadruples** the control force required for a given deflection. Aircraft with a very wide speed range will therefore tend to have excessively light control forces at slow speeds and excessively heavy control forces at the high speed end. One way to counteract this is with a modification to the trailing edge of the control surface, because this has a powerful effect on the BL. The beveled trailing edge (Ref. 12 and Fig. 17.19) has been used with success, as on the Mooney. The V^2 law can also be defeated by connecting the control stick to a **tab** on the control surface trailing edge rather than to the control surface itself. Many large aircraft use this scheme. Or the tab can be geared to the control surface through fixed linkage so that it continuously moves either in the opposite or in the same direction as the control surface. If the tab is geared to move in the opposite direction, the control forces will be lightened and vice versa.

Everything else being equal, a long and narrow aileron has the lightest control forces because this bring the center of pressure closest to the hinge line. The problem with such an aileron is that it doesn't leave as much room for a flap.

An interesting example of control reversal that anyone can experience is the reaction of an aircraft to winds coming from **behind** it. (Tail slides don't count because not everyone has access to a super aerobatic machine that can do it safely.) The author knows of one sad example where a pilot started up the engine of a lightweight taildragger aircraft while there was a relatively strong wind blowing from the tail to the nose. Force of habit made the pilot pull back on the stick before adding power but this caused the tail to start to come up. The addition of power and stick forward completed the 180° out of phase work by the pilot and the plane ended up on its nose.

Some taildraggers such as the Cub do not carry much weight on their tailwheels and so it is easy to wheel land them and then leave enough power on to keep the tail up while taxiing down the runway to the nearest turnoff, maximizing visibility and minimizing the time the aircraft is on the runway. Doing this on a taxiway while taxiing **downwind** results in quite a different story. Power and forward elevator and rudder will bring the nose up again because the prop blast overpowers the wind. The ailerons probably aren't in the prop blast and will need to be adjusted based on the actual crosswind component. As soon as the power is removed, though, the elevator and rudder controls **reverse** in their effect and the pilot better be ready on the brakes as an aid to directional control.

"**Active** controls" are now being used on some transport aircraft and on some low level bombers to reduce bending loads and improve the ride in turbulent air; their use will probably expand a great deal in the future. With ailerons, for instance, the idea is to let a computer sense a momentary wind gust and actively adjust the aileron deflection so as to keep the overall wing lift constant.

Summary

An aircraft possesses static stability if it tends to return to wings-level flight, at the trim speed, when momentarily disturbed in pitch or bank or yaw. For pitch (longitudinal) stability, this is equivalent to requiring that the slope of the pitching moment curve for the whole airplane about its cg be negative about the $M_{CG} = 0$ trim point.

Longitudinal stability for a conventional tail-last configuration normally requires a download on the tail, resulting in a wing lift force greater than the weight and therefore, effectively, an added drag force called trim drag. If the rear surface is large enough, or if it is equal in size (a tandem configuration), or if it is even greater in size than the wing (a canard configuration),

Fig. 17.19 Beveled trailing edge on an aileron

longitudinal stability can be obtained with both surfaces generating positive lift.

The neutral point is the *cg* location for which an airplane has neutral stability; a stable aircraft must have its *cg* ahead of this point, although stick-fixed and stick-free neutral points often do not coincide. The neutral point is the *ac* for the whole airplane. The static margin describes the amount of longitudinal stability by giving the distance from the neutral point as a fraction of the effective or mean wing chord.

Tractor power is a destabilizing influence, effectively moving the neutral point forward. The fuselage is also a destabilizing influence. If the pitching moment curve becomes zero or has a positive slope at high angles of attack, the possibility of a stable, deep stall is present; swept-winged/T-tailed aircraft and canard aircraft with a too-powerful canard are likely candidates. For all configurations, the rule is that the rearmost surface must never be allowed to stall.

The size of the tail required for static stability depends on the wing and the wing-to-tail distance; these contributions are summed up in the tail volume coefficient, which should be greater than about 0.2. Larger tail volume coefficients allow a greater *cg* range.

For all configurations, longitudinal stability decreases and eventually becomes negative as the aircraft's *cg* is moved aft. The stall speed also decreases as the *cg* is moved aft because the wing doesn't have to provide as much extra lift to compensate for the download on the tail (in a conventional configuration) or the less-efficient canard is less heavily loaded (in a canard configuration).

Yaw stability is provided by the vertical tail surface; its effectiveness depends on its area and the distance from the wing of this area. A dorsal fin to the vertical stabilizer adds to the area and reduces the aspect ratio, helping to make sure it doesn't stall.

Roll stability is obtained with dihedral or sweepback; too much of either or both causes a wiggle around the longitudinal axis called Dutch roll. Low wing aircraft need more dihedral than high wing aircraft because the latter obtain a pendulum-type stability contribution.

The amount of longitudinal stability possessed by a given airplane can be measured as the "stick force per *g*" or even "stick force per kt"; doing this for different *cg* locations allows one to estimate the location of the neutral point.

The effectiveness and the force required for a given control surface can be altered by changing the hinge line location, the area ahead of the hinge line, or the shape of the leading or trailing edges of the control surface. Adverse aileron yaw can be minimized by Frise ailerons and differential deflection (more up than down).

Flutter is avoided by mass balancing of control surfaces (so that they tend to float when no air is moving past them) and by minimizing slack or slop in the control system and in any tab system.

If a control surface is attached to a too-flexible structure, control reversal can occur in which the effect of the deflected structure is greater than the effect of the deflected control surface. The effect of control movement away from neutral also reverses when the relative wind is coming from behind the aircraft—as often happens when taxiing. Active control is present when a computer deflects control surfaces to minimize load factors or other effects of turbulence, without input from the pilot.

Symbol Table

$C_{L,w}$	lift coefficient for the front wing (which would be the canard on that type of aircraft)
$C_{L,T}$	lift coefficient for the rear wing (which would be the horizontal tail surface on a conventional configuration aircraft)
M_{CG}	pitching moment for the whole aircraft about the airplane's *cg*
x	distance from the *ac* of the front wing to the airplane's *cg*
s	distance from the *ac* of the front wing to the *ac* of the rear wing
S_W	surface area of the front wing
S_T	surface area of the rear wing
α_W	angle of attack of the front wing
α_T	angle of attack of the rear wing
a_W	slope of the lift curve for the front wing
a_T	slope of the lift curve for the rear wing
x_{NP}	distance from the *ac* of the front wing to the point (the neutral point) where the aircraft has neutral stability
$\bar{c}$	mean geometric chord (MGC) of the front wing for a simple rectangular wing or, for a tapered or swept wing, it is the aerodynamically equivalent average chord, the mean aerodynamic chord (MAC)

Review Questions

1. Lateral stability is normally obtained by designing ___ into the wing structure.

2. Longitudinal or pitch stability requires that the overall pitching moment curve have a slope that is (negative, zero, positive).

3. Unstable pitch-up on a rearward-swept wing occurs only at angles of attack ___.

4. The aspect ratio of the vertical stabilizer is often decreased by adding a ___ ___. One of the resulting benefits is the extension of the effectiveness of the surface to large ___ angles.

5. An aircraft in which a computer moves the controls (without any input from the pilot) so as to minimize turbulence-induced air loads and altitude and pitch variations is said to have ___ controls.

6. Yaw stability can be increased without changing the tail if a ___ ___ is added.

7. When the *cg* of an aircraft is at the ___ ___, an aircraft has neither negative nor positive pitch stability.

8. Which of the following are true statements?
   a. Sweepback increases yaw stability.
   b. Sweepback increases pitch stability.
   c. Sweepback increases roll stability.
   d. Sweepback moves the *ac* rearward.
   e. Sweepback tends to improve stall characteristics.
   f. Sweepback increases the wing's aspect ratio.

9. The trim angle of attack is that angle for which the moment coefficient relative to the ___ is ___ ; a longitudinally stable aircraft will have a moment coefficient that is ___ at angles greater than this angle.

10. Power-on stability is always worse than power-off stability for a ___ configuration.

11. A flying wing can be made stable in pitch by sweeping the wing ___.

12. If a stable aircraft with a conventional tail suddenly loses its horizontal tail surfaces while in level flight, what will be the response of the aircraft?

13. The most important factor in determining whether your new aircraft design has an adequately large tail is the ___ coefficient.

14. Lateral (roll) stability is obtained in most aircraft through the use of ___ ; a ___ type of aircraft requires the most.

15. The primary reason that rearward sweep has been used rather than forward sweep in the past is the ___ problem.

16. One good reason to use low aspect ratio tail surfaces is to make sure that they do not ___.

17. The ___ ___ is the location of the *cg* relative to the *np*, expressed in terms of the *MAC*.

18. The fuselage is a ___ influence. (stabilizing or destabilizing)

19. A stall that is stable (and unrecoverable without a spin 'chute or JATO or something) is called a ___ stall; the tail configuration that has led to this problem in recent years is the ___ ; the wing configuration that has led to this problem in recent years is the ___ .

20. "Tuck-under" is due to the ___ movement of the *ac* as an aircraft's speed approaches the speed of sound. (forward or rearward)

21. What determines the forward limit for the *cg* on an aircraft? Is this as efficient a location as a more rearward one?

22. ___ stability is long-term static stability.

23. Suppose the *cg*-to-tail distance on a certain aircraft is increased by 10% while the span of the wing is increased by 10% through simply extending the current wingtips. Will pitch stability of the aircraft be affected and, if so, how?

24. The ___ ___ can be considered to be the location of the aerodynamic center for the whole aircraft.

25. The stall speed will ___ (increase or decrease) as the *cg* is moved forward because the download on the tail will ___. (increase or decrease)

26. In a conventional wing/tail arrangement, an aircraft can be longitudinally stable while still possessing a lifting tail if ___.

27. Would you expect the ailerons to have a greater or lesser maximum angular deflection than the rudder? Why?

28. What are the advantages of active controls?

29. Why does a transport-type aircraft normally have a larger tail volume coefficient than a light aircraft?

30. How is the Dutch roll illness treated by an aircraft designer?

31. Under what circumstances will pulling the stick or wheel back cause the nose of the aircraft to move down (away from the pilot)?

32. Aileron control forces can be lightened by ___ the trailing edge of the ailerons—at some cost in pressure drag.

33. On a large aircraft without powered controls, the control lines will probably go to a ___ on the control surface rather than to the control surface itself.

34. How can the drag and effectiveness of control surfaces be improved without changing them (assuming this hasn't been accomplished already).

35. Why is it a bad idea to trying to use the ailerons to pick up a low wing while flying at speeds close to the stall, especially when flying old aircraft?

36. The primary preventive against control surface flutter is the ___ ___ of the surface.

37. Flutter can be caused by ___ in the control system.

38. Control forces can be lightened by placing part of the control surface ahead of the ___; this is called ___ balancing.

39. The ___ aileron minimizes aileron yaw by using an ___ nose shape that creates extra drag when the surface (increases, decreases) the lift of that wing.

40. The other design method for minimizing adverse yaw is with differential aileron deflection—i.e., more ___ deflection than ___. (up or down)

41. Loading an aircraft so as to move the *cg* forward will **increase**
   a. the stick force per *g*
   b. the stick force per kt
   c. lateral stability
   d. longitudinal stability
   e. all of the above

42. Does forward sweep make a positive or negative contribution to yaw stability? (Hint: Try redrawing Fig. 17.9.)

43. When sweepback causes too much roll stability, the solution is often ___ for the wing.

44. If you fold an 8½"×11" sheet of paper down the middle in the long direction and give it a toss, it won't glide satisfactorily. In the traditional paper airplane, you complete the construction by adding a couple of triangular folds at the nose end and then bend over the wings with folds in the longitudinal direction. How do these last two steps impart stability to the plane?

45. Too much dihedral or dihedral effect relative to ___ stability produces the oscillatory motion called ___ roll.

46. Moving the *cg* to the rear (up until the ___ ___ is reached) always (increases, reduces) the stall speed.

47. What control deflection on the B-2 bomber would correct for a yaw to the left, without doing anything else?

48. A flying wing is stable if the ___ decreases going from the front to the rear of the machine.

49. Pitch-up at the stall is reflected in a moment curve with a ___ slope at that point.

50. When a pilot trims an airplane and then lets it fly itself, he or she is relying on ___ static stability.

51. The horizontal surface that must never be allowed to stall, in general, is the ___ one.

52. The fuselage has a (positive, negative) effect on the pitch stability; it has a (positive, negative) effect on the yaw stability.

53. The drag force resulting from making the wing lift more than the weight of the airplane in a conventional configuration is known as ___ drag.

54. The trim speed of an aircraft is the speed for which the ___ ___ is equal to zero.

55. A conventional aircraft, with a standard tail and not a T-tail, trimmed for cruising flight, will immediately nose down when power is reduced. The probable reason is that
    a. the effective *cg* has moved rearward
    b. lift has become less than the weight and has moved aft
    c. the effective angle of attack of the wing has been reduced
    d. the downwash on the elevators from the propwash is reduced

56. An aircraft loaded so as to bring the *cg* to the rear of the certified rear limit may produce an airplane that
    a. tends to land on its nosewheel
    b. has a higher than normal stall speed
    c. has an unrecoverable stall
    d. has a longer than normal takeoff roll
    e. all of the above

Problems

1. What is the static margin for the paper airplane when it has a download of 150 lb on its tail? Answer: $0.21\bar{c}$

2. Calculate the tail volume coefficient for the paper airplane.

3. Determine the trim point for the paper airplane if the tail has a download of 75 pounds. Answer: $\alpha_W = -0.9°$, $\alpha_T = -0.625°$

4. Determine the trim point for the canard paper airplane if the canard is lifting 600 pounds in cruise. Answer: $x = 7.5$ ft

5. A Skylane stalls at about 56 kt at a gross weight of 2950 lb when the *cg* is at its aft limit. Taking s to be 14 ft, estimate the stall speed when the *cg* is 2 ft farther forward, assuming adequate elevator power to stall the aircraft is available. Answer: About 60 kt

6. Where should the *cg* of the canard paper airplane be located for neutral stability? (Hint: Use the general stability expression.) Answer: $x = 9.6$ ft

7. What percentage of the wing area must the horizontal tail of the paper airplane have if it is to possess neutral stability with zero tail lift?

8. For angles of attack between 2° and 16° and a typical wing aspect ratio of 6.0, what is the corresponding variation in the angle of attack of the air impinging on the tail, because of downwash? Answer: $-1°$ to $-8°$

9. In Fig. 17.10, what are the tail loads if the *cg* is moved forward to coincide with the wing's *ac*? Answers: -250 lb, -29 lb, -23 lb, -61 lb

10. What is the stall speed for the *cg* location of problem 9? Answer: 40 kt

References

1. Wooldridge, E. T., *Winged Wonders: The Story of the Flying Wings*, Smithsonian Institution Press, 1983. (The history of Northrop's flying wings.)

2. Stinton, Darrol, *The Design of the Airplane*, Van Nostrand Reinhold Company, 1983. (An unusually well-written and interesting text by an author with sterling credentials as a test pilot for England's CAA and as an aeronautical engineer/designer.)

3. Lippisch, Alexander, *The Delta Wing: History and Development*, Iowa State University Press, 1981. (A definitive description of the tailless aircraft designed by a prominent German engineer. His best known tailless aircraft was the rocket-propelled ME 163.)

4. Downie, Don and Downie, Julia, *Complete Guide to Rutan Homebuilt Aircraft*, Tab Books Inc., 2nd Edition, 1984. (Popular account of Burt Rutan's designs through the round-the-world Voyager project.)

5. Lennon, Andy, *Canard—A Revolution in Flight*, Aviation Publishers, 1984. (A popular history of canard aircraft through the Beechcraft Starship.)

6. Horten, von Reimar and Selinger, Peter F., *Nurflügel*, H. Weishaupt Verlag, Graz, 1985. (Flying wing designs of the Horten brothers, from 1933 to about 1950, are described in this partial translation into English.)

7. Wolko, Howard S. (editor), *The Wright Flyer—An Engineering Perspective*, Smithsonian Institution Press, 1987. (Topics include the stability and control, the propulsion system, and the structural design of the Wright Flyers.)

8. Norton, F. H., "A Study of Longitudinal Dynamic Stability in Flight," NACA TR 170, 1923.

9. Jones, R. T., *Modern Subsonic Aerodynamics*, Martin Hollmann, California, 1988.

10. Chambers, Joseph R., Yip, Long P., and Moul, Thomas M., "Wind-Tunnel Investigation of an Advanced General Aviation Canard Configuration," NASA TM 85700, 1984. (Available from NTIS as N84-21539.)

11. Garrison, Peter E., "Balancing Act," *Flying*, May, 1988. (An interesting pilot-oriented discussion of various aspects of longitudinal stability and control.)

12. Purser, Paul E. and McKee, John W., "Wind-Tunnel Investigation of a Plain Aileron with Thickened and Beveled Trailing Edges on a Tapered Low-Drag Wing," NACA Advanced Confidential Report (ACR) L-525, January, 1943. (If properly sealed, beveled-edge ailerons require substantially lower deflection forces at high speed than plane ailerons, but with presumably some pressure drag penalty.)

13. Weick, Fred E. and Wenzinger, Carl J., "Wind-Tunnel Research Comparing Lateral Control Devices, Particularly at High Angles of Attack; I.—Ordinary Ailerons on Rectangular Wings," NACA TR 419, 1932. (Aileron effectiveness at the stall was of particular concern at this time because ailerons were aggravating the tendency of a spin to develop after a stall; this first installment compared ailerons with different length/chord ratios mounted on a model, untwisted wing.)

14. Weick, Fred E. and Noyes, Richard W., "Wind-Tunnel Research Comparing Lateral Control Devices, Particularly at High Angles of Attack; II.—Slotted Ailerons and Frise Ailerons," NACA TR 422, 1932. (None of the tested ailerons gave satisfactory rolling control beyond the stall angle; the Frise ailerons had about half the adverse yaw of the other types but differential aileron motion, i.e., more up than down, was even better.)

15. NTSB Report No. NTSB-AAR-78-1, January 5, 1978.

16. Crane, Richard, "Beyond the Frisbee," *The Physics Teacher*, November, 1986. (A brief description of the aerodynamics of the Aerobee.)

17. Bihrle, William, Jr., Barnhart, Billy, and Pantason, Paul, "Static Aerodynamic Characteristics of a Typical Single-Engine Low-Wing General Aviation Design for an Angle-of-Attack Range of $-8°$ to 90°," NASA CR-2971, 1978.

18. Wainfan, Barnaby, "Wind Tunnel: Here are ways that designers optimize control feel," *Kitplanes*, March, 1992.

19. Vincenti, Walter G., "Establishment of Design Requirements: Flying-Quality Specifications for American Aircraft, 1918–1943," Chapter 3 in *What Engineers Know and How They Know it*, The Johns Hopkins University Press, 1990.

Chapter 18

Aerodynamic Simulation: Tunnels and Computers

Although one might argue that the difficulties of measuring drag are less important than stability and control (because drag affects performance, but not safety), the simple truth is that whether or not a plane is ever built depends very much on the drag data obtained in the wind tunnel, and how well one can make potential customers believe it.[1]

It is more reliable to compare two different airfoils by data obtained from the same computer program than by tests from different wind tunnels.[2]

A. Background and Accomplishments

Hovering birds, lifted by steady winds, inspired would-be airman, including the Wright brothers, to hope that they too could do aerodynamic testing in full-size aircraft without the hazards of motion relative to the ground. Alas, the requisite strong, steady winds were not available, even at Kitty Hawk. Later, for the Wrights' first airfoil tests, the "steady" wind was generated by furious pedaling—because they had mounted two airfoils to be compared on a horizontal wheel ahead of the bicycle. More recently, airfoils have been mounted perpendicular to the wing of an aircraft to obtain airfoil data; the P-51 Mustang was used in this manner during early testing of transonic effects and more recently an F-106 has been used. The top of a speeding car (by Burt Rutan) or truck has been used; disadvantages of this simulation technique include the unsteadiness of low level winds, interference from the supports and the vehicle, vibration from the vehicle, and speeding tickets. Finally, the oldest simulation device is the rotating arm with an airfoil or model at the end of the arm; disadvantages include the difficulty of model observation and instrumentation and the unrealistic immersion of the model in its own wake.

And so we come to the winner: a wind tunnel where the model and the instrumentation are stationary and the air is set in motion past them. Relative motion between the model and the air is all that counts except if ground effects are being investigated, and then a "ground board" will have to be placed in continuous motion through some sort of belt arrangement.

Englishman Frank Wenham is generally credited with designing and operating the first wind tunnel, way back in 1871. Dividends from his tunnel included the discovery that L/D could be better than then-current theories predicted and that wings with large aspect ratios were better than those with low aspect ratios. However, the first wind tunnel to be built for the express purpose of supporting an ongoing flight program was that constructed by the Wrights in 1901. Their glider No. 2, based on existing data, did not perform as predicted and this constrained them to do their own airfoil research. Glider No. 3 achieved an L/D of about 8 and, with a 12-horsepower gasoline engine of their own design added, became the world's first successful airplane in December 1903. Europe soon regained the lead in aircraft design and in wind tunnels, however, and the lead was not regained by the United States until after World War II.

Many have been the accomplishment of wind tunnels in the United States, even though we were late to the starting gate. The first NACA (National Advisory Committee on Aeronautics) wind tunnel, just a copy of an existing English tunnel, became operational in 1920. But a new and innovative design, the Variable Density Tunnel (VDT), was built inside a big tank so it could be pressurized up to 20 atmospheres to permit the testing of small models under realistic Reynolds numbers. The VDT became operational in 1923 and by 1933 had been used to study and characterize the famous 4-digit family of airfoils (NACA 0006 to NACA 6721). NACA's 1927 Propeller Research Tunnel (PRT) was used to measure landing

---

[1]William H. Rae and Alan Pope, *Low-Speed Wind Tunnel Testing* (New York: Wiley, 1984).

[2]Richard Eppler, *Airfoil Design and Data* (Berlin: Springer-Verlag, 1990).

gear drag—and this drag turned out to be so large that retractable landing gears soon became the norm for high speed aircraft. A second major success, in 1928, was the famous, Fred Weick-developed NACA Cowling for radial engines, which not only greatly reduced the drag of this large-frontal-area engine but also did a better job of cooling it. With the NACA Cowl the Lockheed Vega jumped in speed from 157 mi/hr to 177 mi/hr. A full-scale tunnel, with a 30 ft wide by 60 ft tall test section (called the "30×60 ft full-scale tunnel"), became operational in 1931 and was used to test practically every high performance aircraft used by the U.S. in World War II. A high speed wind tunnel, with a 2 ft test section and nearly Mach 1 speeds, became operational in 1934 and was instrumental in the development of the high speed airfoils needed for propellers; they are still in use.

War clouds caused NACA to expand its research facilities beyond Langley, Virginia (which was limited in available space and available electric power) to Moffet Field south of San Francisco (now NASA Ames Research Center) and to Cleveland Hopkins Airport for engine/propeller research (now NASA Lewis Research Center). The new, enlarged inventory of wind tunnels (**a**) solved the compressibility dive problem of the Lockheed P-38 Lighting fighter aircraft by devising dive recovery flaps that improved the flow over the tail, (**b**) solved a rumbling noise caused by the belly air scoop on the North American P-51 Mustang by extending the leading edge of the scoop below the fuselage boundary layer, (**c**) verified the validity of the Area Rule for reducing drag on transonic aircraft, giving the F-102 what it needed to go supersonic in level flight and, in more recent times, (**d**) has been instrumental in the development of the supercritical airfoil and winglets. Currently there are about a thousand wind tunnels in the world, and approximately a quarter of them are in the United States.

B. Types of Wind Tunnels

Every wind tunnel contains a test section, a means for persuading air or some other gas to flow through the test section, instrumentation for the model in the test section, and flow improving devices. The test section is usually rectangular in cross-section for ease in working inside and is usually taller than it is wide to minimize the effect of the floor and ceiling on flow streamlines. Test section sizes vary from a few inches in some hypersonic tunnels to the giant 80 ft × 120 ft full scale wind tunnel at Ames, but a common average size is 7 ft×10 ft. The test section appears just after a converging part of the tunnel in a subsonic or transonic tunnel but it must be in a diverging section following a nozzle for supersonic or hypersonic tunnels (Chapter 6, Section A). Because the proper shape of the nozzle depends on the exact Mach number needed, the nozzle is often formed of heavy steel plate with heavy-duty screw jacks adjusting their shape. Most tunnels use a **closed** test section, but some open or nearly open test sections are also commonly employed.

The power source for a continuous flow wind tunnel is typically an electric motor or motors driving either a fan (for low speeds) or a multi-stage compressor (for supersonic

speeds). The size of the electric motor(s) varies from a few horsepower for a low Reynolds number airfoil testing tunnel to 135,000 hp for the 80 ft×120 ft tunnel (be sure to shut off the city before turning it on!). The power requirements for supersonic and hypersonic flow are so great that such tunnels are often designed for intermittent flow; the power source is then usually a high pressure tank before the test section (a "blow-down" tunnel), a vacuum tank after the test section (an "in-draft" tunnel), or both. However, methane combustion and contained explosions have also been used to generate extreme speeds.

If high quality flow is to be obtained in the test section, various flow straightening and flow smoothing devices must be installed. Fine mesh screens and honeycomb sections typically precede the contraction cone leading to the test section in a subsonic tunnel or the nozzle for higher speeds. The lowest turbulence, and the greatest power losses, result from using many screens and a large contraction ratio. The turning vanes in a closed circuit tunnel must be carefully designed to promote uniform flow without excessive drag. The test section is followed by an expanding cone region (the "diffuser") where the flow speed decreases; then, at the lowest speed, it is ready to meet the fan blades, if the tunnel is of the continuous flow type.

A continuous flow tunnel can be either the **open-circuit** type or the **closed-circuit** type. In the open-circuit tunnel, the gas follows a straight path from entrance to test section to exit; if the tunnel is inside a larger building, the gas mixes with the surrounding air and makes its own way back to the entrance. An open-circuit tunnel (Fig. 18.1) is cheaper to build and removes the problem of recirculating exhaust in a propulsion test or smoke in a smoke tunnel, but it is harder to get good flow quality, more power is required to operate it, and more noise is produced. The closed-circuit type usually is designed with a single-return path, using turning vanes at each corner, but double-return and annular-return tunnels exist. Most research tunnels are closed-circuit types with a closed test section.

Many special purpose wind tunnels have been built. For example, because badly spinning aircraft have been such a persistent, pernicious problem for both Hollywood and real-life pilots, **spin tunnels** have been kept busy since the 1920s. In these tunnels, the air is directed upwards, a radio-controlled model with pro-spin controls is tossed into the flow, the air speed in the tunnel is adjusted to keep the model stationary while the spin mode is studied, and recovery effectiveness is

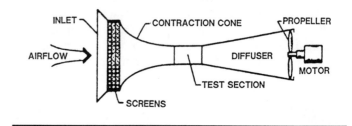

Fig. 18.1 An open-circuit type of wind tunnel (from Ref. 6)

Pic. 18.1 Smoke tunnel installed in a full-scale tunnel building; smoke is discharged from a line of orifices in a vertical plane in such a manner that smoke filaments are made to pass through the tunnel, indicating the nature of the flow around various bodies; this tunnel is used for demonstrations and for the production of films used in flow studies. (NACA photo provided by Robert Baals, NASA Langley)

assessed for various anti-spin control movements. A strictly secondary possible application for a spin tunnel is for sky diving practice — with a safety net below, please.

In-flight icing is still a major problem, especially for propeller-driven aircraft which cruise at icing altitudes, and **ice tunnels** address this need. NASA Lewis has a 6 ft×9 ft icing tunnel that refrigerates the air to –40° F and then an atomizer sprays droplets of water into the air flow upstream of the test section. The tunnel walls are insulated with 3" of fiberglass to keep heat out. Figure 18.2 presents the design features of BFGoodrich Aerospace's new closed-circuit, low speed wind tunnel in Uniontown, Ohio, designed primarily for research into de-icing protection (Ref. 31). The five horizontal spray bars which produce the supercooled water-vapor icing cloud are located in the settling chamber just before the 11:1 contraction duct leading to the 22"×44" test section. Temperatures in the test section range from –22°F to +35°F and speeds to about 225 kt are available.

Smoke tunnels have been extremely useful for flow visualization for many decades; even supersonic flows can be made visible with smoke (Pic. 18.1 and Ref. 9). **Full-scale**

tunnels have a test section large enough to completely enclose most single-engine aircraft without undue wall interference; applications include the verification of scale (Reynolds number) corrections for data gathered in other tunnels, the determination and reduction of interference drag, and landing configuration studies. Figure 18.3 diagrams the test set-up when a full-scale tunnel was used in a free-flight test technique with a scale model (Ref. 30). Two pilots are shown flying a VariEze model in the test section of NASA Langley's 30- by 60-ft wind tunnel. A flexible cable supplies compressed air for the turbine-air motor that drives the propeller as well as for the electro-pneumatic servos that position the control surfaces. Data acquisition transducers are mounted within the model.

Cryogenic tunnels are relatively new; they are able to control Reynolds number and Mach number somewhat independently. For example, NASA Langley's new cryogenic tunnel, designated the National Transonic Facility (NTF), was built in the early 1980s and attains Reynolds numbers from 1 million to over 100 million and Mach numbers up to about 1. Test gas temperatures vary from –300° F to +175° F and pressures vary from near vacuum to 9 atmospheres. In opera-

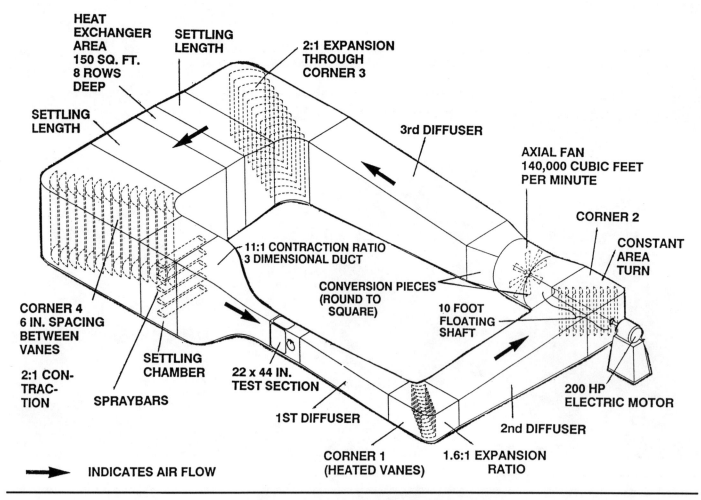

Fig. 18.2 A closed-circuit wind tunnel placed in operation in 1988 by BFGoodrich Aerospace, Uniontown, Ohio, for doing research in icing protection (Reprinted by permission of BFGoodrich Aerospace and Aviation Week & Space Technology)

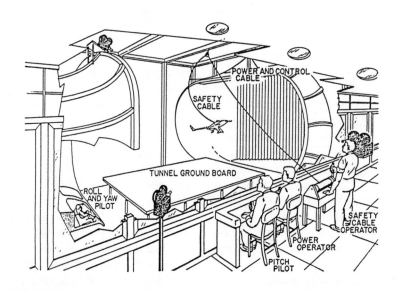

Fig. 18.3 Schematic of a 34% scale model of the VariEze being free-flight tested in the open-throat test section of the NASA Langley 30- by 60-foot wind tunnel (from Ref. 30)

Pic. 18.2   A VariEze mounted in NASA Langley's 30- by 60-ft full-scale wind tunnel (Photo courtesy of NASA Langley)

tion, the working fluid is nitrogen gas and cooling is accomplished by injecting liquid nitrogen into the stream.

The electrical energy given to a drive motor is mostly converted into heat energy in the working fluid and so typical high speed tunnels require active cooling to augment the inherent heat losses in the walls, to maintain reasonable tunnel temperatures. This extra cooling may be accomplished by cooling the walls with water, by cooling hollow turning vanes, or by replacing some of the hot air with cool outside air by means of an air exchanger. The air in a supersonic tunnel must be dried to prevent moisture from condensing as the tunnel air expands and cools in the nozzle. Finally, in a **hypersonic** tunnel, air must be heated to prevent it from liquefying as it expands and cools in front of the test section.

At the highest speeds, **hybrid** tunnels in which a model is fired into the airstream are used. If a model is fired at 8000 ft/s into a Mach 3 supersonic tunnel, for example, the relative Mach number is 15. Useful data have been acquired during the necessarily intermittent operation of such tunnels, assisted immensely by high speed electronics and high speed photography.

Controlled transonic speeds were as difficult to obtain in the wind tunnel as in flight. Wall interference becomes especially critical as the flow begins to choke near Mach 1; the normal shock waves forming on the model are reflected from the walls to interact multiple times with the model. Using an open test section causes the shock wave to be reflected also, but as an expansion wave, and this isn't an acceptable simulation either. A compromise, a partially open test section with slots in the flow direction (the "slotted-wall tunnel"), was so successful that it won the Collier Trophy for NACA's John Stack and his associates in 1951. A perforated wall is also used.

C. Model Support and Data Acquisition

Airfoils are normally mounted between the walls of the tunnel or between endplates inserted inside a larger three-dimensional section. Models other than airfoils, though, have to be supported out in the middle of the airstream. Wires were tried first and the model was hung upside down so the lift added to the weight, increasing the tension in the wires — but the drag of the wires was sometimes ten times that of the model, with accompanying air flow distortion. So most low speed models are mounted to the tunnel floor through one or more stream-lined struts. To minimize the drag contribution of the strut and yet not introduce indeterminate interference drag, shielding for the support strut is extended from the mounting point to within a few inches of the model. The test sections of early supersonic tunnels were too small to permit this mounting scheme and so a "sting" support — a hollow tube extending from the rear of the fuselage — has become common. Sting mounts are particularly well suited for those aircraft having jet engine exhausts at the rear of the fuselage.

An ideal support system disturbs the flow not at all. Such is the promise of magnetic suspension. Magnetic fields can be used to supply the six forces and moments to a non-magnetic model containing a magnetic core (or, better, a superconduct-ing solenoid) and the strengths of the required fields are direct-ly related to the forces and moments acting on the model. The feasibility of magnetic suspension was demonstrated in France over 30 years ago but only two systems are in active use at this time because of model size constraints, limited angle-of-attack range, and low control force capability. Support interference problems are most severe in the transonic wind tunnel; there-fore a magnetic suspension system is being studied for NASA Langley's National Transonic Facility, a new cryogenic tunnel. The newly discovered materials that superconduct at the much "warmer" temperatures provided by liquid nitrogen (rather than those provided by liquid helium) may play an important part in such a large-scale magnetic suspension system (Refs. 21 and 22).

An aircraft is subjected to three mutually perpendicular forces (lift, drag, and side force) and to three moments (pitch-ing, rolling, and yawing); therefore wind tunnel balances must be capable of separating and measuring these six components. The wind tunnel balance which carries the load outside the tunnel to measure them is known as an external balance. The internal balance counters the loads within the model and reports on its activities through the support strut or sting. Both types are in common use. Careful calibration of a wind tunnel's balances is essential if the data obtained are to be meaningful; Reference 1 tells us that even an expert crew will require at least three months for an initial calibration.

The forces themselves are usually measured with an electric wire strain gauge; in this type of gauge, the force stretches a very fine wire, changing its electrical resistance. Great sensitivity can be obtained by making the gauge one element in a balanced bridge network and using electronic amplification. The electrical signal that results from forces is converted electronically to a digital number for computer analysis. The advent of relatively inexpensive and powerful mini- and microcomputers has revolutionized data analysis and reduction for even the smallest tunnels. By using additional sensors for positions and angles, the various aerodynamic coef-ficients can be calculated and graphically displayed in real time — before the next data point is taken — rather than having to wait days or weeks.

Formerly, pressure measurements were made by connect-ing many orifices on the model with equally many manometer tubes. When the flow condition and the manometer reading stabilized, they were manually read or photographed. This is still an informative method for an instructional wind tunnel but most tunnels have updated to an electrically driven multiport valve (a **scanivalve**) that sequentially connects the pressure orifices to a pressure transducer (converting pressure to an electrical voltage) and then to an analog-to-digital converter, so the data acquisition computer gets what it needs (0's and 1's). The pressures can be integrated in real time, as you did by hand in Chapter 8 with flat plate airfoils, to obtain and display immediately the lift and drag and moment coefficients. Then desirable intermediate points where the data are changing rapidly, as well as questionable points, can be identified and immediately checked.

Airspeeds are measured in a number of ways. The fun-damental tool is the pitot tube discussed in Chapter 7. Another tool is hot-wire anemometry in which a very fine wire is main-tained at a constant temperature with the help of electronic feedback circuits; the amount of current required is directly related to the air flow speed past the wire because that deter-mines the heat loss rate. A newer, non-invasive technique is to use a laser beam (**laser velocimetry**). The beam is split and then refocused at the point where the flow velocity is to be measured. An interference pattern consisting of dark and light fringes is formed and photodetectors determine how long it takes seed particles to cross between fringes. Natural dust particles can provide the needed particles at low speeds; at higher speeds the flow may need to be "seeded" with droplets of atomized oils. Flow visualization may also be accomplished, based on this velocity information, if a large region around a model is scanned.

Flow visualization continues to be a vital part of the infor-mation gained from wind tunnel studies. Smoke tunnels make streamlines visible, immediately identifying regions of separated flow; small, slow speed instructional smoke tunnels are relatively easy to build — even supersonic ones. (See refer-ences at the end of the chapter.) As in flight tests, strips of string or yarn can be taped to a model to make cross-flows and separation visible. China clay, a suspension of white clay in kerosene, can be painted on a model just before a tunnel is started and the drying pattern later photographed to yield some of the same information. However, visualization studies of BL transition usually use a 40W motor oil or a sublimating chemi-cal. A laser beam can be used to map the velocity field around a model or, when split into an illuminating beam and a reference beam, for making holographs containing three-dimensional information regarding the density variations. At high speeds, the Schlieren system is the most common optical system for viewing and photographing density variations; in this system, a parallel beam of light passes through the test section and is focused on a knife edge, followed shortly by the viewing screen or photographic film. Regions of varying density refract (bend) the light so that it doesn't pass through the focal point; the knife edge eliminates rays that have been deflected in one direction, resulting in a brightness variation proportional to density gradients. If white light is broken into a spectrum with a prism, a picture can be obtained in which density variations show up as color variations on a background of uniform color (as was done in the Shell films referenced in Chapter 6).

D. Translating Wind Tunnel Data

Pic. 18.3 An eight-bladed model of a single-rotation advanced turboprop undergoes blade deflection tests in an 8 ft by 6 ft transonic wind tunnel (Photo courtesy of NASA Lewis)

We don't fly airplanes in wind tunnels any more than we fly airfoils or even wings in the atmosphere; we fly complete fuselage-wing-tail combinations in a genuinely variable atmosphere. Many difficult and even problematic corrections must be applied to typical wind tunnel data before conclusions relative to this "real world" can be drawn.

Scale effect errors have been with us since the earliest wind tunnel studies suggested erroneously that very thin airfoils gave the best lift to drag ratio for aircraft; the magnitude of the problem and the lack of consensus regarding the proper way to make corrections has led to the development of full-scale tunnels, despite their enormous cost. The sensitivity of airfoils to Reynolds number effects is well known but all parts of an aircraft are involved. A **transition strip** or BL trip is commonly employed to trip the BL at the point where it would be expected to transition to turbulence under full-scale conditions. A typical transition strip uses a 0.2" wide strip of carefully-sized grit. A torn strip of heating ("duct") tape has been successfully used in both the wind tunnel and in flight to simulate the contamination of a laminar airfoil with inset debris or rain. Even when data are taken at varying speeds to determine Reynolds number effects, the result may be as much Reynolds number effects of the tunnel as Reynolds number effects of the model.

Laminar airfoils could not be tested in wind tunnels until it was realized that the small-scale turbulence produced by the propeller, guide vanes, and tunnel wall vibrations were raising the effective Reynolds number to a value much greater that the stated test values. Specifically, this is why early reports on airfoils cannot be taken at face value, especially those in the Reynolds number range below a million. Langley's Low Turbulence Pressure Tunnel, first operational in 1941, provided low turbulence and realistic Reynolds numbers for the first time, leading directly to the widely used 6 series of laminar flow airfoils. The problem of turbulence has surfaced again in recent years because the new-found interest in low Reynolds number airfoils (those with a Reynolds number between about 1×10^4 and 2×10^6) has revealed great sensitivity of the laminar bubble, transition/reattachment, bubble bursting, and hysteresis phenomena to tunnel turbulence level.

Wall corrections are a second vital correction to wind tunnel test data. Because the model reduces the cross-sectional area seen by the air ("solid blockage"), the result is an increase in the dynamic pressure sensed by the model. Solid blockage is minimized by making a relatively small model and then mathematically correcting the data. There is also a problem with the flow outside the wake being artificially speeded up because of the decreased flow speed in the wake; this is known as "wake blockage." The ceiling and floor of the tunnel prevent the curved flow over a lifting wing and the downwash following it from being equal to that enjoyed by the full-scale aircraft; coefficients must be corrected for this effect and it must be recognized that the effects on stability measurements and on the proper angle of incidence for the stabilizer may be large. Another correction has to be made for the fact that the static pressure decreases through the test section (the "buoyancy effect"), tending to draw the model downstream much as a cork in water tends to be pushed upward. A good, expensive solution to these problems is to dynamically alter the shape of the walls until free streamline flow is obtained (the "adaptive" wall).

There exists a distinct lack of published data on how well the corrected wind tunnel data correlates with flight test data. Part of this is due to the fact that the wind tunnel is used mostly to design an aircraft rather than for improving existing aircraft; notable exceptions are where a new aircraft has exhibited unexpected stability or control problems or very much poorer performance than expected. For airfoils, the correlation is clouded by construction problems (both of the airfoil and the full-scale wing), planform effects, and interference/trim drag effects (especially for the new aft-loaded airfoils). Secondly, wind tunnel/flight correlations are often considered proprietary information by the companies involved.

Aeroelasticity, the deflection of an aircraft's structure under aerodynamic loads, causes changes in the full-scale aerodynamic coefficients which are very difficult to simulate in the wind tunnel. This has been given as one of the problems with wind tunnel data obtained for the Concorde, for example.

In any case, learning about potential problems and searching for the optimum configuration will continue to be cheaper in the wind tunnel than in the free air. An interesting and well-publicized exception, or at least a complimentary approach for relatively small aircraft, may be the proof-of-concept aircraft that are being built by companies such as Scaled Composites, using the foam and fiberglass and Kevlar and graphite technology developed for homebuilt aircraft. Scaled Composites has built flying prototypes such as the AD-1 Skew Wing, the Next Generation Trainer for a military competition, the Starship, and the Advanced Technology Turboprop Transport Aircraft. With this construction technology, a wing or a tail can be modified in a matter of days and test flights resumed.

It is estimated that the venerable twin-engine DC-3 transport of the 1930s required about 100 hours of wind tunnel time; the early jet transports required about 5000 hours; the space shuttle about 100,000 hours. The nature of the trend is apparent: exponentially increasing.

E. The Electronic Wind Tunnel: CFD

Even though the use of wind tunnels is cost-effective (their cost is not going up as rapidly as aircraft costs), the moneys involved in the construction, instrumentation, and running of wind tunnels are enormous. The ubiquitous digital computer increasingly is assuming a complementary role in what is generally referred to as computational fluid dynamics (CFD). A personal computer is more than adequate in the design of at least low speed airfoils, although even then only the talented airfoil artist is really successful in manipulating the shapes to produce the pressures which determine the flow velocities and the extent of laminar flow and separation. Typically, the computer is now being used to define the most promising profiles before they go into the tunnel, rather than using the cut-and-try techniques previously in vogue. For amateurs and for instructional purposes, at least, two computer-based airfoil design programs are in the public domain and generally available. The Oshkosh Airfoil Program is basically an inviscid flow calculation (i.e., one ignoring the viscosity of the air and the BL it produces) but it is capable of designing airfoils for cruise conditions as well as showing streamline flow (Ref. 23). The next level of sophistication, an early version of Richard Eppler's airfoil program, is in general use for low speed airfoil design (Refs. 24 and 32). The really difficult problem for an airfoil program is the need to correctly predict and describe the transition point and the degree of separation over the usable range of angles of attack. It is still not possible to model turbulent flow completely.

Fluid flow is most precisely predicted by solving the Navier-Stokes fluid equations, first published in 1823. Technically described as being non-linear, second-order, partial differential equations with over 60 terms, their solution for a complete aircraft (requiring perhaps a billion points around the aircraft) is well beyond the practical capability of even the current generation of supercomputers. Approximations to the full Navier-Stokes equations are now commonly employed to model airfoils and, with additional simplifications, complete configurations. The desire to solve the Navier-Stokes equations more exactly is, to a large extent, driving the push to develop the next generation of supercomputers. Flow visualization via computer graphics is considered to be an integral part of CFD, so that insight into what is happening can guide the number crunching.

Jet engine manufacturers are vigorously and successfully using CFD in the design of more efficient turbines. The United States Air Force is very interested in modeling high angle-of-attack, separated vortex flow because this is the key area for enhanced maneuverability. The major airframe manufacturers are using CFD to evaluate and develop alternative configurations including wing and nacelle design, winglets, and canards. Finally, because hypersonic flow is still very difficult to model in the wind tunnel, most of the design of the National Aerospace Plane (NASP), a hypersonic transport project, will have to be accomplished in the electronic wind tunnel.

To lead the way in computer power for CFD, a national facility, the Numerical Aerodynamic Simulation (NAS) facility became operational at NASA Ames in 1987.

References

1. Rae, William H. and Pope, Alan, *Low-Speed Wind Tunnel Testing*, Second Edition, John Wiley & Sons, Inc., 1984. (This is the basic reference to the design, instrumentation, testing, and required data corrections for subsonic wind tunnels.)

2. Pope, Alan and Goin, Kennith L., *High-Speed Wind Tunnel Testing*, John Wiley & Sons, Inc., 1965. (Although in need of updating, this remains a basic reference for transonic, supersonic, and hypersonic wind tunnels.)

3. Perry, A. E., *Hot-wire Anemometry*, Oxford University Press, 1982. (Describes the use of a hot-wire filament for measuring local airspeeds.)

4. Baals, Donald D. and Corliss, William R., "Wind Tunnels of NASA," NASA SP-440, 1981. (Highly recommended for its very readable description of the history and development and victories achieved by NACA/NASA's wind tunnels.)

5. Ludington, Charles Townsend, *Smoke Streams*, Coward-McCann, Inc., 1943. (If you can find it in a library, this is a useful, fun, little old book with many pictures from a smoke tunnel, accompanied by a chatty discussion of their significance.)

6. "Wind Tunnels," *Air & Space*, National Air and Space Museum, Smithsonian Institution, Vol. 1, No. 3, May–June, 1978, pp. 3–6. (Included are some neat photos and plans for a demonstration wind tunnel for use with an overhead projector.)

7. Merzkirch, Wolfgang, *Flow Visualization*, Second Edition, Academic Press, 1987. (Techniques described include the addition of foreign particles such as smoke, optical methods using variations in the index of refraction and interferometry, and the tracing of heat and energy additions.)

8. Moran, Jack, *An Introduction to Theoretical and Computational Aerodynamics*, John Wiley & Sons, Inc., 1984. (A college textbook at the junior or senior level.)

9. Mueller, Thomas J., "Flow Visualization of Subsonic and Supersonic Flows (The Legacy of F. N. M. Brown)," AD-A059443, June, 1978. (Available from NTIS; see the end of Chapter 11.)

10. Ray, Edward J., Ladson, Charles L., Adcock, Jerry B., Lawing, Pierce L. and Hall, Robert M., "Review of the Design and Operational Characteristics of the 0.3-meter Transonic Cryogenic Tunnel," NASA TM-80123, Sept. 1979. (Available as N79-32159 from NTIS.)

11. Elson, Benjamin M., "Technical Survey: Computational Fluid Dynamics," *Aviation Week & Space Technology*, August 29, 1983, pp. 50ff.

12. "Some Simple Apparatuses for Studying the Dynamics of Airflow and Waterflow," *Scientific American*, October, 1955, pp. 124ff.

13. "How to Make an Aerodynamic Smoke Tunnel," *Scientific American*, May, 1955, pp. 118ff.

14. Strong, C. L., "How to Build a Wind Tunnel that Achieves Supersonic Speeds with a Vacuum System," *Scientific American*, October, 1966, pp. 120ff.

15. Erikson, Gary E., "Water Tunnel Flow Visualization: Insight into Complex Three-Dimensional Flowfields," *Journal of Aircraft*, Vol. 17, No. 9, 1980, pp. 656–662. (Describes the use of a water tunnel for flow visualization of vortices and forebody influences for fighter aircraft.)

16. Saltzman, Edwin J. and Ayers, Theodore G., "Review of Flight-to-Wind Tunnel Drag Correlation," *Journal of Aircraft*, Vol. 19, No. 10, 1982, pp. 801ff.

17. Testa, Al J., "Homebuilt Wind Tunnel Research," *Sport Aviation*, April, 1974, pp. 25ff.

18. Covault, Craig, "Models, Varied Tails Aid Stall-Spin Work," *Aviation Week & Space Technology*, February 17, 1975, pp. 62–64. (Describes NASA Langley's stall-spin research, including the use of the spin tunnel.)

19. Chapman, Dean R., Mark, Hans and Pirtle, Melvin W., "Computers vs. Wind Tunnels for Aerodynamic Flow Simulations," *Astronautics and Aeronautics*, April, 1975, pp. 22–35.

20. Graves, Randolph A., Jr., "Computational Fluid Dynamics—The Coming Revolution," *Astronautics and Aeronautics*, March, 1982, pp. 20ff.

21. Phillips, Edward H., "Magnetic Suspension Studied for Large-Scale Wind Tunnel," *Aviation Week and Space Technology*, April 11, 1988, pp. 41ff.

22. Boyden, Richmond P., Kilgore, Robert H., Rcheng, Ping, and Britcher, Colin P., "Super Magnets for Large Tunnels," *Aerospace America*, June, 1988, pp. 36–40. (magnetic suspension of models)

23. Jones, Robert T., *Modern Subsonic Aerodynamics*, Aircraft Designs, Inc., California, 1988. (A listing of the Oshkosh Airfoil Program is included.)

24. Eppler, Richard and Somers, Dan M., "A Computer Program for the Design and Analysis of Low-Speed Airfoils," NASA-TM-80210, 1980. (Available from NTIS as N80-29254). Eppler, Richard and Somers, Dan M., "Supplement to: A Computer Program for the Design and Analysis of Low-Speed Airfoils," NASA-TM-81862, 1980. (Available from NTIS as N81-13921).

25. Lawing, Pierce L., Kilgore, Robert A. and Dress, David A., "Magnets Promise Productivity," *Aerospace America*, March, 1989. (Magnetic suspension and balance systems could cut the cost of testing by as much as half, this report states.)

26. Munk, Max M. and Miller, Elton W., "The Variable Density Wind Tunnel of the National Advisory Committee for Aeronautics," NACA Report No. 227, 1925.

27. Krothapalli, Anjaneyulu and Smith, Charles A., "Recent Advances in Aerodynamics: Proceedings of an International Symposium held at Stanford University, August 22–26, 1983, Springer-Verlag, New York, 1983. (Recent papers on experimental wind tunnel techniques, including adaptive walls and laser velocimetry.)

28. Dong, Zhao, Zhongqi, Wang, Xiaodi, Zhang, Jialin, Qin, and Xianzhong, Lang, "Color Helium Bubble Flow-Visualization Technique," *Journal of Aircraft*, Vol. 26, No. 6, June, 1989, p. 499ff.

29. Peñaranda, Frank E. and Freda, M. Shannon, "Aeronautical Facilities Catalogue, Volume 1, Wind Tunnels," NASA RP-1132, 1985. (Detailed survey with some historical information as well.)

30. Satran, Dale R., "Wind-Tunnel Investigation of the Flight Characteristics of a Canard General Aviation Airplane Configuration," NASA-TP-2623, 1986. (Available from NTIS as N87-10039.)

31. Phillips, Edward H., "New Goodrich Wind Tunnel Tests Advanced Aircraft De-Icing Systems," *Aviation Week & Space Technology*, October 3, 1988.

32. Eppler, Richard, *Airfoil Design and Data*, Springer-Verlag, 1990. (Description of the theory and practice of airfoil design using the Eppler program plus a data catalog with many airfoils designed for specific applications.)

33. Baals, Donald A., "Baals 7×10 Inch Wind Tunnel," 1989. (These plans for a homebuilt wind tunnel, suitable for an instructional environment but also for some low speed research, are available from the author at NASA Langley Research Center, Mail Stop 464, Hampton, VA 23665-5225. With a 1/4 horsepower electric motor, speeds up to 40 kt in this closed-throat, non-return tunnel are possible.)

Pic. 18.4 The Fred Weick–designed Ercoupe featured the first tricycle gear for lightplanes, a high-visibility cockpit, a two-control control system that prevented spins, and a distinctive twin tail.

Pic. 18.5 This is a modern example of the 1930's homebuilt Pietenpol design; gobs of rudder are needed to coordinate the roll into a turn (i.e., there is a great deal of adverse yaw).

Chapter 19

Aircraft Design Considerations

The achievements of the Wright brothers appear more remarkable the deeper we understand their technical work. . . . What they could not solve with theory and analysis they figured out with careful testing and observations. The standards they set as aeronautical engineers remain unsurpassed.[1]

At ERCO my main project would be to put the ideas we had developed in the W-1 and W-1A into an airplane that would be unusually simple and easy to fly and free from the difficulties associated with stalling and spinning. . . . (Berliner) argued that we should design the ERCO plane so that it could get in and out of the smallest ordinary airports, but not for extremely small places that would be used very seldom. A smaller wing could then be used . . . and extra speed obtained for cruising flight. . . . In selecting the airfoil section for the wing, I went over all of the latest work by Eastman Jacobs and his cohorts in Langley's variable-density tunnel section, and by cross-plotting, determined that the NACA 43013 airfoil would be the optimum shape for our needs. Knowing the NACA's test data for this airfoil enabled us to calculate the approximate wing area of our proposed airplane and to make a reliable series of performance calculations. Climb performance computations with different spans led us to select 30 feet as the span. . . . Next, we went to two separate fins and rudders at the tips of the stabilizer, which would make them entirely free of the slipstream. . . . With the engine sloped three degrees to the right, we found that the plane, with wings level, would trim as desired in a full throttle climb.[2]

A. Mission Definition

Aerodynamics plays a major role in the design of aircraft. Therefore a brief discussion of design considerations, one that builds on material presented in previous chapters, is in order. Surely it is only natural and right that one who has gained some insight into the dynamics of interacting with air should feel a yen to apply this knowledge. It should be noted, though, that aerodynamicists, by themselves, are expected to design aircraft of incredible beauty and with magnificent performance—albeit without adequate provision for structural integrity, powerplants, or people. Structural engineers, by themselves, are expected to design aircraft of great strength—and all of the

good aerodynamic performance possessed by typical bridges. Powerplants engineers, by themselves, are expected to design aircraft with beautifully enclosed, efficiently cooled, powerful engines—attached to a fuselage and wing that pale in comparison. And so all real aircraft are flying compromises between the demands of aerodynamics, propulsion, and structure.

The payload, by definition, justifies the existence of an aircraft and thus many designs begin here. Payload includes the number of occupants and the cargo to be carried, both as regards their volume and their weight. Most transport aircraft, for example, are designed around the fastest and most efficient way to convey a given number of passengers and their luggage over a given distance.

Next, the speed, endurance, and range of the aircraft must be defined. For designs such as ultralights or military aircraft or airliners, there may not be much flexibility in choosing either the minimum or the maximum speeds. Endurance and range are most significant for business and transport aircraft, least

[1]F.E.C. Culick and Henry R. Jex, *The Wright Flyer: An Engineering Perspective* (Washington: Smithsonian Institution Press, 1987).
[2]Fred E. Weick and James R. Hansen, *From the Ground Up: The Autobiography of an Aeronautical Engineer* (Washington: Smithsonian Institution Press, 1988).

significant for sport aircraft. The Voyager presents a marvelous example of how different an aircraft may end up looking when its design is driven, above all, by the range requirement.

Finally, no aircraft design will succeed unless it can be built at reasonable cost. Both the quantity and types of materials used as well as the man-hours required for construction are involved.

Somehow, almost every aircraft ever built ends up heavier, slower, more expensive, and with less range than originally predicted by the designer—just another manifestation of Murphy's Law.

B. Powerplant Selection

It is an old axiom in aircraft design that the process can safely begin only **after** a suitable engine has been located. Few individuals or companies can afford to simultaneously develop a new engine and a new airframe; the Wrights were an impressive exception. No engine with less than 100 hp is being built in the U.S. at present and all aircraft engines are many times more expensive than other engines of equivalent horsepower. Thus the design of the Quickie, intended to provide usable cross-country performance for one person at minimum cost (about $5000 in 1980 dollars), began only after a reliable industrial engine of 18 hp was located. (Jet engines, on the other hand, are expected to grow into higher power versions as the design matures.

Two cycle engines, adapted from snowmobiles and with horsepower ratings ranging from about 20 to 65 hp, have been developed to a satisfactory level of reliability for ultralights. For aircraft with higher landings speeds, four cycle engines are still preferred but are heavier and considerably more expensive to purchase and maintain. A few automobile engine conversions have been developed into reasonably reliable aircraft engines but none have yet attained widespread use; the initial costs of such engines is not much less than certificated aircraft engines although overhaul and parts costs may be much less. An aircraft engine has special requirements because of the thrust loads and bending moments produced by the propeller on the crankshaft. Vibration is an insidious enemy, always trying to shake loose a baffle or ignition lead or throttle cable. The author has had two forced landings caused by induction parts breaking loose and blocking the air flow to the carburetor.

If fuel prices rise sufficiently in the future, new airliners will be designed with the newly efficient turboprop powerplants rather than with pure jet engines. Military aircraft will continue to demand greater thrust and lower fuel consumption from the same size and weight engine, while adding vectored thrust to enhance maneuverability. The propulsion challenge for the National Aerospace Plane, an aircraft which is intended to take off and land on a conventional runway while travelling at hypersonic speeds in orbit, is a daunting one, to say the least.

The only universal rule in powerplant design seems to be that users will want **more** power after the aircraft is built and flying.

C. Configuration Choices

Configuration choices include (a) tricycle or tailwheel, (b) straight wing, or swept forward, or swept back, (c) a fuselage-mounted horizontal stabilizer or a T-tail, (d) tractor or pusher powerplant(s), (e) a horizontal stabilizer with an elevator or a stabilator, and (f) a tandem, canard, 3-surface, or joined-wing configuration—or no tail at all as in the flying wing.

The little wheel ended up at the tail for most aircraft designed through about 1945. A tailwheel is lighter than a nosewheel, is intrinsically more rugged, and produces less drag, everything else being equal. The switchover from a "conventional" gear to the tricycle gear became necessary when concrete replaced sod for runways, when the landing speeds increased to the point that the poor visibility over the nose and the intrinsic directional instability on the ground became unacceptable. For light aircraft, the changeover can be largely credited to the influence of the Fred Weick–designed Ercoupe (Pic. 19.1). After World War II, Cessna's tailwheel C-170 and C-180 gained nosewheels and became the popular C-172 and C-182; Piper converted its tailwheel Pacer to the Tri-Pacer and then designed the popular Cherokee series with a nosewheel from the start. Yet some of the most interesting and useful and fun aircraft still sport tailwheels and continue to attract the more-than-casual pilot who is willing to take on the challenge of their well-earned "have ground-looped or will" reputation.

Wings are swept either for center of pressure reasons, for increased aerobatic capability and easier access on biplanes, or to minimize high speed drag. The canard configuration has been most successful in a pusher configuration—but if a long drive shaft is to be avoided, the wing must be swept back to counter the rear-cg contribution of the engine installation. Another reason to use a swept wing is the potential for moving the wing carry-through structure to a position that doesn't interfere as much with the utilization of interior space. In any case, low speed and landing characteristics of a swept-wing aircraft must be made acceptable through pitch limiting, vortilons, outboard slats, and so forth.

Forward-swept wings require extra strength and rigidity (and weight) to compensate for divergent twist (more lift generates more twist which generates more lift which generates more twist . . .). The advantages, for a subsonic light aircraft, are the excellent outside visibility for the pilot and the fact that the stall naturally tends to begin close to the wing root, with the ailerons stalling last, even for a tapered-wing planform. The blow-by-blow account of the design of a forward-swept light aircraft (scheduled to fly in 1992 or 1993) is described in Ref. 37.

Most horizontal tail surfaces that have moved to the top of the vertical tail have done so for aesthetic or practical reasons unrelated to aerodynamics. Transport aircraft began to use T-tails when the engines were moved from the wing to the rear of the fuselage. It has been argued that T-tails generate less cruise drag because they are out of the wing wake and the vertical fin can be smaller because of the end plate effect of the horizontal surface; however the vertical fin must then be strengthened to counter the additional bending loads and also the takeoff performance may be significantly compromised for a propeller-driven aircraft because the elevator is out of the propwash and unable to rotate the aircraft to the climb angle as early in the takeoff roll. The major aerodynamic problem for swept-wing, T-tailed aircraft has been the possibility of a stable ($\sum M_{CG} = 0$) state at very high angles of attack, for which the T-tail can become **immersed** in the wake of the wing (the "deep stall" problem).

Engines for propeller-driven aircraft most commonly find themselves at the front of the aircraft (the "tractor" configuration). This places a heavy item at the front where it helps keep the *cg* forward of the neutral point. It also may be easier to cool the engine because it gets first shot at the energetic oncoming air. However, on a single-engine aircraft, a forward propeller envelopes the cockpit and the wing behind it in the high velocity air that is swirling back from it, causing extra drag and noise and vibration — although in recent years the drag penalty has been shown to be relatively small. A propeller at the front of an aircraft is more likely to be a hazard to people if they get close or if the plane gets away while being hand-propped. A pusher configuration has its own problems: the propeller tends to be less efficient and more highly loaded structurally because it is getting air in pulses and at different angles as it first meets the air coming over one side of the wing and then over the other; the propeller is much more susceptible to foreign object damage from loose fasteners, from ice chunks breaking loose from the airframe, and from debris thrown up by the tires during the takeoff or landing roll. In the pusher-configured Long-EZ, the canopy can be left open during engine runup because there is very little perceptible increase in air flow with power increase; the front cockpit is noticeably quieter than the rear; and the varying propeller sounds with air turbulence are very noticeable. The fuselage is carefully designed structurally to ensure that the powerplant doesn't move forward and crush the passenger compartment in the event of a crash. Perhaps the greatest plus for the pusher configuration is that the addition of power **adds** to the longitudinal stability rather than subtracting from it; canard-type aircraft in particular may need this.

The one-piece horizontal surface (stabilator or slab tail) became popular in the early days of transonic and supersonic flight because it increased the effectiveness of the surface at these difficult speeds. However a stabilator has been used on much slower aircraft, dating from 1909 (the Santos-Dumont *Demoiselle* with a **one**-piece tail on a universal joint) to the present (the large clan of Piper Cherokees, Warriors, Cadets and derivatives). A stabilator can be made smaller and lighter than a combination horizontal stabilizer/elevator because it is aerodynamically more efficient — but the weight and complexity that is lost externally is often regained internally; to keep the control forces low, the hinge must be close to the aerodynamic center but then significant forward weight must be added to the stabilator inside the fuselage to mass-balance it about the hinge point for flutter prevention.

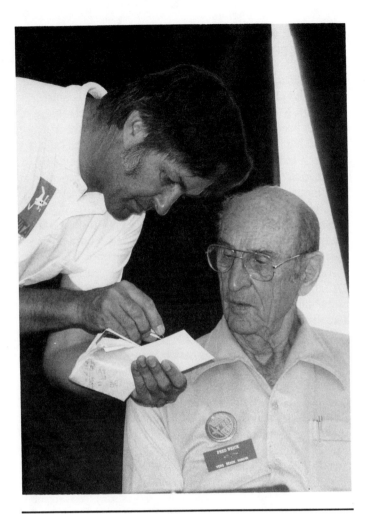

Pic. 19.1 Aircraft designers Burt Rutan (Long-EZ, Voyager, Pond Racer, and others) and Fred Weick (Ercoupe, Pawnee, Cherokee, and others) in a huddle at EAA Oshkosh 1988

The most recent tandem wing design is the Quickie (see the first chapter and Pic. 11.3). In its single-seat form it offers glorious visibility, delightfully light and effective controls, very efficient 100 mi/hr cross-country performance, and the safety advantages inherent in its unstallable, unspinnable design. The original industrial engine has proved reasonably reliable but unable to provide enough climb capability, especially with the inevitable weight increases. The original canard airfoil was not safe in rain or with a load of insect debris or even with a paint stripe across the leading edge because they all caused the BL to thicken and separate at mid-chord; vortex generators solve the problem but add drag. Ground handling has been the second big problem for these aircraft; if the rear wing is lifting significantly during the takeoff or landing roll, the effectiveness of the tailwheel in providing directional stability is severely compromised. A Quickie that is properly built and adjusted can be flown safely by a pilot who has become familiar with its unusual ground handling characteristics, though. Takeoff is normally made with **full** back stick (elevator surface **down**) because the canard has to develop considerable lift to counter the pitching moment of the rear wing and do its share in countering gravity; when liftoff speed is reached, the plane suddenly "levitates" into the air, at which point the stick pressure is relaxed to gain best angle-of-climb speed—although the plane will continue to climb under full control even with full back stick. The aircraft can be safely flown with full back stick, even in steep banks, because the wing (containing the ailerons) remains well below its stall angle of attack. Induced drag builds up on the canard very quickly as the load factor increases in a bank, with a correspondingly large loss of speed. (This proved to be a problem when the canard configuration was used in a soaring aircraft.) It is a little difficult, psychologically, for a pilot to flare for landing at the proper altitude in a Quickie because one's own tail is only a few inches from the ground upon landing. Two-place derivatives of the Quickie design such as the Q-200 and the Dragonfly have solved the engine problem and obtained excellent cruising performance but have continued the history of difficult ground handling characteristics.

Pic. 19.2 Long-EZ landing at Oshkosh at EAA Oshkosh, 1988. Author is in front seat, partner and primary builder Terry Brumbaugh is in rear. (Photo by Aero Vistas, Inc., Chino Valley, Arizona)

Canard aircraft designs have been present throughout the history of powered aircraft, from the 1905 Wright Flyer to the 1930s Focke-Wulf *Ente* to the 1940s XP-55 Curtiss Ascender to the 1980s Beech Starship. But the conventional large wing/small tail completely dominated aircraft design until the 1970s. The man most responsible for the resurgence of the canard is designer Burt Rutan (Pic. 19.1) who, inspired by the Swedish Viggen fighter, designed his VariViggen as a proof-of-concept for his stall-proof canard ideas. Attendees at the week-long EAA (Experimental Aircraft Association) Convention in Oshkosh, Wisconsin, in August of 1972 were thoroughly startled by the appearance and full-aft-stick/steep turn performance of this strange-looking aircraft. Rutan's next canard design, the 2-place VariEze, appeared in 1976, simultaneously introducing the moldless fiberglass/foam sandwich construction technique. The VariEze offered 185 mi/hr or greater performance on a 100 hp engine; Rutan's plans included the necessary tutorials for learning to use this inside-to-out, no-special-tool-required fiberglass and foam method, which resulted in a relatively fast-building, super-slick aircraft. In 1980 the VariEze was sized up to include a larger fuselage, a 115 hp engine, and large fuel tanks for long range; this aircraft is the Long-EZ. The VariEze/Long-EZ was a spectacularly successful design for homebuilding; hundreds of these aircraft have been completed and flown. Over 80 flew to Oshkosh from all over the country for the 1991 EAA Convention.

The Long-EZ the author has flown (Pic. 19.2) is a uniquely fun aircraft. Roll control is outstandingly light and effective, closer to that of a true aerobatic aircraft than to the typical factory-built aircraft. At moderate roll rates there is no significant adverse yaw, so the person in the rear cockpit can fly the aircraft well even without rudder pedals. Pitch control is light but the stick force increases rapidly with increasing load factor, as it should, to discourage overloading the airframe. With one person, the tanks can be filled to give a range of around 1600 miles. On takeoff, the canard is not getting significant propwash and won't rotate the aircraft to bring the wing into a lifting mood until 60 kt or so is reached, a distinct disadvantage on a soft field. Properly built and loaded, the aircraft simply won't stall its rear wing or lose directional or roll control, no matter what the pilot does. The view from the front seat is reasonably good in cruising flight but the canard blanks the normal angle of view when landing and this takes a little getting used to. When the nosewheel comes down after landing (which seems to require a lot of travel because the angle is large and the pilot is well forward of the main landing gear), the wing abruptly loses lift and the plane displays strong directional stability. It has not been possible to fit steering capability into the nose and so all steering on the ground is accomplished via differential braking; this makes the brakes a critical failure item, but the same system is used on many other homebuilt aircraft as well as on some certificated aircraft.

Although the Long-EZ doesn't have good short or rough field capability, its cruise cross-country performance is far greater than that offered by present-day certificated 2-place aircraft. Even faster speed could be obtained if the main gear could be retracted as well as the nose gear, but there is little room in the fuselage until the fuselage-mounted engine becomes two engines on the wings, as on the Starship. Flaps cannot be added to the wings to improve takeoff and landing speeds because the canard wouldn't be able to counter the additional pitching moment. Most people find the semi-supine seating very comfortable once they are in, and this helps bring the equivalent flat plate area down, but entry and exit isn't as easy as it might be.

Because the pilot sits so far forward of the airplane's *cg* in the VariEze and Long-EZ, the unoccupied *cg* is almost directly over the main gear. The result is that the aircraft must be parked with its nose gear retracted and the nose resting on a hard rubber bumper—adding an element of humility to the sea of fork-winged plastic beauties that return to Oshkosh every year. When ready to leave, the pilot lifts the nose by pulling up on the canard, cranks down the nose gear, and climbs in. One very nice feature of a pusher aircraft with a retractable nose gear is that a failure to extend the nose gear usually results in just a scraping and sandpapering of the bottom of the nose. (Although plans for the VariEze and Long-EZ are no longer available, many derivatives have surfaced, some with four seats and some with retractable gear.)

Originally it was suggested that foam and fiberglass would not have good energy absorption capability in a crash; the evidence so far suggests the opposite is true.

The Beechcraft Starship is a high performance, twin-turboprop business aircraft that received its initial airworthiness certificate in 1988 (Pic. 19.3). Its canard/strake/swept wing/winglets/pusher configuration reveals its debt to the Long-EZ design. (Beech prefers to call its winglets "wing tipsails.") Certification was difficult because both the composite construction method and the canard configuration were new to the certificating agency. Wing flaps can be used on this aircraft through the stratagem of sweeping the canard back for cruise and forward for landing. The safety advantages of the canard configuration will be proven with the Starship, it is hoped. Certainly its futuristic shape will forever alter our ideas of what an aircraft should look like.

Another aircraft aiming for the same general market is the Piaggio Avanti (Pic. 19.4); it is also a twin-turboprop, pusher design. It differs from the Starship in having a fixed canard as well as a main wing and a horizontal tail with elevator; i.e., it is a **3-surface** design. Lift on the canard moves the center of pressure forward so the main wing can be in **back** of the passenger compartment rather than in the middle of it. Performance appears to be somewhat better than that of the Starship. Expect to see more of these 3-surface aircraft designs in the future.

Another new configuration is the joined wing aircraft, for which a rearward swept wing is attached directly to tips of a forward swept tail. The primary advantage of such an arrangement is thought to be structural, which means that the empty weight of a joined-wing aircraft should be lower than that for other configurations, for a given payload. An ultralight joined-wing aircraft has been demonstrated and flown at Oshkosh but a large aircraft, intended as a proof-of-concept prototype, is being funded by NASA.

Pic. 19.3 The Beechcraft Model 2000 Starship I, a beautiful new 6- to 7-passenger, turboprop-powered business aircraft; cruise speeds to 336 kt are available. (Photo courtesy of Beech Aircraft Corporation)

Pic. 19.4 The Piaggio P180 Avanti, a beautiful new 6- to 9-passenger, turboprop-powered business aircraft with cruise speeds to 400 kt (Photo courtesy of Rinaldo Piaggio, Genoa, Italy)

A final new/old configuration is the flying wing. The attraction of the flying wing has always been its elimination of the weight and drag of the fuselage and tail while distributing the payload **within** the lifting surface. Of course an aircraft must be **very** large before the wing itself is large enough to hold people and cargo entirely within its natural contours. The flying wing can be made longitudinally stable by using techniques such as wing sweep, wing twist, and reflexed airfoils (turned up at the rear to give a positive pitching moment). Flying wings have traditionally had relatively poor takeoff and landing performance because they don't have a long moment arm for their elevons to compensate for the pitching moment produced by

Pic. 19.5 The Piper Malibu, a 6-seat business aircraft with cruise speeds to 225 kt on 350 hp (Photo courtesy of Piper Aircraft Corporation)

Pic. 19.6 The Glasair III, a 2-seat homebuilt design with cruise speeds to 245 kt on 300 hp (Photo courtesy of Stoddard-Hamilton Aircraft, Inc.)

flaps. In Germany, in the 1930s and 1940s, the Horten brothers designed and flew some relatively successful flying wings, including a jet-powered fighter. In the United States, Jack Northrop was a great proponent of the flying wing and independently developed flying wing bombers for the U. S. Air Force in the 1940s. The XB-35 made its first flight in 1946 with the help of four 3000-horsepower piston engines; the YB-49 flew one year later with the help of eight 4000-pound thrust jet engines. None of these aircraft ever quite made it into full-scale production, for varying reasons. Now Northrop is building the B-2 "Stealth" bomber as a flying wing because this configuration has intrinsically low radar reflectivity through its lack of vertical surface area, as well as good range from its high L/D. Fly-by-wire and active controls give the B-2 the requisite stability and maneuverability.

D. Toward Minimum Drag

In Chap. 13 we calculated the drag of the whole airplane as a drag coefficient times the dynamic pressure times a reference area. The reference area, we noted, was for convenience normally the wing area even though the greater part of the drag in cruising flight might be due to the fuselage. Even if separation can be avoided in the flow around the fuselage, there will be skin friction drag that is nearly proportional to how much actual surface area the air sees. This is called the **wetted area** of the plane because it is the area that would get wet if the plane was hung from a cable and dipped in water. The appropriate skin friction drag coefficient that should be used with the wetted area to estimate the total skin friction drag force, $C_{D, skin}$, has a value of about 0.003 to 0.005 for a typical single-engine aircraft. For larger aircraft the skin friction coefficient is smaller and for very small aircraft the coefficient is larger because of the dependence of BL thickness on RN.

From wetted area considerations alone, the optimum fuselage length would be just enough to accommodate the desired payload. But the fuselage must not taper too abruptly or the flow separates and causes considerable pressure drag. Also, the tail surfaces will have to be very large (with the resultant increase in wetted area) if they are placed too close to the wing. There is a theoretical optimum for the length/diameter ratio (called the **fineness ratio**) for a cylindrical approximation to a fuselage. At low RN, a fineness ratio of 8 is predicted to have the minimum overall drag; this decreases to about 3 at very high RN.

This suggests that, for a 2-seat aircraft, a tandem seat configuration should have significantly lower drag than a side-by-side configuration, everything else being equal. However, many people prefer side-by-side seating to allow face-to-face communication, as well as the extra panel space for full instrument flight rules equipment. Surprisingly, a well-designed side-by-side aircraft gives up very little in speed to a tandem seater of equivalent payload.

A type of drag that has received significantly more attention in recent years is **cooling drag**, the effective drag force due to forcing air through the cooling fins or radiator of the engine. This ranges from about 10% to as large as 30% of the total drag in cruising flight! Some production aircraft (e.g., the Mooney) have been speeded up very noticeably simply by reducing cooling drag. The basic idea is to convert most of the dynamic pressure of the cooling air to additional static pressure in a high pressure, high volume "plenum chamber" before it reaches the cooling fins, and afterwards encourage the air to speed up again before exiting into the slipstream. The heat added to the air helps to speed it up.

Among the business-type, single-engine aircraft currently being manufactured in the U.S., apparently only one can claim to be a new design. This is the Piper Malibu (Pic. 19.5), a 6-place business aircraft boasting a pressurized cabin and a maximum cruise of 225 kt at 25,000 ft with a 350 hp engine. Although it is a very efficient design, possessing very long wings ($b = 46.0$ ft)

Pic. 19.7 The Lancair 320, a 2-seat homebuilt design with cruise speeds to 210 kt on 160 hp (Photo courtesy of Neico Aviation Inc.)

Pic. 19.8 The Questair Venture, a 2-seat homebuilt design with cruise speeds to 242 kt on 280 hp (Photo courtesy of Questair)

and a relatively low span loading ($S = 175$ ft^2, AR = 10.6, and $W/b = 100$), it still uses only aeronautical knowledge available by the mid 1930s.

As this is being written, three homebuilt designs vie for the honor of being the "most efficient" 2-seat, cross-country design: the Glasair (Pic. 19.6), the newer Lancair (Pic. 19.7), and the even newer Questair Venture (Pic. 19.8). All three of these aircraft feature fully retractable gear; the Glasair uses 150 to 300 hp engines, the Lancair 115 and 150 hp engines, and the Venture a derated 300 hp engine. The Glasair and Lancair use foam/fiberglass composite construction but the Venture is a traditional stressed-skin aluminum aircraft.

The Glasair was the first homebuilt design to feature a pre-molded fiberglass/foam fuselage and wing section, yielding an immediately smooth exterior; the design has been refined and extended during the past ten years of its existence. The Glasair and Lancair both use new laminar airfoils.

The Venture uses NACA 230 series airfoils, presumably for their very low pitching moments. Largely designed by Jim Griswold (who is also credited with the design management of the very efficient Piper Malibu), the Venture has an unusually low fineness ratio but possesses a normal tail volume coefficient for good stability and reaches speeds of 300 mi/hr on about 280 hp. NASA's new outer wing cuffs are used along with two Rao-designed "saw cuts" (vertical slots) in the leading edge to make it very spin-resistant. The slots pump energetic vortices over the top of the wing when it is at a high angle of attack. (Wing cuffs by themselves weren't enough because of the complex stall pattern of the 10.4 aspect-ratio wing and the leading

edge stall characteristic of the 230 series airfoils. See Ref. 35 for details.)

In the 1989 CAFE 400 race in California, a highly-tuned VariEze demonstrated a calculated equivalent flat plate area of only 1.33 ft^2 at 158 mi/hr, a Lancair 235 an area of 1.59 ft^2, and the Venture an area of 1.38 ft^2 at a speed of 270 mi/hr.

Beyond laminar airfoils and smooth contours, minimum interference drag, minimum cooling drag, and minimum wetted area, is there anything else that can be done to reduce drag? Yes, it appears that the drag of the fuselage can be reduced through the use of **riblets** and **LEBU** (Large Eddy Break-Up) devices.

Riblets are very tiny, flow-aligned, V-shaped grooves carved into the surface of a fuselage at or beyond the point where the BL has become **turbulent**. At jet speeds, the grooves should have a spacing and a height of only about 0.001" to 0.003". Alignment with the air flow is not particularly critical; there is little degradation in effectiveness up to a yaw angle of about 15° and benefit is entirely lost only at about 30°. The practical achievement of fuselage grooving has been made possible through the use of pre-grooved, self-adhesive vinyl sheets. Skin friction reductions of 6% to 8% have been reported, which translates into significant fuel savings for a transport aircraft. Riblet sheets with 0.0045" spacing and height were used successfully on the 1987 champion racing yacht, *Stars and Stripes*, and are now commercially available.

Riblet size is critical because a grooved surface can double the wetted area. Riblets apparently work by maintaining a more vertically stable turbulent layer close to the surface, so there is less energy transfer outside this layer into more distant layers of the turbulent BL.

LEBUs, on the other hand, are used in the **outer** edge of the turbulent BL to break up large vortices that transfer energy all the way from the surface. Their application to transport aircraft probably will take the form of rings around the fuselage. Reductions of up to 10% in skin friction appear to be possible.

Because riblets and LEBUs operate on different parts of the turbulent BL, they can be combined to maximize skin friction reduction. It may be more difficult to design effective LEBUs for high RN applications, though.

E. Preliminary Performance Estimates

The most popular homebuilt aircraft using traditional riveted-aluminum construction, presently, is Van Aircraft's RV series. Family members include the RV-3 (single-seat), RV-4 (2 tandem seats), and RV-6/RV-6A (2 side-by-side seats); all are tailwheel designs except for the RV-6A. With a relatively clean airframe, moderate wing loading, and good power loading, the aircraft provides both good cruise performance and good short takeoff/landing performance.

The designer's specifications for the 2-place RV-4 are provided in Table 19.1.

$b = 23.0$ ft	$S = 110$ ft^2	$W = 1500$ lb
Power: 150 hp, with fixed pitch propeller		
$V_{MAX} = 201$ mi/hr		
$V_{STALL} = 54$ mi/hr		
Takeoff distance = 450 ft		
Rate of climb = 1650 ft/min		

Table 19.1 Sea Level Standard Performance of the RV-4

Suppose that we had designed this aircraft. What sort of performance could we expect? Clearly, the first step is to estimate the equivalent flat plate area. The most accurate method is to (a) add up the wetted areas of all exposed surfaces and multiply this sum by the appropriate skin friction coefficient (depending on the smoothness of the surface and protuberances), (b) add an estimated increment for pressure drag (regions of separated flow in cruise), (c) add an increment or percentage for interference drag (depending on wing location and degree of filleting), and (d) add 10% or more for cooling drag. References 15 and 37 in this chapter and Ref. 10 of Chap. 13 provide examples of this approach various types of aircraft.

A second approach, which we use here, is to make preliminary estimates based on average experimental data from other, similar designs. Reference 15, for example, suggests that a single-engine aircraft will have an f approximately given by

$$f = 0.0055 S_{WET} \tag{19.1}$$

The wetted area of a flat plate must be just twice the surface area of one side—but for a wing or tail member with a finite thickness, Ref. 15 provides the following approximation

$$S_{WET, WINGS \& TAIL} = A_{EXPOSED}\left[1.977 + 0.52\left(\frac{t}{c}\right)\right] \tag{19.2}$$

where $A_{EXPOSED}$ is the exposed, projected area, t is the thickness and c is the chord; also, for the fuselage,

$$S_{WET, FUSELAGE} = 3.4\left(\frac{A_{TOP} + A_{SIDE}}{2}\right) \tag{19.3}$$

Using a 3-view drawing of the RV-4, I obtain

$$S_{WET, RV-4} \approx S_{WET, WINGS} + S_{WET, TAIL} + S_{WET, FUSELAGE}$$
$$\approx 200 \text{ ft}^2 + 20 \text{ ft}^2 + 160 \text{ ft}^2 = 380 \text{ ft}^2$$

Then, from Eq. 19.1,

$$f_{RV-4} \approx \left(0.0055\right)\left(380 \text{ ft}^2\right) = 2.1 \text{ ft}^2$$

Since the top speed at SL is primarily determined by f (because the induced drag term is small), we can neglect the second term in Eq. 13.45 for the moment and solve for the speed:

$$V_{MAX, SL} \approx \left[\frac{2 \times P_{AVAIL}}{\rho f}\right]^{1/3} \tag{19.4}$$

For a propeller efficiency of 0.8 (about right for the compromise that a fixed-pitch propeller enforces), the available power at sea level is

$$P_{AVAIL} = \eta P_{RATED} = \left(0.8\right)\left(150 \text{ hp} \times \frac{550 \text{ ft-lb/s}}{hp}\right)$$
$$= 66,000 \text{ ft-lb/s}$$

Substituting this into Eq. 19.4 yields

$$V_{MAX, SL} \approx \left[\frac{2 \times 66,000}{(0.002377)(2.1)}\right]^{1/3} = 298 \text{ ft/s} = 203 \text{ mi/hr}$$

which is closer to the designer's value than it has any right to be.

Next we wish to estimate the stalling speed of the RV-4. It uses a NACA 5-digit, 230-series airfoil and the NACA "airfoil bible" (Ref. 4 in Chap. 10) suggests a maximum lift coefficient of 1.6 for a RN of 3.0×10^6, which will turn out to be the appropriate RN for the RV's average chord of

$$\bar{c} = \frac{S}{b} = \frac{110 \text{ ft}^2}{23.0 \text{ ft}} = 4.78 \text{ ft}$$

Reference 15 suggests that a wing will obtain about 90% of the airfoil's maximum lift coefficient (which should take care of the download on the tail) and that plain flaps (which the RV-4 uses) will add about 0.9 times the ratio of the flapped wing area to the reference wing area. The flapped wing area on the RV-4 is about 46 ft^2 so we estimate that

$$C_{L, MAX} \approx (0.9)(1.6) + \left(\frac{46}{110}\right)(0.9) = 1.82$$

The equivalent stalling speed then depends on the weight and the wing area (from Eq. 12.16):

$$V_{STALL, SL} = \sqrt{\frac{2W}{\rho C_{L, MAX} S}}$$
$$= \sqrt{\frac{2(1500)}{(0.002377)(1.82)(110)}} = 79.4 \text{ ft/s} = 54 \text{ mi/hr}$$

which happens to be just the same as the designer's value.

Estimating the SL rate of climb will require the generation of P_{REQ} and P_{AVAIL} curves for the RV-4, as Fig. 13.10 did for the C-152. First, though, we must estimate the value of e, the wing

Pic. 19.9 A brand-new RV-4 leaps off the ground on its first takeoff.

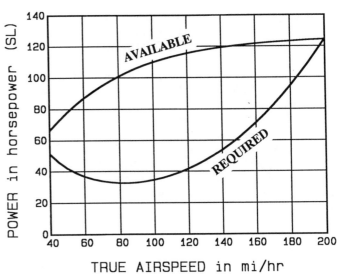

Fig. 19.1 Calculated power curves for the RV-4, SL conditions

efficiency factor. Since the RV-4 has a very low aspect ratio (using Eq. 12.1):

$$\text{Aspect Ratio} = \frac{b^2}{S} = \frac{(23 \text{ ft})^2}{110 \text{ ft}^2} = 4.8$$

we can argue for a low-end value of $e = 0.7$.

Such a low aspect ratio undoubtedly extracts a drag penalty in climb and high altitude performance, but that is largely overcome with the very good power-to-weight ratio. The good news is that the RV-4 has good stall characteristics even though it uses an airfoil with a nasty leading edge stall (Fig. 10.7) and an untwisted wing. Evidently the favorable stall pattern of its low AR, rectangular planform overwhelms the adverse factors. (The high AR, tapered-wing Venture, on the other hand, has record-breaking high altitude performance but had to resort to the heroics detailed in Ref. 35 to obtain satisfactory wing stall characteristics with basically the same airfoil.)

Now that f and e have been estimated, power curves for standard sea level conditions are easily drawn with the help of the propeller theory of section 13B and Eq. 13.45 (Fig. 19.1).

Note that, as remarked earlier, the simple polar approximation to the drag curve can't hope to realistically model the power required near the stall, so that curve in Fig. 19.1 should turn up much more abruptly before and at the stall speed.

Careful examination of Fig. 19.1 shows a maximum excess power of about 75.7 hp, so the maximum rate of climb at SL, from Eq. 14.17, can be estimated to be

$$\text{Rate of Climb} = \frac{\text{excess power}}{\text{weight}}$$

$$= \frac{(75.7 \text{ hp})(550 \text{ ft–lb/s/hp})}{1500 \text{ lb}} = 27.8 \text{ ft/s} = 1665 \text{ ft/min}$$

which is in close agreement with the designer's figures once again.

Finally, let's try to estimate the takeoff roll under SL standard conditions. We might start with the calculated average acceleration for the C-152 of 6 ft/s² (Fig. 14.2) and estimate a

correction for the RV-4's lighter weight and greater power. The acceleration is more or less inversely proportional to the mass (and the weight) and directly proportional to the power and the thrust. Thus we might estimate that

$$a_{\text{RV–4}} = \left(6 \text{ ft/s}^2\right)\left(\frac{1670 \text{ lb}}{1500 \text{ lb}}\right)\left(\frac{150 \text{ hp}}{110 \text{ hp}}\right) = 9.11 \text{ ft/s}^2$$

If we use a liftoff speed of 15% above the stall speed of 54 mi/hr (79.2 ft/s), application of Eq. 14.10 yields

$$\text{Takeoff Distance} \approx \frac{V_{\text{LIFTOFF}}^2}{2\,a} = \frac{(1.15 \times 79.2)^2}{2\,(10.0)} = 455 \text{ ft}$$

References

1. AIAA Professional Study Series: *Case Studies in Aircraft Design* (Available from The American Institute of Aeronautics and Astronautics, 370 L'Enfant Promenade, SW, Washington, DC 20077-0820.)

    1. Northrup F-5 (1978)
    2. Boeing 727 (1978)
    3. Harrier
    4. Concorde
    5. Rockwell International Sabreliner-65
    6. Lockheed C-5
    7. de Havilland STOL Aircraft
    8. Gossamer Condor and Albatross (1980)
    9. F-16 Fly-by-wire Flight Control System
    10. Gulfstream III (1980)

2. Bond, Dan and Doo, Johnny, "Basic Diffuser Design," *Sport Aviation*, Vol. 38, No. 9, September, 1989. (Some concepts underlying the design of low-drag cooling ducts.)

3. Crawford, Donald R., *A Practical Guide to Airplane Performance and Design*, Crawford Aviation, California, 1979. (The performance aspect of aircraft design, especially for homebuilt aircraft.)

4. Crawford, Donald R., *Airplane Design*, Crawford Aviation, California, 1986. (Reprint of a series that appeared first in *Kitplanes* magazine.)

5. Heinemann, Edward H. and Rausa, Rosario, *Ed Heinemann: Combat Aircraft Designer*, Naval Institute Press, Maryland, 1980. (The autobiography of the designer of the Douglas Dauntless and Skyraider and many other well-known aircraft.)

6. Heinemann, Edward H., Rausa, Rosario, and Van Every, K. E., *Aircraft Design*, The Nautical & Aviation Publishing Co. of American, Inc., 1985. (Non-technical, well-illustrated description of design by the well-known Douglas Aircraft designer.)

7. Hollmann, Martin, *Modern Aircraft Design*, Aircraft Designs, Inc., California, 1985. Also Volume 2, 1988. (Conceptual and structural design of composite aircraft for the homebuilder.)

8. Hooker, Sir Stanley, *Not much of an Engineer*, Airlife Publishing Ltd., Shrewsbury, England, 1984. (Autobiography of a famous British aircraft designer.)

9. Huenecke, Klaus, *Modern Combat Aircraft Design*, Naval Institute Press, Airlife Publishing Ltd., 1987. (Well-illustrated, mostly qualitative treatment.)

10. Johnson, Clarence L., *Kelly—More Than My Share of It All*, Smithsonian Institution Press, Washington, D.C., 1985. (Autobiography by the famous Lockheed designer, whose "Skunk Works" produced such famous aircraft as the F-80, F-104, U-2, and SR-71.)

11. Nicolai, Leland M., *Fundamentals of Aircraft Design*, School of Engineering, University of Dayton, Ohio, 1975. (A college text with emphasis on high speed, turbine-powered aircraft.)

12. Pazmany, L., *Light Airplane Design*, L. Pazmany, Pazmany Aircraft Corporation, California, 1986. (A description of the preliminary design of the author's PL-1 and Pl-2 metal aircraft.)

13. Pazmany, Ladislao, *Landing Gear Design for Light Aircraft*, Pazmany Aircraft Corporation, California, 1986. (Well-illustrated; useful.)

14. Penrose, Harald, *Architect of Wings*, Airlife Publishing Ltd., Shrewsbury, England, 1985. (Biography of Roy Chadwick, the designer of the British Lancaster bomber.)

15. Raymer, Daniel P., *Aircraft Design: A Conceptual Approach*, American Institute of Aeronautics and Astronautics, Inc., Washington, D.C., 1987. (This is a recent and very useful text by an experienced aircraft designer; the emphasis is on the conceptual design phase in which the configuration, size, weight, and performance of an a new design are estimated. Design examples discussed are a single-seat aerobatic homebuilt and a lightweight supersonic fighter.)

16. Roskam, Jan, *Airplane Design*, Part I: Preliminary Sizing of Airplane, Part II: Preliminary Configuration Design and Integration of the Propulsion System, Roskam Aviation and Engineering Corporation, Kansas, 1985. (College text by an aeronautical engineer at the University of Kansas.)

17. Stinton, Darrol, *The Design of the Airplane*, Van Nostrand Reinhold Company, New York, 1983. (A chock-full, readable preliminary design text by a practicing designer and test pilot.)

18. Teichmann, Frederick K., *Airplane Design Manual*, Pitman, New York, 1958. (A classic old college text, here in its fourth edition.)

19. Thurston, David B., *Design for Flying*, McGraw-Hill Book Company, 1978. (An interesting, non-technical book by a former Grumman engineer.)

20. Torenbeek, Egbert, *Synthesis of Subsonic Aircraft Design*, Martinus Nijhoff Publishers, Holland, 1982. (The emphasis here is on the layout, aerodynamic design, propulsion and performance of large aircraft. Some good data on current designs are included.)

21. Warner, Edward P., *Airplane Design: Performance*, McGraw-Hill Book Company, Inc., 1936. (This second edition of a 1927 text, commonly available in aviation libraries, provides a delightful historical insight into aircraft design a half-century ago.)

22. Weick, Fred E. and Hansen, James R., *From the Ground Up: The Autobiography of an Aeronautical Engineer*, Smithsonian Institution Press, Washington, D.C., 1988. (This is the attractive story of the practical man responsible for the design of the Ercoupe and various Piper aircraft.)

23. Whitford, Ray, *Design for Air Combat*, Jane's Publishing Inc., New York, 1987. (Qualitative with many illustrations and good explanations for the design features of modern fighter aircraft.)

24. Woko, Howard S., editor, *The Wright Flyer: An Engineering Perspective*, Smithsonian Institution Press, Washington, D.C., 1987. (Modern analyses of structural, propulsive, aerodynamics, and stability aspects of the world's first successful airplane.)

25. Wood, K. D., *Aircraft Design*, Third Edition, Johnson Publishing Company, Colorado, 1968. (Various aspects of layout design, with copious data in the appendices.)

26. Wright, Orville, *How We Invented the Airplane: An Illustrated History*, Dover Publications, Inc., New York, 1953 and 1988. (The central piece here is Orville's paper in connection with a 1920 legal action. Many historic photos are provided, as is an essay by Fred C. Kelly, the Wrights' "official" biographer.)

27. Selberg, Bruce P. and Rokhsaz, Kamran, "Aerodynamic Tradeoff Study of Conventional, Canard, and Trisurface Aircraft Systems," *Journal of Aircraft*, Vol. 23, No. 10, October, 1986.

28. Lange, Roy H., "Review of Unconventional Aircraft Design Concepts," *Journal of Aircraft*, Vol. 25, No. 5, May, 1988.

29. Rokhsaz, Kamran and Selberg, Bruce P., "Three-Surface Aircraft—Optimum vs Typical," *Journal of Aircraft*, Vol. 26, No. 8, August, 1989.

30. "Special Report: Advanced Fighter Technology," *Aviation Week & Space Technology*, June 23, 1986.

31. "Special Section: Drag Reduction," *Aerospace America*, January, 1988, pp. 14–27, 48–49. (Natural laminar flow, laminar flow control, riblets, and LEBUs.)

32. Pazmany, L., "Groovy Drag Reduction," *Kitplanes* magazine, May, 1988. (The design of riblets.)

33. Walsh, Michael J., Sellers, William L., III, and McGinley, Catherine B., "Riblet Drag at Flight Conditions," *Journal of Aircraft*, Vol. 26, No. 6, June, 1989.

34. Philips, Edward H., "Modified LEBU Devices Could Cut Skin-Friction Drag," *Aviation Week & Space Technology*, August 21, 1989.

35. Ross, H. M., Yip, L. P., Perkins, J. N., Vess, R. J., and Owens, D. B., "Wing Leading-Edge Droop/Slot Modification for Stall Departure Resistance," *Journal of Aircraft*, Vol. 28, No. 7, July, 1991. (Venture wing development: outer-wing cuff plus saw-cut vortex generators.)

36. *Piaggio Avanti, Beech Starship Offer Differing Performance Characteristics*, Aviation Week & Space Technology, October 2, 1989.

37. Roncz, John G., *Sport Aviation*: "Designing Your Homebuilt," February, 1990; "Sizing Your Wings," March, 1990; "Wing Incidence & Tail Size," April, 1990; "Forward Sweep & The Great Tire Crisis," May, 1990; "Questions and Answers," June, 1990; "Tail Incidence," August, 1990 and September, 1990; "Ground Effect," December, 1990; "Canards and Other Unsolved Mysteries," January, 1991; "Evolution of a Homebuilt Design," February, 1991. (John here substitutes a spreadsheet for traditional computer programming, and plays it like a piano—but you still better know what you're doing to use it. Lots of interesting observations and tidbits, all delivered in the inimitable Roncz style.)

Appendix A: Answers to Review Questions

Chapter 1. **1.** It is by virtue of the opposite (and equal) push they receive when they push air down (i.e., giving it a downward momentum it didn't possess before). **2.** Stabilize the aircraft in a side-to-side sense. **3.** left **4.** forward **5.** horizontal and vertical stabilizing and control surfaces **6.** They can't influence the air ahead of them so they need a more powerful control surface. **7.** right aileron up, left aileron down **8.** It is a **rate** control rather than a proportional control; additional back (nose-up) elevator pressure and deflection **9.** More drag and lowered approach and landing speeds **10.** control deflection: wheel turned to the right or stick pushed to the right; aileron deflection: right aileron up, left aileron down **11.** its camber or curvature; increasing **12.** Spoilers free the whole trailing edge of the wing for use as a flap surface. **13.** a lifting surface (usually small) in front of the primary horizontal surface (the wing) **14.** A narrow forward extension of the wing along the fuselage. **15.** Control the flow of air around the tip of the wing between upper and lower surfaces. **16.** This is when the control stick or wheel produces electrical signals which indirectly cause control surface deflection through a computer and electric motors rather than through a direct mechanical linkage. **17.** spoiler: left spoiler up; ailerons: left aileron is up and right aileron is down. For a properly rigged airplane, there is no control deflection when in a constantly-banked turn with a medium angle of bank (around 30°); at smaller angles the natural stability of the aircraft tends to reduce the bank angle and at large angles there is usually an over-banking tendency. **18.** wing **19.** power **20.** span

Chapter 2. **1.** magnitude; direction **2.** distance, velocity, force **3.** end **4.** in the same or opposite directions **5.** components **6.** use of the Pythagorean theorem **7.** arctan(y component/x component) **8.** Velocity is a vector, and the use of the word means that its direction as well as its magnitude is being considered; for speed, only the magnitude is being considered. **9.** (a) Second velocity was after the first but for the same time, (b) Second velocity was after the first, for the same time, but due to movement of the air while the aircraft was stationary, (c) simultaneous motion of the aircraft relative to the air. **10.** (a) Wind speed relative to air speed is the determining factor as to whether we can compensate for the wind. (b) It is going to be harder to compensate for a direct crosswind than for a nearly direct headwind or tailwind, as far as going along (or backward from) the destination. (c) large ratio for (wind speed)/(airplane speed) and ($\theta_C - \theta_W$) close to ± 90° **11.** wind speed less than airplane speed **12.** airplane speed greater than wind speed and then from Eq. 2.8, $V_G = \pm\sqrt{[V_A{}^2 + V_W{}^2]}$ **13.** increased, decreased **14.** cruise **15.** short trip: use 5000 ft; long trip: use 8000 ft

Chapter 3. **1.** the point, not necessarily within the object, at which the force of gravity can be considered to act (rather than on the actual distributed mass) **2.** when the mass distribution dimensions of a body are comparable with the distance from the attracting body **3.** weight; third **4.** *cg.* **5.** The heavier object has comparably greater inertia so the acceleration is the same. **6.** In a vertical dive the air drag force just balances the weight vector; there is no acceleration and so this is an equilibrium situation. **7.** If it were in front, both the moments due to the gear would be negative and it would be unstable toward assuming a nose-dragging attitude. **8.** positive (CW) moment **9.** reduced maximum (±) lift by the control surface **10.** No, only if the plane is not accelerating vertically; e.g., the ground exerts a greater force than the weight of the aircraft during a hard landing. **11.** the glider's weight

Chapter 4. **1.** The temperature of a single atom or molecule isn't meaningful because we can't measure individual energies or speeds; we can measure average energies for a group of atoms or molecules. **2.** linear **3.** perpendicular to the bounding surface **4.** equal pressures inside and outside **5.** pressure **6.** density, absolute temperature **7.** lb/ft^2, sl/ft^3, R **8.** Weight varies with position on the earth's surface. **9.** c_V **10.** square root of the absolute temperature **11.** nitrogen and oxygen; about 98% **12.** the transition to a zero lapse rate **13.** with both **14.** static equilibrium **15.** no water vapor and no air motion **16.** pressure altitude **17.** An altimeter must accurately measure pressures, independent of the ambient temperature. **18.** weather conditions **19.** Height is inversely proportional to the density of fluid; atmospheric pressure is supporting the weight of the column of fluid and so a less dense fluid will be taller for a given atmospheric pressure. **20.** See text. **21.** the same (assuming at same temperature) **22.** See text. **23.** less than **24.** greater than **25.** yes for both **26.** cooler, more

Chapter 5. **1.** shear **2.** large **3.** coefficient of viscosity **4.** temperature **5.** kinematic viscosity **6.** kinematic viscosity **7.** turbulent **8.** laminar, turbulent **9.** RN **10.** smoothness and waviness; turbulence **11.** laminar **12.** viscosity **13.** kinematic viscosity **14.** streamlines **15.** mass **16.** speed **17.** isolated **18.** pressure, kinetic **19.** total, stagnation **20.** total **21.** 4 **22.** oncoming air **23.** less **24.** greater, less **25.** the same as **26.** zero, zero **27.** more, stagnation point **28.** Kutta-Joukowski **29.** It has a zero or a small angle of attack. **30.** natural, resonance **31.** clockwise **32.** pressure, viscous, turbulent **33.** pressure, viscous **34.** turbulence **35.** low, backward, pressure **36.** pressures, less than **37.** pressure, viscous **38.** false, RN is the determining factor; example: low speeds such as for a falling dust particle in air **39.** No. Like frictional forces, the viscous forces only oppose motion and are proportional to the motional speed. They would violate energy conservation if they were greater than the driving forces doing the work on the object. **40.** wake **41.** turbulent, pressure **42.** smaller **43.** density, four (same RN) **44.** smoothness and waviness **45.** thickness **46.** one-half **47.** increasing **48.** Air close to the airfoil and in the BL has stalled and then separated. **49.** separation and transition to turbulence **50.** c **51.** near the upper leading edge **52.** greater than **53.** The droplet is in the wake or is near the separation point for a separated BL. **54.** induced

Chapter 6. **1.** greater than the speed of sound **2.** compressible, Mach no. **3.** pressure, density **4.** less than 1 **5.** normal, decreases, doesn't change   **6.** always subsonic   **7.** M_D **8.** greater than   **9.** less (Fig. 6.3)   **10.** vortex generators   **11.** shock   **12.** bow   **13.** oblique   **14.** supercritical   **15.** the drag divergence Mach number   **16.** sharp, greater than   **17.** wave   **18.** kinetic   **19.** Transonic, supersonic, 5, hypersonic   **20.** The loss in efficiency for the propeller tips (and perhaps the airframe) because of Mach effects is a greater problem than that of getting more power out of an engine.

Chapter 7.   **1.** the difference between total and static air pressure (the dynamic pressure)   **2.** air density greater than sea level standard (Eq. 7.8)   **3.** The quantity (total pressure – ambient pressure) will be greater than it should be because the ambient pressure seen by the indicator is lower than it should be; therefore the indicated airspeed will be higher than it should be.   **4.** In this case the quantity will be less than it should be because the indicator senses a constant ambient pressure when it should be sensing a decreasing value; therefore the indicated airspeed is less than it should be.   **5.** The total pressure must have remained constant because (total pressure – ambient pressure) evidently increased as the ambient pressure was decreasing during the climb; if the total pressure line became blocked except for a vent hole, the indicated airspeed would approach zero as the total pressure approached the ambient pressure. It is even conceivable that a venturi effect could reduce the total pressure that is sensed to a value below the ambient pressure, causing negative indicated airspeeds.   **6.** the aircraft at 20,000 ft, by inspection of Eq. 7.8, solved for V_∞ **7.** only when the ambient pressure is greater than SL standard; when ambient pressure is less than SL standard   **8.** indicated, true   **9.** The ground speed at which takeoff airspeed is reached depends on reaching the proper dynamic pressure; this depends on the wind and on the density altitude; the airspeed indicator should be monitored and relied on for takeoff information (assuming no icing!).   **10.** the air temperature   **11.** higher than, greater than   **12.** The pressure difference depends on the geometry of the venturi system and the dynamic pressure, so the instrument should work similarly well at similar **indicated** airspeeds, independent of air density.   **13.** 80 kt indicated air speed and SL density, by inspection of the figures   **14.** normal shock wave because the air speed in the probe must be zero, and therefore surely subsonic

Chapter 8.   **1.** all chord-wise flow with no span-wise flow; allow us to determine the effects of airfoil **shapes**, isolated from the effects of the properties of the air and the wing shape   **2.** symmetric airfoil   **3.** develops lift at zero angle of attack (it also causes the minimum drag to occur at a positive value for the lift coefficient rather than at the unusable zero lift value) and, especially for thin airfoils, yields larger maximum lift coefficients because the upper nose curvature can be reduced   **4.** the **location** of the maximum thickness point along the chord line   **5.** angles of attack; flight Mach number and RN   **6.** whether the angle of attack is too close to the stalling angle of attack, not the angle of attack relative to the horizon   **7.** z and x **8.** zero for a symmetric wing, negative for a positively cambered airfoil (assuming no wing twist)   **9.** near the leading edge; bugs, ice, "hangar rash" (handling dents)   **10.** high speed and aerobatic aircraft for wings and for tail members on others   **11.** dynamic pressure; flight Mach number and RN   **12.** +1, which corresponds to the minimum relative airspeed of zero at a stagnation point   **13.** the direction of the relative wind   **14.** level flight   **15.** b **16.** perpendicular to and pointing toward the surface for pressures greater than ambient; perpendicular to and pointing away from the surface for pressures less than ambient   **17.** center of pressure, zero   **18.** pressure; viscous or skin friction; energy or momentum; wake rake   **19.** energy, surface area, coeffi-

cient   **20.** chord   **21.** coefficient, angle of attack, RN and Mach number   **22.** dimensional   **23.** a nearly constant pressure region, ending at the trailing edge

Chapter 9.   **1.** (a) forward from $0.48c$ to $0.33c$, (b) forward from $0.33c$ to $0.29c$, (c) rearward from $0.29c$ to $0.39c$ **2.** rearward; leading edge   **3.** none, the lower surface has negative lift at the design lift coefficient   **4.** about 7 times greater (based on the length of the arrows in Fig. 9.1)   **5.** increase, in general (i.e., less negative)   **6.** angle of attack, ac **7.** about 20% as much   **8.** about 85%   **9.** moment; symmetrical   **10.** RN, separated   **11.** a, a_0 **12.** lift, polar   **13.** large   **14.** less (the minimum drag coefficient is larger even though the total drag force will be less because of lesser surface area, S **15.** zero, camber   **16.** negative, zero   **17.** negative, zero   **18.** shock stall; reaches the trailing edge   **19.** fully supersonic flow prevails   **20.** minimum   **21.** $0.25c$, $0.5c$ **22.** before   **23.** design   **24.** it moves from close to the leading edge to noticeably below the leading edge

Chapter 10.   **1.** NACA airfoil from about 1935, max camber of 1.8% of chord is located at 15% of chord; thickness is 24% of chord   **2.** NACA airfoil from about 1932, the max camber of 2% of chord is located at 40% of the chord; thickness is 18% of chord   **3.** NACA airfoil from about 1944, intended to supported laminar flow over front part, minimum pressure point is at 30% of chord, low drag range extends over + and –0.1 from the design lift coefficient of 0.2, thickness of 12% of chord   **4.** Wortmann airfoil from 1962, flapped, 13.1% thick, 17% chord flap   **5.** NASA airfoil from the 1990s, intended for medium speed applications, the second series with a design lift coefficient of 0.2 and a thickness of 11% of chord   **6.** turbulent flow airfoils (they were tested with a BL trip)   **7.** scale (RN) effects   **8.** lift, camber, pitching moment   **9.** propeller sections   **10.** six   **11.** NLF (**natural** laminar flow)   **12.** lift coefficient   **13.** low speed   **14.** Felix Wortmann   **15.** Whitcomb   **16.** flatness, downward curvature (positive camber)   **17.** (pitching) moment   **18.** 0.0001, drag   **19.** an oil film   **20.** thickness   **21.** 13%   **22.** decreased, thinnest   **23.** thinnest   **24.** structural   **25.** pressure distribution   **26.** 2412   **27.** RAF-6, Clark Y   **28.** maximum lift coefficients   **29.** two to nine, one   **30.** 0.4   **31.** greater maximum lift coefficient and a smaller (less negative) pitching moment for the 230 series   **32.** 66-216 because the minimum pressure point is farther back   **33.** high speed natural laminar flow, low Reynolds number, supercritical

Chapter 11.   **1.** lift   **2.** transitions (to a turbulent BL)   **3.** camber   **4.** very smooth and wave-free   **5.** very large   **6.** 2.0, 2.5   **7.** adverse pressure, distance   **8.** man-powered   **9.** roof-top   **10.** smooth, free from waves   **11.** rain, lift, drag   **12.** fluid, leading edge   **13.** transition (to help it resist separation)   **14.** sixty, one hundred   **15.** sublimating   **16.** curved plate   **17.** tripping   **18.** trailing edge, rounded, lift   **19.** bubble, bursts   **20.** angle of attack   **21.** glaze   **22.** pitch-up, roll   **23.** 1.1   **24.** stall; maximum lift, lift/drag ratio   **25.** vortex   **26.** the LS(1)-0013 because camber shifts the lift curve to the left   **27.** hysteresis, low RN

Chapter 12.   **1.** induced   **2.** pitch up   **3.** roll counterclockwise   **4.** rotational   **5.** lift   **6.** heavy, low, minimum   **7.** climb, high   **8.** same as, greater than (a bit of algebra here)   **9.** viscous or frictional, increased, increased (because of the smaller AR)   **10.** straight up and straight down and for other short term "zero-g" maneuvers   **11.** greater than, slow (high angle of attack—Eq. 12.2)   **12.** shape   **13.** downwash   **14.** about the same   **15.** tip, tapered, lesser, washout

16. bending moment **17.** rectangular **18.** tight turning (trying to stay in a narrow thermal), glide **19.** middle of the fuselage, high wing, fillets **20.** higher (because the smaller version would have a smaller maximum lift coefficient because of RN effects) **21.** Start at wing root, progress gradually to tip to retain lateral (aileron) control through the stall and get a pre-stall buffet of the tail and elevator. **22.** stall strips **23.** pitch-up **24.** rearward; snap rolls **25.** curves from top to bottom **26.** flaps, slots and suction **27.** Increase overall maximum lift coefficient while tending to prevent tip stall. **28.** spoilers, flaps **29.** decrease, four **30.** roots, structural, composite **31.** upwash, downwash, upwash, downwash **32.** nose-down (–) **33.** induced, distributes the savings most uniformly among the birds **34.** fences **35.** blown **36.** structural and weight **37.** supercritical airfoils, sweep, proplets, composite materials **38.** zero **39.** aspect ratio **40.** lift, angle of attack **41.** elliptical, 1 **42.** increase, four (Eq. 12.6) **43.** decrease, increase **44.** more, more, better, poorer **45.** vortices, tip **46.** tight turns **47.** lift force = weight **48.** indicated, calibrated and equivalent airspeeds, weight **49.** wing loading, maximum lift coefficient **50.** root **51.** slot **52.** twist, stall strips **53.** sweepback, tips **54.** snag (leading edge discontinuity) **55.** wool (yarn) tufts **56.** upper surface blowing, engine **57.** pitch (nose) down **58.** angles of attack (yielding poor pilot visibility) **59.** drag, downwash **60.** Fowler, Fowler **61.** airfoil **62.** angle of attack **63.** flap, dead spot **64.** component, perpendicular **65.** oblique **66.** fly-by-wire **67.** instability **68.** supersonic **69.** vortilons; being easier to make and install and causing less drag in cruise **70.** sheared, 60°, 25%, aileron **71.** No difference in the angle of attack but the split flap generates considerably more drag at that lift coefficient. Thus the plain flap would be more likely to be useful for takeoffs while the split flap would give steeper approaches for landing. **72.** forward-swept wing and then the rectangular planform **73.** The upward-deflected aileron stalls at a higher angle of attack than the basic section, and the downward-deflected aileron at a lower angle of attack. **74.** $\alpha = 0°, \beta = 40°$. **75.** vortex

Chapter 13. **1.** center of gravity **2.** moments, angular **3.** mass, moment of inertia **4.** effective, perpendicular **5.** (linear) momentum **6.** three, three **7.** acceleration **8.** motion, lift **9.** energy **10.** transferred **11.** gravitational, potential **12.** Power **13.** Newton's 3rd **14.** thrust, speed **15.** drag, speed **16.** angle of attack **17.** power, power **18.** 80% **19.** sonic **20.** static thrust **21.** thrust, power **22.** induced, lift **23.** area, 1.0 **24.** maximum, minimum **25.** PE is being transformed into kinetic energy (turbulence) and heat energy of the air **26.** The torque developed and the angular speed (Eq. 13.15) **27.** The effect is the same as a lower propeller efficiency so both e and f would be calculated to be worse than they really are; i.e., e too small and f too large. **28.** stall of the propeller blades **29.** radians and rads/s **30.** About 28% more since the drag coefficient of an actual flat plate is about 1.28 rather than 1.00. **31.** induced; parasite

Chapter 14 **1.** weight, thrust **2.** (rolling) friction **3.** (profile) parasite drag, induced drag **4.** ground **5.** thrust, liftoff, (aerodynamic) drag **6.** zero, zero **7.** angle of attack **8.** excess, weight **9.** best angle, best rate, density altitude **10.** controllability (lateral stability) **11.** 8000 ft **12.** power **13.** rate of climb, maximum or cruise speed **14.** range, glide **15.** laminar **16.** windmilling propeller **17.** airspeed, steep angle, rate **18.** frictional, weight **19.** power, thrust **20.** induced, friction **21.** time **22.** angle, rate, less than **23.** decreases, decrease **24.** low **25.** increases, induced, slower, parasite **26.** increases, doesn't change, doesn't change **27.**

decreases, doesn't change **28.** airspeed, descent **29.** f **30.** increased **31.** side, forward **32.** right **33.** more upward **34.** excess speed **35.** potential, kinetic **36.** d **37.** a **38.** static friction between the tires and the ground from braking, rolling friction on tires with or without braking, and aerodynamic drag **39.** static, sliding **40.** weight

Chapter 15. **1.** lift, weight **2.** b, c, d (and e?) **3.** Yes, just have a sufficiently powerful elevator and apply abrupt control movement. **4.** steeper curve on the down side and a shorter time of flight, just as for a thrown object with significant drag **5.** maintain a zero-lift angle of attack throughout the maneuver **6.** Time is directly proportional to the entry speed (see calculation in Sec. B), so would double. **7.** no density dependence in our calculation; the decreased drag and reduced gravitational force at higher altitudes would make the times a trifle longer **8.** Power increases the time because the thrust can be used to equal (or nearly equal) the drag. **9.** negative lift, to counteract the negative zero-lift pitching moment of the wing **10.** induced **11.** equal to its weight **12.** one **13.** positive because the force on the fuselage is **up**; if the wings are stalled, though, they try to continue their trip to the ground and will tend to bend **down** as the fuselage tries to stop them (as in a negative-g maneuver); this explains why one set of wires on a biplane is called the "flying wires" and the other set is called the "landing wires." **14.** Significant downward bending moments compared with 1-g upward bending moments at touchdown for a smooth landing. **15.** zero **16.** one **17.** centripetal acceleration **18.** lateral **19.** the most downwind point (Fig. 15.10) because then the steepest bank for the maneuver will be determined immediately, at the entry point **20.** In straight-and-level flight, there is no forward or rearward force acting on the pilot while the gravitational force (his weight, a downward force) is just balanced by the upward force of the seat on him. In a terminal velocity dive, on the other hand, there are **no** forces on the pilot in the plane through the seat but the forward force of the pilot's weight is now just balanced by a rearward force exerted by the seat belt harness. **21.** zero (i.e., it can't be stalled) **22.** flying wires: lower fuselage to upper wing, landing wires: upper fuselage to lower wing **23.** g, straight down, just like a rock **24.** centripetal **25.** yes — in a descending, constant angle-of-attack turn **26.** same **27.** if one wing stalls first, especially if the wingtips tend to stall first; power effects can also be significant if power is used **28.** greater than one, greater than **29.** Unload the wings (i.e., reduce the load factor to less than one with forward or relaxed pressure on the stick). **30.** induced **31.** span (because a small span loading goes along with a large induced drag term to begin with—Eq. 15.18) **32.** speed **33.** less than, apparent **34.** true ground **35.** (slightly) less than **36.** increases, weight **37.** doesn't change **38.** increases **39.** increases **40.** doesn't affect **41.** 0 **42.** greater than +1

Chapter 16. **1.** limit **2.** tail slide **3.** wind gust considerations and flutter **4.** the wings are then capable of providing greater accelerations **5.** one and one-half times **6.** increase or decrease the angle of attack **7.** the inside wing has less lift and more drag (and sometimes answer e) **8.** because jogging tends to lower the blood pressure **9.** nose slice, wing rock **10.** negative, positive **11.** dynamic, rotor blades **12.** deep, T **13.** stick shaker, stick pusher **14.** rotary, spin **15.** lower fuselage **16.** drooped **17.** flat **18.** power, stick, rudder, opposite to spin rotation **19.** descent or vertical, yaw **20.** up; increases (Fig. 16.4) **21.** one **22.** Limit elevator power to prevent a stall, use the canard configuration to prevent the wing from stalling, or design a spin-resistant wing configuration.

322

APPENDIX A

23. barrel, positive, two   24. slow, longitudinal, 1, 1   25. load factor, equivalent airspeed   26. loss of peripheral vision, darkening vision, loss of vision, blackout   27. the seat reclines   28. "grunt" (M-1) maneuver   29. The person farthest from the cg experiences the greater load factor extremes because any angular acceleration about the cg is added to the acceleration of the cg.

Chapter 17.   1. dihedral   2. negative   3. greater than the angle for onset of stall along the wing   4. dorsal fin; yaw   5. (fly-by-wire) active   6. dorsal fin   7. neutral point   8. a, b, c, and d   9. cg, zero, negative   10. tractor   11. rearward   12. Normally the aircraft will nose **down** and the structure will fail with negative load factors because the tail usually has a download on it.   13. tail volume   14. dihedral, low wing   15. aeroelastic divergence   16. stall   17. static margin   18. destabilizing   19. deep, T-tail, canard   20. rearward   21. adequate elevator power for landing with power off; no, because the download on the tail is a maximum at forward cg 22. dynamic   23. no effect because the effects cancel, as shown by inspection of the tail volume coefficient formula   24. neutral point   25. increase, increase   26. The horizontal tail has a sufficiently large surface area.   27. lesser because they are typically a higher aspect ratio surface and we don't want them to stall   28. Reduce maximum loads in turbulence.   29. to ensure a wide loading envelope (i.e., a long range of safe cg locations)   30. Decrease the dihedral effect or increase the tail effectiveness.   31. at the stall, in a stable aircraft   32. beveling   33. tab   34. sealing the gaps   35. The extra drag of the lowered aileron tends to cause a roll in that direction.   36. mass balance   37. slack   38. hinge line, aerodynamic   39. Frise, unsymmetrical, decreases   40. up, down   41. all of them   42. negative   43. anhedral   44. move the cg back relative to the cp 45. yaw, Dutch   46. neutral point, reduces   47. deflect both outboard drag rudder/spoilers (up and down equally)   48. lift   49. positive   50. stick-free   51. rearmost   52. negative, negative (since the fuselage's cg is behind its cp)   53. trim   54. pitching moment   55. downwash from propeller is reduced   56. has an unrecoverable stall

Appendix B: Standard Atmosphere Tables (Geometric Altitude)

Altitude ft	Pressure lb/ft²	Density sl/ft³	Temperature R	Temperature °F	Speed of Sound ft/s	Speed of Sound kt	μ (Viscosity) 10^{-7} lb-s/ft²
−10000	3002.0	0.003155	554.3	94.68	1154.2	683.9	3.934
−9900	2991.9	0.003146	554.0	94.32	1153.8	683.6	3.932
−9800	2981.7	0.003138	553.6	93.96	1153.5	683.4	3.930
−9700	2971.7	0.003129	553.3	93.61	1153.1	683.2	3.928
−9600	2961.6	0.003120	552.9	93.25	1152.7	683.0	3.926
−9500	2951.6	0.003112	552.6	92.89	1152.4	682.7	3.924
−9400	2941.6	0.003103	552.2	92.54	1152.0	682.5	3.922
−9300	2931.6	0.003095	551.9	92.18	1151.6	682.3	3.920
−9200	2921.6	0.003086	551.5	91.82	1151.2	682.1	3.918
−9100	2911.7	0.003078	551.1	91.47	1150.9	681.9	3.916
−9000	2901.8	0.003069	550.8	91.11	1150.5	681.6	3.914
−8900	2891.9	0.003061	550.4	90.75	1150.1	681.4	3.912
−8800	2882.1	0.003052	550.1	90.40	1149.7	681.2	3.910
−8700	2872.3	0.003044	549.7	90.04	1149.4	681.0	3.909
−8600	2862.5	0.003036	549.4	89.68	1149.0	680.8	3.907
−8500	2852.7	0.003027	549.0	89.32	1148.6	680.5	3.905
−8400	2843.0	0.003019	548.6	88.97	1148.3	680.3	3.903
−8300	2833.3	0.003010	548.3	88.61	1147.9	680.1	3.901
−8200	2823.6	0.003002	547.9	88.25	1147.5	679.9	3.899
−8100	2813.9	0.002994	547.6	87.90	1147.1	679.7	3.897
−8000	2804.3	0.002985	547.2	87.54	1146.8	679.4	3.895
−7900	2794.7	0.002977	546.9	87.18	1146.4	679.2	3.893
−7800	2785.2	0.002969	546.5	86.83	1146.0	679.0	3.891
−7700	2775.6	0.002961	546.1	86.47	1145.6	678.8	3.889
−7600	2766.1	0.002952	545.8	86.11	1145.3	678.5	3.887
−7500	2756.6	0.002944	545.4	85.76	1144.9	678.3	3.885
−7400	2747.1	0.002936	545.1	85.40	1144.5	678.1	3.883
−7300	2737.7	0.002928	544.7	85.04	1144.1	677.9	3.881
−7200	2728.3	0.002920	544.4	84.69	1143.8	677.7	3.879
−7100	2718.9	0.002912	544.0	84.33	1143.4	677.4	3.877
−7000	2709.5	0.002903	543.6	83.97	1143.0	677.2	3.875
−6900	2700.2	0.002895	543.3	83.61	1142.6	677.0	3.873
−6800	2690.9	0.002887	542.9	83.26	1142.3	676.8	3.872
−6700	2681.6	0.002879	542.6	82.90	1141.9	676.5	3.870
−6600	2672.3	0.002871	542.2	82.54	1141.5	676.3	3.868
−6500	2663.1	0.002863	541.9	82.19	1141.1	676.1	3.866
−6400	2653.9	0.002855	541.5	81.83	1140.8	675.9	3.864

324

APPENDIX B

Altitude ft	Pressure lb/ft²	Density sl/ft³	Temperature R	Temperature °F	Speed of Sound ft/s	Speed of Sound kt	μ (Viscosity) 10^{-7} lb-s/ft²
−6300	2644.7	0.002847	541.1	81.47	1140.4	675.7	3.862
−6200	2635.6	0.002839	540.8	81.12	1140.0	675.4	3.860
−6100	2626.4	0.002831	540.4	80.76	1139.6	675.2	3.858
−6000	2617.3	0.002823	540.1	80.40	1139.3	675.0	3.856
−5900	2608.3	0.002815	539.7	80.05	1138.9	674.8	3.854
−5800	2599.2	0.002807	539.4	79.69	1138.5	674.5	3.852
−5700	2590.2	0.002800	539.0	79.33	1138.1	674.3	3.850
−5600	2581.2	0.002792	538.6	78.98	1137.7	674.1	3.848
−5500	2572.2	0.002784	538.3	78.62	1137.4	673.9	3.846
−5400	2563.3	0.002776	537.9	78.26	1137.0	673.6	3.844
−5300	2554.3	0.002768	537.6	77.91	1136.6	673.4	3.842
−5200	2545.5	0.002760	537.2	77.55	1136.2	673.2	3.840
−5100	2536.6	0.002752	536.9	77.19	1135.9	673.0	3.838
−5000	2527.7	0.002745	536.5	76.84	1135.5	672.8	3.836
−4900	2518.9	0.002737	536.1	76.48	1135.1	672.5	3.834
−4800	2510.1	0.002729	535.8	76.12	1134.7	672.3	3.832
−4700	2501.3	0.002721	535.4	75.76	1134.4	672.1	3.830
−4600	2492.6	0.002714	535.1	75.41	1134.0	671.9	3.828
−4500	2483.9	0.002706	534.7	75.05	1133.6	671.6	3.826
−4400	2475.2	0.002698	534.4	74.69	1133.2	671.4	3.824
−4300	2466.5	0.002691	534.0	74.34	1132.8	671.2	3.822
−4200	2457.8	0.002683	533.7	73.98	1132.5	671.0	3.820
−4100	2449.2	0.002675	533.3	73.62	1132.1	670.7	3.819
−4000	2440.6	0.002668	532.9	73.27	1131.7	670.5	3.817
−3900	2432.0	0.002660	532.6	72.91	1131.3	670.3	3.815
−3800	2423.5	0.002653	532.2	72.55	1130.9	670.1	3.813
−3700	2415.0	0.002645	531.9	72.20	1130.6	669.8	3.811
−3600	2406.5	0.002638	531.5	71.84	1130.2	669.6	3.809
−3500	2398.0	0.002630	531.2	71.48	1129.8	669.4	3.807
−3400	2389.5	0.002623	530.8	71.13	1129.4	669.2	3.805
−3300	2381.1	0.002615	530.4	70.77	1129.0	668.9	3.803
−3200	2372.7	0.002608	530.1	70.41	1128.7	668.7	3.801
−3100	2364.3	0.002600	529.7	70.06	1128.3	668.5	3.799
−3000	2356.0	0.002593	529.4	69.70	1127.9	668.3	3.797
−2900	2347.6	0.002585	529.0	69.34	1127.5	668.0	3.795
−2800	2339.3	0.002578	528.7	68.99	1127.1	667.8	3.793
−2700	2331.0	0.002570	528.3	68.63	1126.8	667.6	3.791
−2600	2322.8	0.002563	527.9	68.27	1126.4	667.4	3.789
−2500	2314.6	0.002556	527.6	67.92	1126.0	667.1	3.787
−2400	2306.3	0.002548	527.2	67.56	1125.6	666.9	3.785
−2300	2298.1	0.002541	526.9	67.20	1125.2	666.7	3.783
−2200	2290.0	0.002534	526.5	66.85	1124.9	666.5	3.781
−2100	2281.8	0.002526	526.2	66.49	1124.5	666.2	3.779

Altitude ft	Pressure lb/ft²	Density sl/ft³	Temperature		Speed of Sound		μ (Viscosity)
			R	°F	ft/s	kt	10⁻⁷ lb-s/ft²
−2000	2273.7	0.002519	525.8	66.13	1124.1	666.0	3.777
−1900	2265.6	0.002512	525.4	65.78	1123.7	665.8	3.775
−1800	2257.6	0.002505	525.1	65.42	1123.3	665.6	3.773
−1700	2249.5	0.002497	524.7	65.06	1123.0	665.3	3.771
−1600	2241.5	0.002490	524.4	64.71	1122.6	665.1	3.769
−1500	2233.5	0.002483	524.0	64.35	1122.2	664.9	3.767
−1400	2225.5	0.002476	523.7	63.99	1121.8	664.7	3.765
−1300	2217.5	0.002469	523.3	63.64	1121.4	664.4	3.763
−1200	2209.6	0.002461	522.9	63.28	1121.0	664.2	3.761
−1100	2201.7	0.002454	522.6	62.92	1120.7	664.0	3.759
−1000	2193.8	0.002447	522.2	62.57	1120.3	663.7	3.757
−900	2186.0	0.002440	521.9	62.21	1119.9	663.5	3.755
−800	2178.1	0.002433	521.5	61.85	1119.5	663.3	3.753
−700	2170.3	0.002426	521.2	61.50	1119.1	663.1	3.751
−600	2162.5	0.002419	520.8	61.14	1118.8	662.8	3.749
−500	2154.7	0.002412	520.5	60.78	1118.4	662.6	3.747
−400	2147.0	0.002405	520.1	60.43	1118.0	662.4	3.745
−300	2139.3	0.002398	519.7	60.07	1117.6	662.2	3.743
−200	2131.6	0.002391	519.4	59.71	1117.2	661.9	3.741
−100	2123.9	0.002384	519.0	59.36	1116.8	661.7	3.739,
0	2116.2	0.002377	518.7	59.00	1116.5	661.5	3.737
100	2108.6	0.002370	518.3	58.64	1116.1	661.3	3.735
200	2101.0	0.002363	518.0	58.29	1115.7	661.0	3.733
300	2093.4	0.002356	517.6	57.93	1115.3	660.8	3.731
400	2085.8	0.002349	517.2	57.57	1114.9	660.6	3.729
500	2078.3	0.002342	516.9	57.22	1114.5	660.3	3.727
600	2070.7	0.002335	516.5	56.86	1114.1	660.1	3.725
700	2063.2	0.002329	516.2	56.50	1113.8	659.9	3.723
800	2055.8	0.002322	515.8	56.15	1113.4	659.7	3.721
900	2048.3	0.002315	515.5	55.79	1113.0	659.4	3.719
1000	2040.9	0.002308	515.1	55.43	1112.6	659.2	3.717
1100	2033.4	0.002301	514.7	55.08	1112.2	659.0	3.715
1200	2026.0	0.002295	514.4	54.72	1111.8	658.7	3.713
1300	2018.7	0.002288	514.0	54.36	1111.5	658.5	3.711
1400	2011.3	0.002281	513.7	54.01	1111.1	658.3	3.709
1500	2004.0	0.002274	513.3	53.65	1110.7	658.1	3.707
1600	1996.7	0.002268	513.0	53.29	1110.3	657.8	3.705
1700	1989.4	0.002261	512.6	52.94	1109.9	657.6	3.703
1800	1982.2	0.002254	512.3	52.58	1109.5	657.4	3.701
1900	1974.9	0.002248	511.9	52.22	1109.1	657.1	3.699
2000	1967.7	0.002241	511.5	51.87	1108.7	656.9	3.697
2100	1960.5	0.002234	511.2	51.51	1108.4	656.7	3.695
2200	1953.3	0.002228	510.8	51.16	1108.0	656.5	3.693

Altitude ft	Pressure lb/ft^2	Density sl/ft^3	Temperature R	Temperature °F	Speed of Sound ft/s	Speed of Sound kt	μ (Viscosity) 10^{-7} lb-s/ft^2
2300	1946.2	0.002221	510.5	50.80	1107.6	656.2	3.691
2400	1939.0	0.002214	510.1	50.44	1107.2	656.0	3.689
2500	1931.9	0.002208	509.8	50.09	1106.8	655.8	3.687
2600	1924.8	0.002201	509.4	49.73	1106.4	655.5	3.685
2700	1917.8	0.002195	509.0	49.37	1106.0	655.3	3.683
2800	1910.7	0.002188	508.7	49.02	1105.7	655.1	3.681
2900	1903.7	0.002182	508.3	48.66	1105.3	654.9	3.679
3000	1896.7	0.002175	508.0	48.30	1104.9	654.6	3.677
3100	1889.7	0.002169	507.6	47.95	1104.5	654.4	3.675
3200	1882.7	0.002162	507.3	47.59	1104.1	654.2	3.673
3300	1875.8	0.002156	506.9	47.23	1103.7	653.9	3.671
3400	1868.9	0.002149	506.5	46.88	1103.3	653.7	3.669
3500	1862.0	0.002143	506.2	46.52	1102.9	653.5	3.667
3600	1855.1	0.002136	505.8	46.16	1102.5	653.2	3.665
3700	1848.2	0.002130	505.5	45.81	1102.2	653.0	3.663
3800	1841.4	0.002124	505.1	45.45	1101.8	652.8	3.661
3900	1834.5	0.002117	504.8	45.09	1101.4	652.6	3.659
4000	1827.7	0.002111	504.4	44.74	1101.0	652.3	3.657
4100	1821.0	0.002105	504.1	44.38	1100.6	652.1	3.655
4200	1814.2	0.002098	503.7	44.03	1100.2	651.9	3.653
4300	1807.5	0.002092	503.3	43.67	1099.8	651.6	3.651
4400	1800.8	0.002086	503.0	43.31	1099.4	651.4	3.649
4500	1794.1	0.002079	502.6	42.96	1099.0	651.2	3.647
4600	1787.4	0.002073	502.3	42.60	1098.7	650.9	3.645
4700	1780.7	0.002067	501.9	42.24	1098.3	650.7	3.643
4800	1774.1	0.002061	501.6	41.89	1097.9	650.5	3.641
4900	1767.5	0.002054	501.2	41.53	1097.5	650.2	3.639
5000	1760.9	0.002048	500.8	41.17	1097.1	650.0	3.637
5100	1754.3	0.002042	500.5	40.82	1096.7	649.8	3.635
5200	1747.7	0.002036	500.1	40.46	1096.3	649.5	3.632
5300	1741.2	0.002030	499.8	40.10	1095.9	649.3	3.630
5400	1734.7	0.002023	499.4	39.75	1095.5	649.1	3.628
5500	1728.2	0.002017	499.1	39.39	1095.1	648.9	3.626
5600	1721.7	0.002011	498.7	39.03	1094.8	648.6	3.624
5700	1715.3	0.002005	498.3	38.68	1094.4	648.4	3.622
5800	1708.8	0.001999	498.0	38.32	1094.0	648.2	3.620
5900	1702.4	0.001993	497.6	37.97	1093.6	647.9	3.618
6000	1696.0	0.001987	497.3	37.61	1093.2	647.7	3.616
6100	1689.6	0.001981	496.9	37.25	1092.8	647.5	3.614
6200	1683.3	0.001975	496.6	36.90	1092.4	647.2	3.612
6300	1676.9	0.001969	496.2	36.54	1092.0	647.0	3.610
6400	1670.6	0.001963	495.9	36.18	1091.6	646.8	3.608
6500	1664.3	0.001957	495.5	35.83	1091.2	646.5	3.606

Altitude ft	Pressure lb/ft^2	Density sl/ft^3	Temperature		Speed of Sound		μ (Viscosity) 10^{-7} lb-s/ft^2
			R	°F	ft/s	kt	
6600	1658.0	0.001951	495.1	35.47	1090.8	646.3	3.604
6700	1651.8	0.001945	494.8	35.11	1090.4	646.1	3.602
6800	1645.5	0.001939	494.4	34.76	1090.0	645.8	3.600
6900	1639.3	0.001933	494.1	34.40	1089.7	645.6	3.598
7000	1633.1	0.001927	493.7	34.05	1089.3	645.4	3.596
7100	1626.9	0.001921	493.4	33.69	1088.9	645.1	3.594
7200	1620.7	0.001915	493.0	33.33	1088.5	644.9	3.592
7300	1614.6	0.001909	492.6	32.98	1088.1	644.7	3.590
7400	1608.5	0.001903	492.3	32.62	1087.7	644.4	3.588
7500	1602.3	0.001898	491.9	32.26	1087.3	644.2	3.586
7600	1596.2	0.001892	491.6	31.91	1086.9	644.0	3.584
7700	1590.2	0.001886	491.2	31.55	1086.5	643.7	3.582
7800	1584.1	0.001880	490.9	31.19	1086.1	643.5	3.579
7900	1578.1	0.001874	490.5	30.84	1085.7	643.3	3.577
8000	1572.1	0.001868	490.2	30.48	1085.3	643.0	3.575
8100	1566.1	0.001863	489.8	30.13	1084.9	642.8	3.573
8200	1560.1	0.001857	489.4	29.77	1084.5	642.6	3.571
8300	1554.1	0.001851	489.1	29.41	1084.1	642.3	3.569
8400	1548.2	0.001845	488.7	29.06	1083.7	642.1	3.567
8500	1542.3	0.001840	488.4	28.70	1083.3	641.9	3.565
8600	1536.4	0.001834	488.0	28.34	1083.0	641.6	3.563
8700	1530.5	0.001828	487.7	27.99	1082.6	641.4	3.561
8800	1524.6	0.001823	487.3	27.63	1082.2	641.2	3.559
8900	1518.8	0.001817	486.9	27.27	1081.8	640.9	3.557
9000	1512.9	0.001811	486.6	26.92	1081.4	640.7	3.555
9100	1507.1	0.001806	486.2	26.56	1081.0	640.5	3.553
9200	1501.3	0.001800	485.9	26.21	1080.6	640.2	3.551
9300	1495.5	0.001794	485.5	25.85	1080.2	640.0	3.549
9400	1489.8	0.001789	485.2	25.49	1079.8	639.8	3.547
9500	1484.0	0.001783	484.8	25.14	1079.4	639.5	3.545
9600	1478.3	0.001778	484.5	24.78	1079.0	639.3	3.542
9700	1472.6	0.001772	484.1	24.42	1078.6	639.1	3.540
9800	1466.9	0.001767	483.7	24.07	1078.2	638.8	3.538
9900	1461.3	0.001761	483.4	23.71	1077.8	638.6	3.536
10000	1455.6	0.001756	483.0	23.36	1077.4	638.3	3.534
10100	1450.0	0.001750	482.7	23.00	1077.0	638.1	3.532
10200	1444.4	0.001745	482.3	22.64	1076.6	637.9	3.530
10300	1438.8	0.001739	482.0	22.29	1076.2	637.6	3.528
10400	1433.2	0.001734	481.6	21.93	1075.8	637.4	3.526
10500	1427.6	0.001728	481.2	21.57	1075.4	637.2	3.524
10600	1422.1	0.001723	480.9	21.22	1075.0	636.9	3.522
10700	1416.5	0.001717	480.5	20.86	1074.6	636.7	3.520

Altitude ft	Pressure lb/ft²	Density sl/ft³	Temperature		Speed of Sound		μ (Viscosity) 10⁻⁷ lb-s/ft²
			R	°F	ft/s	kt	
10800	1411.0	0.001712	480.2	20.51	1074.2	636.5	3.518
10900	1405.5	0.001706	479.8	20.15	1073.8	636.2	3.516
11000	1400.1	0.001701	479.5	19.79	1073.4	636.0	3.514
11100	1394.6	0.001696	479.1	19.44	1073.0	635.8	3.512
11200	1389.2	0.001690	478.8	19.08	1072.6	635.5	3.509
11300	1383.7	0.001685	478.4	18.72	1072.2	635.3	3.507
11400	1378.3	0.001680	478.0	18.37	1071.8	635.0	3.505
11500	1372.9	0.001674	477.7	18.01	1071.4	634.8	3.503
11600	1367.6	0.001669	477.3	17.66	1071.0	634.6	3.501
11700	1362.2	0.001664	477.0	17.30	1070.6	634.3	3.499
11800	1356.9	0.001658	476.6	16.94	1070.2	634.1	3.497
11900	1351.5	0.001653	476.3	16.59	1069.8	633.9	3.495
12000	1346.2	0.001648	475.9	16.23	1069.4	633.6	3.493
12100	1341.0	0.001643	475.5	15.87	1069.0	633.4	3.491
12200	1335.7	0.001637	475.2	15.52	1068.6	633.1	3.489
12300	1330.4	0.001632	474.8	15.16	1068.2	632.9	3.487
12400	1325.2	0.001627	474.5	14.81	1067.8	632.7	3.485
12500	1320.0	0.001622	474.1	14.45	1067.4	632.4	3.482
12600	1314.8	0.001617	473.8	14.09	1067.0	632.2	3.480
12700	1309.6	0.001612	473.4	13.74	1066.6	632.0	3.478
12800	1304.4	0.001606	473.1	13.38	1066.2	631.7	3.476
12900	1299.3	0.001601	472.7	13.02	1065.8	631.5	3.474
13000	1294.1	0.001596	472.3	12.67	1065.4	631.2	3.472
13100	1289.0	0.001591	472.0	12.31	1065.0	631.0	3.470
13200	1283.9	0.001586	471.6	11.96	1064.6	630.8	3.468
13300	1278.8	0.001581	471.3	11.60	1064.2	630.5	3.466
13400	1273.7	0.001576	470.9	11.24	1063.8	630.3	3.464
13500	1268.7	0.001571	470.6	10.89	1063.4	630.1	3.462
13600	1263.6	0.001566	470.2	10.53	1063.0	629.8	3.460
13700	1258.6	0.001561	469.8	10.18	1062.6	629.6	3.457
13800	1253.6	0.001556	469.5	9.82	1062.2	629.3	3.455
13900	1248.6	0.001551	469.1	9.46	1061.8	629.1	3.453
14000	1243.6	0.001546	468.8	9.11	1061.4	628.9	3.451
14100	1238.7	0.001541	468.4	8.75	1061.0	628.6	3.449
14200	1233.7	0.001536	468.1	8.39	1060.6	628.4	3.447
14300	1228.8	0.001531	467.7	8.04	1060.2	628.1	3.445
14400	1223.9	0.001526	467.4	7.68	1059.8	627.9	3.443
14500	1219.0	0.001521	467.0	7.33	1059.4	627.7	3.441
14600	1214.1	0.001516	466.6	6.97	1059.0	627.4	3.439
14700	1209.3	0.001511	466.3	6.61	1058.6	627.2	3.437
14800	1204.4	0.001506	465.9	6.26	1058.2	626.9	3.434
14900	1199.6	0.001501	465.6	5.90	1057.8	626.7	3.432

Altitude ft	Pressure lb/ft²	Density sl/ft³	Temperature		Speed of Sound		μ (Viscosity) 10⁻⁷ lb-s/ft²
			R	°F	ft/s	kt	
15000	1194.8	0.001496	465.2	5.55	1057.4	626.5	3.430
15100	1190.0	0.001491	464.9	5.19	1057.0	626.2	3.428
15200	1185.2	0.001486	464.5	4.83	1056.5	626.0	3.426
15300	1180.4	0.001482	464.1	4.48	1056.1	625.7	3.424
15400	1175.7	0.001477	463.8	4.12	1055.7	625.5	3.422
15500	1171.0	0.001472	463.4	3.77	1055.3	625.3	3.420
15600	1166.2	0.001467	463.1	3.41	1054.9	625.0	3.418
15700	1161.5	0.001462	462.7	3.05	1054.5	624.8	3.415
15800	1156.8	0.001458	462.4	2.70	1054.1	624.5	3.413
15900	1152.2	0.001453	462.0	2.34	1053.7	624.3	3.411
16000	1147.5	0.001448	461.7	1.99	1053.3	624.1	3.409
16100	1142.9	0.001443	461.3	1.63	1052.9	623.8	3.407
16200	1138.2	0.001439	460.9	1.27	1052.5	623.6	3.405
16300	1133.6	0.001434	460.6	0.92	1052.1	623.3	3.403
16400	1129.0	0.001429	460.2	0.56	1051.7	623.1	3.401
16500	1124.4	0.001424	459.9	0.20	1051.3	622.9	3.399
16600	1119.9	0.001420	459.5	−0.15	1050.9	622.6	3.397
16700	1115.3	0.001415	459.2	−0.51	1050.5	622.4	3.394
16800	1110.8	0.001410	458.8	−0.86	1050.0	622.1	3.392
16900	1106.2	0.001406	458.5	−1.22	1049.6	621.9	3.390
17000	1101.7	0.001401	458.1	−1.58	1049.2	621.7	3.388
17100	1097.2	0.001396	457.7	−1.93	1048.8	621.4	3.386
17200	1092.8	0.001392	457.4	−2.29	1048.4	621.2	3.384
17300	1088.3	0.001387	457.0	−2.64	1048.0	620.9	3.382
17400	1083.9	0.001383	456.7	−3.00	1047.6	620.7	3.380
17500	1079.4	0.001378	456.3	−3.36	1047.2	620.4	3.378
17600	1075.0	0.001373	456.0	−3.71	1046.8	620.2	3.375
17700	1070.6	0.001369	455.6	−4.07	1046.4	620.0	3.373
17800	1066.2	0.001364	455.2	−4.42	1046.0	619.7	3.371
17900	1061.8	0.001360	454.9	−4.78	1045.6	619.5	3.369
18000	1057.5	0.001355	454.5	−5.14	1045.1	619.2	3.367
18100	1053.1	0.001351	454.2	−5.49	1044.7	619.0	3.365
18200	1048.8	0.001346	453.8	−5.85	1044.3	618.7	3.363
18300	1044.5	0.001342	453.5	−6.20	1043.9	618.5	3.361
18400	1040.2	0.001337	453.1	−6.56	1043.5	618.3	3.358
18500	1035.9	0.001333	452.8	−6.92	1043.1	618.0	3.356
18600	1031.6	0.001328	452.4	−7.27	1042.7	617.8	3.354
18700	1027.4	0.001324	452.0	−7.63	1042.3	617.5	3.352
18800	1023.1	0.001320	451.7	−7.98	1041.9	617.3	3.350
18900	1018.9	0.001315	451.3	−8.34	1041.5	617.0	3.348
19000	1014.7	0.001311	451.0	−8.70	1041.0	616.8	3.346
19100	1010.5	0.001306	450.6	−9.05	1040.6	616.6	3.344
19200	1006.3	0.001302	450.3	−9.41	1040.2	616.3	3.341

Altitude ft	Pressure lb/ft^2	Density sl/ft^3	Temperature R	Temperature °F	Speed of Sound ft/s	Speed of Sound kt	μ (Viscosity) 10^{-7} lb-s/ft^2
19300	1002.1	0.001298	449.9	−9.76	1039.8	616.1	3.339
19400	997.9	0.001293	449.6	−10.12	1039.4	615.8	3.337
19500	993.8	0.001289	449.2	−10.48	1039.0	615.6	3.335
19600	989.7	0.001285	448.8	−10.83	1038.6	615.3	3.333
19700	985.5	0.001280	448.5	−11.19	1038.2	615.1	3.331
19800	981.4	0.001276	448.1	−11.54	1037.8	614.9	3.329
19900	977.4	0.001272	447.8	−11.90	1037.3	614.6	3.326
20000	973.3	0.001267	447.4	−12.25	1036.9	614.4	3.324
20100	969.2	0.001263	447.1	−12.61	1036.5	614.1	3.322
20200	965.2	0.001259	446.7	−12.97	1036.1	613.9	3.320
20300	961.1	0.001254	446.3	−13.32	1035.7	613.6	3.318
20400	957.1	0.001250	446.0	−13.68	1035.3	613.4	3.316
20500	953.1	0.001246	445.6	−14.03	1034.9	613.1	3.314
20600	949.1	0.001242	445.3	−14.39	1034.5	612.9	3.312
20700	945.1	0.001237	444.9	−14.75	1034.0	612.7	3.309
20800	941.2	0.001233	444.6	−15.10	1033.6	612.4	3.307
20900	937.2	0.001229	444.2	−15.46	1033.2	612.2	3.305
21000	933.3	0.001225	443.9	−15.81	1032.8	611.9	3.303
21100	929.3	0.001221	443.5	−16.17	1032.4	611.7	3.301
21200	925.4	0.001217	443.1	−16.53	1032.0	611.4	3.299
21300	921.5	0.001212	442.8	−16.88	1031.6	611.2	3.297
21400	917.6	0.001208	442.4	−17.24	1031.1	610.9	3.294
21500	913.8	0.001204	442.1	−17.59	1030.7	610.7	3.292
21600	909.9	0.001200	441.7	−17.95	1030.3	610.4	3.290
21700	906.1	0.001196	441.4	−18.31	1029.9	610.2	3.288
21800	902.2	0.001192	441.0	−18.66	1029.5	609.9	3.286
21900	898.4	0.001188	440.7	−19.02	1029.1	609.7	3.284
22000	894.6	0.001184	440.3	−19.37	1028.6	609.5	3.281
22100	890.8	0.001180	439.9	−19.73	1028.2	609.2	3.279
22200	887.0	0.001176	439.6	−20.08	1027.8	609.0	3.277
22300	883.3	0.001171	439.2	−20.44	1027.4	608.7	3.275
22400	879.5	0.001167	438.9	−20.80	1027.0	608.5	3.273
22500	875.8	0.001163	438.5	−21.15	1026.6	608.2	3.271
22600	872.0	0.001159	438.2	−21.51	1026.2	608.0	3.269
22700	868.3	0.001155	437.8	−21.86	1025.7	607.7	3.266
22800	864.6	0.001151	437.5	−22.22	1025.3	607.5	3.264
22900	860.9	0.001147	437.1	−22.58	1024.9	607.2	3.262
23000	857.2	0.001143	436.7	−22.93	1024.5	607.0	3.260
23100	853.6	0.001140	436.4	−23.29	1024.1	606.7	3.258
23200	849.9	0.001136	436.0	−23.64	1023.6	606.5	3.256
23300	846.3	0.001132	435.7	−24.00	1023.2	606.2	3.253
23400	842.7	0.001128	435.3	−24.35	1022.8	606.0	3.251
23500	839.1	0.001124	435.0	−24.71	1022.4	605.8	3.249

Altitude ft	Pressure lb/ft²	Density sl/ft³	Temperature R	Temperature °F	Speed of Sound ft/s	Speed of Sound kt	μ (Viscosity) 10^{-7} lb-s/ft²
23600	835.5	0.001120	434.6	−25.07	1022.0	605.5	3.247
23700	831.9	0.001116	434.2	−25.42	1021.6	605.3	3.245
23800	828.3	0.001112	433.9	−25.78	1021.1	605.0	3.243
23900	824.7	0.001108	433.5	−26.13	1020.7	604.8	3.240
24000	821.2	0.001104	433.2	−26.49	1020.3	604.5	3.238
24100	817.6	0.001100	432.8	−26.85	1019.9	604.3	3.236
24200	814.1	0.001097	432.5	−27.20	1019.5	604.0	3.234
24300	810.6	0.001093	432.1	−27.56	1019.0	603.8	3.232
24400	807.1	0.001089	431.8	−27.91	1018.6	603.5	3.230
24500	803.6	0.001085	431.4	−28.27	1018.2	603.3	3.227
24600	800.1	0.001081	431.0	−28.62	1017.8	603.0	3.225
24700	796.7	0.001078	430.7	−28.98	1017.4	602.8	3.223
24800	793.2	0.001074	430.3	−29.34	1016.9	602.5	3.221
24900	789.8	0.001070	430.0	−29.69	1016.5	602.3	3.219
25000	786.3	0.001066	429.6	−30.05	1016.1	602.0	3.217
25100	782.9	0.001063	429.3	−30.40	1015.7	601.8	3.214
25200	779.5	0.001059	428.9	−30.76	1015.3	601.5	3.212
25300	776.1	0.001055	428.6	−31.11	1014.8	601.3	3.210
25400	772.7	0.001051	428.2	−31.47	1014.4	601.0	3.208
25500	769.4	0.001048	427.8	−31.83	1014.0	600.8	3.206
25600	766.0	0.001044	427.5	−32.18	1013.6	600.5	3.204
25700	762.7	0.001040	427.1	−32.54	1013.2	600.3	3.201
25800	759.3	0.001037	426.8	−32.89	1012.7	600.0	3.199
25900	756.0	0.001033	426.4	−33.25	1012.3	599.8	3.197
26000	752.7	0.001029	426.1	−33.60	1011.9	599.5	3.195
26100	749.4	0.001026	425.7	−33.96	1011.5	599.3	3.193
26200	746.1	0.001022	425.4	−34.32	1011.0	599.0	3.190
26300	742.9	0.001018	425.0	−34.67	1010.6	598.8	3.188
26400	739.6	0.001015	424.6	−35.03	1010.2	598.5	3.186
26500	736.3	0.001011	424.3	−35.38	1009.8	598.3	3.184
26600	733.1	0.001007	423.9	−35.74	1009.3	598.0	3.182
26700	729.9	0.001004	423.6	−36.09	1008.9	597.8	3.180
26800	726.7	0.001000	423.2	−36.45	1008.5	597.5	3.177
26900	723.5	0.000997	422.9	−36.81	1008.1	597.3	3.175
27000	720.3	0.000993	422.5	−37.16	1007.7	597.0	3.173
27100	717.1	0.000990	422.2	−37.52	1007.2	596.8	3.171
27200	713.9	0.000986	421.8	−37.87	1006.8	596.5	3.169
27300	710.8	0.000982	421.4	−38.23	1006.4	596.3	3.166
27400	707.6	0.000979	421.1	−38.58	1006.0	596.0	3.164
27500	704.5	0.000975	420.7	−38.94	1005.5	595.8	3.162
27600	701.3	0.000972	420.4	−39.30	1005.1	595.5	3.160
27700	698.2	0.000968	420.0	−39.65	1004.7	595.3	3.158

Altitude ft	Pressure lb/ft²	Density sl/ft³	Temperature		Speed of Sound		μ (Viscosity) 10⁻⁷ lb-s/ft²
			R	°F	ft/s	kt	
27800	695.1	0.000965	419.7	−40.01	1004.3	595.0	3.155
27900	692.0	0.000961	419.3	−40.36	1003.8	594.8	3.153
28000	689.0	0.000958	419.0	−40.72	1003.4	594.5	3.151
28100	685.9	0.000955	418.6	−41.07	1003.0	594.2	3.149
28200	682.8	0.000951	418.2	−41.43	1002.6	594.0	3.147
28300	679.8	0.000948	417.9	−41.79	1002.1	593.7	3.144
28400	676.8	0.000944	417.5	−42.14	1001.7	593.5	3.142
28500	673.7	0.000941	417.2	−42.50	1001.3	593.2	3.140
28600	670.7	0.000937	416.8	−42.85	1000.8	593.0	3.138
28700	667.7	0.000934	416.5	−43.21	1000.4	592.7	3.136
28800	664.7	0.000931	416.1	−43.56	1000.0	592.5	3.133
28900	661.7	0.000927	415.8	−43.92	999.6	592.2	3.131
29000	658.8	0.000924	415.4	−44.28	999.1	592.0	3.129
29100	655.8	0.000921	415.0	−44.63	998.7	591.7	3.127
29200	652.9	0.000917	414.7	−44.99	998.3	591.5	3.125
29300	649.9	0.000914	414.3	−45.34	997.9	591.2	3.122
29400	647.0	0.000910	414.0	−45.70	997.4	591.0	3.120
29500	644.1	0.000907	413.6	−46.05	997.0	590.7	3.118
29600	641.2	0.000904	413.3	−46.41	996.6	590.4	3.116
29700	638.3	0.000901	412.9	−46.76	996.1	590.2	3.114
29800	635.4	0.000897	412.6	−47.12	995.7	589.9	3.111
29900	632.5	0.000894	412.2	−47.48	995.3	589.7	3.109
30000	629.7	0.000891	411.8	−47.83	994.8	589.4	3.107

Appendix C: **List of Symbols**

α (a) angle of attack: the angle of the chord line (or other fixed line on the airfoil) relative to the direction of distant, oncoming, undisturbed air (the relative wind)
(b) angular acceleration

$\alpha_{2\text{-D}}$ effective angle of attack of an airfoil

$\alpha_{3\text{-D}}$ effective angle of attack of a wing

α_T angle of attack of the rear wing

α_W angle of attack of the front wing

β bank angle

ε wing downwash angle

ϕ wind correction angle

γ (a) ratio of c_P to c_V (for air, value is 1.400)
(b) angle of climb or (if negative) angle of glide

η propeller efficiency

Λ angle of sweep for a swept wing

μ (a) coefficient of viscosity
(b) coefficient of friction

ν kinematic viscosity

θ an angle

θ_M Mach angle (half-vertex angle for a Mach wave)

ρ density (mass density)

∞ as a subscript: refers to a fluid property before it is affected by a bounding surface

Δt an arbitrarily small time interval

a (a) acceleration
(b) the lift slope, i.e., the slope of the lift curve

a_0 the lift slope in the linear region near $\alpha = 0°$

a_T slope of the lift curve for the rear wing

a_W slope of the lift curve for the front wing

A area; cross-sectional area

ac aerodynamic center, the point on an airfoil where the moment coefficient is nearly constant (usually at about $0.25c$ subsonically and about $0.50c$ supersonically)

AR aspect ratio (of a wing)
$= (\text{span})^2/(\text{wing area}) = b^2/S$

b wing span

BL boundary layer

c wing or airfoil chord (from the leading edge to the trailing edge)

$\bar{c}$ mean geometric chord of the front wing for a simple rectangular wing, or the mean aerodynamic chord (MAC), which is the aerodynamically equivalent chord, for a tapered or swept wing

c_d the airfoil drag coefficient

c_ℓ the airfoil lift coefficient

c_{ℓ_i} design lift coefficient for an airfoil

c_m the airfoil moment coefficient

c_P specific heat of a gas at constant pressure

c_V specific heat of a gas at constant volume

C_{Di} induced drag coefficient

C_{Li} airplane design lift coefficient

$C_{L,W}$ lift coefficient for the front wing (which would be the canard on a canard-type of aircraft)

$C_{L,T}$ lift coefficient for the rear wing (which would be the horizontal tail surface on a conventional configuration aircraft)

C_P the pressure coefficient, which gives the local air pressure at a given point on the surface of an airfoil in terms of a fraction of the dynamic pressure

cg the center of gravity of an object (where the force of gravity effectively acts)

cm the center of mass of an object

cp the center of pressure, the point where the resultant aerodynamic force on an airfoil equivalently acts

d (a) diameter
(b) distance

$d_\perp$ perpendicular moment arm

D the drag force on an airfoil, wing, or aircraft

D_i	induced drag force	0	as a subscript: refers to SL standard value
e	(a) wing efficiency factor (b) airplane efficiency factor	p	(a) the pressure exerted by a fluid (b) the linear momentum of an object
$e\mathrm{AR}$	effective aspect ratio	p_T	total pressure
f	equivalent flat plate area	P	power
F	force	PE	gravitational potential energy
F_D	the force component in the direction of the motion	q	dynamic pressure, often local dynamic pressure
F_F	force of friction (rolling, static, or kinetic)	q_∞	flight dynamic pressure (based on V_∞)
F_R	the resultant or equivalent force on an airfoil due to the distributed pressures on it	R	(a) resultant aerodynamic force; usually resolved into equivalent perpendicular components called lift and drag (b) radius of a loop (c) radius of a turn
g	acceleration due to the gravitational force in the absence of restraining air drag or other forces	RN	Reynolds number
h	(a) height above the surface of the earth (b) height difference of the liquids in a U-tube manometer	s	(a) thickness of a thin layer of fluid (b) distance form the *ac* of the front wing to the *ac* of the rear wing
I	moment of inertia of an object	S	the surface area of an airfoil or wing (projected area, as in a shadowgraph—not the actual area)
J	advance ratio for a propeller		
k	gas constant for dry air	S_W	surface area of the front wing
k_W	gas constant for gaseous water molecules	S_T	surface area of the rear wing
KE	kinetic energy, linear or angular	StAt	U.S. Standard Atmosphere (1976)
L	(a) the lift force on an airfoil, wing, or aircraft (b) the angular momentum of an object	T	thrust force produced by a propeller or jet engine
m	mass	T_T	total temperature, the temperature at a stagnation point
M	(a) local Mach number or flight Mach number (b) moment of a force about a designated point (c) the pitching moment on an airfoil or wing or aircraft, measured about some designated point	U_E	equivalent wind gust speed
		v	speed (of molecules of air)
M_∞	flight Mach number	V	(a) speed (b) volume
M_{CG}	pitching moment for the whole aircraft about the airplane's *cg*	V_∞	(a) fluid speed relative to a surface before it is affected by the surface (b) the speed of an aircraft relative to undisturbed air (i.e., the true airspeed or flight speed)
M_{CR}	critical Mach number, when local fluid flow first becomes supersonic		
$M_{CR,\,AIRFOIL}$	critical Mach number for an airfoil	$V_{\infty,\,STALL}$	true airspeed at the stall
$M_{CR,\,WING}$	critical Mach number for a wing	V_A	(a) speed of sound in air (b) design maneuvering speed
M_D	drag divergence Mach number, where drag increases rapidly due to shock wave formation and the separation it causes	V_{CAL}	calibrated airspeed (the reading of the airspeed indicator when there is no instrument or position error)
M_{LE}	the pitching moment on an airfoil, measured about the leading edge	$V_{CAL,\,STALL}$	calibrated airspeed at the stall
M_{TE}	the pitching moment on an airfoil, measured about the trailing edge	V_E	equivalent airspeed (the airspeed that produces the same dynamic pressure at altitude as does a true airspeed of the same value under SL density conditions)
n	load factor (L/W)		

$V_{E,STALL}$ equivalent airspeed at the stall

V_C design cruising speed

V_D design dive speed

V_{IND} indicated airspeed

V_{S+} positive load factor stall speed

V_{S-} negative load factor stall speed

V_W (a) wind speed
         (b) the wave speed in a given fluid

W work done by a force

x (a) a characteristic dimension (often the length of a
           bounding surface in fluid flow)
        (b) a location along the chord line of an airfoil,
           measured from the leading edge and
           expressed in fractions of the chord length
        (c) distance from the ac of the front wing to the
           airplane's cg

x_{NP} distance from the ac of the front wing to the point
           where the aircraft has neutral stability
           (the neutral point)

z a location perpendicular to the chord line of an airfoil,
           measured from the chord line and in fractions of
           the chord length (used in specifying the coord-
           inates of an airfoil, + for the upper surface, − for
           the lower surface)

Appendix D: **Computer Program for Calculating Cruise Speeds**

```
REM   Program to solve Eq. 15.18 for the true aircraft speed corresponding to
REM      a given f (flat plate area), e (efficiency factor), W (weight in
REM      pounds), b (span in ft), PRated (rated power in units of hp),
REM      eta (propeller efficiency), percent power, bank angle,
REM      and the density altitude in the StAt
REM   This example program uses the parameters derived for the C-152 (Fig. 13.7)
REM   Change the parameters in the DATA statement to model a different aircraft
REM      Program language is Microsoft's QuickBASIC (Provided with MS-DOS 5.0)


READ f, e, W, b, PRated                         'Read in performance parameters
      DATA 5.39, 0.716, 1670, 33.33, 110
      REM   Change data as desired for other airplanes
CLS : PRINT                                     'Clear screen; space 1 line

REM ***** INPUT OPERATING CONDITIONS FOR THE A/C MODELED WITH ABOVE PARAMETERS
      INPUT "Percent of available power being used"; PUsed: PRINT
      INPUT "Estimated propeller efficiency in PERCENT"; eta: PRINT
      INPUT "For density altitude (in ft)"; DAltitude: PRINT
          Rho = .002377 * EXP(−.0000297 * DAltitude)    'air density in sl/ft^3
                    'Note: this calculation of air density is good to 10,000 ft or so;
                    '    input the air density directly if desired
      INPUT "Bank angle (in degrees)"; BetaDeg: PRINT  'supply bank angle (0 if level)
          BetaRad = BetaDeg * 3.14159 / 180          'convert bank angle to radians

REM ***** DEFINE THE POWER REQUIRED EQUATION (Eq. 15.18)
      COEF1 = .5 * f * Rho
      COEF2 = (1 / e) * (2 / 3.14159) * (W / b) ^ 2 / (Rho * COS(BetaRad) ^ 2)
      DEF FnPowerReq (V) = COEF1 * V ^ 3 + COEF2 / V

REM ***** CALCULATE THE POWER AVAILABLE
      PAvail = (eta / 100!) * 550! * (PUsed / 100!) * PRated

REM ***** SOLUTION OF THE POWER EQUATION INVOLVES STARTING WITH THE SPEED
REM *****    FOR NO INDUCED DRAG (which is too LARGE), DECREMENTING THE SPEED
REM *****    UNTIL A too SMALL SPEED IS OBTAINED, AND THEN HOMING IN ON THE
REM *****    ACTUAL VALUE BETWEEN THESE TWO SPEEDS.
REM ***** THE SOLUTION OBTAINED IS THE cruise SPEED; IF THE ITERATION IS
REM *****    RE-STARTED below THE CRUISE SPEED, THE SOLUTION WILL BE THE
REM *****    MINIMUM SPEED CLOSE TO (or below) THE STALL SPEED.
REM ***** THE PROCEDURE MAY NOT CONVERGE IF THE minimum AND cruise SPEEDS ARE
REM *****    VERY CLOSE; CHOOSING A MUCH SMALLER DeltaV SHOULD SOLVE THAT
REM *****    PROBLEM. ALSO, THE PROCEDURE WILL NOT CONVERGE IF NO SOLUTION
REM *****    EXISTS, WHICH WOULD BE THE CASE FOR INSUFFICIENT POWER AVAILABLE.

      V1 = (2 * PAvail / (f * Rho)) ^ (1 / 3)  'Start V based on only flat plate
      DeltaV = −10: Iteration% = 0      'drag and then count down to correct value
      F1 = FnPowerReq(V1) − PAvail             'Starting difference, PAvail & PReq
```

```
   IF F1 = 0 THEN V = V1: GOTO PrintAnswer
            'Previous statement takes care of odd chance that stumble onto answer
   DO UNTIL F1 * F2 < 0
     IF Iteration% < > 0 THEN V1 = V2: F1 = F2
     V2 = V1 + DeltaV: F2 = FnPowerReq(V2) - PAvail      'iterate below answer
     IF F2 = 0 THEN GOTO PrintAnswer                     'stumble onto answer?
     Iteration% = 1
   LOOP               'Normally will iterate until we overshoot the solution speed

REM *****    NOW V1 IS THE SPEED ABOVE THE CRUISE SPEED & V2 THE SPEED BELOW IT
REM *****         WILL HOME IN ON THE EXACT VALUE (V) BETWEEN THESE VALUES
   DO
     V3 = V1 - F1 * (V2 - V1) / (F2 - F1)
     IF V3 < 0 THEN PRINT "No solution is possible.": END
     F3 = FnPowerReq(V3) - PAvail
     IF V3 = V1 THEN V = V1: GOTO PrintAnswer            'stumble onto answer?
     IF V3 = V2 THEN V = V2: GOTO PrintAnswer            'stumble onto answer?
     IF F3 = 0 THEN V = V3: GOTO PrintAnswer             'stumble onto answer?
     IF F1 * F3 < 0 THEN
          V2 = V3: F2 = F3
       ELSE
          V1 = V3: F1 = F3
     END IF
   LOOP UNTIL ABS(F2 - F1) < .0000001       'Convergence (precision) requirement

REM *** PR1$ & PR2$ provide the formatting for the answer
PrintAnswer:
   PRT1$ = "  For a density altitude of ##,### ft and a bank angle of ## deg,"
   PRT2$ = "The calculated true airspeed is    ### ft/s = ### mi/hr = ###.# kt"
   PRINT : PRINT USING PRT1$; DAltitude; BetaDeg
   PRINT : PRINT USING PRT2$; V; V * 60 / 88; V / 1.6889

END
```

Sample Output #1

```
Percent of available power being used? 75
Estimated propeller efficiency in PERCENT? 80
For density altitude (in ft)? 4000
Bank angle (in degrees)? 0
For a density altitude of  4,000 ft and a bank angle of  0 deg,
The calculated true airspeed is    175 ft/s = 119 mi/hr = 103.3 kt
```

Sample Output #2

```
Percent of available power being used? 75
Estimated propeller efficiency in PERCENT? 80
For density altitude (in ft)? 2500
Bank angle (in degrees)? 45
For a density altitude of  2,500 ft and a bank angle of 45 deg,
The calculated true airspeed is    158 ft/s = 108 mi/hr =  93.6 kt
```

Appendix E: **Mathematics Review**

In science we often use symbols as a shorthand representation for physical properties. For example, the average speed for a given travel distance is defined as the distance divided by the required time. In symbols this can be written as

$$v_{AV} = \frac{d}{t} \tag{E.1}$$

An equality such as this is maintained whenever **both** the left and right sides of the equation are multiplied by the same quantity. Similarly, the equality is maintained whenever **both** sides are divided by the same quantity — but only if that quantity is not zero. Thus we can multiply both sides of Eq. E.1 by t and then switch sides to obtain

$$d = v_{AV}\, t \tag{E.2}$$

where we have used the fact that $\dfrac{t}{t} = 1$.

Next, dividing both sides of Eq. E.2 by v_{AV} and then interchanging sides yields

$$t = \frac{d}{v_{AV}} \tag{E.3}$$

Adding or subtracting the same quantity from each side of an equation also maintains the equality. For example, an equation for the distance travelled when the speed is **not** constant is

$$d = v_0 t + \frac{1}{2} a t^2 \tag{E.4}$$

where v_0 is the speed when timing begins and a is the acceleration (i.e., a is the actual value of the acceleration if the acceleration is constant but it is the average value of the acceleration if the acceleration changes during the time interval). If we wish to solve Eq. E.4 for a, we would first subtract $(v_0 t)$ from each side to isolate the term with the unknown quantity (a):

$$d - v_0 t = v_0 + \frac{1}{2} a t^2 - v_0 t = \frac{1}{2} a t^2 \tag{E.5}$$

Then we multiply each side by 2 and divide each side by t^2 to further isolate a:

$$\frac{2(d - v_0 t)}{t^2} = \frac{(2)\,(\tfrac{1}{2} a t^2)}{t^2} = a$$

and switch sides:

$$a = \frac{2\,(d - v_0 t)}{t^2} \tag{E.6}$$

Note that we used parentheses around the quantity $(d - v_0 t)$ to indicate that the whole value, and not just the first term, is multiplied by the 2. (When you are manipulating equations in this manner, you must be careful to always subtract or add from the whole term and not just from the numerator or other part of the term.)

Another algebraic challenge arises when we discover that we have to combine two equations with two unknowns into one equation with one unknown. Consider Eqs. 5.18 and 5.19 in Chap. 5:

$$A_1 V_\infty = A_2 V_2 \tag{5.18}$$

$$p_\infty + \frac{1}{2}\rho V_\infty^2 = p_2 + \frac{1}{2}\rho V_2^2 \tag{5.19}$$

(A_1 and A_2 are areas, V_∞ and V_2 are speeds, p_∞ and p_2 are pressures, and ρ — the Greek letter rho — is the density.)

Suppose that we wish to calculate V_∞ and we know values for only A_1, A_2, p_∞, p_1, and p_2. It is no use to solve just Eq. 5.18 for V_∞ because we don't have a value for V_2. It is also of no use to solve just Eq. 5.19 for V_∞ because that equation also contains the unknown quantity V_2. The proper approach is to solve the equations **simultaneously**; i.e., solve one equation for the second unknown (V_2) and substitute this expression in the other equation so as to eliminate the second unknown and thereby obtain a new equation with a single unknown.

We first solve the simpler equation (Eq. 5.18) for V_2:

$$V_2 = \left(\frac{A_1}{A_2}\right) V_\infty \tag{E.7}$$

Next we substitute this expression for V_2 in Eq. 5.19:

$$p_\infty + \frac{1}{2}\rho V_\infty^2 = p_2 + \left(\frac{1}{2}\rho\right)\left[\left(\frac{A_1}{A_2}\right) V_\infty\right]^2$$
$$= p_2 + \frac{\rho A_1^2 V_\infty^2}{2 A_2^2} \tag{E.8}$$

We wish to solve this equation for V_∞ but it appears on both sides of the equation. So the next step is to place both terms in V_∞ together on the same side and then factor out the two multiplying terms, so as to isolate V_∞.

$$p_\infty = p_2 + \frac{\rho A_1^2 V_\infty^2}{2 A_2^2} - \frac{1}{2}\rho V_\infty^2$$
$$= p_2 + \frac{1}{2}\rho \left(A_1^2/A_2^2 - 1\right) V_\infty^2$$

Then, subtracting p_2 from both sides of this equation,

$$p_\infty - p_2 = \frac{1}{2}\rho \left(\frac{A_1^2}{A_2^2} - 1\right) V_\infty^2$$

Interchanging sides of this equations gives

$$\frac{1}{2}\rho \left(\frac{A_1^2}{A_2^2} - 1 \right) V_\infty^2 = p_\infty - p_2$$

Dividing through by the multiplying factor $\frac{1}{2}\rho \left(\frac{A_1^2}{A_2^2} - 1 \right)$ yields

$$V_\infty^2 = \frac{2(p_\infty - p_2)}{\rho \left(A_1^2/A_2^2 - 1 \right)}$$

and taking the square root of both sides of this equation completes the algebraic solution for V_∞:

$$V_\infty = \sqrt{\frac{2(p_\infty - p_2)}{\rho \left(A_1^2/A_2^2 - 1 \right)}}$$

which is just the expression given in the text (Eq. 5.20) — but without these steps!

Note that the last equation was obtained from the preceding one by taking the square root of each side. As for additions/subtractions and multiplications/divisions, an equality is maintained whenever both sides of equation are raised to any power, as long as it is the same for both.

We commonly use the square root symbol ($\sqrt{}$) to indicate the square root of a quantity, where the square root is defined as the quantity which gives the original quantity back again when multiplied by itself; i.e.,

$$\sqrt{x} \times \sqrt{x} = x \qquad (E.9)$$

However, writing the square root as an **exponent** of ½ is a desirable alternative because it implies the generality of algebraic manipulations that are possible with numbers raised to integer or fractional exponents:

$$\sqrt{x} \equiv x^{1/2} \qquad (E.10)$$

(Note that three lines ($\equiv$) are often used to define two quantities that are equal by **definition**.

Now, when we square both sides of this equation, we obtain

$$\sqrt{x} \times \sqrt{x} = x = x^{1/2} \times x^{1/2} = x^{1/2+1/2} = x^1 = x \qquad (E.11)$$

which illustrates the rule that exponents add when one number, raised to a given power, is multiplied times the same number, raised to a (possibly) different power.

A quantity raised to a **negative** power is equal to 1 over the same quantity raised to the positive value of that power. Thus

$$x^{-1} \equiv \frac{1}{x} \quad \text{and} \quad x^{-2} \equiv \frac{1}{x^2} \quad \text{etc.} \qquad (E.12)$$

A little thought will now convince you that dividing by a quantity raised to a given power is equivalent to multiplying by the same quantity raised to the negative of that power. As an example of the use of these exponent rules, consider the derivation of important Eq. 14.27 in Chap. 14:

$$\frac{V}{P_{REQ}} = \left(\frac{2W}{\rho C_L S} \right)^{1/2} \div \left[\left(\frac{2W^3}{\rho S} \right)^{1/2} \left(\frac{C_D}{C_L^{3/2}} \right) \right]$$

To simplify this, we need to combine like terms on the right side of the equation. First, separate out all the variables with their individual exponents:

$$\frac{V}{P_{REQ}} = \frac{2^{1/2} W^{1/2} \rho^{-1/2} C_L^{-1/2} S^{-1/2}}{2^{1/2} W^{3/2} \rho^{-1/2} S^{-1/2} C_D C_L^{-3/2}}$$

Next, change the divisor terms to multiplying terms by taking the negative of signs of their exponents:

$$\frac{V}{P_{REQ}} = \left(2^{1/2} W^{1/2} \rho^{-1/2} C_L^{-1/2} S^{-1/2} \right)$$
$$\times \left(2^{-1/2} W^{-3/2} \rho^{1/2} S^{1/2} C_D^{-1} C_L^{3/2} \right)$$

Finally, **add** the exponents for similar variables:

$$\frac{V}{P_{REQ}} = \left(2^{1/2-1/2} \right) \left(W^{1/2-3/2} \right) \left(\rho^{-1/2+1/2} \right) \left(S^{-1/2+1/2} \right)$$
$$\times \left(C_L^{-1/2+3/2} \right) C_D^{-1}$$
$$= (1)\left(W^{-1} \right)(1)(1)\left(C_L^1 \right)\left(C_D^{-1} \right) = \frac{1}{W}\left(\frac{C_L}{C_D} \right)$$

which is just the result given in Eq. 14.27. In following this activity, you may need to recall that fractions add by first finding a common denominator: $\frac{1}{2} - \frac{3}{2} = \frac{(1-3)}{2} = \frac{-2}{2} = -1$

Decimal fractions are defined almost as easily as integer fractions. Thus $x^{0.32}$ can be defined as the quantity that is equal to x^{32} when multiplied times itself one hundred times. Fortunately, all these exponential values are easily calculated with the help of a scientific calculator (one with a y^x function key).

Table E.1 contains common mathematical symbols used with equations. The "implies" symbol ($\Rightarrow$) is often used to indicate that the following equation is obtained from the previous one by algebraic manipulations such as the ones we have just accomplished.

The **proportional** symbol ($\propto$) deserves some little additional attention. A variable is said to be proportional to another variable if changing the second variable also changes the first variable; often this is taken to be equivalent to being **linearly** (or directly) proportional: doubling the second variable also doubles the first variable, halving the second variable also halfs the first variable, etc. For example, for motion at constant speed, the distance travelled is directly proportional to the time spent travelling:

$$d \propto t \qquad (E.13)$$

The proportional symbol must be used rather than the equal symbol because the distance travelled also depends on constant(s) or variable(s) that are not indicated (in this case, the variable of time is omitted).

Air drag, for a non-lifting body, varies as the **square** of the speed through the air, so we can write

$$D \propto V^2 \qquad (E.14)$$

to emphasize this dependence (even though drag also depends on the shape and size of the object and the density of the air).

SYMBOL	MEANING		GREEK LETTER	NAME
=	is equal to		α	alpha
$\approx$	is approximately equal to		β	beta
$\neq$	is not equal to		δ	delta
$\equiv$	is equal by definition to		ε	epsilon
<	is less than		ϕ	phi
>	is greater than		γ	gamma
$\leq$	is less than or equal to		μ	mu
$\geq$	is greater than or equal to		ν	nu
$\propto$	is proportional to		π	pi
$\Rightarrow$	implies that		θ	theta
$\pm$	plus or minus		ρ	rho
$\times$	times		ω	omega
$\div$	divided by		Δ	delta (capital)
∞	infinity		Λ	lambda (capital)
<<	is much less than		Σ	sigma (capital)
>>	is much greater than			
$\therefore$	Therefore	**Table E.1** Mathematical Symbol Definitions and Some Greek Letters		

Drag due to lift, on the other hand, is proportional to the **inverse** of the speed:

$$D_{\text{DUE TO LIFT}} \propto \frac{1}{V} \qquad (\text{E.15})$$

The great value and utility of these proportionalities is that important physical consequences can be drawn immediately. For air drag, we can now conclude that, if we wish to increase our aircraft's cruising speed by 30% (1.3 times), then the non-lifting drag will increase by a factor of 1.69 (1.3 × 1.3) times (which is a 69% increase) while the drag-due-to-lift will decrease to 77% of its previous value (1/1.3 = 0.77). Clearly, if we only want to go faster, our major concern should be the increase in non-lifting drag.

You are strongly urged to inspect equations for their physical significance **and** for their intuitive reasonableness. Look at each variable on the right side of an equation and ask yourself if this dependence of the variable on the left side with that on the right makes sense — or are you going to have to correct some long-held intuitional error? Remember that increasing any variable in the numerator increases the number and increasing any variable in the denominator decreases the number. Try asking yourself the result of doubling the value of each variable in turn, and see how this correlates with your own personal experimental data bank.

As another interesting example, the power required for flight by a bird is roughly proportional to the bird's weight raised to the 7/6 power:

$$P_{\text{REQ}} \propto W^{7/6} \qquad (\text{E.16})$$

while the flapping power available increases much more slowly with increased weight:

$$P_{\text{AVAIL}} \propto W^{2/3} \qquad (\text{E.17})$$

Consequently, birds can only grow so big (up to about 30 lb) before they can no longer fly — and so we don't have to snivel and scurry into a cave every time we heard the flutter of wings.

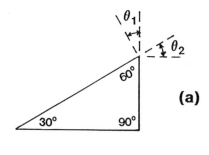

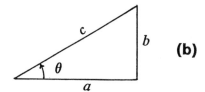

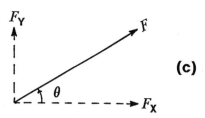

Fig. E.1 Figures to be used to describe properties of angles and triangles

Working with airplanes means working with angles. Figure E.1 provides sketches for reference. All three sketches show a right triangle (one with a 90° angle) but, except as noted, the properties given below are properties of the **angles** themselves and so apply to any triangle or any angle formed by two straight lines.

(a) The **sum** of the interior angles of any triangle is 180°. Note in Fig. E.1a that 30° + 60° + 90° = 180°. This property of a triangle allows us to determine a third angle if we already know two of them.

(b) Two angles are equal if their sides are either parallel or perpendicular to each other. (A side that is an extension of another side is a special case of parallel sides.) In Fig. E.1a, angle θ_1 must equal 30° because the sides of the lines defining it are parallel to the lines defining the 30° angle inside the triangle; the angle θ_2 must equal 30° also, because the sides of the lines defining it are perpendicular to the sides defining the 30° angle inside the triangle.

(c) **Pythagorean theorem** for a right triangle (Fig. E.1b):

$$c^2 = d^2 + b^2 \qquad (E.18)$$

THEOREM OF PYTHAGORAS

This allows us to solve for a third side if the other two are known. It is an extremely useful theorem. c, the longest side of a **right** triangle, is called the **hypotenuse** of the triangle.

(d) Elementary trigonometric functions (Fig. E.1b):

$$\text{sine } \theta \equiv \sin \theta = \frac{\text{opposite side}}{\text{hypotenuse}} = \frac{b}{c} \qquad (E.19)$$

$$\text{cosine } \theta \equiv \cos \theta = \frac{\text{adjacent side}}{\text{hypotenuse}} = \frac{a}{c} \qquad (E.20)$$

$$\text{tangent } \theta \equiv \tan \theta = \frac{\text{opposite side}}{\text{adjacent side}} = \frac{b}{a} \qquad (E.21)$$

These trig functions are properties of the angles themselves, even though they are defined with the help of a right triangle. Special cases:

if $\theta = 90°$, then $\sin \theta = 1$, $\cos \theta = 1$, and $\tan \theta = 0$

if $\theta = 90°$, $\sin \theta = 1$, $\cos \theta = 0$, and $\tan \theta = \infty$

The inverses of these elementary functions are also sometimes seen:

$$\cot \theta \equiv \frac{1}{\tan \theta}, \quad \sec \theta \equiv \frac{1}{\cos \theta}, \quad \csc \theta \equiv \frac{1}{\sin \theta} \quad (E.22)$$

(e) So long as we restrict ourselves to angles less than or equal to 90°, every number between 0 and 1 defines a unique angle through either the cosine or the sine function, and every number between 0 and infinity defines a unique angle through the tangent function. These are the inverse trigonometric functions (as opposed to the inverses of the functions just given) and are the mandatory method for analytically determining an angle (as opposed to geometric techniques). A calculator does all the work for you; all that is necessary is to enter the number and then press the INV key before pressing the trig function key—assuming that the function is defined for that value.

The symbolic representation of the statement "θ is the angle whose tangent is x" is either

$$\theta = \arctan x \quad \text{or, alternately,} \quad \theta = \tan^{-1} x \qquad (E.23)$$

The first form is the one used in this text. As an example, since $\sin 30° = 0.5$, it is also true that $30° = \arcsin 0.5$.

If you are serious about any of the physical sciences, you should transfer these definitions of sine, cosine, and tangent into the "quick recall" part of your memory. Try drawing a little sketch of a triangle whenever you have a scrap of time and write down the definitions on it. Try triangles with very small angles and then ones with an angle close to 90° to help recall the values of the functions for the special cases. It is also sometimes helpful to remember that

$$\sin 30° = \cos 60° = 0.5,$$

$$\sin 45° = \frac{1}{\sqrt{2}} \approx 0.707,$$

$$\cos 45° = \frac{1}{\sqrt{2}} \approx 0.707, \text{ and}$$

$$\tan 45° = 1.$$

(e) Applications to vectors (Fig. E.1c). In this text, angles are most often used in connection with velocities or forces or accelerations—all of which have a particular direction associated with them as well as a particular magnitude or size. Figure E.1c shows a force acting upwards at an angle θ to the horizontal; this could be the force of the propeller on the propeller shaft in climbing flight, for example. That same force, though, is equivalent in its effects to the two separate forces F_X and F_Y because they form a right triangle. From the definition of the sine of an angle,

$$\sin \theta = \frac{F_Y}{F} \quad \Rightarrow \quad F_Y = F \sin \theta \qquad (E.24)$$

and

$$\cos \theta = \frac{F_X}{F} \quad \Rightarrow \quad F_X = F \cos \theta \qquad (E.25)$$

where F_X and F_Y are said to be the x and y **components** of the vector F. These relationships are used so often that you are urged to study them sufficiently that you are able to reliably write down the second form of these equations immediately, without having to start with the first form.

On some lucky days, we start with knowledge of the components; then we can recover the angle θ from the trig functions and the resultant vector magnitude from the Pythagorean theorem. If we need to sum the effects of two forces (or two velocities, etc.), this is usually accomplished by adding components in two mutually perpendicular directions, and then recovering the vector as before, since vectors can't be simply added numerically unless their directions are the same. Chapter 2 explores some of the consequences of these concepts in the analytical determination of ground speeds and winds aloft.

Appendix F: Problem Solving Hints

Once you are comfortable with the mathematical notation discussed in the preceding appendix, you are ready to begin calculating meaningful aerodynamic quantities. A scientific calculator (or a little computer program) removes most of the pain from the procedure — if you manage to escape the pitfalls about to be described.

A recommended calculator would have scientific notation capability, trigonometric functions, the square root ($\sqrt{}$) function, the square (x^2) function, the inverse ($\frac{1}{x}$) function, the general exponent (y^x) function, a built-in value for π, parentheses, and at least one memory for storage of intermediate values.

When numbers become very large or very small, it is most untidy to have to write down all the zeroes. Therefore scientific notation, which records numbers as a simple number followed by a decimal multiplier (or decimal point locator), is universally used by scientists and engineers. For example, air has a standard sea level viscosity of 0.0000003737 lb s/ft² (pound-seconds per square foot). It is far better to write this as 3.737×10^{-7} lb s/ft², for both aesthetic and practical reasons. The 10^{-7} indicates that the decimal point belong 7 places to the **left** of its position in the simple number. (It is customary in scientific notation to place the decimal point immediately after the first digit but it is not incorrect to write the number as 37.37×10^{-8} lb s/ft² or as 0.3737×10^{-6} lb s/ft², if you so desire.) As a second example, the speed of light is about 186,000 mi/s (miles per second); this would normally be written as 1.86×10^5 mi/s. Note that conversion to scientific notation involves just counting the number of places that you move the decimal point and using this as the power of ten, except using the negative of the number if you had to move the decimal point to the right.

The essence of scientific notation is to consider a number such as 1.86×10^5 as a **single** number rather than as the multiplication of one number by another (which the $\times$ symbol might well suggest). Therefore you should enter numbers in scientific notation into your calculator as a single number rather than as two numbers multiplied together, as in $(1.86) \times (1 \times 10^5)$. The usual procedure is to enter the number with the decimal point into the calculator first and then the power of ten with the help of the EE (Enter Exponent) key. If the exponent is negative, use the CHS (Change Sign) key or the +/- key (whichever notation is used by your calculator) before or after entering the value of the exponent. Thus the sequence of keystrokes for entering the speed of light in mi/s would be **1.86**

EE 5. To enter 3.737×10^{-7} you would key in either the sequence **3.737 EE CHS 7** or **3.737 EE 7 CHS.**

Your goal in using your calculator should be to efficiently obtain the correct answer without losing any precision in the process. The precision part implies keeping all the numbers used **inside** the calculator right up until the final answer pops out on the display. If you try to write down intermediate numbers and later try to re-enter them, you will tend to make random errors and, if you don't enter the complete number, will contribute to a round-off error that may become quite large and cause needless worry regarding the correctness of the procedure. For similar reasons, enter π as 3.14159 rather than as 3.14, if your calculator is not sufficiently enlightened to have it built in.

Parenthesis keys [(and)] and memory storage make it possible to keep all numbers in the calculator until the end. Practice and forethought are still required to make it happen, though. Following are some examples for practice.

Suppose you need to calculate the quantity x from the following expression:

$$x = 43.6 \left[7.3 - \frac{(3.2)\,(1.6)}{(.89)\,(1.49)} \right]$$

In general, your calculations should always begin with the innermost calculation and proceed outward; in this case it means starting with the second term inside the brackets:

3.2 × 1.6 ÷ .89 ÷ 1.49 = CHS + 7.3 = × 43.6 =

(which should cause the answer, which rounds off to 150, to appear in your calculator's display). (Note again that some calculators use a +/- legend rather than the CHS legend.) If you have parenthesis capability, an alternate keystroke sequence is this one:

7.3 – (3.2 × 1.6 ÷ .89 ÷ 1.49) = × 43.6 =

Still another method uses the $\frac{1}{x}$ key:

.89 × 1.49 = $\frac{1}{x}$ × 3.2 × 1.6 = CHS + 7.3 = × 43.6 =

Note well that there is no need to use your = key unless you must complete the previous operation before doing the next one. Specifically, when you have a succession of multiply/divide or add/subtract operations, you need the = function only when you change from one type to the other.

The parenthesis capability can also be used to multiply all the components of a denominator, as they are written:

3.2 × 1.6 ÷ (.89 × 1.49) = CHS + 7.3 = × 43.6 =

The size of an angle is most commonly given in degrees, where 360° has been very arbitrarily chosen to represent one complete circle (for which any blame or credit rests with the Babylonians). Thus 90° is one-quarter of a turn (also known as a **right angle**) and 180° is one-half of a turn. Pilots know all about this because they often talk about "doing a 180" when their present course has carried them into worsening weather and they want out.

The **radian** is a natural measure of the size of an angle because it is defined as the ratio of the length of a circular arc traced out by the lines defining the angle divided by the radius of the arc. For a full turn, the arc length is just the circumference of a circle and so

$$360° = \frac{\text{circumference of a circle}}{\text{radius of the circle}} = \frac{2\pi r}{r} = 2\pi \text{ radians}$$

or $1 \text{ radian} = \frac{360°}{2\pi} = \frac{180°}{\pi} \approx 57.3°$

Note that the radian is a much larger unit for measuring angles than is the degree. The radian is also not a true unit like feet or seconds because it is defined as the **ratio** of two distances, but we often write the word down with the number to remind us that radians and not degrees are being used.

Radian measure is sometimes used in aerodynamics when dealing with angles of attack. Calculators typically default to assuming that angles are being given in degrees (with an option for radians) but computer languages generally require all angles to be given in radians before using a trig function, and they return the angle in radians when an inverse function is used; inspect Appendix C for an example. Forgetting this detail in computer programs causes thousands of hours of heartburn for programmers every week, at least.

Some calculator **functions** operate on the current number in the display without requiring the use of the = key and without affecting a pending operation. Thus

$$\left[\frac{1}{7} + \sin(30°)\right]^2$$

can be computed with this key sequence:

7 $\frac{1}{x}$ + 30 **sin x** = x^2 (Answer: 0.413)

in which the $\sin x$ function acts on the 30 in the display while maintaining the + operation as a pending operation, to be consummated when the = key is pressed.

The y^x function is very useful in working some of the problems in this text. With this function you can compute the value of any number raised to any power. First the number is entered into the calculator, then the y^x key is pressed, then x is entered, and finally the = key is pressed to obtain the answer. For example, the value of $(8)^{1/3}$ can be accomplished through these keystrokes:

8 y^x .3333333 =

or, better, using parentheses,

8 y^x (1 ÷ 3) =

with a value of 2.

As well as knowing how to use your calculator, successful problem solving requires a **structured** approach. Generally it is best to begin by writing down all the information that is given, using the proper symbols for each quantity. For example, suppose you are told that a certain airport has a current pressure altitude of 3500 ft and an outside air temperature of 115 °F, and you are to calculate the air density.

Step 1. Write down the known facts:

pressure altitude = 3500 ft

$T = 115\,°F$

Step 2. Look for an equation that relates the desired quantity to the known quantities. In this case, Eq. 4.5 in Section A of Chap. 4 is what we seek:

$p = \rho k T$

where p = pressure, ρ = density, $k \equiv$ gas constant = 1716 lb ft / sl / R.

Step 3. Solve the equation for the unknown:

$\rho = \frac{p}{kT}$

Step 4. Write down all necessary quantities with their correct units. In this case we have the pressure altitude rather than the pressure but the connection is through the Standard Atmosphere tables of Appendix B; so

pressure altitude = 3500 ft $\Rightarrow p = 1862\,\text{lb/ft}^2$

$k = 1716$ lb ft / sl / R

$T = 115\,°F = (115 + 460)R = 575\,R$

Note that we had to convert the temperature unit from degrees Fahrenheit to the absolute unit of Rankines (R), using Eq. 4.2. (Now don't go around cussing out your kindly old professor for this state of affairs. Our everyday units are chosen for convenience and utility and for historical reasons but the relations between these physical quantities always require **consistent** units that may be quite different. In this text, in the examples and in the problems, properties are given with the units that would normally be used in measuring them and it is your responsibility to convert them—whenever necessary—before using them in an equation, just as in the real world.)

Step 4. Write down the equation in symbols followed by the equation with all the numbers, followed by the calculated answer—followed by THE PROPER UNIT!

$$\rho = \text{density} = \frac{p}{kT} = \frac{1862}{(1716)(565)} \approx 0.001887\,\text{sl/ft}^3$$

(As was done in the example above, I normally choose to omit the units when I am writing down the intermediate numbers in an equation because they have been written down in the previous step and they just get in the way when using the calculator—but this means that you must **know** the unit for the answer when using the set of consistent units given in the text. The alternative is to substitute the units along with the numbers and use algebra to cancel and simplify; this works quite well in the above example but can be more than a little tedious and sometimes requires an additional conversion.)

It cannot be emphasized too strongly that the most likely reason for a wrong answer to a problem lies in a failure to convert to the proper units! Distances and speeds are normally given in miles or knots but almost all equations require them to be expressed in units of ft and ft/s, respectively, for example. The important exception to this rule occurs when an equation uses the **ratio** of two like quantities (such as speed over speed, etc); then any units can be used because any conversion constants will apply to both numerator and denominator and cancel out anyway.

Your calculator will present you with answers containing eight digits or so. Most of these digits are pure garbage and common decency requires that they be discarded. Determining which digits are "significant" is a complex business, so in this text some approximate rules are suggested.

The number of significant figures in a number is the number of meaningful digits **after** any leading zeroes (because they are simply decimal place markers); when scientific notation is used, all of the written digits should be significant.

The important point in the approximate rule for significant figures is that the number of significant figures in the **least** precise variable is what usually governs the precision of the answer. (An exception is when two nearly equal numbers are subtracted; then the number of significant figures in the answer is less—possibly much less—than the smallest number in the variables used. In any case, the precision of the answer can never be **better** than that of the least precise number used in its calculation.)

In the previous sample problem, the pressure altitude (3500 ft) appears to have four significant figures, k (1716 lb ft / sl /R) has about four, and T (115 °F) has three. The answer should therefore have (at most) three significant figures; however, in this text I have used an amended rule which adds an additional digit if the first digit is a **1** (... because then the digit is very close to zero and it doesn't seem right to take a chance on throwing out a possibly significant digit). Therefore the answer to the sample problem was given with four significant figures. Scientific notation can be used to make the number of significant figures clear (for example, $\rho = 1.887 \times 10^{-3}$ sl / ft^3) and this is a very good practice.

Writing the pressure altitude as 3500 ft implies **four** significant figures, but in all likelihood the precision is at best ±20 ft, and so the value only possesses three significant figures. Writing it as 3.50×10^3 ft or as 3500 ft ± 20 ft would make this clear. To arrive at an estimate for the possible **range** of values for the answer due to the limited precision in the numbers used, you can try the worst possible combinations of worst possible values and see what you calculate. Maximum worst case values in the numerator combined with minimum worst case values in the denominator produce the worst-case maximum value for the answer, for example.

In sum, when you solve problems in this text, assume that the given quantities are good to three significant figures and round off your final answer to three significant figures (four if the first significant figure is a 1). But be sure you don't lose any precision through round-off errors in intermediate calculations!

Help stamp out naked numbers! An answer is not complete and is **not correct** until the correct unit is appended! Units make all the difference in the world; just ask yourself if you are willing to work for 1000 a day (pennies or bubble gum drops, perhaps?).

One last plea: Respect the mathematical sanctity of the equals symbol. Do not use it unless it is really true. Think of the two lines as being even more important than the don't-cross two lines you see on highways. The equal symbol has been violated already too many times. For example, every year I see something like this:

$$\tan \theta = \frac{4.7}{8.1} = 0.58 = 30$$

The last = symbol is WRONG, WRONG, WRONG. How could 0.58 possibly equal 30°? Instead, use one of these forms:

$$\tan \theta = \frac{4.7}{8.1} = 0.58 \quad \Rightarrow \quad \theta = 30°$$

or

$$\tan \theta = \frac{4.7}{8.1} = 0.58 \quad \text{so} \quad \theta = 30°$$

Last reminder: If your calculated answer differs significantly from the given answer, the chances are 50:50 that you forgot to convert one of the known quantities, or your numerical answer, to the proper unit.

Index

1. page 10, third paragraph and fourth line, should read as follows: "usually given in units of pounds of **weight** per horsepower."

2. page 50, Table 4.3: p_s for 52° should be shown as 27.61 rather than as 27/61; T for p_s = 100.6 (middle column) should be shown as 90 rather than as 80.

3. page 63, next to last paragraph: the conversion factor from kt to ft/s should be 1.689 rather than 1.688.

4. page 80, Fig. 6.2a. The distance $V_W \Delta t$ should be shown as the **radius** of the circle rather than as the height of the circle, in agreement with the text:

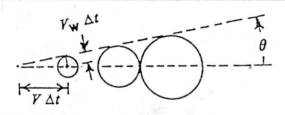

5. page 91, Fig. 7.3, and page 93, Fig. 7.4: to match the notation in the text, the horizontal axis should be labelled as V_{IND} rather than as V_I.

6. page 101, right column, calculation of M_{TE}: (0.576c) should be (0.575c) [from 1.000c–0.425c]

7. page 111, for $\alpha = 0°$, the section lift coefficient is **0.47** (not 0.047).

8. page 132, Problem 4: Answer is 0.40

9. page 140, Fig. 11.9: The "smooth" data are represented by the dashed curve and the "tripped" data by the solid curve.

10. page 164, left column, line 10, should read as follows: "big advantage of forward sweep over rearward sweep is that"

11. page 179, Problem 27: Answer should be given as: "For 10°, L = 7708 lb, D = 144 lb."

12. page 182, bottom left column, should read as follows:

Therefore one complete revolution (360°) is equal to $\frac{2\pi r}{r} = 2\pi$ radians and one radian is equal to $\frac{180°}{\pi}$ or about 57.3°

13. page 212, Eq. 14.12 should read as follows:

$$\sum F_{PERP} = L + T \sin \alpha - W \cos \gamma = 0 \qquad (14.12)$$

14. page 214, lines 3→14 should read as follows:

"The top curve in Fig. 14.12 shows that the reduction in weight has not only increased the maximum climb rate, as expected, but the speed to obtain this has decreased. It works the other direction, too: If you overload an aircraft the optimum climb speed will also increase.

Figure 14.13 explores the effect of density altitude on the rate of climb, using the theoretical dependence of P_{REQ} and P_{AVAIL}. The increase in density altitude has a powerful effect, as expected from Eq. 14.17 and power curves such as Figs. 13.10 and 13.11. One useful fact pointed out by Fig. 14.13 is that the maximum rate of climb is a nearly linear function of density altitude:"

15. page 225, Figs. 14.23 and 14.24. To match the discussion in the text, the figures and their captions should be interchanged.

16. page 228, second column, should read as follows:

as the flare begins the drag is no longer equal to [W sin(γ)]

17. page 234, the equation in the lower left column should read as follows:

$$s = \frac{V_F^2 - V_0^2}{2a} = \frac{0^2 - 81.6^2}{(2)(-0.2 \times 32.2)} = 517 \text{ ft}$$

18. page 247, lower left column, should read as follows:

In the general expressions of Eqs. 15.23 and 15.24, I have substituted a general symbol for the **true** airspeed, V_A, in place of the symbol V_{A2} to remind us that these equations are valid everywhere in the turn.

19. page 254, left column, should read as follows: (on reverse side)

20. page 278, Problem 7: "Use the stall speed given in Problem 6."